AF808912

Compendium of Blueberry, Cranberry, and Lingonberry Diseases and Pests

SECOND EDITION

Edited by

James J. Polashock
USDA-ARS, Genetic Improvement of Fruits and Vegetables Laboratory
Chatsworth, NJ

Frank L. Caruso
Cranberry Station
University of Massachusetts-Amherst, East Wareham

Anne L. Averill
Department of Environmental Conservation
University of Massachusetts, Amherst

Annemiek C. Schilder
Department of Plant, Soil and Microbiological Sciences
Michigan State University, East Lansing

APS PRESS
The American Phytopathological Society
St. Paul, Minnesota, U.S.A.

Front cover images: highbush blueberries (Courtesy W. O. Cline); cranberry rake (inset; Courtesy E. Saalau Rojas)

Back cover images: rabbiteye blueberries (top left; Courtesy W. O. Cline); cranberry blossoms (top right; Courtesy E. Saalau Rojas); cranberry harvest (bottom left; Courtesy E. Saalau Rojas); lowbush blueberries (bottom right; Courtesy C. Slemmons)

Reference in this publication to a trademark, proprietary product, or company name by personnel of the U.S. Department of Agriculture or anyone else is intended for explicit description only and does not imply approval or recommendation to the exclusion of others that may be suitable.

Library of Congress Control Number: 2016947338
International Standard Book Numbers:
Print: 978-0-89054-536-2
Online: 978-0-89054-538-6
Mobi: 978-0-89054-537-9
ePub: 978-0-89054-545-4

© 1995, 2017 by The American Phytopathological Society
First edition published 1995
Second edition published 2017
Second printing 2017

All rights reserved.
No portion of this book may be reproduced in any form, including photocopy, microfilm, information storage and retrieval system, computer database, or software, or by any means, including electronic or mechanical, without written permission from the publisher.

Printed in the United States of America on acid-free paper.

The American Phytopathological Society
3340 Pilot Knob Road
St. Paul, Minnesota 55121, U.S.A.

Preface

The first edition of this book was published as *Compendium of Blueberry and Cranberry Diseases* in 1995. The pathology, entomology, crop production, and breeding for these crops have advanced significantly in the past 21 years. In addition, there is significant acreage now planted with lingonberries, and it was decided to include this *Vaccinium* sp. along with highbush, lowbush, and rabbiteye blueberries and cranberry in this second edition. Insect and mite pests were also added to this edition. Consequently, this edition is entitled *Compendium of Blueberry, Cranberry, and Lingonberry Diseases and Pests, Second Edition* to represent these enhancements. As in the first edition, the purpose of this new edition is to assist in the diagnosis of diseases and pests in the field, classroom, and diagnostic clinic by providing a current and comprehensive reference source for the diseases, insect pests, and abiotic issues affecting these *Vaccinium* spp. It is intended to be utilized by plant pathologists, entomologists, crop specialists, growers, diagnostic clinicians, master gardeners, regulatory specialists, crop consultants, extension educators, researchers, and others who have a stake in the recognition and management of diseases, pests, and other maladies affecting these crops throughout the world. This edition updates all of the previous disease information from the first edition, includes diseases not included in the first edition, and provides many new photos to assist in disease diagnosis. In addition, this second edition includes descriptions of the major arthropod pests of these crops, their life cycles and behavior, and descriptions of the damage they cause, as well as their involvement as vectors of certain pathogens. Many new authors brought their expertise to the writing of new or updated sections. Recommendations for management of diseases and pests in this volume are general in nature and primarily focus on those cultural practices that can be used in all areas where the crops are grown. Specific fungicide/insecticide recommendations were not deemed appropriate, since these quickly become obsolete and vary from one geographic region to another.

This new edition includes an appendix that identifies common names of diseases of blueberry, cranberry, and lingonberry, along with their pathogens. The same content is available on the website of The American Phytopathological Society in the section "Common Names of Plant Diseases." Any updates that are subsequently made to the taxonomy and nomenclature will be reflected on the website. Throughout the text, the common names of diseases and pests are generally provided, along with the currently accepted scientific names. Authorities for the scientific names of pathogens and pests are included in the appendix. Within the text, plants are mostly referred to by their common names; scientific names are also provided for those plants whose common names may be unfamiliar to readers.

We thank Rose Gergerich for her enthusiasm and support for this project in its early stages, as well as Margery Daughtrey for her undaunting diligence in moving the project forward when it reached speed bumps along the way. The list of authors and photo contributors is quite long and we sincerely appreciate the contributions of each and every one of them. A list of the second edition authors follows. There are a few contributors that we especially thank. These include Bill Cline, Patty McManus, Peter Oudemans, Bob Martin, Yannis Tzanetakis, Cesar Rodriguez-Saona, Martha Sylvia, and Sheila Fitzpatrick. Thanks also to the two anonymous reviewers for their reading of all sections of the project. We are especially thankful to the authors and editors of the first edition, who set the stage for this second edition.

James J. Polashock
Frank L. Caruso
Anne L. Averill
Annemiek C. Schilder

Second edition authors:

Seanna L. Annis, School of Biology and Ecology, University of Maine, Orono, Maine, U.S.A.
H. Alejandro Arevalo, BASF Corporation, Research Triangle Park, North Carolina, U.S.A.
Anne L. Averill, Department of Environmental Conservation, University of Massachusetts, Amherst, Massachusetts, U.S.A.
Phillip M. Brannen, Department of Plant Pathology, University of Georgia, Athens, Georgia, U.S.A.
Marin T. Brewer, Department of Plant Pathology, University of Georgia, Athens, Georgia, U.S.A.
Lori M. Carris, Department of Plant Pathology, Washington State University, Pullman, Washington, U.S.A.
Frank L. Caruso, Cranberry Station, University of Massachusetts-Amherst, East Wareham, Massachusetts, U.S.A. (retired)
Nora J. Catlin, Cornell Cooperative Extension of Suffolk County, Riverhead, New York, U.S.A.
Chung Jan Chang, Department of Plant Pathology, University of Georgia, Griffin, Georgia, U.S.A.
William O. Cline, Department of Plant Pathology, North Carolina State University, Castle Hayne, North Carolina, U.S.A.
Judith A. Collins, School of Biology and Ecology and Cooperative Extension, University of Maine, Orono, Maine, U.S.A.
Anne DeMarsay, Department of Plant Biology & Pathology, Rutgers University, New Brunswick, New Jersey, U.S.A.
Carolyn J. DeMoranville, Cranberry Station, University of Massachusetts-Amherst, East Wareham, Massachusetts, U.S.A.
Tim G. Dittl, Madison, Wisconsin, U.S.A.
Peggy L. Dixon, Science and Technology Branch, Agriculture and Agri-Food Canada, St. John's, Newfoundland and Labrador, Canada
Frank A. Drummond, School of Biology and Ecology and Cooperative Extension, University of Maine, Orono, Maine, U.S.A.
Sheila M. Fitzpatrick, Agriculture and Agri-Food Canada, Agassiz, British Columbia, Canada
Thomas A. Forge, Agriculture and Agri-Food Canada, Summerland, British Columbia, Canada
Andres France, INIA Quilamapu, Chillan, Chile

Carlos Garcia-Salazar, Michigan State University Extension, West Olive, Michigan, U.S.A.
Jim F. Hancock, Department of Horticulture, Michigan State University, East Lansing, Michigan, U.S.A.
Eric J. Hanson, Department of Horticulture, Michigan State University, East Lansing, Michigan, U.S.A.
Paul D. Hildebrand, Agriculture and Agri-Food Canada, Kentville, Nova Scotia, Canada (retired)
N. Kirk Hillier, Department of Biology, Acadia University, Wolfville, Nova Scotia, Canada
Bradley I. Hillman, Department of Plant Biology & Pathology, Rutgers University, New Brunswick, New Jersey, U.S.A.
Dan L. Horton, Entomology Department, University of Georgia, Athens, Georgia, U.S.A.
Rufus Isaacs, Department of Entomology, Center for Integrated Plant Systems, Michigan State University, East Lansing, Michigan, U.S.A.
Donald Y. Kobayashi, Department of Plant Biology & Pathology, Rutgers University, New Brunswick, New Jersey, U.S.A.
Oscar E. Liburd, Entomology and Nematology Department, University of Florida, Gainesville, Florida, U.S.A.
Mark Longstroth, Michigan State University Extension, East Lansing, Michigan, U.S.A.
Robert R. Martin, USDA-ARS, Horticultural Crops Research Unit, Corvallis, Oregon, U.S.A.
Patricia S. McManus, Department of Plant Pathology, University of Wisconsin, Madison, Wisconsin, U.S.A.
Jane S. Mika, Cranberry Station, University of Massachusetts-Amherst, East Wareham, Massachusetts, U.S.A.
Peter V. Oudemans, Department of Plant Biology & Pathology, Rutgers University, New Brunswick, New Jersey, U.S.A.
Kim D. Patten, WSU Long Beach Research and Extension Unit, Washington State University, Long Beach, Washington, U.S.A.
James J. Polashock, USDA-ARS, Genetic Improvement of Fruits and Vegetables Laboratory, Chatsworth, New Jersey, U.S.A.
Dean Polk, Rutgers Agricultural Research and Extension Center, Bridgeton, New Jersey, U.S.A.
Jay W. Pscheidt, Department of Botany and Plant Pathology, Oregon State University, Corvallis, Oregon, U.S.A.
Melodie L. Putnam, Oregon State University Plant Clinic, Corvallis, Oregon, U.S.A.
D. C. Ramsdell, Michigan State University, East Lansing, Michigan, U.S.A. (retired)
Willy E. Renderos, Agriculture and Agri-Food Canada, Kentville, Nova Scotia, Canada
Paul S. Robbins, USDA-ARS Subtropical Insects Research Unit, Fort Pierce, Florida, U.S.A.
Cesar Rodriguez-Saona, Department of Entomology, Rutgers University, New Brunswick, New Jersey, U.S.A.
Hilary A. Sandler, Cranberry Station, University of Massachusetts-Amherst, East Wareham, Massachusetts, U.S.A.
Harald Scherm, Department of Plant Pathology, University of Georgia, Athens, Georgia, U.S.A.
Dan E. Schiffhauer, Ocean Spray Cranberries, Inc., Chatsworth, New Jersey, U.S.A.
Annemiek C. Schilder, Department of Plant, Soil and Microbiological Sciences, Michigan State University, East Lansing, Michigan, U.S.A.
Barbara J. Smith, USDA-ARS Thad Cochran Southern Horticultural Research Laboratory, Poplarville, Mississippi, U.S.A.
Cheryl A. Smith, University of New Hampshire Cooperative Extension, Durham, New Hampshire, U.S.A.
Bernadine C. Strik, Department of Horticulture, Oregon State University, Corvallis, Oregon, U.S.A.
Martha M. Sylvia, Cranberry Station, University of Massachusetts-Amherst, East Wareham, Massachusetts, U.S.A.
Ioannis E. Tzanetakis, Department of Plant Pathology, University of Arkansas, Fayetteville, Arkansas, U.S.A.
Timothy J. Waller, Department of Plant Biology & Pathology, Rutgers University, New Brunswick, New Jersey, U.S.A.
Elke Weibelzahl, Entomology and Nematology Department, University of Florida, Gainesville, Florida, U.S.A. (retired)
Erik J. Wenninger, Department of Plant, Soil and Entomological Sciences, University of Idaho, Kimberly, Idaho, U.S.A.
John C. Wise, Department of Entomology, Michigan State University, East Lansing, Michigan, U.S.A.
David E. Yarborough, Department of Plant, Soil, and Environmental Sciences, University of Maine, Orono, Maine, U.S.A.

Contents

Introduction

The Taxonomy, Botany, and Culture of *Vaccinium* spp.

Blueberries, cranberries, and lingonberries belong to the genus *Vaccinium* in the heath family (Ericaceae). Indigenous peoples have harvested many of the wild, edible *Vaccinium* spp. for thousands of years. Some evidence suggests that native people of eastern North America may have intentionally burned stands of lowbush blueberries to renew their vigor. Farmers in the Cape Cod area of Massachusetts (United States) began the cultivation of *Vaccinium* spp. in the early nineteenth century by building dikes and ditches to control the water level in native cranberry stands. Lingonberries are native to the circumpolar boreal region and are harvested from the wild; they are not widely cultivated.

Vaccinium fruits are used in a wide variety of ways. The first widespread use of cranberries was in a sauce served at Christmas and the U.S. holiday of Thanksgiving. Juice products were introduced during the 1930s and are now the dominant cranberry product. Cranberry juice is marketed alone and in mixtures with other types of juices. Other cranberry products include dried fruit, wines, and liqueurs. Lingonberries are processed and mainly used as preserves. Blueberries are sold fresh or processed as frozen, juice, puree, or dried berries. Processed fruits are used in desserts, yogurt, ice cream, baked goods, prepared muffin and pancake mixes, and breakfast cereal. Blueberry syrups, jams, and preserves are also produced but in more limited quantities.

The genus *Vaccinium* is now divided into many sections, four of which are agriculturally important: true blueberries (subgenus *Cyanococcus*), cranberries (*Oxycoccus*), lingonberries (*Vitis-idaea*), and bilberries (*Myrtillus*). Wild representatives of cranberries, lingonberries, and bilberries are found in both Europe and North America, while true blueberries are solely North American in origin.

The commonly cultivated cranberry species is *V. macrocarpon* of section *Oxycoccus*. This species is cultivated primarily in North America in Wisconsin, Massachusetts, New Jersey, Washington, and Oregon (United States) and in Quebec, British Columbia, and Nova Scotia (Canada). Production was about 395,000 t from 16,390 ha (in 2014) in the United States and 86,300 t from 6,100 ha in Canada (in 2011). Cranberries are harvested from September to November, depending on the production region and market. About 94% of total cranberry production in North America is sold for processing. There were about 600 ha of cranberry cultivated in South America in 2005. The small-fruited cranberry (*Vaccinium oxycoccos*) is harvested from about 10,800 ha in Asia, but yields are relatively low, with 33,500 t harvested in 2005.

Lingonberries, *V. vitis-idaea*, are mainly harvested from native stands in Scandinavia and northern Europe, United States (Alaska), and Canada. Lingonberries are being cultivated to a very limited extent (estimated 25 ha) in Oregon and Washington (United States) and in Germany. The lingonberry is known by many other common names worldwide, including lingen, alpine cranberry, dry-ground cranberry, foxberry, moss cranberry, mountain cranberry, northern mountain cranberry, cowberry, partridgeberry, red whortleberry, and rock cranberry. Lingonberries flower twice per year, but often only the later, autumn crop is harvested.

The most important species of blueberries cultivated worldwide are northern highbush (*V. corymbosum*), lowbush (mainly *V. angustifolium*, but also *V. myrtilloides*), southern highbush (complex hybrids based largely on *V. corymbosum* and *V. darrowii*), and rabbiteye (*V. virgatum* [syn. *V. ashei*]), all native to eastern North America. The lowbush blueberry is harvested from managed or cultivated (pruned, sprayed, fertilized) native stands; fruit are commonly marketed as "wild blueberries." Various other species are harvested from the wild in many regions of the world, including *V. uliginosum*, *V. myrtillus*, *V. deliciosum*, *V. ovalifolium*, and *V. membranaceum*.

There were 60,000 ha of lowbush blueberry reported in Canada in 2011, but the production area has been increasing, particularly in New Brunswick. There were 8,660 ha of lowbush blueberry in Maine, the predominant producing state in the United States. Only half of the total area of lowbush blueberries is harvested annually because of alternate-year pruning practices. Total lowbush blueberry production in North America in 2014 was 151,275 t, with approximately 99% sold for processing.

The highbush blueberry is the most commercially important blueberry worldwide, with a production of about 562,465 t from 110,835 ha in 2014. Total planted area in North America was approximately 45,000 ha, producing 330,150 t. In 2014, 67% of total world highbush blueberry production was marketed fresh. Growth has remained strong in many production regions, but particularly in North America. Southern highbush blueberry cultivars have greatly expanded the regions in which blueberries are grown worldwide.

The estimated area of rabbiteye blueberries in North America was 4,270 ha in 2003. This type of blueberry is grown mainly in the southeastern United States.

The planted area of organic highbush blueberry worldwide increased fourfold from 2006 to an estimated 4,156 ha in 2011; the largest area was planted in the United States and Chile. The certified organic blueberry area in the United States increased from an estimated 194 ha in 2003 to 1,665 ha in 2011. The greatest growth has occurred in the western United States, which accounted for 26% of the total planted highbush blueberry area, but 64% of the total organic area planted in 2011.

Positive findings from health-related research on blueberries, including age-related diseases and oxidative stress, have been used in marketing efforts. Demand for blueberries in the early 2000s has exceeded supply.

Fresh blueberry fruit are available all year in North America from "local" production and imports from the production regions in the Southern Hemisphere. In North America, the blueberry season begins in Florida (United States) in March, and follows in California, Georgia and North Carolina, southern Oregon, eastern Washington and New Jersey, western Oregon, northwestern Washington, Michigan (United States), and British Columbia (Canada), in that order. In Europe, the blueberry season spans the period from January (Spain) to October (Poland). In the Southern Hemisphere, blueberry production begins in late August (Australia) and follows in South Africa, Argentina (late September), Chile, and New Zealand, ending in late March.

Cultivars

The lowbush blueberry and cranberry industries are both based on genotypes native to North America. Lowbush blueberry cultivars exist but are rarely planted; commercial fields are developed by enhancing naturally established native stands. A number of cultivars have been released by Agriculture and Agri-Foods Canada (Kentville, Nova Scotia), but none of them have been widely planted.

Cranberries are commercially planted, but some of the important cultivars today were selected from the wild in the nineteenth century, including McFarlin, Ben Lear, Howes, Early Black, and Searles. The hybrid cultivar Stevens accounts for the largest share of planted area in North America and is still considered the most common cultivar for new plantings. The next most predominant cultivars in North America are Early Black, Howes, and Ben Lear. In new plantings, cultivars Stevens, Grygleski #1, Crimson Queen, and HyRed are common.

More than 30 hybrid cultivars of rabbiteye blueberries have been released. Historically, the most important cultivars have been Tifblue and Brightwell from Georgia (United States), Bluegem and Bonita from Florida (United States), and Powderblue and Premier from North Carolina (United States). Newer plantings include cultivars Ochlockonee and Titan from Georgia.

More than 100 hybrid cultivars of northern highbush blueberry have been developed, although relatively few are commercially important. 'Bluecrop' from the U.S. Department of Agriculture (USDA) remains the dominant cultivar across all production regions, being planted on one-third of the total highbush area in the United States; however, this cultivar has not been widely planted on new acreage. Other important cultivars include Croatan (North Carolina, United States); Duke, Elliott, and Legacy (USDA); and Reka (New Zealand). Cultivars being widely planted include Aurora, Draper, and Liberty (Michigan, United States) and various proprietary cultivars.

Interspecific hybridization has greatly increased the range of blueberry production. Half-high hybrids were developed from *V. corymbosum* × *V. angustifolium* that could be grown in very cold regions with dependable snow cover. The most commonly planted cultivars are Northblue and Chippewa from the University of Minnesota. The wild diploid *V. darrowii* and other native species found in the southern United States have proven useful in breeding the southern highbush hybrids that require less chilling than northern highbush types. Some of the most successful southern highbush cultivars have been Emerald, Jewel, and Star (Florida, United States). Other cultivars being widely planted include Biloxi (USDA); Kestrel, Farthing, and Flicker (Florida); and Rebel and Camellia (Georgia, United States).

Evolutionary Background

The genus *Vaccinium* is widespread, with high densities of species being found in the Himalayas, New Guinea, and the Andean region of South America. The origin of the group is thought to be South America. Estimates of species numbers vary from 150 to 450 in 30 sections. The primary mode of speciation in the section *Cyanococcus* is thought to be through unreduced gametes. Cultivated *V. corymbosum* and *V. angustifolium* are tetraploids ($2n = 48$); *V. virgatum* is a hexaploid ($2n = 72$); and *V. macrocarpon*, *V. myrtilloides*, and *V. vitis-idaea* are diploids ($2n = 24$). The tetraploid highbush blueberry *V. corymbosum* has been shown to be genetically an autopolyploid.

Species boundaries have been difficult to resolve in the section *Cyanococcus* because of polyploidy, overlapping morphologies, and a general lack of chromosome differentiation. In the first detailed taxonomy of the group, Camp described nine diploid, twelve tetraploid, and three hexaploid species, but Vander Kloet reduced this list to six diploid, five tetraploid, and one hexaploid species. He included all the crown-forming species (diploid, tetraploid, and hexaploid) into *V. corymbosum* with three chromosome levels. Most horticulturists and blueberry breeders believe that the variation patterns in *V. corymbosum* are distinct enough to retain Camp's diploids *V. elliottii* and *V. fuscatum*, tetraploid *V. simulatum*, and hexaploids *V. virgatum* (syn. *V. ashei*) and *V. constablaei*.

V. macrocarpon is an endemic of eastern North America and is thought to be the most primitive species in the section *Oxycoccus*. Its closest relatives are diploid, tetraploid, and hexaploid races of *V. oxycoccos* that have circumboreal distribution. Tetraploid *V. oxycoccos* has been shown to be genetically an autopolyploid, although it likely carries genes from *V. macrocarpon*. *V. myrtillus* is very similar to *V. scoparium* and may have been derived in the Rocky Mountains of North America. There has been little speculation about the origin of *V. vitis-idaea*, but it must be closely related to *V. myrtillus*, since hybrids have been discovered between these two species at numerous locations across northern Europe.

Botany

All *Vaccinium* spp. are acidophilic. Cranberries grow naturally on hummocks in acidic peat bogs and wetlands of pH 4.0–5.0. Lingonberries grow in densely wooded areas, heaths, grass moorlands, raised bogs, and rocky exposed cliffs in the circumpolar region. Highbush and rabbiteye blueberries are commonly found in acidic swamps and bottomlands but are usually held above water level on hummocks for most of the year. Lowbush blueberries are located primarily on rocky heaths with limited moisture.

Stature is one of the most striking differences among the major types of *Vaccinium* spp. In nature, cranberry shoots typically are 0.1–0.15 m high, although prostrate runners can extend several meters across the peat. Lingonberry plants typically grow no taller than 0.5 m. Lowbush blueberry plants grow 0.30–0.60 m high, highbush blueberry plants reach 1.8–4.0 m, and rabbiteye plants may grow to 6 m tall.

All *Vaccinium* spp. are woody perennials. Cranberries and lingonberries are evergreen, while the cultivated blueberries are deciduous. Cultivated *Vaccinium* varieties are asexually propagated through stem or rhizome cuttings or by micropropagation. All *Vaccinium* spp. are colonized with ericoid mycorrhizae.

Cranberry leaves are either oblong or oval shaped, 1–1.5 cm long and 0.3 cm wide. The leaves remain on the plant for up to 2 years before senescing. Senescing leaves fall to the surface of the bog and contribute to the organic or duff layer. Leaves are dark green and glossy during the growing season, changing to reddish brown during the dormant season. Cranberry plants produce various types of shoots (Fig. 1). Runners are horizontal stems that are 0.3–2 m long and spread profusely over the bed. Short, vertical branches of cranberry are called uprights.

Uprights originate from axillary buds on the runners or from older uprights and grow for several years. Uprights typically grow 5–8 cm per year, depending on light intensity and nutrition. Uprights can be either fruiting or vegetative (nonfruiting). Cranberry roots are very fine and fibrous and lack root hairs. Roots are readily formed on decumbent stems when covered with a moist medium, such as sand. Most cranberry roots are in the upper 10–20 cm of soil, but some may extend 45 cm deep.

Cranberry flower initiation takes place on the terminal buds of some uprights during mid- to late summer when days are shortening. Uprights tend to biennially bear; if they flower and fruit in a given year, they are more likely to be vegetative the following year.

Cranberry uprights produce terminal buds that are vegetative (simple) or flowering (mixed). Flower buds are larger and more round than vegetative buds and may contain from one to seven flowers as well as leaves and a growing point.

In early spring in the Northern Hemisphere, vegetative growth resumes after winter dormancy. At the hook stage, the flower pedicels are visible on fruiting uprights. Bloom begins at early to mid-June and lasts 3–6 weeks, depending on the growing region and cultivar. Blossoms and fruit are usually borne on uprights, but under certain conditions in some cultivars, blossoms and fruit may be produced on runners.

Each flower is borne individually, and there are one to seven flowers per upright. Normally, vegetative growth continues distal to the flowers. However, sometimes the upright forms an umbrella bloom, in which case the flowers terminate the upright. The normal flower consists of an inferior ovary with four cells, a calyx of four sepals, and a corolla of four petals that are pinkish white and deeply cleft. The petals curve back when the flower is fully open. Eight stamens surround the pistil in a whorl and shed tetrad pollen grains from a terminal pore in the anther (Fig. 1).

The cranberry is self-fertile; however, some evidence indicates that cross-pollination may improve fruit set and berry size. Mixtures of cranberry cultivars, or off-types, are not unusual in commercial beds, although this is not desired from a management perspective. Agitation of the flower by wind is inadequate to produce a commercial crop. Honeybees are introduced for pollination in commercial plantings.

Cranberry fruit are borne singly at three to seven nodes on upright shoots. Fertilization of the ovule occurs within about 72 h after pollen grain germination. Unfertilized flowers may hang on to the upright for more than 3 weeks. The percentage of blossoms that set fruit is relatively low in cranberry, i.e., 10–60%. Fruit set may be limited by an overproduction of flowers, insufficient pollination, frost in the spring, nutritional imbalance, and/or application of certain fungicides. Final fruit size depends largely on the number of developed seeds. The ovary and calyx fuse to form a true berry that varies in shape and color. The berry consists of a relatively thin epidermis and four locules containing 0–50 seeds, depending on set. A waxy cuticle covers the epidermis and contributes to the ability of the cranberry to resist moisture loss after harvest. Although the thickness of this waxy cuticle has been found to differ among cultivars, it is not related to the postharvest quality of the fruit. A typical fruit weight is about 1–1.5 g.

The total amount of anthocyanins is 25–100 mg per 100 g in cranberry, and the anthocyanins cyanidin-3-monogalactoside, peonidin-3-monogalactoside, cyanidin-3-monoarabinoside, and peonidin-3-monoarabinoside are predominant.

Cranberry fruit go through several stages of color development, including green, white, and red. However, they grow more linearly than blueberry fruit; once berry growth begins, it continues at a relatively constant rate for 4–6 weeks. Cranberry fruit mature 60–120 days after bloom, depending on the cultivar.

After harvest, the cranberry vine enters dormancy and requires more than 1,000 h of chilling (0–7°C) before normal growth resumes the following spring. Cold hardiness of cranberry is affected by the stage of dormancy or growth. Flower buds can survive temperatures of –30 to –35°C before being damaged, although desiccation can be a complicating factor.

Highbush and rabbiteye blueberries have very fine, fibrous roots that lack root hairs. Lowbush blueberries produce roots adventitiously from rhizomes (Fig. 2). Thus, lowbush blueberries are maintained in solid beds without aisles, whereas highbush and rabbiteye types are planted in fields with aisles between the rows of plants. Root growth begins in the spring when soil temperatures reach about 6°C; often this coincides with bud swell. Root growth continues through late spring but slows during fruit development and maturation. Most root growth occurs from after harvest throughout the autumn, until soil temperatures fall below 6°C. Rabbiteye and southern highbush blueberries may have more than two flushes of root growth in regions with a long growing season. Most highbush and rabbiteye blueberry roots are found within the dripline of the bush and in the upper 0.5 m of soil. Lowbush roots extend 6–25 cm below the soil surface.

Blueberries produce simple buds—vegetative buds, producing a leafy shoot, and flower buds, producing only an inflorescence (Figs. 2–4). Flower buds are relatively large and almost spherical, whereas vegetative buds are pointed, smaller, and scalelike. Buds are found mainly on 1-year-old wood; latent buds are also on older wood. Vegetative bud break may take place before or after floral bud break, depending on the cultivar. Blueberry shoots have a determinant, sympodial (zigzag) growth pattern caused by periodic apical abortions (black tip). Each abortion terminates a flush of growth. The adjacent bud

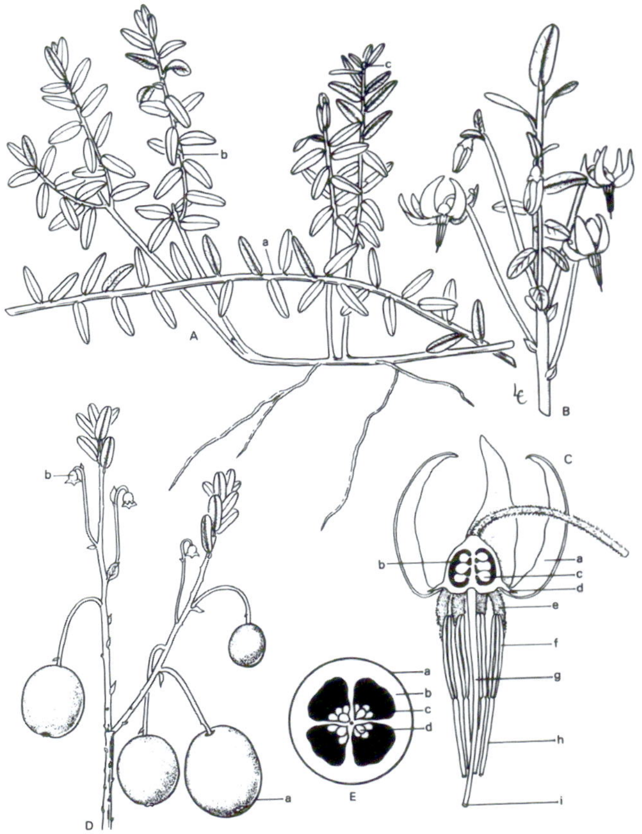

Fig. 1. Morphology of cranberry. **A,** Plant habit; note the procumbent stem (a), upright (b), and terminal bud (c). **B,** Flowering shoot. **C,** Longitudinal section of a flower: petals (a), ovary (b), ovule (c), sepal (d), filament (e), anther (f), style (g), pollen delivery tube (h), and stigma (i). **D,** Fruiting uprights with a mature berry (a) and an unfertilized ovary (b). **E,** Cross section of a berry: epidermis (a), flesh (b), locule (c), and seeds (d). (Reproduced, by permission of Prentice Hall, from Galletta and Himelrick, 1990)

may then remain dormant or enlarge and initiate a new shoot extension. The number of flushes depends on the cultivar, species, and vigor. Some lowbush types undergo only a single flush per season, whereas highbush blueberries generally have one to three flushes of growth, depending on the cultivar and length of the growing season. New canes develop from the crown of the blueberry plant in early spring or from older wood higher up in the bush. These shoots, usually called whips, are extremely vigorous and often are the renewal wood for subsequent years' production.

Flower bud initiation takes place under the short day length during late summer and early autumn, usually on the tip of the current season's shoot. Initiation proceeds basipetally and the number of floral buds per shoot is affected by the cultivar, climate, and production practices. Flower bud development continues until temperatures become too low in the autumn. In general, northern highbush cultivars have a higher chilling requirement (more than 800 h at 0–7°C) than southern highbush or rabbiteye cultivars (generally less than 300 h of chilling). Northern highbush blueberries are more cold hardy than rabbiteye blueberries; their flower buds can survive temperatures of –30 to –35°C when fully dormant. Cold damage may happen within buds at higher temperatures if plants are in the acclimation or deacclimation phase of dormancy. Most rabbiteye flower buds are injured by temperatures lower than –15°C. Lowbush blueberries are considered the most winter hardy of the domesticated *Vaccinium* spp.; however, much of their hardiness is because of their low stature and consequent protection by snow cover.

In most production systems, blueberry plants flower in the spring, with flowers at the tip of 1-year-old wood and the tip of the cluster opening first (Fig. 5). In areas with limited chilling hours below 7°C, many southern highbush types flower and fruit on the same season's growth. The length of the bloom period varies with the cultivar but can be affected by climatic factors, such as heat units accumulated. Each flower bud contains six to twelve potential flowers. The flowers of cultivated blueberries are epigynous and usually inverted and are pollinated by insects. Flowers consist of sepals (calyx), petals (a four- or five-lobed, urn-shaped corolla), 10 stamens, and a pistil. Commercial growers usually use honeybees for pollination, although sonicating bees, such as bumblebees, are more efficient. Blueberry pollen is dehisced as tetrads through pores at the end of long pollen delivery tubes (Fig. 1). Flower receptivity to pollination is 5–8 days in northern highbush blueberry, 4–6 days in lowbush blueberry, and 6 days in rabbiteye blueberry. Although northern highbush blueberries are self-fertile, cross-pollination increases fruit set and fruit size for many cultivars. Rabbiteye and southern highbush blueberries have low self-fertility and are generally grown in alternating rows of cross-compatible cultivars. Lowbush clones are generally not self-fertile, but cross-pollination is rarely a problem because they grow naturally in mixed stands. Self-infertility in *Vaccinium* spp. is probably caused by a combination of self-incompatibility and late-acting inbreeding depression.

Blueberry fruit are held on corymbs or racemes. The blueberry fruit is a true berry, consisting of an ovary with up to 100 seeds. The fruit of some species, particularly rabbiteye blueberry, contain stone cells that contribute to fruit grittiness.

The fruit development period is 42–90 days for northern highbush blueberry, 55–60 days for southern highbush blueberry, 70–90 days for lowbush blueberry, and 60–135 days for

Fig. 2. Morphology of lowbush blueberry. **A,** Plant habit; note the underground rhizome (a) and upright stems (b). **B,** Dormant shoots with vegetative buds (a) and flower buds (b). **C,** Flowering shoot. **D,** Flowering branch. (Reproduced, by permission of Prentice Hall, from Galletta and Himelrick, 1990)

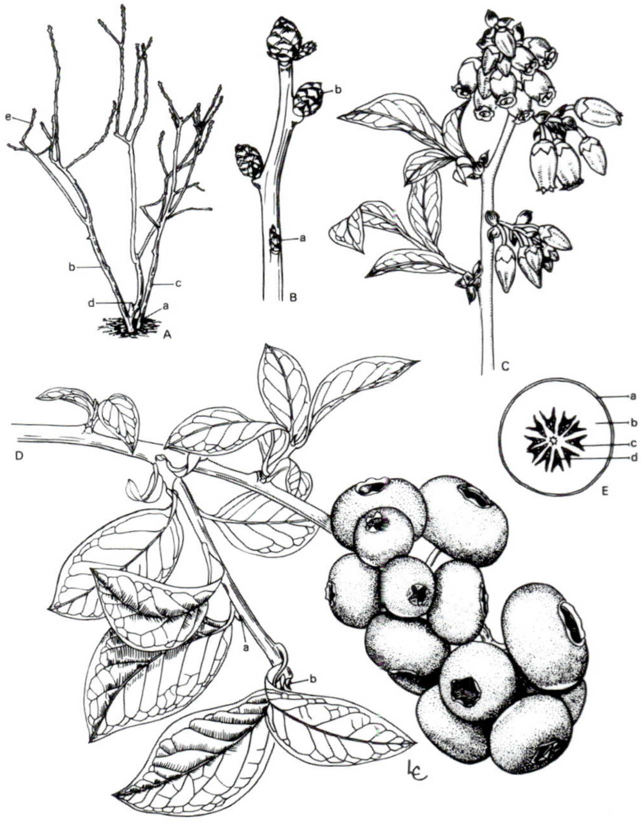

Fig. 3. Morphology of highbush blueberry. **A,** Dormant plant habit: crown area (a), major cane (b), renewal canes (c), pruning cut (d), and lateral shoot (e). **B,** Dormant lateral shoot with vegetative buds (a) and flower buds (b). **C,** Flowering lateral shoot. **D,** Fruiting branch with an aborted apical shoot tip (a) and an axillary bud (b). **E,** Cross section of a berry: epidermis (a), pulp (b), locule (c), and placenta and seeds (d). (Reproduced, by permission of Prentice Hall, from Galletta and Himelrick, 1990)

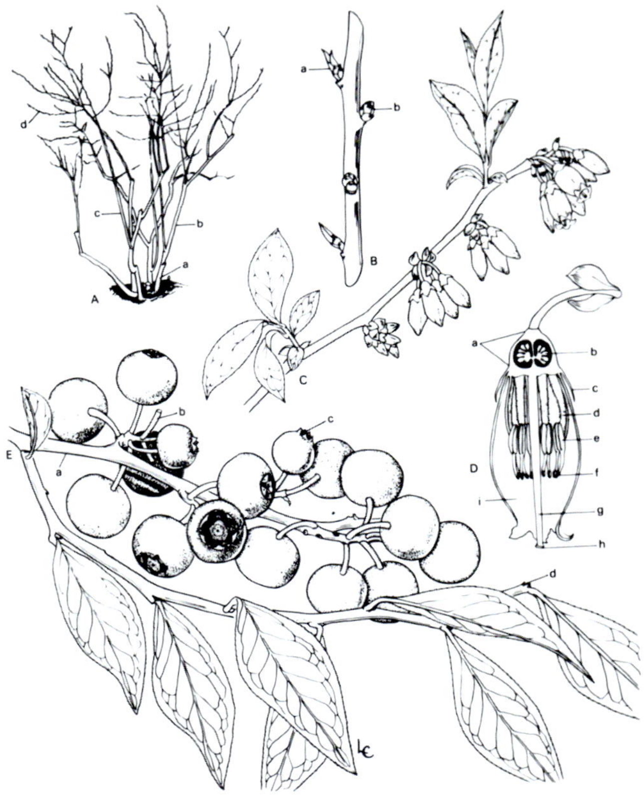

Fig. 4. Morphology of rabbiteye blueberry. **A,** Dormant plant habit: crown area (a), major cane (b), renewal cane (c), and lateral shoot (d). **B,** Dormant lateral shoot with vegetative buds (a) and flowering buds (b). **C,** Flowering lateral shoot. **D,** Cross section of inverted blueberry flower: ovary (a), ovule (b), sepal (c), filament (d), anther (e), pore (f), style (g), stigma (h), and corolla (i). **E,** Fruiting branch: peduncle (a), pedicel (b), calyx (c), and aborted shoot tip (d). (Reproduced, by permission of Prentice Hall, from Galletta and Himelrick, 1990)

rabbiteye blueberry, depending on the cultivar, weather, and plant vigor. Blueberry fruit exhibit a double sigmoid growth curve with rapid gain in dry weight during stage I, seed development and little change in volume or weight during stage II, and rapid increase in dry weight during stage III. During stage III, sugars and anthocyanins accumulate. Berries go through several phases of color development, from immature green to translucent greenish white, greenish pink, blue-red, and finally a complete blue. Up to 50% of the increase in berry volume happens during the shift from greenish pink to blue.

Anthocyanin pigments are generally located in the epidermis and hypodermis. The total amount of anthocyanins is 85–270 mg per 100 g in blueberry fruit. The predominant anthocyanins in blueberry are delphinidin-monogalactoside, cyanidin-monogalactoside, petunidin-monogalactoside, malvidin-monogalactoside, and malvidin-monoarabinoside. The light blue fruit color of many blueberry cultivars results from the combination of dark blue pigments overlaid by a translucent waxy bloom. Individual fruit typically weigh 0.5–4 g, depending on the cultivar.

Lingonberry is a low-growing, woody, perennial, evergreen plant that spreads by rhizomes. Leaves are ovate (6–25 mm long) and arranged alternately on the stems. The white or pink flowers are urn shaped. Flower morphology is similar to that of blueberry. Fruit are small, red berries weighing 0.2–0.5 g.

Lingonberries flower on 1-year-old wood, similar to blueberry. However, lingonberry plants may have two bloom periods, March to April and July to August, with two potential fruit harvest seasons. A pollenizing cultivar is required for fruit set and good fruit production.

Fig. 5. Bloom in 'Duke' blueberry showing early fruit set on the upper cluster, open flowers on the apical clusters, and late pink stage on the lower clusters. Photo taken on April 6, Aurora, Oregon, United States. (Courtesy B. C. Strik; © APS)

The chemical content of blueberry and cranberry fruit parallels that of other berry crops. On average, ripe blueberry fruit are about 80% water, 0.7% protein, 0.5% fat, 1.5% fiber, and 17.3% carbohydrate. The predominant reducing sugars in blueberries are glucose and fructose, which represent 8.2% of the fresh weight; sucrose represents 2.4%. The fruit of cranberry are composed of 2.7% glucose, 0.74% fructose, and 0.14% sucrose. The primary organic acid in blueberries is citric acid (1.2%), while cranberries have high levels of quinic (1.3%), malic (0.9%), and benzoic (0.6%) acids. Blueberries contain an average of 22.1 mg of vitamin C per 100 g of fresh weight, compared with 7.5–10.5 mg per 100 g for cranberries. Berries from all *Vaccinium* spp. have relatively high levels of polyphenolic acids; caffeic and para-coumaric predominate in blueberries.

Culture

Cranberry production has evolved from the management of native plantings on peat bogs in eastern production areas in North America. In many sites in eastern North America, cranberries, growing in peat bogs, have been modified to increase yield and facilitate management. New planting sites are developed in areas with a soil and microclimate similar to native plantings. Cranberries are best grown in soil with a pH of 4.2–5.5. In the Pacific coastal region of the United States, cranberry beds are more commonly established using a sand layer over organic soil (Washington) or over a clay subgrade soil (Oregon). Cranberry beds or bogs are surrounded by perimeter ditches and dike systems with an adjacent water reservoir. These all are connected to irrigation systems that provide frost protection to new spring growth and ripe berries in the autumn and water to plants during the growing season.

Planting is done from January to May, depending on the production region. To establish a new planting, vine cuttings are spread over the surface of the planting area and then mechanically pushed into a wet sand layer to a depth of 5–10 cm. Planting material originates from the prunings of established beds or from beds that are mowed or renovated. The cranberry bed is kept moist to enhance rooting of the cuttings, which usually takes place in 2–4 weeks. Sprinkler irrigation is used most commonly in cranberry beds to satisfy the water requirements of the crop, for frost protection in the spring and autumn, and for evaporative cooling of the vines on hot days.

The surface of the bed is usually well covered by runners by the end of the second year. A small yield may be harvested the year after planting, but year three is more typically the first production year, with production increasing until the bed is mature in year five. A mature cranberry bed or bog consists of a solid bed of vines with no aisles.

Three to four honeybee hives are placed near cranberry plantings at about 5–10% bloom to promote good pollination. Berries are harvested with or without the use of flooding—wet harvested or dry harvested, respectively. During wet harvest, the cranberry bed is flooded to just above the top of the plant canopy. A water reel or "egg beater" machine is driven through the flooded bog (Fig. 6). Bars on the reel are set to rotate just above the canopy, thus agitating the water and forcing the berries to abscise from the pedicel. The floating berries are corralled on the downwind end of the bed using booms and are then lifted into a truck using a conveyor belt or a berry pump. A few other types of wet-harvest machinery, including combs and soft paddles that dislodge berries, are used less commonly.

Dry harvest has evolved from the hand scoops used in the early 1900s to a Furford picker (Furford Picker Co., Grayland, WA), which has horizontal knives to prune runners and tines or rakes to pick fruit. Prunings and fruit are commonly collected in burlap bags. Plant material is separated from the berries on a mechanical shaker, and fruit are transported to a storage or packing facility. Dry harvest is less common than wet harvest.

Cranberry yield is still measured in barrels (bbl), the containers used in early production. One barrel weighs 100 lb. An average yield is 95–210 bbl/acre, but yields as high as 700 bbl/acre have been documented.

About 95% of commercially harvested cranberries are processed. Cranberries to be sold as fresh fruit are generally bounced over barrier boards: sound berries bounce and travel forward to a conveyor belt, while soft berries are collected below. Small and large berries are then separated over a wire screen and carried along moving belts for further sorting for color and defects. Electronic color sorting is rapidly replacing hand sorting. Water-harvested berries are first dumped on a long, inclined mesh belt that passes through a dryer to prepare them for sorting. Flotation tanks are also used in some instances to sort both wet- and dry-harvested fruit.

Cranberry fruit can be stored for months without significant loss in quality. Fruit were once maintained for fresh marketing in large ventilated rooms at low ambient temperatures, but refrigerated storage at 2–4°C is now universal. Berries destined for processing are generally frozen within 12–24 h of harvest. Most cranberries are stored in bulk, but some are prepackaged in perforated polyethylene bags or unsealed cardboard boxes.

Fig. 6. Wet harvest of cranberries in Oregon (United States). (Courtesy B. C. Strik; © APS)

Cranberries are pruned during the winter using a machine that nonselectively cuts runners and some uprights. The frequency of pruning varies by region; alternate-year pruning is common in the Pacific coastal region of North America. A thin layer of sand (<3 cm) is spread over the surface of the bed every 3–4 years to stimulate new rooting of stems.

In eastern and Wisconsin (United States) production regions, a winter flood is used to protect plants against winter cold injury and desiccation. A winter flood is applied any time from late November to early January, depending on the region and temperature. Vines are encased and protected in a layer of ice throughout the winter. The flood is usually removed in March or April to prevent damage to the vines from oxygen deficiency. Flooding is used in all production regions as a pest management tool. A flood can be applied for a short period of time during the growing season or after harvest to control various insect pests.

Lingonberries are most commonly harvested from wild, unmanaged native stands. In the limited number of commercial, established plantings, lingonberries are grown in light, well-drained, acidic soil, high in organic matter. Some plantings are established on raised beds to improve drainage, and organic matter is often incorporated into the soil before planting. Plants are established 20–45 cm apart in rows that are 0.6–1 m apart. Plants fill in the rows as they spread by rhizomes, forming dense matted rows in 4 or 5 years. An organic mulch, such as sawdust, is often added to the soil surface after planting and replenished as needed. Plantings are irrigated and pruned by mowing alternate rows to about 5 cm above the soil level every 3–6 years.

Berries can be harvested when they are light red to dark red. Lingonberries ripen to a dark red after being stored for several weeks under refrigeration. Berries that ripen during the summer will senesce, if not harvested, before the autumn crop is mature. Lingonberries are most commonly harvested by hand using a rake that is similar to the ones used for lowbush blueberries. Machine harvest is possible but not common. Yields are 1–30 t/ha. Berries can be stored for 3–5 weeks under refrigeration.

Northern highbush blueberries are the main type of blueberry planted in the northern regions of the world, whereas southern highbush cultivars are the most common type planted in warmer regions of the Northern and Southern Hemispheres. Rabbiteye blueberries are less widely planted than northern or southern highbush blueberries, but they offer a late fruiting season that is an advantage for targeting key high-value fresh markets.

Blueberries grow best on well-drained soils that are high in organic matter and have a pH of 4.2–5.5. However, in many regions, blueberries are successfully grown through the modification of traditional cultural practices. For example, blueberries are now grown on sandy soil, with or without organic material incorporated, using acidified irrigation water. Soil that is not naturally acidic may be modified by adding elemental sulfur. In most production regions, blueberry plants are grown on raised beds, about 0.3–0.6 m high, to improve soil drainage in the root zone.

Many older blueberry plantings in North America were established at an in-row spacing of 1.2 m, with 3 m between rows. However, research and subsequent positive grower experience has led to a change to higher density plantings worldwide; the most common in-row spacing is now 0.75–0.9 m, with 3–3.4 m between rows. Plantings are established, usually with 2-year-old container-grown plants but in many regions also with younger plants, during the autumn or spring. An organic material, commonly sawdust, is often incorporated prior to planting and also may be applied as a surface mulch after planting; this layer is then maintained for the life of the planting. Many fields in the western region of North America are now established using porous, polyethylene ground cover (weed mat) as a weed barrier in the row (Fig. 7). Most fields have a permanent grass cover in the aisles between rows.

While most of the older blueberry plantings in North America are irrigated using sprinklers, use of drip irrigation became very common in new plantings in the early 2000s. Drip irrigation systems are also commonly used to apply fertilizer (fertigation). In some regions, growers may also install overhead irrigation systems to help protect against frost in the spring and heat in the summer.

Cultivated blueberry plants usually require 4–8 years to reach full production, depending on the climate. Yield for highbush blueberries varies with the production region. In North America, typical yields for well-managed, mature highbush blueberry fields are 7–20 t/ha, depending on the production region. Some high-yielding cultivars, such as Legacy, can reach yields as high as 36 t/ha in Oregon (United States). Growers use domesticated honeybees at rates of two to eight colonies per hectare to augment natural bumblebee populations and enhance pollination.

In North America, most highbush blueberry fruit destined for the processed market are harvested by machine. A large change since 2005 has been the increased use of machine harvest for fresh market fruit; this trend is expected to continue in most production regions of North America because of the high cost and unavailability of labor. However, in 2014, machine-harvested fruit were not suited for more distant fresh markets.

Most of the machine harvesters used are an over-the-row rotary type in which nylon beaters or fingers on revolving drums on either side of the picking tunnel vibrate to dislodge ripe fruit (Fig. 8). Fruit fall onto catcher plates, are carried to the top of the machine, pass under a fan to remove most debris, and then fall onto a sorting belt before falling into flats or lugs. For fields that are entirely harvested by machine, losses have been as high as 25% of total yield in untrellised plantings. In many production regions of the world and in western North America, blueberries that may be harvested by machine are now trellised (Fig. 9), improving recovery by 3–8% of total yield. Trellises are used in many hand-picked fields also.

With a properly adjusted rotary machine and selection of cultivars, growers may sell machine-harvested fruit in the fresh market; however, currently, this fruit is sold to less distant markets since it has a shorter shelf life than hand-picked fruit.

Fruit for the fresh market may be either packed into its final container in the field or picked into larger containers and then transported quickly to a cleaning, sorting, packing, and storage facility. Mechanical color sorters or hand sorting is done to remove any unripe or damaged fruit. Fruit are quickly cooled using forced air. For longer-term storage, berries are kept under a controlled atmosphere. Most berries for the fresh market are now sold in plastic clamshell containers and shipped via refrigerated truck, airplane, or ship to distant markets.

Fruit for processing are transported quickly to a processing plant, where it is either individually quick frozen, bulk frozen, pureed, juiced, or freeze dried, depending on the facility and end use or market. The proportion of the highbush blueberry crop in North America that is processed varies from year to year but is approximately 50%. Most of the rabbiteye crop is marketed fresh.

Blueberries are pruned annually in most production regions with the severity varying depending on the type of blueberry grown and the amount of vegetative vigor. In most northern production regions, highbush blueberries are pruned annually mainly during the dormant season with little, if any, summer pruning. Southern highbush blueberries, however, are often pruned after harvest by hand or machine (hedging), especially in areas with a long growing season. Rabbiteye blueberries are often pruned during the summer by hand-tipping vigorous shoots on several occasions to stimulate the production of fruit buds for the next year's crop; machine pruning is common in the southeastern United States. Plants to be machine harvested

Fig. 8. Littau over-the-row harvester (Littau Harvester, Stayton, OR) in a field of mature 'Elliott' blueberry in Oregon (United States). (Courtesy B. C. Strik; © APS)

Fig. 7. 'Duke' northern highbush blueberry field in Oregon (United States) showing the use of porous, polyethylene ground cover (weed mat) in the row for weed management. (Courtesy B. C. Strik; © APS)

Fig. 9. Mature northern highbush blueberry field in Oregon (United States) showing the trellis used to improve machine harvest efficiency. (Courtesy B. C. Strik; © APS)

are pruned to a more upright habit with a narrow crown to allow close fitting of the machine harvester's catcher plates and thus improve machine harvest efficiency or minimize fruit loss. For hand harvest, plants are pruned to maintain branches within easy picking height, and low-fruiting branches or canes are removed.

In lowbush blueberry, native stands are brought into commercial production through management practices that improve planting density. The cultural management of lowbush blueberries has changed significantly since the late 1990s, including increased use of fertilization, irrigation, bee hives for pollination, herbicides for weed management, and management of soil pH, all of which have led to increased yield. Only half of the total area of lowbush blueberries is harvested annually because of alternate-year pruning practices. Most growers mow the field to about 1 cm above the soil level during the winter. During the following growing season, upright, unbranched shoots grow and produce flower buds; these yield during the following year. Some growers still burn fields to prune during the winter, which offers some advantages for pest control but has a higher cost than mowing. Average yields were 5.5 t/ha in the United States in 2014 but, in a well-managed field, yields of 11 t/ha can be achieved.

Lowbush blueberries are harvested all at one time, often using handheld scoops with tines to remove berries from the bush (Fig. 10). However, machine harvest of lowbush blueberries increased in all production areas to 40% in Maine (United States) and to 25–80% in Canada, depending on the province, by 2005 (Fig. 11). The most common type of machine used is a Bragg reel-type harvester (Doug Bragg Enterprises, Collingwood, Nova Scotia, Canada). In Quebec, Canada, walk-behind machines are used. Almost all of the lowbush crop is processed.

Fig. 10. Hand rake used to pick lowbush blueberries in Nova Scotia, Canada. (Courtesy B. C. Strik; © APS)

Fig. 11. Machine harvest of lowbush blueberries in Nova Scotia, Canada. (Courtesy B. C. Strik; © APS)

Selected References

Aalders, L. E., and Hall, I. V. 1961. Pollen incompatibility and fruit set in lowbush blueberries. Can. J. Genet. Cytol. 3:300-307.

Abbott, J. D., and Gough, R. E. 1987. Seasonal development of highbush blueberry roots under sawdust mulch. J. Am. Soc. Hortic. Sci. 112:60-62.

Austin, M. E., and Brightwell, W. T. 1984. Rejuvenating the rabbiteye blueberry cultivars Tifblue and Woodard. J. Am. Soc. Hortic. Sci. 109:74-77.

Ballington, J. R., Kirkman, W. B., Ballinger, W. E., and Maness, E. P. 1988. Anthocyanin, aglycone, and aglycone-sugar content in the fruits of temperate North American species of four sections in *Vaccinium*. J. Am. Soc. Hortic. Sci. 113:746-749.

Beaudry, R. M., Cameron, A. C., Shirazi, A., and Dostal, D. L. 1992. Modified atmospheric packaging of blueberry fruit: Effect of temperature on package oxygen and carbon dioxide. J. Am. Soc. Hortic. Sci. 117:436-441.

Brazelton, D., and Strik, B. C. 2007. Perspective on the U.S. and global blueberry industry. J. Am. Pomol. Soc. 61(3):144-147.

Camp, W. H. 1945. The North American blueberries with notes on other groups of Vacciniaceae. Brittonia 5:203-275.

Eck, P. 1988. Blueberry Science. Rutgers University Press, New Brunswick, NJ.

Eck, P. 1990. The American Cranberry. Rutgers University Press, New Brunswick, NJ.

Galletta, G. J., and Ballington, J. R. 1996. Blueberries and cranberries. Pages 1-107 in: Fruit Breeding. Vol. II: Vine and Small Fruits. J. Janick and J. N. Moore, eds. John Wiley & Sons, New York.

Galletta, G. J., and Himelrick, D. G., eds. 1990. Small Fruit Crop Management. Prentice-Hall, Englewood Cliffs, NJ.

Gough, R. E., and Shutak, V. G. 1978. Anatomy and morphology of cultivated highbush blueberry. R. I. Agric. Exp. Stn. Bull. 423.

Hancock, J. F., and Draper, A. D. 1989. Blueberry culture in North America. HortScience 24:551-556.

Hancock, J. F., Beaudry, R. M., and Luby, J. J. 1995. Fruits of the Ericaceae. In: Encyclopedia of Fruit Science, Food Technology and Nutrition. R. Macrae, R. Robinson, and M. Sadler, eds. Academic Press, London.

Krebs, S. L., and Hancock, J. F. 1988. The consequences of inbreeding on fertility in *Vaccinium corymbosum* L. J. Am. Soc. Hortic. Sci. 113:914-918.

Luby, J. J., Ballington, J. R., Draper, A. D., Pliszka, K., and Austin, M. E. 1991. Blueberries and cranberries (*Vaccinium*). Pages 391-456 in: Genetic Resources of Temperate Fruit and Nut Crops. J. N. Moore and J. R. Ballington, eds. International Society for Horticultural Science, Wageningen, the Netherlands.

Moon, J. W., Hancock, J. F., Draper, A. D., and Flore, J. A. 1987. Genotype differences in the effect of temperature on CO_2 assimilation and water use efficiency in blueberry. J. Am. Soc. Hortic. Sci. 112:170-173.

Penhallegon, R. 2006. Lingonberry Production Guide for the Pacific Northwest. Oregon State University Extension Service. PNW 583-E.

Roberts, R. H., and Struckmeyer, B. E. 1942. Growth and fruiting of the cranberry. Proc. Am. Soc. Hortic. Sci. 40:373-379.

Strik, B. 2007. Berry crops: Worldwide area and production systems. Pages 3-49 in: Berry Fruit: Value-Added Products for Health Promotion. Y. Zhao, ed. CRC Press, Taylor and Francis Group, Boca Raton, FL.

Strik, B. C. 2014. Organic blueberry production systems—Advances in research and industry. Acta Hortic. 1017:257-267.

Strik, B., and Yarborough, D. 2005. Blueberry production trends in North America, 1992 to 2003 and predictions for growth. HortTechnology 15:391-398.

Vander Kloet, S. P. 1988. The Genus *Vaccinium* in North America. Publication 1828. Agriculture Canada, Ottawa.

(Prepared by B. C. Strik and J. F. Hancock)

Part I. Infectious Diseases

Diseases Caused by Fungi

Fungi and oomycetes known to infect blueberry, cranberry, and lingonberry are also listed in the appendix. Readers will note that in both the appendix and the text, alternative names of fungi are identified exclusively as "synonyms," rather than as specific life cycle stages (e.g., "anamorph," "teleomorph"). This practice reflects the policy for naming fungi issued in 2011 by the International Botanical Congress: namely, that the use of multiple names for the same fungus be abandoned. (Readers interested in learning more about this policy are referred to the 2011 article by Hawksworth [see Selected Reference].) Going forward, accommodating this new policy may mean that the most familiar names of some fungi will change. Changes to the names of fungi, as well as other updates to taxonomy and nomenclature, will be made in the Common Names of Diseases lists for blueberry, cranberry, and lingonberry, which are available on the website of The American Phytopathological Society.

Selected Reference

Hawksworth, D. L. 2011. A new dawn for the naming of fungi: Impacts of decisions made in Melbourne in July 2011 on the future publication and regulation of fungal names. IMA Fungus 2:155-162.

Blueberries (Highbush and Rabbiteye)

Alternaria Leaf Spot and Fruit Rot

A leaf spot disease of highbush blueberry caused by an *Alternaria* sp. was first observed in North Carolina in 1973. *Alternaria* spp. had been isolated from blueberry stems and leaves by several researchers, but they were considered secondary invaders. The fungus has since been widely recognized as a pathogen of blueberry, is worldwide in distribution, and is likely to be found wherever blueberries are grown. The disease was described on southern highbush blueberry in Argentina in 2004 and on highbush blueberry in China in 2006. The fungus also was identified as the principal cause of postharvest fruit decay of blueberries in North Carolina (United States) in 1971 and has been reported to cause a fruit rot in most blueberry-growing regions. The development of semiautomated sorting and packing equipment has increased the incidence of this and other postharvest rots because of increased handling and the contamination of surfaces with spores during the packing process.

Symptoms

Leaf lesions develop as circular to irregularly shaped, light brown to gray spots 1–5 mm in diameter surrounded by a brownish red border (Fig. 12). Lesions continue to enlarge when humidity is high but remain as small flecks when humidity is low. Flecking also occurs on stems.

Symptoms of fruit infection develop when the fruit ripen. The calyx end of the berry is covered with closely appressed, greenish black fungal mycelium and spores (Fig. 13). After storage at room temperature for 10–20 h, fruit may become "leaky." Postharvest infections appear as fluffy gray-green mycelium, often (but not exclusively) on the stem end of the berry. If berries are handled wet or are not kept cool after packing, this mycelium may completely engulf the berry by the time commercially packed fruit reach the consumer.

Causal Organism

This leaf spot and fruit rot is caused by *Alternaria tenuissima*. Conidiophores (4–5 × 30–50 µm) are yellow-brown, simple, and smooth and have three to five septa, with slightly swollen tips. Conidia are dark, ovoid, obclavate, and pyriform and are muriformly divided with one to seven cross septa and one to three vertical septa; they are 7–15 × 15–55 µm (mean 10.8 × 35.8 µm). Beaks (spore tails) are cylindrical, 2–3 µm in diameter, and up to 15 µm long (Fig. 14). Colonies in culture are initially gray, becoming dark olive green with sporulation.

Fig. 12. Alternaria leaf spot, caused by *Alternaria tenuissima*, on blueberry leaves. (Courtesy R. D. Milholland; © APS)

Disease Cycle and Epidemiology

Alternaria leaf spot is a minor leaf disease that develops primarily during prolonged periods of cool, wet weather, when spores are produced in abundance. In most cases, only lower leaves are affected but, in rare cases, severe defoliation occurs. *A. tenuissima* overwinters in and on twigs and in debris on the ground. The fungus also readily colonizes freeze- and herbicide-injured shoots. The optimal temperature for growth and spore germination is 28°C; however, disease development is optimal at 20°C. In postharvest handling and packing facilities, spores from field-infected berries rapidly contaminate inspection belts and other handling surfaces. Healthy berries coming into contact with these surfaces pick up the spores, and even dry berries may become infected through wet stem scars.

Management

Control of Alternaria leaf spot and preharvest fruit rot is essential to the delivery of high-quality blueberries to the market. Fungicides provide some control, but the primary means of controlling Alternaria fruit rot is through cultural practices, such as timely harvesting, dry handling, and rapid postharvest cooling. High levels of Alternaria fruit rot in the field are an indicator of berries that have been left on the bush too long. In North Carolina, highbush blueberries must be picked every 4–5 days and rabbiteye blueberries should be picked every 8–10 days. To avoid postharvest fruit rots, it is critical that berries are dry during harvesting and packaging. Because fungal spores need moisture to germinate and infect fruit, handling berries when they are wet with rain or dew greatly increases the chance of postharvest rots. Picking buckets, packing lines, and inspection belts should be cleaned frequently to reduce the chance of contamination by fungal spores and other microorganisms. Rapid forced-air cooling of fruit after harvest is necessary to dry the stem scars, retard fungal growth, and preserve the quality of harvested fruit. Cultivars that have a wet stem scar are more prone to postharvest infection and should be avoided.

Fig. 13. Alternaria fruit rot, caused by *Alternaria tenuissima*, on blueberry. (Courtesy W. O. Cline)

Fig. 14. Conidia of *Alternaria tenuissima*, the causal organism of Alternaria leaf spot and fruit rot. (Courtesy W. O. Cline)

Selected References

Cline, W. O. 1996. Postharvest infection of highbush blueberries following contact with infested surfaces. HortScience 31:981-983.

Luan, Y. S., Feng, L., Xia, X. Y., and An, L. J. 2007. First report of *Alternaria tenuissima* causing disease on blueberry in China. Plant Dis. 91:464.

Milholland, R. D. 1973. A leaf spot disease of highbush blueberry caused by *Alternaria tenuissima*. Phytopathology 63:1395-1397.

Milholland, R. D., and Jones, R. K. 1972. Postharvest decay of highbush blueberry fruit in North Carolina. Plant Dis. Rep. 56:118-122.

Wright, E. R., Rivera, M. C., Esperón, J., Cheheid, A., and Rodríguez Codazzi, A. 2004. Alternaria leaf spot, twig blight, and fruit rot of highbush blueberry in Argentina. Plant Dis. 88:1383.

(Prepared by R. D. Milholland; Revised by W. O. Cline)

Anthracnose Fruit Rot

Blueberry anthracnose is one of the most economically important diseases of highbush, southern highbush, and rabbiteye blueberries. The disease is found in most growing areas of North America as well as several other continents. It has been reported most recently from Australia, Japan, New Zealand, Norway, Spain, and Switzerland. In addition, the primary causal organism, *Colletotrichum fioriniae*, has been widely reported on other host species. While the disease can affect blossoms and shoots, its most destructive symptom is a rot of ripe fruit that may develop before or after harvest, from which its other common name, ripe rot, is derived. Anthracnose fruit rot is less common on lowbush blueberries. *C. fioriniae* also causes a ripe rot of other fruit, nut, and vegetable crops, including almond, apple, avocado, citrus, cranberry, grape, mango, papaya, pecan, peppers, and strawberry.

Although anthracnose fruit rot is widespread on commercial blueberries, losses vary greatly from region to region and from season to season. There is also significant variability in susceptibility among commercial cultivars. In Michigan (United States), for example, overall crop losses may range from 10 to 20%, depending on the season. The incidence of infection typically increases in successive pickings; in New Jersey (United States), losses exceeding 50% have been observed in the third picking. Ripe fruit may show symptoms of infection in the field, but most economic losses occur after harvest, when rot symptoms from latent infections develop on fruit in storage and spread to healthy fruit. Losses from postharvest rot may approach 100% under poor storage conditions.

Symptoms

Fruit rot is the most common and economically important stage of blueberry anthracnose. Fruit infections begin during bloom but infections remain latent and symptomless until the fruit ripen. The first symptom of infection may be shriveling of the berries on the bush. On overripe fruit, acervuli may form and exude sticky, salmon-colored conidial masses during warm, humid weather. Symptom development, however, may be delayed until fruit have been harvested and stored for a week or more (Fig. 15). In storage, fruit rot typically begins at the blossom end of a berry, which softens and becomes sunken.

Additional symptoms include blossom blight, but this symptom is rarely seen in the field except on susceptible cultivars under very humid conditions. Shoot tip blight (Fig. 16) and leaf spots may also develop during the growing season. Leaf lesions vary from small, brown, circular to irregularly shaped spots to large, black, poorly defined necrotic lesions. Shoot tip blight is common on succulent, rapidly growing shoots and may be more abundant on susceptible cultivars such as Coville and Bluecrop.

Causal Organisms

Colletotrichum fioriniae is the species most often associated with anthracnose fruit rot on blueberries. The former name for this pathogen is *Colletotrichum acutatum* (syn. *Glomerella acutata*). *C. fioriniae* is a coelomycete fungus that produces fusiform, one-celled, hyaline conidia averaging about 3 × 14 µm. Conidia are produced on conidiophores in acervuli just below the epidermis of infected tissue and emerge as orange- to salmon-colored, sticky masses. On a berry, the acervuli sometimes grow in concentric rings, giving a lesion a bull's-eye appearance. In culture, the fungus is often chromogenic when grown on standard growth media, such as potato dextrose agar. Blueberry isolates are typically deep red to light orange on the underside of the colony and white on top, although some include black or dark gray rays or sections.

C. gloeosporioides (syn. *Glomerella cingulata*) was considered the causal organism of anthracnose fruit rot in the original literature and is still found on rabbiteye and southern highbush blueberries in the southeastern United States but occurs less frequently than *C. fioriniae*. *C. gloeosporioides* produces masses of sticky, salmon-colored conidia in acervuli formed on infected fruit and in culture. The conidia are hyaline, oblong-elliptical, and variable in size (average 4.7 × 10–28 µm) and shape but are usually rounded at the ends (Fig. 17). Some isolates of *C. gloeosporioides* from rotted fruit produce perithecia of *G. cingulata*. Perithecia are subspherical; asci (55 × 70 µm) are subclavate; ascospores are 3.5 × 12–22 µm. Fruit infected by *C. gloeosporioides* display symptoms similar to fruit infected by *C. fioriniae*.

Disease Cycle and Epidemiology

The fungus overwinters as mycelium in the bud scales surrounding floral and vegetative buds, blighted twigs, and fruit pedicels from the previous season. Conidia are dispersed from the overwintering sites to flower clusters and developing fruit (Fig. 18) either by splash dispersal or by epiphytic growth on the plant surface. Because fruit are susceptible to infection at all stages of development, the level of infection depends mainly on inoculum availability, cultivar susceptibility, and microclimate. Floral bud scales that persist on fruit clusters are believed to be an important source of initial inoculum because

Fig. 15. Symptoms of anthracnose fruit rot, caused by *Colletotrichum fioriniae*, on harvested fruit. All blueberries were produced in commercial settings but with various levels of disease management. Note the pint on the right has nearly 100% anthracnose fruit rot. (Courtesy P. V. Oudemans; © APS)

Fig. 16. Anthracnose shoot tip blight on blueberry, caused by *Colletotrichum fioriniae*. (Courtesy P. V. Oudemans; © APS)

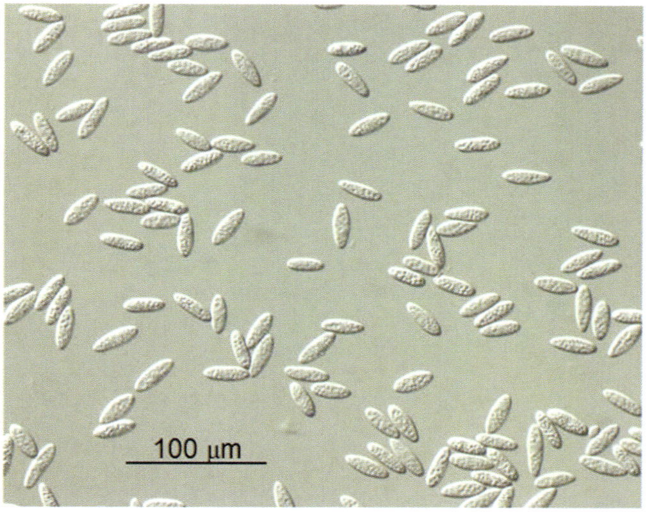

Fig. 17. Conidia characteristic of *Colletotrichum fioriniae*, the causal organism of blueberry anthracnose. (Courtesy P. V. Oudemans; © APS)

Fig. 18. Blueberry fruit naturally infected with *Colletotrichum fioriniae*, the causal organism of anthracnose fruit rot, with pedicels sporulating under laboratory conditions. (Courtesy P. V. Oudemans; © APS)

of their proximity to developing fruit. The rate of bud scale dehiscence differs among cultivars and is related to field resistance; the faster these bud scales drop, the lower the incidence of fruit infection. Conidia germinate, produce germ tubes with melanized appressoria, and penetrate the cuticle through an infection peg. An infection hypha grows from the appressoria and ceases development until the fruit ripens. Infection requires a minimum of 8 h of continual leaf wetness at 25°C, with longer wetness durations needed at higher and lower temperatures. At 10°C, about 36 h of wetness are needed for infection. Sporulation on fruit is a major source of secondary inoculum. The optimal temperature for hyphal growth of the fungus is 26°C. Losses are most serious when there are prolonged periods of warm, wet weather during bloom and just before harvest.

Management

The highbush cultivars Bluecrop, Bluetta, Cape Fear, and Coville are highly susceptible to the ripe rot fungus, whereas Elliott, Legacy, and Reveille exhibit resistance to infection. 'Duke' is considered only moderately resistant in controlled trials but exhibits significant field resistance under commercial growing conditions in New Jersey. Rabbiteye cultivars are generally far less susceptible, although differences among cultivars are evident. 'Powderblue', for instance, develops less ripe rot than 'Premier'.

Preharvest control of ripe rot on susceptible cultivars is best achieved with applications of an effective fungicide at 7- to 10-day intervals, beginning at full bloom. Postharvest disease development can be reduced by timely harvests, rapid postharvest cooling, and sanitation of sorting lines.

Fungicide applications should be initiated during early bloom. Broad-spectrum fungicides (such as captan, fluazinam, and ziram) and more selective fungicides (such as the strobilurins) are very effective. Demethylation inhibitor (DMI) fungicides are variable in performance against these pathogens. Delayed dormant sprays with lime sulfur or captan reportedly help reduce overwintering inoculum.

Selected References

Daykin, M. E., and Milholland, R. D. 1984. Infection of blueberry fruit by *Colletotrichum gloeosporioides*. Plant Dis. 68:948-950.

DeMarsay, A. 2005. Anthracnose fruit rot of highbush blueberry: Biology and epidemiology. Ph.D. dissertation. Rutgers University, New Brunswick, NJ.

Hartung, J. S., Burton, C. L., and Ramsdell, D. C. 1981. Epidemiological studies of blueberry anthracnose disease caused by *Colletotrichum gloeosporioides*. Phytopathology 71:449-453.

Polashock, J. J., Ehlenfeldt, M. K., Stretch, A. W., and Kramer, M. 2005. Anthracnose fruit rot resistance in blueberry cultivars. Plant Dis. 89:33-38.

Smith, B. J., Magee, J. B., and Gupton, C. L. 1996. Susceptibility of rabbiteye blueberry cultivars to postharvest diseases. Plant Dis. 80:215-218.

Verma, N., MacDonald, L., and Punja, Z. K. 2007. Environmental and host requirements for field infection of blueberry fruits by *Colletotrichum acutatum* in British Columbia. Plant Pathol. 56:107-113.

Wharton, P. S., and Schilder, A. C. 2008. Novel infection strategies of *Colletotrichum acutatum* on ripe blueberry fruit. Plant Pathol. 57:122-134.

Yoshida, S., Tsukiboshi, T., Shinohara, H., Koitabashi, M., and Tsushima, S. 2007. Occurrence and development of *Colletotrichum acutatum* on symptomless blueberry bushes. Plant Pathol. 56:871-877.

(Prepared by R. D. Milholland;
Revised by A. DeMarsay and P. V. Oudemans)

Armillaria Root Rot

Armillaria root rot is an important disease worldwide; it affects more than 500 species of woody plants. The causal fungus can attack fruit trees, strawberry, raspberry, blackberry, and grape. To date, *Armillaria* spp. have been reported as pathogens of highbush blueberry only in the United States and Italy. Many observations of the disease have probably gone unreported, particularly because the symptoms are easily confused with other problems. Although the disease is uncommon, it can cause serious damage in fields where it is found, and it is very difficult to control.

The pathogen is sometimes referred to as the honey fungus, oak root fungus, oak fungus, and shoestring fungus. The disease is most likely to be found on sites where an oak forest has been cleared for planting.

Symptoms

Infected bushes may display a range of symptoms. Plants low in vigor may appear to be suffering from nutritional disorders or a pH imbalance and may slowly decline in vitality over several years. Infected plants have small, chlorotic leaves and may be more susceptible to drought stress and winter injury than healthy plants. Branches may wilt suddenly, and the entire plant may die within a short time period.

Affected bushes may be distributed in patches around focal points of fungal inoculum. The disease may be confined to one area of a field (Fig. 19) or it may be interspersed throughout the field, depending on the distribution of the primary inoculum.

The most reliable diagnostic procedure is to scrape away the outer bark of the trunk and larger roots at or slightly below the soil line to examine the base of the affected bush. The fungus produces white mycelial fans (Fig. 20) between the bark and the hardwood, and striations may be present. Plants killed several months before examination may also have the mycelium of secondary invading fungi. Rhizomorphs—black, rootlike structures that also resemble shoestrings (Fig. 21)—may be attached to the roots or trunk, or they may be found growing freely in the soil.

The fungus may produce mushroom fruiting bodies (Fig. 22) at the base of affected plants during late summer to early autumn if temperature and moisture conditions are right. These mushrooms are typically yellowish brown (honey colored), with a conspicuous ring around the stem below the cap. Because the mushrooms are frequently not produced, the mycelial fans and rhizomorphs are more reliable diagnostic features of the disease.

Causal Organisms

Very little is known about the identity of the *Armillaria* spp. that cause root rot in blueberry. The most important species is

Fig. 19. Highbush blueberry bushes affected by Armillaria root rot, caused by an *Armillaria* sp., in the field. (Courtesy A. M. C. Schilder; © APS)

Fig. 20. Mycelial fan of an *Armillaria* sp. in an Armillaria root rot-infected highbush blueberry plant. (Courtesy A. M. C. Schilder; © APS)

Fig. 21. Rhizomorphs of an *Armillaria* sp. attached to an Armillaria root rot-infected blueberry root. (Courtesy T. D. Miles; © APS)

probably *A. mellea*, although *A. ostoyae* and *A. gallica* also have been isolated from infected bushes. All belong to the order Agaricales in the phylum Basidiomycota.

Mushroom caps are 2–25 cm in diameter; however, they are usually produced in a dense clump and tend to be on the smaller end of this scale. The fruiting bodies darken with age; fresh mushrooms are easier to identify. It is easier to culture the fungus from fresh hyphal fans than from wood. The fungus often produces light to dark brown rhizomorphs in culture. Polymerase chain reaction is necessary to confirm the species.

Disease Cycle and Epidemiology

Armillaria root rot is most likely to occur in recently cleared fields that have sandy, well-drained soil. *Armillaria* spp. survive in the soil on the debris of susceptible host plant roots. Blueberry roots that come in contact with this colonized food base are colonized by direct mycelial contact or from rhizomorphs. The fungus spreads further by root-to-root contact between infected and healthy bushes. The fungus can also survive on wood chip mulches. Spores disseminated from the mushrooms (basidiospores) are probably not important in the spread of the pathogen within a field.

If the fungus remains confined to the roots, plants die slowly over several years. If the fungus invades the trunk, the plants are girdled and killed in a short time period.

Management

When a forested location is to be used for planting, particularly where oak is an important species, the soil should be thoroughly disked and as many root fragments as possible should be removed. The area should be left fallow for at least 3 years so that the amount of inoculum of *Armillaria* spp. is significantly reduced. Soil fumigants or sterilants such as carbon disulfide are effective; however, if they do not penetrate deeper than 50 cm, some inoculum may remain in the soil. *Armillaria* spp. may be significantly weakened by the treatments, however, and may be attacked by *Trichoderma* spp. and other antagonistic fungi in the soil, thus further reducing inoculum potential. If the area to be planted is large, a few bushes planted in various locations 1 or 2 years after clearing might serve as a test of the degree of soil inoculum.

Infected bushes should be removed and burned. A bush from beyond the disease perimeter that appears healthy should also be examined and should be removed if rhizomorphs and

Fig. 22. Basidiocarps of an *Armillaria* sp. at the base of an Armillaria root rot-infected highbush blueberry plant. (Courtesy F. L. Caruso; © APS)

mycelial fans are present at the root collar. Soil fumigation of areas where infected bushes have been removed helps to prevent recurrence of the disease in replanted areas. Wood chip mulch should be removed from healthy bushes around the disease centers and should not be replaced as long as there are rhizomorphs growing in the chips or at the mulch–soil line.

No information is available on cultivar susceptibility to Armillaria root rot. Most cultivars are probably highly susceptible to the fungus.

Selected References

Eglitis, M., Gould, C. J., and Johnson, F. 1966. Fungi found on Ericaceae in the Pacific coastal area. Wash. State Agric. Exp. Stn. Bull. 675.

Fulton, R. H. 1958. New or unusual small fruit diseases and disease-like occurrences in Michigan. Plant Dis. Rep. 42:71-73.

Munnecke, D. E., Kolbezen, M. J., Wilbur, W. D., and Ohr, H. D. 1981. Interactions involved in controlling *Armillaria mellea*. Plant Dis. 65:384-389.

Prodorutti, D., Palmieri, L., Gobbin, D., and Pertot, I. 2006. First report of *Armillaria gallica* on highbush blueberry (*Vaccinium corymbosum*) in Italy. Plant Pathol. 55:583.

Shaw, C. G., III, and Kile, G. A. 1991. Armillaria root disease. U. S. Dep. Agric. For. Serv. Agric. Handb. 691.

(Prepared by F. L. Caruso)

Botryosphaeria Stem Blight

Botryosphaeria stem blight, commonly referred to as dieback, is a destructive disease of highbush and rabbiteye blueberries, primarily in the southeastern United States extending north into New Jersey. Losses are most severe in young fields, where plants often become infected and die in the first 2 years. In older fields, stem blight is most commonly seen as a dieback on one or more canes in a bush. A dramatic increase in stem blight incidence over the past 50 years has been associated with an increase in mechanized harvesting and mechanized pruning and a shift to more susceptible cultivars. Stem blight is also a major disease of low-chill southern highbush cultivars grown in the Gulf south. Besides blueberry, the disease affects other fruit crops, including apple, blackberry, peach, and citrus, as well as many other woody plants in temperate and tropical zones around the world.

Symptoms and Host Susceptibility

The point of initiation of most stem blight infections is a wound on stems or canes. During the early stages of infection, characteristic symptoms are yellowing and reddening or drying of the leaves on one or more branches. A rapid wilt of leaves on individual branches is often followed by death of the entire plant as the fungus spreads downward through vascular tissue to the base of the plant. The most conspicuous symptom, called flagging, is a dead branch with leaves still attached in close proximity to living branches (Fig. 23). A stem blight-infected stem has a uniform, light brown discoloration in the wood extending down the infected side of the stem from a few centimeters to the entire length (Fig. 24). Twig infections may be confused with winter injury or other twig diseases. The infection can also develop in wounds at the base (crown) of the bush in susceptible cultivars, resulting in rapid plant death without the typical flagging symptoms associated with infections on individual stems. Younger plants die rapidly within 1–2 years of planting, and the mortality rate is highest when infection develops at or near the crown (Fig. 25).

Fig. 23. Botryosphaeria stem blight on blueberry, caused by *Botryosphaeria dothidea*. (Courtesy R. D. Milholland; © APS)

Fig. 24. Cross section of a blueberry stem infected with *Botryosphaeria dothidea*, the causal organism of Botryosphaeria stem blight, showing brown discoloration of the infected wood. (Courtesy P. V. Oudemans; © APS)

Stem blight infections usually become evident soon after harvest, and new infections can be observed throughout the summer months. Infections are often associated with wounds caused by mechanical damage, contact herbicide injury, insect damage, or late-season cold injury on succulent shoots that happened during the previous growing season. Temperatures below –18°C (0°F) may cause cracking in the forks on blueberry stems, which has resulted in wound-related epidemics the following spring. The greatest incidence of stem blight in New Jersey was found in years following winter injury. Drought stress also predisposes plants to stem blight. Fields that are irrigated regularly during dry periods have a much lower incidence of death and stem blight symptoms on infected plants than do fields that do not have sufficient and consistent soil moisture. On rabbiteye cultivars, stem blight infection usually is limited to dieback of individual canes or twigs, with an occasional major cane dying back to the ground, and it does not typically cause plant death except in young, vigorous plants. However, many mature rabbiteye plants in south Mississippi (United States) died during the spring following extensive bush damage caused by hurricanes the previous summer. Rabbiteye cultivars also show an increase in disease severity following exposure to stresses such as heat, drought, or cold.

Blueberry cultivars vary widely in their response to infection by *Botryosphaeria dothidea*, and isolates of the pathogen show a broad range of pathogenicity. Most highbush cultivars grown in the southern United States are susceptible to *B. dothidea*. Field observations in North Carolina (United States) indicate that the highbush cultivars Bluechip and Bounty are very susceptible to stem blight. 'Croatan', 'Reveille', 'Harrison', and 'Bladen' and the rabbiteye cultivars Premier and Powderblue are also susceptible; however, plant losses of these cultivars average less than 10–20%. The highbush cultivars Murphy, Cape Fear, and O'Neal have shown field resistance in North Carolina with few plant deaths, although occasionally they do become

Fig. 25. Young blueberry plants showing symptoms of Botryosphaeria stem blight, caused by *Botryosphaeria dothidea*. (Courtesy R. D. Milholland; © APS)

infected. Field observations of rabbiteye cultivars grown in Mississippi revealed a higher mortality rate of Tifblue plants caused by stem blight compared with that of Climax and Woodard.

Causal Organisms

Stem blight, first reported in 1958 to be caused by an undescribed *Botryosphaeria* sp. closely related to, but distinct from, *B. corticis*, was subsequently shown to be caused by *B. dothidea* (syn. *Fusicoccum aesculi*). Pseudothecial locules are usually 120–220 µm in diameter and contain bitunicate asci (15–20 × 60–100 µm). Ascospores (6–10 × 15–24 µm) are oblong to fusoid, hyaline, and aseptate. Conidia (6–7 × 16–22 µm) are hyaline, aseptate, and ovate to broadly ellipsoid and are borne on simple, short to elongate conidiophores. Pseudothecia and pycnidia are scattered or in botryose groups and are subepidermal, becoming erumpent at maturity. Young cultures on potato dextrose agar or oatmeal agar are fluffy and white initially, but later the mycelium turns light to dark gray. The minimum, optimum, and maximum temperatures for mycelial growth are 10, 28, and 35°C, respectively. Two other species in the family Botryosphaeriaceae, *Neofusicoccum ribis* and *Lasiodiplodia theobromae*, have been reported to be the primary causal organisms of stem blight of southern highbush blueberries in Florida (United States).

Disease Cycle and Epidemiology

B. dothidea overwinters in dead and infected stems of blueberry or other woody plants. Inoculum (primarily conidia) is present throughout the growing season in the southeastern United States, with the highest levels in June and July and the lowest levels during the winter. Maturation and release of spores are affected by temperature and rainfall. Ascospores are windborne, but they are relatively minor in importance as an inoculum source in warmer winter hardiness zones. Spores are carried by wind (ascospores) and rain (conidia) from infected stems to wounds on healthy plants. These spores germinate and invade the vascular tissue of the host, causing a pecan-brown discoloration that extends up and down the stem from the infection point, eventually killing the stem. Stems killed by blight drop their leaves after a few weeks and turn dark brown to black. These dead infected stems are noticeably darker than stems dying from other causes. Fungal fruiting bodies (pycnidia) are produced all along the stem just under the bark, and spores are released and spread to wounds on adjacent bushes. These spores are released year-round with the exception of a few weeks during the winter; however, the greatest numbers of infections are found in early summer.

Plants begin to show symptoms about 4–6 weeks after infection. While the fungus infects the stem primarily through wounds, the mode of penetration and stage of plant development influence both the type and extent of symptom development. Infection of succulent stems through wounds results in a rapid breakdown of all tissues. In older plants, infection usually results in the loss of one or more branches but not the entire plant. Other major entrance sites of the fungus into blueberry plants are through infection of cold-injured shoots around the base of the bush and through stem cankers formed by other fungi, such as *B. corticis* and *Phomopsis* spp. Disease development decreases with increasing age of the wound.

Management

Control of stem blight is limited to cultural practices since fungicides are not effective in the field. Dipping blueberry roots in a fungicidal solution prior to planting was shown to protect plants against stem blight development for a few months but not for a long enough time to offer practical field control. Producing healthy nursery plants plus pruning out diseased tissue during January and February are the most effective cultural practices for the control of stem blight and should be an integral part of disease control programs. Other cultural practices that reduce the incidence of stem blight are proper site selection, management of soil nutrients, use of resistant cultivars, and avoiding wounding plants.

To reduce the amount of damage caused by stem blight, both extremely sandy soils that lead to poor growth and black, heavy muck soils that promote excessive growth should be avoided. On soils with high organic content (>5%), new plants can be established without the use of fertilizer. Thorough field preparation prior to planting is essential in recently cleared fields where old stumps, trunks, and branches are buried. This debris promotes high populations of termites, which have been associated with a high incidence of death in young plants because of stem blight.

Resistant cultivars are available and should be planted in areas where stem blight is a problem. Stem blight resistance is a primary objective of most southern blueberry breeding programs. Selection of resistant genotypes has generally relied on the natural infection of seedlings in field trials; however, several screening procedures utilizing potted plants or detached stems are being used to determine the stem blight susceptibility of blueberry germplasm. New blueberry plantings should be established with disease-free plants, and any practices that wound plants should be avoided. To prevent the formation of succulent shoots late in the growing season, fertilizer should not be applied after midsummer, especially on young bushes. This allows bushes to enter a natural dormancy and reduces the chance of cold injury.

The best method of reducing disease incidence in established fields is to remove entire plants or plant parts infected with *B. dothidea*. This prevents disease spread to healthy plants. Prompt removal of all infected stems 15–20 cm below any sign of diseased wood prevents the infection from continuing down into the crown, possibly killing the plant. All dead and infected wood should be removed from the field and burned or shredded to reduce the number of spores released in the field. While infected material can be removed at any time, pruning during the winter months reduces the possibility that the pruning wounds become infected. Since stem blight is most damaging to young plantings, pruning of young plants should be limited to the removal of stem blight-infected canes and heavy pruning to promote rapid growth should be avoided.

Selected References

Cline, W. O., and Schilder, A. M. C. 2006. Identification and control of blueberry diseases. Pages 115-138 in: Blueberries for Growers, Gardeners, and Promoters. N. F. Childers and P. M. Lyrene, eds. Norman F. Childers Publications, Gainesville, FL.

Creswell, T. C., and Milholland, R. D. 1987. Responses of blueberry genotypes to infection by *Botryosphaeria dothidea*. Plant Dis. 71:710-713.

Creswell, T. C., and Milholland, R. D. 1988. Spore release and infection periods of *Botryosphaeria dothidea* on blueberry in North Carolina. Plant Dis. 72:342-346.

Milholland, R. D. 1972. Histopathology and pathogenicity of *Botryosphaeria dothidea* on blueberry stems. Phytopathology 62:654-660.

Witcher, W., and Clayton, C. N. 1963. Blueberry stem blight caused by *Botryosphaeria dothidea* (*B. ribis*). Phytopathology 53:705-712.

Wright, A. F., and Harmon, P. F. 2010. Identification of species in the *Botryosphaeriaceae* family causing stem blight on southern highbush blueberry in Florida. Plant Dis. 94:966-971.

(Prepared by R. D. Milholland; Revised by B. J. Smith)

Botryosphaeria Stem Canker

Botryosphaeria stem canker has been one of the most important limitations on the production of northern highbush blueberries in southeastern North Carolina (United States) for more than 70 years. Blueberries were first planted in Magnolia,

North Carolina, in 1928. Ten years later, all cultivars had declined so badly from stem canker that they had to be removed. The use of infected wood for propagation has contributed to the prevalence and spread of this disease. Disease incidence is often so high in new plantings that either productivity is severely reduced or the planting is destroyed before it comes into full production. The disease is present in both cultivated and wild highbush blueberry as well as in some rabbiteye blueberry plantings in the United States in North Carolina, South Carolina, Georgia, Florida, Alabama, and Mississippi, and it occurs in highbush plantings in New Jersey. The disease is endemic on several native *Vaccinium* spp. in the southeastern United States.

Symptoms

Several stem diseases of blueberries may be confused with certain stages of Botryosphaeria stem canker. Symptoms first appear as small red lesions on succulent stems 7 days after infection. Lesions develop slowly; they become swollen and broadly conical in 6 months in susceptible cultivars (Fig. 26). Symptoms vary with the susceptibility of the plant. On some species or cultivars (e.g., rabbiteye), lesions may develop initially as brown, slightly sunken spots. Large, swollen cankers with deep fissures and cracks develop on susceptible cultivars after 2–3 years, whereas cankers are restricted in size on more resistant cultivars. On highly susceptible cultivars, such as Weymouth and Wolcott, cankers enlarge and may girdle and kill the stem (Fig. 27). On moderately susceptible cultivars, such as Scammell, cankers are often extensive but not conspicuously swollen. Numerous fruiting bodies of the pathogen appear in cankers of susceptible cultivars.

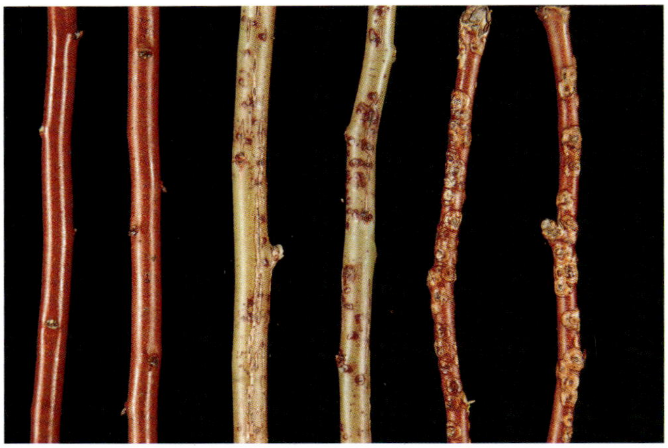

Fig. 26. Lesions of Botryosphaeria stem canker, caused by *Botryosphaeria corticis*, on resistant (left), moderately susceptible (middle), and highly susceptible (right) blueberry stems 6–12 months after infection. (Courtesy R. D. Milholland; © APS)

Fig. 27. Mature blueberry bush with several canes killed by Botryosphaeria stem canker, caused by *Botryosphaeria corticis*. (Courtesy W. O. Cline)

Causal Organism

The causal fungus was first described in 1942 as *Physalospora corticis*; in 1954, the name was changed to *Botryosphaeria corticis*. Pseudothecia are mostly scattered and partly embedded in cankers consisting of a dark stromatic mixture of fungal and host tissue. They are conical and ostiolate, with a dark brown to black wall, and up to 250 µm in diameter. Asci are eight spored, clavate, bitunicate, and 25–28 × 145–165 µm (Fig. 28). Ascospores (9.5–13.5 × 24–34.5 µm) are hyaline, aseptate, ellipsoid to fusoid, and widest in the middle to upper third. Pycnidia are dark brown to black, globose, and up to 450 µm in diameter. Conidia are fusiform, hyaline, aseptate, and widest in the middle to upper third, with a truncate base, and are 5–7.5 × 20.5–34.5 µm. Microconidia may be found in pycnidia on stem cankers and are rod shaped with obtuse ends, hyaline, and 1.5–2 × 4.1–6 µm. Colonies on cornmeal agar are white at first, turning olive green with clumps of loosely aggregated hyphae, and are 28–40 mm in diameter after 7 days at 25°C.

Phylogenetic analysis has shown that isolates from New Jersey are very similar to isolates from North Carolina. The slightly larger ascospores and conidia of *B. corticis* distinguish it from *B. dothidea*, the causal organism of Botryosphaeria stem blight of blueberry.

At least eight pathogenic races of *B. corticis* infect highbush and rabbiteye cultivars. Races 1 and 6 are the most prevalent in North Carolina, and race 6 is the primary race isolated from rabbiteye cultivars grown in Georgia and North Carolina. Race 7 is highly virulent on the cultivars Bluecrop, Croatan, Earliblue, and Wolcott. Race 8 is the most virulent; it causes large cankers on all five cultivars in a set of differential cultivars as well as on most highbush cultivars grown in North Carolina. However, this race is not widespread; it has been detected in only one blueberry planting in North Carolina.

Disease Cycle and Epidemiology

Stems of current-season growth are infected in late spring when ascospores and conidia are released from pseudothecia

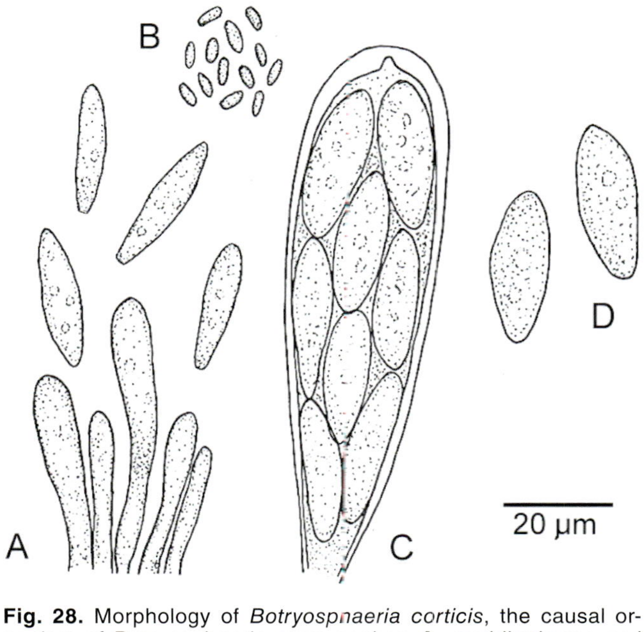

Fig. 28. Morphology of *Botryosphaeria corticis*, the causal organism of Botryosphaeria stem canker: **A,** conidiophores and conidia; **B,** microconidia; **C,** ascus with ascospores; and **D,** ascospores. (Courtesy A. M. C. Schilder; © APS)

and pycnidia during wet weather and disseminated by wind throughout the planting. Temperature affects both the number and the type of lesions produced by *B. corticis* on blueberry stems. The optimum temperature range for fungal growth, sporulation, and spore germination is 25–28°C. At 25°C, ascospores germinate within 24 h. Disease development is limited to small red flecks at 16°C. Alternating temperatures of 16 and 27°C do not appear to restrict stem canker development; however, a constant low temperature (16°C) appears to significantly inhibit disease development in susceptible cultivars.

The infection process—that is, spore germination, penetration, and establishment of *B. corticis* in blueberry stems—takes place within 24 h. Only current-season growth is susceptible to infection by the fungus, a relatively slow-growing organism that is confined to the cortex during the first year. In susceptible cultivars, the fungus causes severe hyperplasia of the cortical parenchyma, phloem, and ray cells. As the fungus continues to grow and invade the wood, cankers develop, and the stem finally becomes girdled and eventually dies. In resistant cultivars, the fungus is restricted to the outer portion of the epidermis, and cankers appear as small, raised lesions.

Management

Resistance to stem canker is essential for the culture of northern and southern highbush blueberries in the southern and southeastern United States. None of the highbush cultivars have resistance to all eight races of *B. corticis*. The northern highbush cultivars Murphy, Croatan, and Bluechip are resistant to the predominant race (race 1) in North Carolina. O'Neal is the only southern highbush cultivar that has shown resistance to stem canker in greenhouse and field testing. Avonblue and Sharpblue have exhibited field resistance in Florida.

Susceptible cultivars can sometimes be grown successfully in eastern North Carolina if the planting stock is free of canker and the planting is isolated from infected plantings. The use of canker-free propagation wood is important in limiting the spread of this disease. Cankers should be removed from blueberry plantings to reduce the inoculum level.

Rabbiteye cultivars are susceptible to *B. corticis* but are seldom killed in commercial plantings. Cultivar susceptibility is largely undetermined. In general, the use of fungicides for stem canker control has been ineffective.

Selected References

Alves, A., Phillips, A. J. L., Henriques, I., and Correia, A. 2007. Rapid differentiation of species of *Botryosphaeriaceae* by PCR fingerprinting. Res. Microbiol. 158:112-121.

Cline, W. O., and Milholland, R. D. 1988. Identification of a new race of *Botryosphaeria corticis* on highbush and rabbiteye blueberry in North Carolina. Plant Dis. 72:268.

Milholland, R. D. 1970. Histology of Botryosphaeria canker of susceptible and resistant highbush blueberries. Phytopathology 60:70-74.

Milholland, R. D. 1972. Factors affecting sporulation and infection by the blueberry stem canker fungus, *Botryosphaeria corticis*. Phytopathology 62:137-139.

Milholland, R. D. 1984. Occurrence of a new race of *Botryosphaeria corticis* on highbush and rabbiteye blueberry. Plant Dis. 68:522-523.

Milholland, R. D., and Galletta, G. J. 1969. Pathogenic variation among isolates of *Botryosphaeria corticis* on blueberry. Phytopathology 59:1540-1543.

Phillips, A. J. L., Oudemans, P. V., Correia, A., and Alves, A. 2006. Characterisation and epitypification of *Botryosphaeria corticis*, the cause of blueberry cane canker. Fung. Diversity 21:141-155.

Phillips, A. J. L., Alves, A., Pennycook, S. R., Johnston, P. R., Ramaley, A., Akulov, A., and Crous, P. W. 2008. Resolving the phylogenetic and taxonomic status of dark-spored teleomorph genera in the Botryosphaeriaceae. Persoonia 21:29-55.

Taylor, J. 1958. Stem canker and related blueberry diseases. N. C. Agric. Exp. Stn. Tech. Bull. 132.

(Prepared by R. D. Milholland; Revised by A. M. C. Schilder)

Botrytis Blight and Fruit Rot

Botrytis blight (or gray mold) is one of the most common diseases found on a wide variety of plants worldwide. The disease was first observed on cultivated blueberry in New Jersey (United States) in 1924. In 1930, Botrytis blossom blight developed on lowbush blueberry in Maine (United States) during a week of foggy, warm weather in late June and became very prevalent. By 1950, Botrytis blight was considered the most important disease of blueberry in the Pacific Northwest. Severe Botrytis blossom blight outbreaks were reported in New Jersey and North Carolina (United States) in 2003, particularly on the cultivar Duke. In the southeastern United States, the disease is of minor importance on highbush blueberry in most years, but it can result in severe losses in plantings of rabbiteye blueberry, particularly of the cultivar Tifblue. In the Midwest, Botrytis blight is a sporadic problem. The disease also is found on highbush blueberry in Canada, Chile, Argentina, and Europe. Yield losses caused by blossom blight can be as high as 30–40%. Postharvest fruit losses caused by Botrytis fruit rot are also widespread. Twig blight reduces the number of flower buds and, thus, yield the following year.

Symptoms

Infections of blossoms, tender green twigs, and young leaves can cause severe damage, especially when rainy weather persists during bloom. Often, the fungus advances from infected flower clusters into the stem, girdling it and killing all flowers and fruit above the infection site (Fig. 29).

Infected twigs at first are brown to black (Fig. 30) and later become bleached to tan or gray. Blossom and twig blight symptoms may be mistaken for freeze injury or Phomopsis twig blight. Infected blossoms may be covered with abundant gray mycelium and conidia. When young leaves are attacked, often by contact with infected flower parts, the first symptom is a chlorotic area that later becomes necrotic and turns brown. Under humid conditions, gray spore masses develop on infected tissues.

Developing berries can also become infected (Fig. 31), but the fruit rot phase usually does not appear until after the berries are harvested. Postharvest decay can be a serious problem.

Fig. 29. Highbush blueberry with Botrytis blossom blight, caused by *Botrytis cinerea*. (Courtesy R. S. Byther; © APS)

Infected berries shrivel slightly and the fungus produces masses of conidia on the surface of the fruit.

Causal Organism

Botrytis blight is caused by the fungus *Botrytis cinerea* (syn. *Botryotinia fuckeliana*). Only the anamorphic state has been reported on blueberry. *B. cinerea* has a wide host range and has pathogenic as well as saprophytic abilities. It is an aggressive colonizer of injured and senescing plant tissues. While this pathogen is generally not considered to be host specific, genetic differences and a certain degree of host specialization exist among isolates. Two genetically distinct groups, which are considered cryptic species, have been identified in France, making *Botrytis cinerea* a species complex.

In culture, *B. cinerea* generally grows rapidly on several media, but growth is variable. On potato dextrose agar, the mycelium is initially whitish but within several days becomes brownish gray. Conidia and sometimes sclerotia are produced in culture (Fig. 32).

Hyphae are branched, septate, and hyaline. Exposure to daylight or ultraviolet light is necessary for sporulation. Conidiophores are long (often more than 16–30 µm × 2 mm, hyaline to dark brown, and branched (Fig. 33 top). Conidiophores terminate in a rounded apical cell with conidia forming as blastospores on very short denticles (Fig. 33 bottom). Conidia (6–9 × 8–14 µm) are one celled, ovate, and hyaline. In mass, they ap-

Fig. 30. Botrytis twig blight, caused by *Botrytis cinerea*, on highbush blueberry. (Courtesy A. M. C. Schilder; © APS)

Fig. 31. Unpollinated ovaries and immature blueberries infected by *Botrytis cinerea*, the causal organism of Botrytis blight. Note the necrotic peduncle. (Courtesy P. V. Oudemans; © APS)

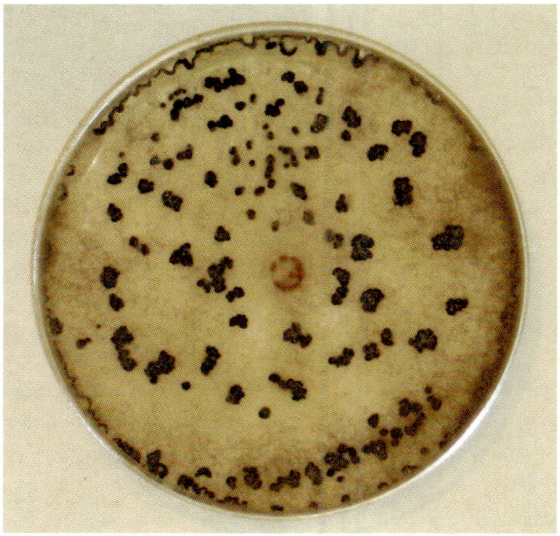

Fig. 32. Culture of *Botrytis cinerea*, the causal organism of Botrytis blight, from blueberry forming sclerotia on potato dextrose agar. (Courtesy T. D. Miles; © APS)

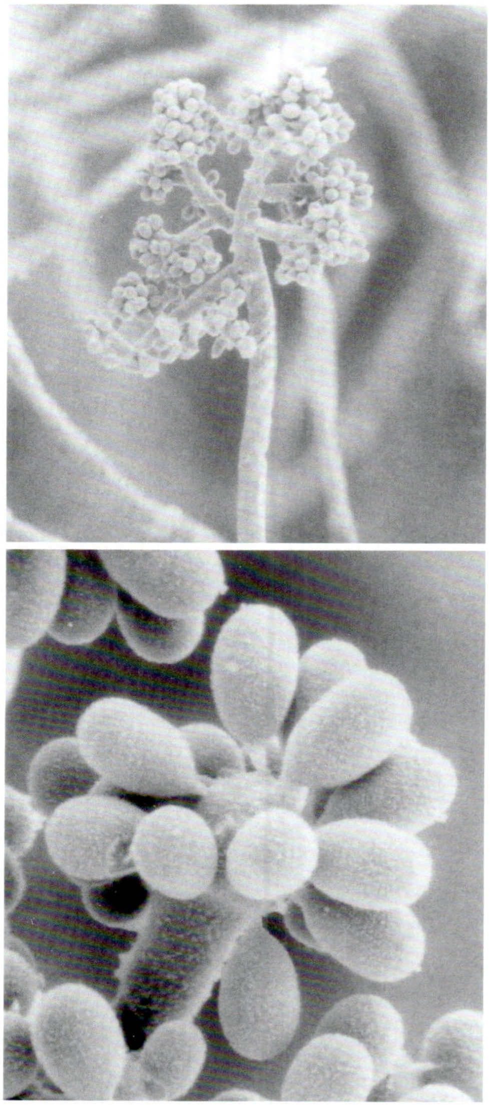

Fig. 33. Scanning electron micrographs of *Botrytis cinerea*, the causal organism of Botrytis blight, conidiophore and conidia (×300) (top) and swollen terminal cell with conidia on denticles (×2,000) (bottom). (Reproduced, by permission of the University of Missouri, from Brown and Brotzman, 1979)

pear grayish brown. Conidia are multinucleate, with one to ten nuclei (average four).

Sclerotia are produced on dead twigs and considered the most important long-term survival structure of the fungus. The flat, black sclerotia are irregularly shaped and up to 5 mm in diameter. They have a black rind and a light interior composed of a dense mass of hyphae. In culture, sclerotia tend to be globose or oblong.

Disease Cycle and Epidemiology

The fungus overwinters as dormant mycelium in or on plant debris or as sclerotia. In the spring, masses of conidia are produced from reactivated mycelia and germinated sclerotia. Conidia are easily detached and carried by wind currents to susceptible blueberry tissue. While the fungus can infect blossoms, twigs, leaves, and fruit, most losses are caused by blossom blight.

Flowers become susceptible to infection just before they open and are most susceptible at full bloom and early senescence. Poor conidium germination on flowers during early stages of development suggests that the availability of exogenous nutrients on flower surfaces is important for infection. Freeze injury may predispose flowers and other tissues to infection.

The optimum temperature for conidium germination is between 20 and 25°C, whereas 16–20°C is the optimum temperature for infection. In water, conidia form a germ tube within 1–3 h, followed by the formation of simple or compound appressoria after 6 h. *B. cinerea* conidia are known to germinate in water or under conditions of high relative humidity (>94%). A period of 3–4 days of high relative humidity is necessary for the infection of open flowers of lowbush blueberry and 6–9 days is necessary for the infection of dormant buds. In the presence of surface water, a minimum of 6 h of wetness is required for the infection of open flowers at 20°C, 8 h is required at 16°C, 13 h is required at 8–12°C, and 24 h is required at 4°C. Sporulation on infected flower clusters may be visible within 3–4 days after infection.

Fruit can also become infected, although ripe fruit apparently are more susceptible than green fruit. In general, the fungus does not progress from infected berries to the stem through the pedicel and peduncle. However, nonpollinated ovaries from which the corolla has abscised are susceptible to infection and can facilitate colonization of the peduncle and destruction of other flowers and fruit in the cluster. Latent fruit infections can manifest themselves during storage after harvest. The ability of *B. cinerea* to grow at temperatures just above freezing makes it difficult to control during storage and shipment.

Management

Planting resistant or less susceptible cultivars is recommended. Highbush blueberry cultivars Elliott and Bluecrop are considered relatively resistant to Botrytis blossom blight, whereas cultivars Duke, Weymouth, Blueray, and Rancocas are susceptible.

Botrytis blight and fruit rot incidence can also be reduced by cultural practices that improve air circulation and promote drying of the plant canopy to minimize conditions favorable for infection. Recommended practices are annual pruning, wide row spacing, controlling weeds, and planting rows in the prevailing wind direction. Pruning out dead twigs and canes and removing dead leaves from the ground in the autumn can also reduce *Botrytis* infections by removing inoculum sources. In addition, open canopies enhance fungicide spray penetration. Avoid establishing plantings in low-lying areas prone to frost or fog. Excessive use of nitrogen fertilizer in the spring should be avoided because it stimulates the growth of succulent, disease-susceptible tissues. To reduce losses to postharvest fruit rot, fruit should be harvested in a timely manner and cooled down rapidly.

While biocontrol agents, such as *Trichoderma harzianum*, *Aureobasidium pullulans*, and *Bacillus subtilis*, are known to be antagonistic to *B. cinerea*, they are rarely deployed for control of Botrytis blight in the field. However, dormant sprays with lime sulfur, copper, or sulfur can reduce the incidence of Botrytis fruit rot.

Effective fungicides applied during bloom control Botrytis blight. Fungicide applications during fruit development may be needed to prevent fruit infection. Fungicides such as iprodione, captan, and ziram are good protectants against *Botrytis* spp., but iprodione is prone to fungicide resistance development. Benomyl and thiophanate-methyl are known to be effective against *B. cinerea* but may not be allowed on blueberries. Fungicides such as fenhexamid, cyprodinil + fludioxonil, and boscalid are very effective against Botrytis blight. Since these fall into different chemical classes, they should be alternated for fungicide resistance management.

Selected References

Brown, M. F., and Brotzman, H. G. 1979. Phytopathogenic Fungi: A Scanning Electron Stereoscopic Survey. University of Missouri Press, Columbia.

Elad, Y., Williamson, B., Tudzynski, P., and Delen, N., eds. 2004. Botrytis: Biology, Pathology, and Control. Springer-Verlag, Heidelberg, Germany.

Ellis, M. B., and Waller, J. M. 1974. *Sclerotinia fuckeliana* (conidial state: *Botrytis cinerea*). Descriptions of Pathogenic Fungi and Bacteria, No. 431. Commonwealth Mycological Institute, Kew, Surrey, England.

Fournier, E., Giraud, T., Albertini, C., and Brygoo, Y. 2005. Partition of the *Botrytis cinerea* complex in France using multiple gene genealogies. Mycologia 97:1251-1267.

Hildebrand, P. D., McRae, K. B., and Lu, X. 2001. Factors affecting flower infection and disease severity of lowbush blueberry by *Botrytis cinerea*. Can. J. Plant Pathol. 23:364-370.

Sciarappa, W. J., and Oudemans, P. 2005. Botrytis blight in highbush blueberry. N. J. Agric. Exp. Stn., Rutgers Coop. Res. Ext. Fact Sheet 512.

Smith, B. J. 1998. Botrytis blossom blight of southern blueberries: Cultivar susceptibility and effect of chemical treatments. Plant Dis. 82:924-927.

(Prepared by P. R. Bristow and R. D. Milholland;
Revised by A. M. C. Schilder)

Cylindrocladium Rot

Cylindrocladium rot of blueberry stems, leaves, and roots primarily affects blueberry cuttings, nursery plants, and plant breeder's seedlings in nurseries. This disease is common in the southeastern United States and has also been seen in Michigan (United States). It is a widespread problem on many hosts in ornamental plant nurseries.

Symptoms

Initial symptoms of Cylindrocladium rot are browning or blackening of the stem near the soil line followed by wilting and death of the plants. Brown leaf spots (Fig. 34) and stem lesions with reddish borders may also be observed. Lesions can girdle the stem, killing all tissue above the girdled area (Fig. 35). Root lesions and root rot are also commonly observed. Orange perithecia develop on infected tissues under humid conditions (Fig. 36).

The disease can spread rapidly in a rooting bed. Dead plants are often found in a circular pattern 0.3–1.2 m in diameter, with disease spreading radially from a point source. When old, contaminated rooting medium is reused, the disease may develop suddenly on all cuttings throughout the bed, and it is not unusual for all the plants to die.

On rooted plants, the disease is most often seen in propagation beds and pot yards that have grown vigorously and are too dense to allow proper air movement between the plants. Spores of the fungus infect leaves and stems and eventually kill entire plants. Often the root system is the last healthy part remaining. The phenomenon of plants dying in the center of a dense block, while those at the edge (with better air circulation) survive, is characteristic of this disease in dense beds of plants.

Causal Organisms

Cylindrocladium rot is caused by the ascomycetous fungus *Calonectria ilicicola* (syns. *Calonectria crotalariae* and *Cylindrocladium parasiticum*). However, other *Cylindrocladium* spp. may also be involved. *Cylindrocladium colhounii* was reported to cause brown leaf spots with purplish margins on lowbush and half-high blueberry plants in a nursery in China and is also suspected to be the cause of blight of blueberry cuttings in a nursery in Michigan.

Fig. 34. Leaf spots are often the first visible sign of Cylindrocladium rot, caused by a *Cylindrocladium* sp., in dense beds of nursery blueberry plants. (Courtesy W. O. Cline)

Fig. 35. Girdling of stems and blight from Cylindrocladium rot, caused by a *Cylindrocladium* sp., of young blueberry plants in rooting trays in the greenhouse. (Courtesy J. Sadowsky; © APS)

Perithecia of *Calonectria ilicicola* are orange to red and have a darker, reddish brown base (Fig. 36). They are subglobose to ovoid and 280–400 × 300–550 µm. Asci are eight spored, clavate, tapering to a long thin stalk, and 12–19 × 90–140 µm (Fig. 37D). Ascospores are hyaline, guttulate, fusoid with rounded ends, and straight or slightly curved. They have one to three septa, depending on the species, and are 4–7 × 30–65 µm (average 6 × 45 µm).

Macroconidiophores are septate, pale brown at the base, and stalked with a penicillate arrangement of phialides. The characteristic stipe extension is septate, straight to flexuous, and 120–240 µm long, terminating in a light-bulb-shaped vesicle (Fig. 37A). Conidia are hyaline, cylindrical, rounded at both ends, and straight, have one to three septa (depending on the species), are 4–7 × 45–90 µm (average 6 × 62 µm), and are held in parallel bundles by colorless slime. Cylindrical and hyaline microconidia, with one septum and 4–7 × 16–47 µm (average 4.5 × 30 µm), may also be present (Fig. 37C). The fungus produces copious amounts of brown chlamydospores and microsclerotia in culture.

On potato dextrose agar, mycelium of the fungus initially is white, later turning a light brown to salmon-buff color. Conidia are produced within 7 days. Perithecia are produced on water agar or sterilized blueberry stems after 14 days at 25°C.

Disease Cycle and Epidemiology

Calonectria ilicicola causes crown rot and blight, damping-off, and leaf spots on a wide range of plants throughout the world. It is considered a high-temperature species. It can grow between 5 and 35°C but the optimum temperature is 25°C. At 25°C, most conidia germinate within 2–4 h. Disease development is rapid; lesions can develop on leaves and stems within 24 h and succulent portions of some stems may be killed within 48 h after inoculation.

Infections in blueberry beds are thought to originate from nearby infected vegetation (e.g., oak, sweet gum, tulip, and poplar trees) and peanut fields in areas where peanuts are widely grown. Initial infection of young shoots may take place in the field or after cuttings are stuck. Spores are splash dispersed to surrounding cuttings, causing the disease to spread rapidly. The fungus can spread by conidia, chlamydospores, or microsclerotia and is known to infest soil and potting–rooting media.

Management

New, clean rooting medium should be used; old media should not be reused. Frames, trays, and pots should be sanitized before reusing them. No cultivars are known to be resistant to Cylindrocladium rot, and 'Jersey' is particularly susceptible to root infection. Propagation beds should not be placed under trees that may be a source of inoculum. Moisture should be managed to reduce the level of relative humidity and water splash as much as possible. Diseased cuttings or plantlets should be removed as quickly as possible and remaining plants should be treated with

Fig. 36. Orange perithecia on a blueberry stem infected by *Calonectria ilicicola*, a causal organism of Cylindrocladium rot. (Courtesy W. O. Cline)

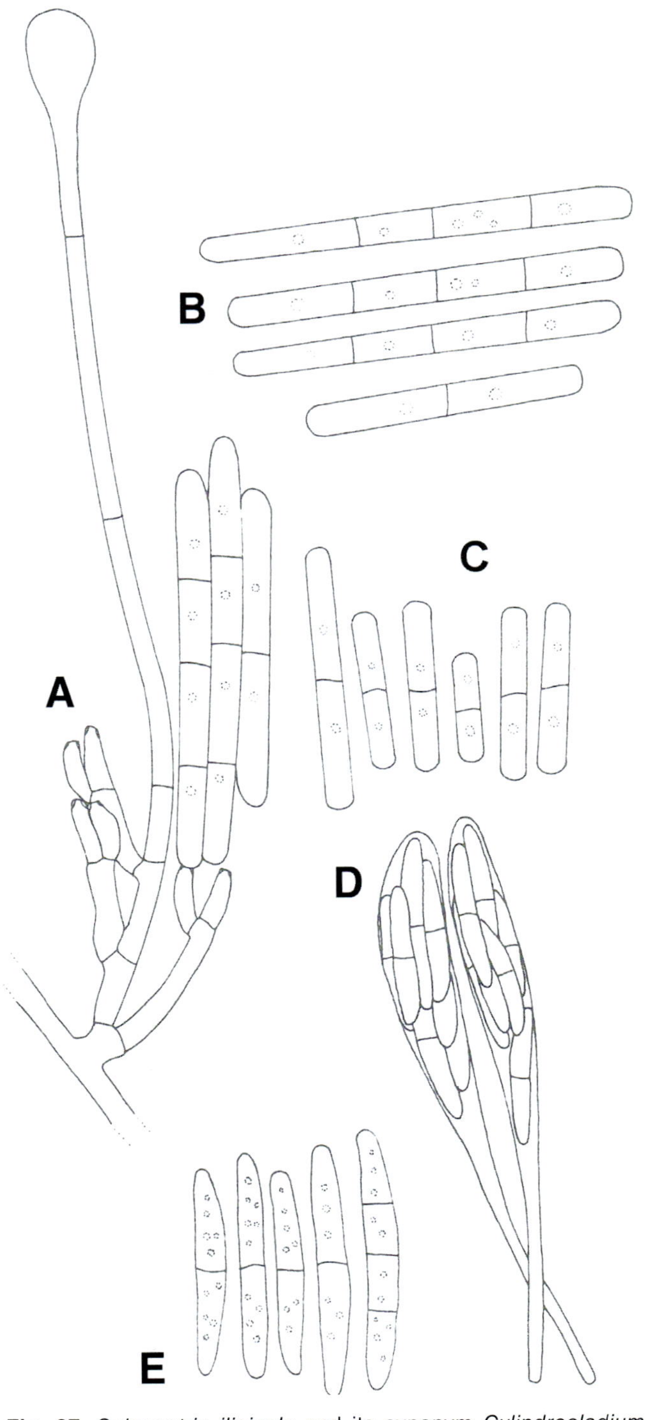

Fig. 37. *Calonectria ilicicola* and its synonym *Cylindrocladium parasiticum*, causal organisms of Cylindrocladium rot. **A,** Macroconidiophore with vesicle. **B,** Macroconidia. **C,** Microconidia. **D,** Asci with ascospores. **E,** Ascospores. (Modified from Crous, 2002; © APS)

an effective fungicide. In fungicide efficacy trials, fludioxonil applied as a drench was superior to thiophanate-methyl or triflumizole. In pot yards and late-stage propagation beds where the plants have already formed roots, this disease is best managed by not overwatering or overfertilizing plants and by maintaining proper plant spacing so that the leaves and stems can dry between waterings.

Selected References

Cline, W. O. 2004. Fungal pathogens associated with blueberry propagation beds in North Carolina. Pages 213-219 in: Proceedings of the Ninth North American Blueberry Research and Extension Workers Conference. C. F. Forney and L. J. Eaton, eds. Haworth Press, Binghamton, NY.

Crous, P. W. 2002. Taxonomy and Pathology of *Cylindrocladium* (*Calonectria*) and Allied Genera. American Phytopathological Society, St. Paul, MN.

Crous, P. W., Wingfield, M. J., and Alfenas, A. C. 1993. *Cylindrocladium parasiticum* sp. nov., a new name for *C. crotalariae*. Mycol. Res. 97:889-896.

Grand, L. F., ed. 1985. North Carolina Plant Disease Index. N. C. Agric. Res. Serv. Tech. Bull. 240.

Haralson, J. C., Brannen, P. M., Scherm, H., and NeSmith, D. S. 2008. Use of fungicides for control of Cylindrocladium root rot in blueberry cuttings, 2007. Plant Dis. Manage. Rep. 2:SMF038.

Luan, Y. S., Feng, L., and An, L. J. 2006. First report of blueberry leaf spot caused by *Cylindrocladium colhounii* in China. Plant Dis. 90:1553.

Milholland, R. D. 1974. Stem and root rot of blueberry caused by *Calonectria crotalariae*. Phytopathology 64:831-834.

Milholland, R. D. 1974. Histopathology of *Calonectria crotalariae* on highbush blueberry. Phytopathology 64:1228-1231.

(Prepared by A. M. C. Schilder and W. O. Cline)

Double Spot

Double spot was first described on highbush blueberry in 1947. The disease has been reported in North Carolina, Georgia, and Florida (United States), where it may contribute to the defoliation of highbush, southern highbush, and rabbiteye cultivars.

Symptoms

Double spot is characterized by light to dark brown leaf spots that are variable in shape. At first, spots are circular with a dark red border and are 2–3 mm in diameter. A secondary spot develops around the original lesions after midsummer and spreads in an irregular to fan-shaped pattern. The dark border surrounding the original lesion persists, and the large necrotic area surrounds or lies to one side of the original lesion (Fig. 38). Typically one to five pycnidia develop per lesion.

Causal Organism

The causal organism of double spot, the fungus *Dothichiza caroliniana*, was identified in 1947. Black pycnidia develop primarily on the upper (adaxial) leaf surface and within the con-

Fig. 38. Double spot lesions on blueberry leaves in late autumn caused by *Dothichiza caroliniana*. (Courtesy W. O. Cline)

fines of the original lesion. Pycnidia (80–130 × 48–95 µm) do not have a neck or ostiole and vary from globose to pyriform. Conidia (7.0 × 2.2 µm) are hyaline and obtuse.

Disease Cycle and Epidemiology

The fungus overwinters in infected leaves. The spores are disseminated during rainy periods in the spring, and leaf spots develop in mid-May. In years when rains are frequent in May and June, severe defoliation can occur by late summer.

Management

Double spot can be controlled by applications of an effective fungicide immediately after harvest and 2, 4, and 6 weeks after harvest. Differences in susceptibility of cultivars are likely but have not been documented. Among rabbiteye cultivars, Climax is reportedly resistant compared with cultivars Tifblue and Homebell. As with other blueberry leaf spots, summer mowing (hedging) immediately after harvest may be beneficial in the management of double spot.

Selected References

Alfieri, S. A., Jr. 1982. Dothichiza leaf spot of blueberries. Plant Pathol. Circ. No. 242, Fla. Dep. Agric. Consum. Serv.

Baird, R. E., and Ruter, J. M. 1996. First report of *Dothichiza caroliniana* on southern highbush blueberry in Georgia. Plant Dis. 80:223.

Demaree, J. B., and Wilcox, M. S. 1947. Fungi pathogenic to blueberries in the eastern United States. Phytopathology 37:487-506.

Taylor, J., and Clayton, C. N. 1959. Comparative studies on Gloeosporium stem and leaf fleck and Dothichiza leaf spot of highbush blueberry. Phytopathology 49:65-67.

(Prepared by R. D. Milholland; Revised by W. O. Cline)

Exobasidium Leaf and Fruit Spot

Exobasidium leaf and fruit spot has been reported on highbush and rabbiteye blueberries in North Carolina, Mississippi, and Georgia (United States). Although the disease is usually sporadic, its incidence has been increasing in prevalence in the southeastern blueberry-growing regions, causing significant losses in some plantings. Fruit infection, in addition to leaf infection, was first reported in North Carolina in 1998 on highbush blueberry. The symptoms of leaf and fruit spot are distinct from those of red leaf disease, attributed to *Exobasidium vaccinii* in most publications but now known to be caused by a different *Exobasidium* sp.

Losses caused by leaf infection are minimal, but fruit infection is especially problematic in commercial handling and shipping for the fresh market. Infected berries have green, unripe spots that persist into the marketplace and are highly visible in the clear plastic clamshell containers used for retail sales. Although the affected berries are not leaky or decayed, this highly visible defect may result in the rejection of entire pallets of otherwise sound fruit. In the most severe cases, growers have abandoned entire fields at harvest rather than try to sort out spotted fruit.

Symptoms

In the spring, infected leaves develop pale green spots that are sharply defined, circular but sometimes with slightly irregular wavy edges, 2–14 mm in diameter, and slightly thicker than the surrounding leaf tissue (Fig. 39A). A dense, clearly delimited white fungal mat where spores are produced (hymenium) forms on the underside of each spot (Fig. 39B). In early summer, the leaf spots become necrotic, turning reddish brown, and the hymenial layer is no longer present. Leaf spots on highbush blueberry have fairly distinct edges and are somewhat uniform, whereas those on rabbiteye blueberry are more variable and likely to coalesce. Infected, immature berries develop sharply defined, light green circular spots that fail to ripen normally (Fig. 40). On highbush blueberry, fruit spots are 2–10 mm in diameter, circular or rarely irregular, not sunken or protruding, and often with a tiny (<0.5 mm) faint spot visible in the center. Affected tissues in ripe fruit extend 2–3 mm into the berry and are the texture and color of green, unripe fruit. As the berries turn blue, spots remain green or become greenish pink. Symptoms differ slightly on rabbiteye blueberry; spots are the same size as those on highbush blueberry fruit but are more likely to be sunken and tinged with pink or purplish colors. Sparse white fungal growth is visible on the surface of some affected berries. When ripe or nearly so, most infected fruit eventually develop this sparse white surface layer of fungal hyphae and spores.

Fig. 39. Leaf spots on blueberry (**A,** adaxial; **B,** abaxial) caused by *Exobasidium maculosum*, the causal organism of Exobasidium leaf and fruit spot. (Courtesy W. O. Cline)

Fig. 40. Fruit spots on blueberry caused by *Exobasidium maculosum*, the causal organism of Exobasidium leaf and fruit spot (**A,** low magnification; **B,** high magnification). (Courtesy W. O. Cline)

Causal Organism

Exobasidium leaf and fruit spot on highbush and rabbiteye blueberries is caused by *Exobasidium maculosum*. The disease on lowbush blueberry is caused by a very closely related, yet genetically differentiated, population that may or may not be the same species. On blueberry leaves and fruit, the fungus produces cylindrical basidia (5–6.5 × 8.5–11.5 μm) bearing four, five, or six sterigmata. Basidiospores (3–5 × 8–15 μm) are musiform (banana shaped) to fusiform, slightly curved, and hyaline with a conspicuous hilar appendage and are often transversely septate at detachment. Basidiospores may give rise to cylindrical to ellipsoid, aseptate conidia (1–1.5 × 6–9 μm) that are smooth and hyaline. When isolated from infected tissue or single basidiospores onto potato dextrose agar, colonies are moist, wrinkled, superficial and pink to tan, reaching a diameter of 1–1.5 cm in 6–8 weeks when incubated at 23°C. Conidia, similar to those recovered from infected leaves and fruit, are produced in culture by budding from either end of a basidiospore or from other conidia.

Disease Cycle and Epidemiology

Very little is known about the epidemiology of Exobasidium leaf and fruit spot, how the fungus overwinters, or its sources of inoculum. The disease is most often seen in older fields, especially in low areas and wooded corners where air movement is limited. Leaf infection appears on the earliest foliar growth flush that coincides with fruit development and fruit infection, but it does not appear on later-emerging foliage. On lowbush blueberry in Canada, the disease is associated with the lowest-growing species (*Vaccinium angustifolium*) and seems more frequent on bushes in partial shade or underneath taller vegetation.

Management

Resistance is likely in some cultivars. Lime sulfur, applied as a late-dormant application, provides substantive management of both leaf and fruit spots. For in-season applications, the fungicide mixture of pyraclostrobin and boscalid was the most effective treatment in a fungicide trial conducted in Mississippi in 2009. However, trials in Georgia have shown that captan or fenbuconazole, applied during bloom and as early cover sprays, are significantly more effective than the fungicide mixture of pyraclostrobin and boscalid. The efficacy of cultural practices for disease suppression is not known; however, since the disease is associated with areas of high moisture levels and poor air movement, pruning to open up the canopy and improve air circulation may be beneficial. Once fruit are harvested, low levels of infected fruit can be removed with some success by repeated visual culling, especially when aided by electronic color-sorting equipment.

Selected References

Brewer, M. T., Turner, A. N., Brannen, P. M., Cline, W. O., and Richardson, E. A. 2014. *Exobasidium maculosum*, a new species causing leaf and fruit spots on blueberry in the southeastern USA, and its relationship with other *Exobasidium* spp. parasitic to blueberry and cranberry. Mycologia 106:415-423.

Cline, W. O. 1998. An Exobasidium disease of fruit and leaves of highbush blueberry. Plant Dis. 82:1064.

Nickerson, N. L., and Vander Kloet, S. P. 1997. Exobasidium leaf spot of lowbush blueberry. Can. J. Plant Pathol. 19:66-68.

(Prepared by W. O. Cline and M. T. Brewer)

Fusicoccum (Godronia) Canker

Fusicoccum (Godronia) canker is a disease of highbush blueberry in the northern United States and southern Canada. In the United States, it occurs in Michigan, Maine, Massachusetts, Oregon, and Washington and in the mountains of North Carolina. In the Upper Peninsula of Michigan, the severity of this disease is a limiting factor to production. In Canada, it has been reported in British Columbia, Nova Scotia, New Brunswick, and Quebec. The disease has also been reported from Finland, Scotland, and England. Fusicoccum canker has been reported on lowbush blueberry in Nova Scotia but is not economically important. The disease has not been reported on rabbiteye blueberry. On highbush blueberry, disease severity ranges from wilting and death of one or two stems to wilting and death of almost all stems and subsequent death of the plant.

Symptoms of Fusicoccum canker are usually first noted during the summer. When temperatures are warm, leaves wilt on those stems with cankers. Leaves remain on infected stems and eventually turn reddish brown. Leaves on infected stems often exhibit early autumn color. Infections on 1- and 2-year-old stems are observed in the autumn as minute (<0.5 mm in diameter), water-soaked lesions that turn red by December. During the following spring and summer, the lesions expand, coalesce, and form a targetlike canker with a light center and reddish brown or dark brown margin. Cankers are 1–10 cm long (Fig. 41). Usually, the canker is centered on a leaf scar. Most cankers on stems are near ground level but some may be as high as 1 m above the ground. The canker becomes covered with pycnidia that are 0.5–1 mm in diameter (Fig. 41). Minute, hard, black apothecia may be found on older dead wood and dead pruning stubs (3 or more years old) (Fig. 42). In some instances, apothecia may be very difficult to find.

Causal Organism

Shear described *Godronia cassandrae* (syn. *Fusicoccum putrefaciens*) from rotted cranberries in 1917 and determined that the fungus was the causal organism of end rot of cranberry. He also reported the presence of a discomycetous fungus on rotted and mummified berries and speculated on the possibility of its connection with *G. cassandrae*. Apothecia of *G. cassandrae* appear occasionally at the base of 3-year-old stems (usually on dead tissue) (Fig. 42).

Asci (1.6–2.2 × 97 μm) are cylindrical. Ascospores (1.6–2.2 × 52–69 μm) are filiform, septate, and nearly straight. Paraphyses

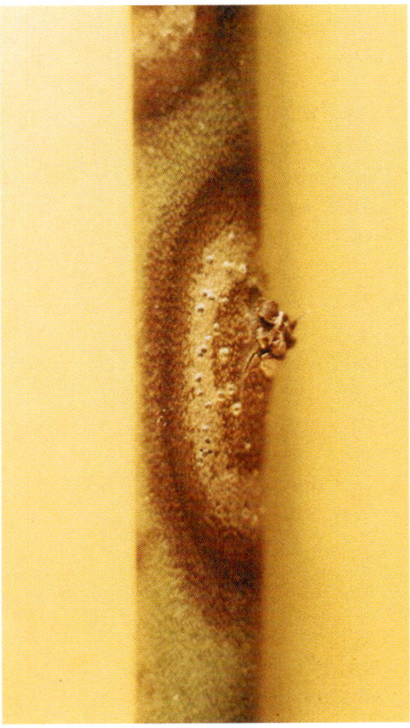

Fig. 41. One-year-old stem of highbush blueberry with a well-defined Fusicoccum canker, caused by *Godronia cassandrae*. Note the pycnidia. (Courtesy D. C. Ramsdell)

Fig. 42. Older, dead highbush blueberry stem with apothecia of *Godronia cassandrae*, the causal organism of Fusicoccum canker, on the surface. (Courtesy D. C. Ramsdell)

Fig. 43. Cankers caused by Fusicoccum canker, caused by *Godronia cassandrae*, at the base of 1-year-old stems of the blueberry cultivar Berkeley. (Courtesy C. A. Smith; © APS)

are filiform and numerous. Brownish pycnidia, abundant in all lesions, are subglobose to pyriform and have thick walls that are more or less roughened. Pycnidia are embedded at first but eventually become erumpent when mature. Pycnidia are sessile or subsessile. Simple or irregularly chambered forms develop, especially on artificial media.

Conidia (1.2–1.9 × 8.4–17.0 μm) of *G. cassandrae* are elliptical to fusiform, hyaline or very faintly yellowish in mass, and pseudoseptate. The fungus is similar to *G. cassandrae* described on cranberry, but the conidia of a cranberry isolate from the eastern United States are straighter and slightly larger compared with conidia from a blueberry isolate.

Disease Cycle and Epidemiology

The fungus overwinters in cankers on stems and crowns (Fig. 43). Conidia are released from pycnidia in stem cankers during rainy periods from bud swell through leaf drop in October. Although most infections occur from bud swell through June, new infections can appear on new growth throughout the growing season. The fungus invades the cortex, cambium, and (to a limited extent) xylem. Infection by ascospores is relatively unimportant in the disease cycle.

Wounding is not a prerequisite for infection. Leaf scars and other natural openings in the bark are common infection sites. Winter injury, frost injury, and cold stress appear to predispose stems to infection. Under controlled conditions, infection takes place in free water within 48 h at 10–22°C. Infection does not take place at or above 30°C.

Management

Sanitation is a critical component of management. Stems with cankers should be pruned, removed from the field, and destroyed prior to bud break. Overhead irrigation should be avoided. Planting on sites prone to frosts should be avoided. Dormant application of lime sulfur is often recommended.

Differences in cultivar susceptibility have been documented. Cultivars Jersey, Earliblue, Johnson, and Bluecrop are highly susceptible. Cultivars Coville, Berkeley, Pioneer, Blueray, Rubel, and Burlington are moderately susceptible, and cultivars Rancocas and Concord are highly resistant. Cultivars with lowbush blueberries as parents exhibit fewer and slower-growing cankers compared with cultivars with only highbush parents.

Selected References

Lockhart, C. L. 1972. Occurrence and pathogenicity of *Godronia cassandrae* f. *vaccinii* on lowbush blueberry in Nova Scotia. Can. Plant Dis. Surv. 52:119-121.

Lockhart, C. L., and Craig, D. L. 1967. Fusicoccum canker of highbush blueberry in Nova Scotia. Can. Plant Dis. Surv. 47:17-20.

McKeen, W. E. 1958. Blueberry canker in British Columbia. Phytopathology 48:277-280.

Parker, P. E., and Ramsdell, D. C. 1977. Epidemiology and chemical control of Godronia (Fusicoccum) canker of highbush blueberry. Phytopathology 67:1475-1480.

Shear, C. L. 1917. End rot of cranberries. J. Agric. Res. 11:35-41.

Shear, C. L., and Bain, H. F. 1929. Life history and pathological aspects of *Godronia cassandrae* Peck (*Fusicoccum putrefaciens* Shear) on cranberry. Phytopathology 19:1017-1024.

Stromeng, G. M., and Stensvand, A. 2001. Susceptibility of highbush blueberry (*Vaccinium corymbosum* L.) cultivars to Godronia canker (*Godronia cassandrae* f. sp. *vaccinii*) in Norway. Gartenbauwissenschaft 66(2)(Suppl.):78-84.

Weingartner, D. P., and Klos, E. J. 1975. Etiology and symptomatology of canker and dieback diseases on highbush blueberries caused by *Godronia* (*Fusicoccum*) *cassandrae* and *Diaporthe* (*Phomopsis*) *vaccinii*. Phytopathology 65:105-110.

(Prepared by D. C. Ramsdell; Revised by C. A. Smith)

Gibbera Twig Blight

Gibbera twig blight is a disease of highbush blueberry. The disease has not been observed in lowbush blueberry. Gibbera twig blight appears to be limited primarily to the northeastern United States and eastern Canada. In the United States, it has been documented at several sites in New Hampshire and has been reported from New York and North Carolina. In Canada, it has been observed in Quebec. Disease severity on infected plants ranges from a few scattered stromata to as great as 40% twig and stem dieback. In addition, infected twigs appear to be predisposed to winter injury.

Symptoms and Signs

The signs and symptoms of Gibbera twig blight are quite striking. The fungus forms raised cushion-shaped black stroma on twigs and stems. Stroma may enlarge to a diameter of 3–4 mm and appear singly or in clusters. Each stroma is usually bordered by a distinct red margin (Fig. 44).

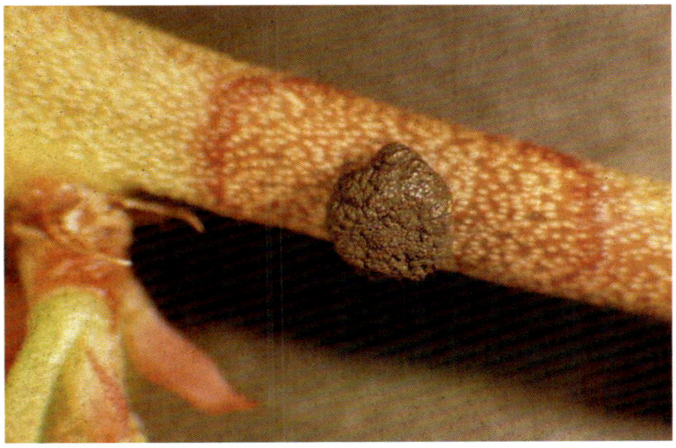

Fig. 44. Stroma of *Gibbera vacciniicola*, the causal organism of Gibbera twig blight, bordered by a distinct ring of reddened tissue on a blueberry stem. (Courtesy C. A. Smith; © APS)

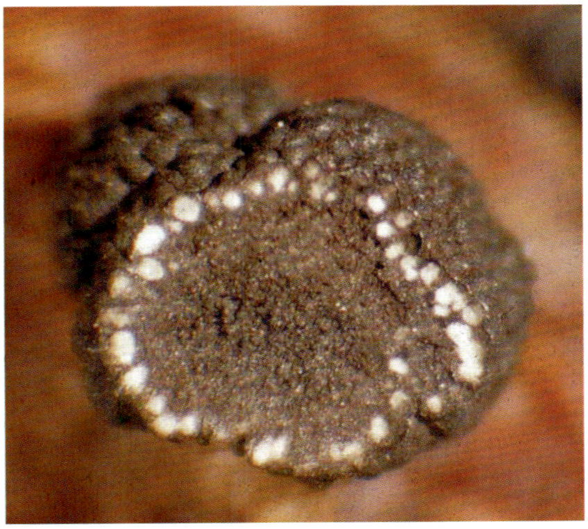

Fig. 45. Cross section of a stroma of *Gibbera vacciniicola*, the causal organism of Gibbera twig blight, on blueberry. White locules near the stromatal surface contain pseudothecia. (Courtesy C. A. Smith; © APS)

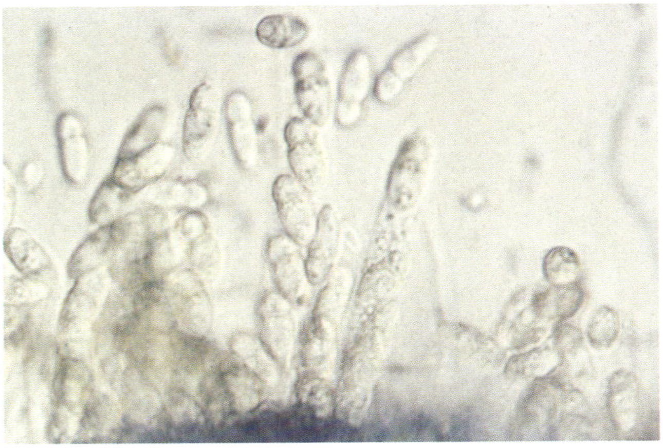

Fig. 46. Asci and ascospores of *Gibbera vacciniicola*, the causal organism of Gibbera twig blight. (Courtesy C. A. Smith; © APS)

Repeated infections over several years often results in stems covered with stromata and red lesions. Heavily infected twigs are often killed when clustered stromata form girdling cankers. Growers have observed increased susceptibility to winter injury on infected stems. Fruit yield is often reduced on infected stems.

Causal Organism

Gibbera vacciniicola is a loculoascomycete. The type material was collected by House in 1912 in Biltmore, North Carolina (United States). The fungus was initially described as *Dothidella vacciniicola* by Dearness and House in 1916. Additional type material was collected and determined by Shear in 1920. In her 1968 treatment of the North American Venturiaceae, Barr reclassified the fungus to *G. vacciniicola*. The fungus forms erumpent stroma on twigs ranging in size from 1 to 4 mm. Pseudothecia form in locules near the stromatal surface (Fig. 45). Ascospores are nearly hyaline to greenish brown, 3.4–6.2 × 12–15 μm, and unequally septate near the upper third of the cell (Fig. 46). Although the conidial state has been described, it is unknown if conidia play a role in disease development.

Disease Cycle and Epidemiology

Few details about the epidemiology of the pathogen are known. Ascospores are released during rainy periods from mid-April to late June, with peak release from mid- to late May. The appearance of tiny, black stromata (<0.25 mm in diameter) from mid-September through October on 1-year-old twigs is the first evidence of infection. By the following April, stroma enlarge to a diameter of 3–4 mm and a red border develops around each stroma.

Management

Diseased twigs and stems should be removed and destroyed. Although fungicide programs have not been developed specifically for control of Gibbera twig blight, fungicides effective for the control of mummy berry (*Monilinia vaccinii-corymbosi*) appear to reduce the incidence of Gibbera twig blight. Fungicide trials indicate two to three applications at labeled intervals beginning just prior to bloom to be most effective.

Differences in cultivar susceptibility have been observed. 'Northland' is highly susceptible, and the disease has also been documented on 'Berkley' and 'Herbert'. The disease has not been reported on other cultivars.

Selected References

Barr, M. E. 1968. Venturiaceae in North America. Can. J. Bot. 46:799-864.

Morrall, R. A. A. 2007. The Canadian Phytopathological Society/Canadian Plant Disease Survey—Disease Highlights. Can. Plant Dis. Surv. 87:62.

Stobbs, L. W. 1998. The Canadian Phytopathological Society/Canadian Plant Disease Survey—Disease Highlights. Can. Plant Dis. Surv. 78:62.

(Prepared by C. A. Smith)

Gloeocercospora Leaf Spot

Gloeocercospora leaf spot was first described on blueberry in 1947. In the United States, the disease has been observed on wild *Vaccinium australe* in Maryland and on highbush and rabbiteye blueberry cultivars in North Carolina. Its occurrence in other blueberry-growing regions has not been reported.

Symptoms and Disease Cycle

Dark brown, circular to angular lesions surrounded by a dark border appear on infected foliage by midsummer (Fig.

47). Lesions resemble the early spots formed by *Dothichiza caroliniana* (see Double Spot above) and are often difficult to distinguish from them before the double spot lesions coalesce.

Sporodochia are formed more frequently on the upper (adaxial) than on the lower (abaxial) leaf surface. When a thin film of water is spread over the lesion, the sporodochia enlarge and appear as clear, glistening, spherical to conical globules.

Causal Organism

Gloeocercospora leaf spot is caused by *Gloeocercospora inconspicua*. Sporodochia (36–60 × 40–90 µm) are slightly embedded in the epidermis. Some are constricted at the base and wider at the top. The body of the sporodochium is an undifferentiated, opaque mass with some intermingling of hyphal strands. Conidia (3 × 30–55 µm) are hyaline, usually curved, and septate (Fig. 48).

Fig. 47. Lesions of Gloeocercospora leaf spot, caused by *Gloeocercospora inconspicua*, on rabbiteye blueberry leaves. (Courtesy R. D. Milholland; © APS)

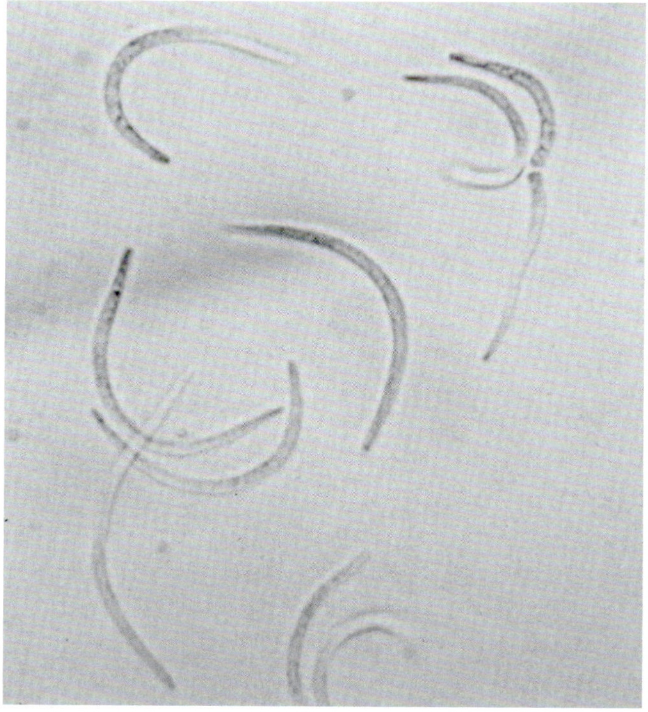

Fig. 48. Conidia of *Gloeocercospora inconspicua*, the causal organism of Gloeocercospora leaf spot, from rabbiteye blueberry. (Courtesy R. D. Milholland; © APS)

Management

Gloeocercospora leaf spot can be effectively controlled with applications of fungicides immediately after harvest and 2, 4, and 6 weeks after harvest.

Selected Reference

Demaree, J. B., and Wilcox, M. S. 1947. Fungi pathogenic to blueberries in the eastern United States Phytopathology 37:487-506.

(Prepared by R. D. Milholland; Revised by W. O. Cline)

Gloeosporium Leaf Spot and Stem Canker

The fungus *Gloeosporium minus* was first described as causing a storage rot of cranberries in 1907, then a stem and leaf fleck disease of blueberry in 1958, and later an anthracnose-type leaf spot of blueberry in 1970 and a stem canker disease of blueberry in 1973. Gloeosporium leaf spot is the most serious foliar disease of highbush blueberry in the southeastern United States. The leaf spot develops in mid- to late June and can cause severe defoliation on susceptible highbush cultivars when inoculum levels are high and periods of warm, wet weather persist.

Symptoms

Gloeosporium leaf spot was first described as a stem and leaf fleck disease. The first symptom is the appearance of small reddish flecks on young leaves and stems of succulent shoots. The leaf flecks do not develop further, and as a result, the enlarging leaves become puckered and malformed. In addition, the fungus produces large, brown, circular to irregularly shaped lesions 5–20 mm in diameter on blueberry leaves (Fig. 49). Stem lesions first appear on current-season growth as dark red, circular to elliptical lesions around leaf scars (Fig. 50). As the lesions enlarge, the affected stems turn brown and eventually become gray and die. The disease results in severe stem dieback, extending up to 50 cm (Fig. 51). Numerous acervuli develop over the entire lesion.

Causal Organism

The genus *Gloeosporium* is considered a synonym of the genus *Marssonina*. Taxonomic reclassification of the pathogen is probably needed; most other fungi in the genus *Gloeosporium* were reclassified by Von Arx in 1970. Acervuli of *G. minus* are

Fig. 49. Highbush blueberry leaves with Gloeosporium leaf spot lesions, caused by *Gloeosporium minus*. (Courtesy R. D. Milholland; © APS)

subepidermal and extrude a mass of hyaline, ellipsoid conidia that are 3–4 × 6–9 µm. The conidiophores are simple and are borne on the surface of the area exposed by the rupture of the epidermis and along the vertical sides of the acervuli.

Disease Cycle and Epidemiology

The severity of Gloeosporium leaf spot depends on the mode of infection and the stage of plant growth. The large anthracnose leaf lesions develop primarily from infection through hydathodes and, to a lesser extent, through wounds. Numerous acervuli appear on both the upper and lower surfaces of leaf lesions that develop during early summer (late June to early July). Stem cankers develop from these leaf infections when the fungus grows through the petiole into the stem and also from bud and leaf scar infections on the stem late in the summer. Invasion through leaf scars results in stem canker development, whereas penetration through open stomata results in small, red lesions that fail to develop further. Minimum, optimum, and maximum temperatures for growth of the pathogen are 12, 25–30, and 36°C, respectively.

Fig. 50. Gloeosporium leaf and stem infections, caused by *Gloeosporium minus* (Gloeosporium leaf spot and stem canker), on highbush blueberry. (Courtesy R. D. Milholland; © APS)

Fig. 51. Gloeosporium stem canker and dieback on highbush blueberry stems, caused by *Gloeosporium minus*. (Courtesy R. D. Milholland; © APS)

Management

Most leaf spot diseases develop from mid- to late June and can cause severe defoliation, which results in reduced plant vigor and reduced flower bud set for the following year. Spraying with fungicides before and immediately after harvest and continuing at 2-week intervals until late summer is the best way to control blueberry leaf spot diseases. Northern highbush cultivars are generally more susceptible than southern highbush and rabbiteye cultivars. Cultivars vary in their susceptibility to this disease; the highbush cultivars Duke and Jersey are very susceptible. In the southeastern United States and other regions with long growing seasons, summer mowing (hedging) is used to manage bush height and reduce the incidence of leaf diseases. For southern highbush cultivars, bushes are mowed off 1–2 m above the ground immediately after harvest. This removes old fruiting twigs and forces a new flush of healthy foliage that will persist throughout the remainder of the growing season, ensuring adequate flower bud set for the following year.

Selected References

Cline, W. O. 2002. Blueberry bud set and yield following the use of fungicides for leaf spot control in North Carolina. Acta Hortic. 574:71-74.

Milholland, R. D. 1970. The effect of leaf exudates on blueberry leaf spot caused by *Gloeosporium minus*. Phytopathology 60:635-640.

Milholland, R. D. 1974. Blueberry stem canker and dieback caused by *Gloeosporium minus*. Phytopathology 64:727-730.

Ojiambo, P. S., Scherm, H., and Brannen, P. M. 2006. Septoria leaf spot reduces flower bud set and yield potential of rabbiteye and southern highbush blueberries. Plant Dis. 90:51-57.

Taylor, J., and Clayton, C. N. 1959. Comparative studies on Gloeosporium stem and leaf fleck and Dothichiza leaf spot of highbush blueberry. Phytopathology 49:65-67.

Von Arx, J. A. 1970. A revision of the fungi classified as Gloeosporium. Bibl. Mycol. 24.

(Prepared by R. D. Milholland; Revised by W. O. Cline)

Leaf Rust

Leaf rust is usually of minor importance; however, epidemics have been reported on lowbush blueberry, where the disease contributes to the leaf spot complex. Occasionally, epidemics occur on highbush blueberry, especially in the states on the east coast of the United States. The disease has been reported in Michigan and in the Pacific Northwest, but it is considered rare. In the lower southeastern states, Florida and Georgia, for example, the disease can be particularly severe on some rabbiteye and southern highbush blueberry varieties. The disease is also taking on international importance, as it has now been reported in Argentina and Spain.

Symptoms

Small, chlorotic to yellow spots, which usually appear by midseason, are the first leaf symptom on blueberry. Spots on the upper surface of the leaf turn reddish brown (Fig. 52), and if the disease is severe, the entire leaf may eventually die as the spots coalesce and become necrotic. Underneath the spots, yellowish orange uredinia develop conspicuously on the abaxial side of infected leaves (Fig. 53). These pustules turn rusty red with age. Infected leaves usually drop prematurely. This disease, like other foliar diseases, can have a direct impact on yield because of early defoliation and reduction in return bloom the following year.

Causal Organism

The pathogen *Pucciniastrum vaccinii* sensu lato is a heteroecious species complex of rust fungi that is found on many ericaceous hosts. There is a need for further taxonomic work, but *P. vaccinii* was divided into two North American species, with *Thekopsora minima* and *Naohidemyces vaccinii* being suggested for the eastern and western North American species, respectively. The aecia of the eastern species are peridermioid (having a peridium that is tonguelike or blisterlike and forms outside the host); those of the western species are uredinoid. Newer publications continue to use the *P. vaccinii* species designation but species should be further clarified. For the purposes of this book, the *P. vaccinii* sensu lato approach to the broad species designation will be used, with the realization that a minimum of two and possibly more species may be represented in this complex.

On blueberry, uredinia are hypophyllous and numerous and often grow in loose clumps in discolored spots. Urediniospores (12–19 × 16–24 μm) are broadly obovate or ellipsoid, with yellowish orange contents (Fig. 54); they are borne on a short stalk and released through a central pore. The cells surrounding the pore are ovoid and smooth. Spore walls are 1–1.5 μm thick, colorless, and slightly roughened or minutely echinulate.

Telia are also hypophyllous, and teliospores form within epidermal cells. Teliospores (7–10 × 14–17 μm) are sessile and ob-

Fig. 52. Reddish brown, necrotic leaf spots on the upper surface of blueberry leaves infected with *Pucciniastrum vaccinii*, the causal organism of leaf rust. (Courtesy J. Clark; © APS)

Fig. 53. Leaf rust uredinia on the lower surface of a blueberry leaf infected with *Pucciniastrum vaccinii*. (Courtesy J. Clark; © APS)

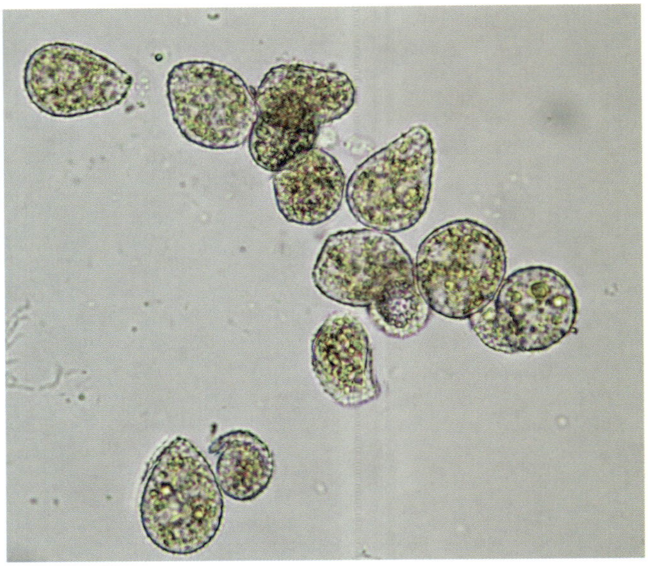

Fig. 54. Urediniospores of *Pucciniastrum vaccinii*, the causal organism of leaf rust. (Courtesy P. M. Brannen; © APS)

Fig. 55. Aecia (uredinoid) of the blueberry leaf rust fungus, *Pucciniastrum vaccinii*, on needles of western hemlock. (Courtesy M. J. Adams; © APS)

Fig. 56. Aecia (peridermioid) of the blueberry leaf rust fungus, *Pucciniastrum vaccinii*, on needles of eastern hemlock. (Courtesy M. J. Adams; © APS)

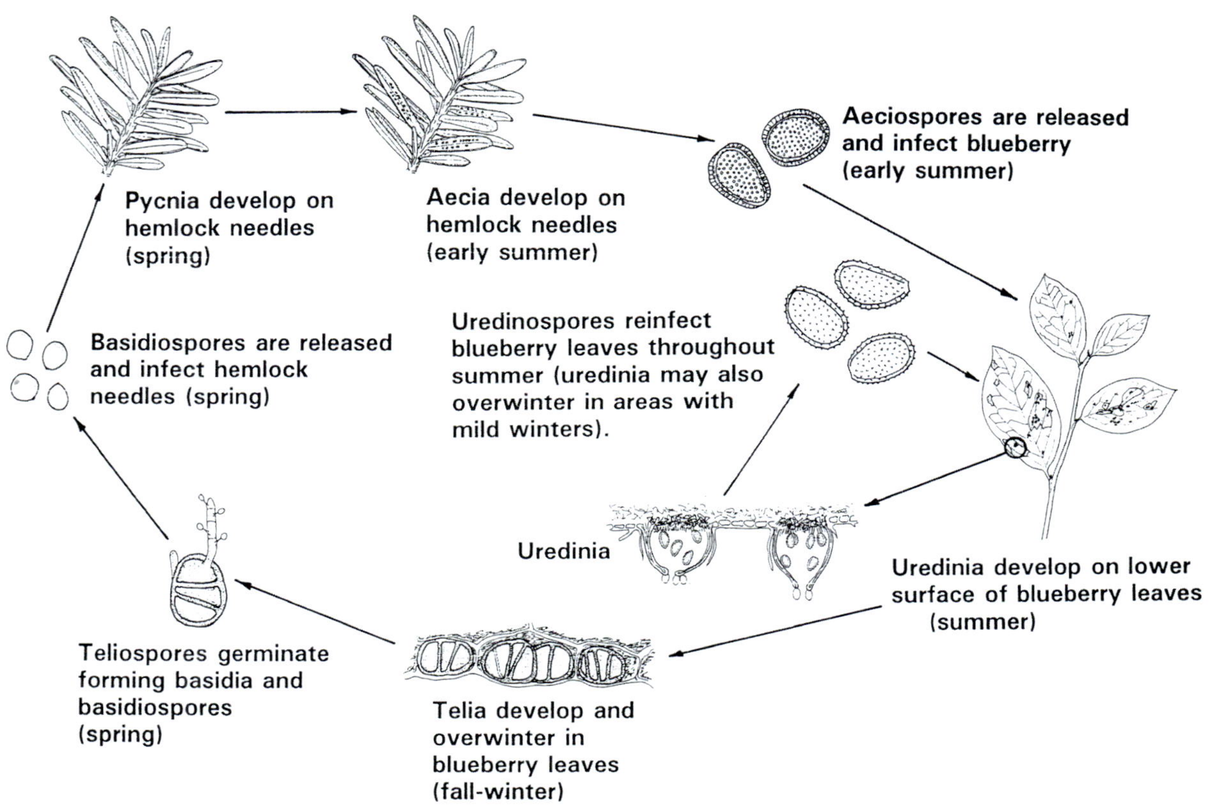

Fig. 57. Disease cycle of leaf rust caused by *Pucciniastrum vaccinii.* (Courtesy P. R. Bristow; © APS)

long to columnar and have two or more vertical cell walls; each cell probably has a single germ pore. Spore walls are lightly colored and uniformly thick. Basidia are external.

On hemlock (*Tsuga* spp.), the alternate host, spermagonia are hypophyllous, subcuticular, numerous, and inconspicuous. Circular aecia form on the lower side of current-season needles in yellow spots. They erupt from beneath the epidermis and form without (*N. vaccinii*; Fig. 55) or with (*T. minima*; Fig. 56) a peridium, depending on the region in North America (as mentioned above). Aeciospores (15–21 × 18–24 µm) are produced in chains and are globoid to broadly ellipsoid; they have a wall (1–1.5 µm thick) that is finely and evenly roughened with flat or rounded warts (*T. minima*) or echinulate (*N. vaccinii*).

Disease Cycle and Epidemiology

P. vaccinii has a macrocyclic life cycle (Fig. 57). In the spring, airborne aeciospores infect newly emerging blueberry leaves. Uredinia form about midsummer on the lower side of leaves. Urediniospores reinfect blueberry in a repeating secondary cycle, which leads to rapid inoculum buildup and disease spread. The latent period may be completed in only 10 days at 20°C. The pathogen does not systemically invade blueberry. Telia form within the epidermal cells of infected leaves, which eventually drop to the ground. Basidia arise from germinating teliospores in dead leaves, and basidiospores infect young hemlock needles in the spring. Inconspicuous spermagonia develop on the lower side of needles, followed by yellow aecia.

Previously, leaf rust was thought to be more prevalent in areas within the native range of hemlock. However, recent reports from both Argentina and Spain note that hemlock is not available, and it is suspected that in regions with mild winters, uredinia may overwinter on the leaves of native evergreen *Vaccinium* spp. or on leaves that remain green, even through the winter. If so, this might account for the presence of the disease in areas, particularly in the south, that are also remote from hemlock. In Maine (United States), the disease has been observed on lowbush blueberry as far as 0.4 km from hemlock. Infrequent outbreaks of the disease in the mid-Atlantic states may be caused by windborne urediniospores from the south.

Management

Numerous fungicides have significant activity against urediniospores of *P. vaccinii*, but not all of these are registered for use on blueberry. Among these, chlorothalonil, mancozeb, and copper sulfate provide fungicidal activity, while propiconazole and azoxystrobin provide fungistatic activity. Pyraclostrobin and fenbuconazole also have activity against rust, which is presumed to be fungistatic. Historically, fungicides have been applied at petal fall and again 10–14 days later to control leaf rust on lowbush blueberry. For rabbiteye and southern highbush varieties, applications are recommended for late spring through summer for management in Georgia, but timing varies by geographic location, as evidenced by recommendations in Florida that include fungicide applications in the spring once leaves first form. In areas where leaves remain throughout the winter, epidemics are likely initiated earlier. Removal of all hemlocks from within a 0.4-km radius of blueberry fields, especially those upwind, may be beneficial. In areas with mild winters, it may also be beneficial to remove all evergreen *Vaccinium* spp. located near blueberry plantings, although this may be impractical in some locations.

Highbush blueberry cultivars vary in their susceptibility to leaf rust, as do rabbiteye and southern highbush blueberry cultivars. Sources of resistance exist within highbush blueberry and possibly within other *Vaccinium* spp. as well. Hybrids between *V. uliginosum* and highbush blueberry are generally susceptible.

Selected References

Barrau, C, de los Santos, B., and Romero, F. 2002. First report of leaf rust of southern high-bush blueberry caused by *Pucciniastrum vaccinii* in southwestern Spain. Plant Dis. 86:1178.

Cummins, G. B., and Hiratsuka, Y. 1983. Illustrated Genera of Rust Fungi, rev. ed. American Phytopathological Society, St. Paul, MN.

Dal Bello, G., and Perelló, A. 1998. First report of leaf rust of blueberry caused by *Pucciniastrum vaccinii* in Argentina. Plant Dis. 82:1062.

Demaree, J. B., and Wilcox, M. S. 1947. Fungi pathogenic to blueberries in the eastern United States. Phytopathology 37:487-506.

Hilborn, M. T. 1955. Blueberry diseases and their control. Pages 27-32 in: Producing Blueberries in Maine. Maine Agric. Exp. Stn. Bull. 479.

Mueller, D. S., Jeffers, S. N., and Buck, J. W. 2005. Toxicity of fungicides to urediniospores of six rust fungi that occur on ornamental crops. Plant Dis. 89:255-261.

Sato, S., Katsuya, K., and Hiratsuka, Y. 1993. Morphology, taxonomy and nomenclature of Tsuga–Ericaceae rusts. Trans. Mycol. Soc. Jpn. 34:47-62.

Ziller, W. G. 1974. The Tree Rusts of Western Canada. Can. For. Serv. Publ. 1329. Department of the Environment, Victoria, British Columbia, Canada.

(Prepared by P. R. Bristow and A. W. Stretch;
Revised by P. M. Brannen)

Mummy Berry

Mummy berry is a widespread disease on all species of cultivated blueberries in North America. Recently, the disease was reported for the first time in central Europe. Under conditions of high disease pressure or when disease management practices are inadequate, yield losses can be severe. Direct yield losses in the field are from shoot blight, flower cluster blight, and fruit mummification. Additional losses may result after harvest from the downgrading or rejection at the packinghouse of fruit loads that do not meet the near-zero tolerance for mummified fruit. The relative importance of these different symptom types depends on the degree of synchronization between the life cycle of the pathogen and the bud break and flowering phenology of its host. For example, on lowbush blueberry in the northeastern United States and Canada and on rabbiteye blueberry in North Carolina (United States), the shoot and flower cluster blight phases are typically most damaging. In contrast, on rabbiteye blueberry in southern Georgia (United States), fruit mummification is more important. Both shoot blight and fruit mummification are prevalent on northern highbush blueberry throughout the production range of this species.

Symptoms

Shoot blight is caused by ascospore infection of young, vegetative tissues soon after bud break. The earliest symptom is the drooping of affected shoots (as early as 1 week after infection), followed by brownish purple discoloration of the upper sides of the midribs and lateral veins of affected leaves. When the leaves are very young at the time of infection, the discoloration may spread across the entire leaf surface within a few days, whereas on somewhat older leaves, the necrotic area often remains restricted to the petiole and proximal end of the lamina. Typically within a few days to 1 week after symptoms first become visible, tufts of grayish conidia appear along the petioles and midribs of affected leaves (Fig. 58). On some *Vaccinium* spp., symptomatic tissues may emit a characteristic odor resembling fermented dark tea leaves. By midseason, blighted tissues desiccate and abscise. The pathogen does not persist in or on infected shoots and does not invade woody tissue.

When ascospore release coincides with the host's bloom period, the peduncle of the inflorescence may become infected, resulting in a blight of the entire flower cluster distally from the point of infection. The pathogen then produces conidial tufts on the infected peduncle or flower pedicels (Fig. 59). The flowers themselves are not subject to infection by ascospores. As indicated above, this phase of the disease is rare in some production areas and common in others.

The conidial infections that lead to fruit mummification are initiated during bloom. Affected fruit remain externally symptomless during the period of green fruit development, although bisection reveals a dense, white mycelial matrix localized within their locules (Fig. 60). As the infected fruit mature, they turn cream to salmon pink and eventually tan or whitish gray (Fig. 61), shriveling and hardening at the same time. Mummified fruit drop to the soil, usually before or during commercial harvest. The fruit epidermis and cortex deteriorate during the autumn and winter to reveal a black pseudosclerotium composed primarily of fungal material.

Causal Organism

Mummy berry is caused by *Monilinia vaccinii-corymbosi*, a member of the Disjunctoriae group within the discomycete genus *Monilinia*. The fungus is oligotrophic in that it is capable of infecting several wild and cultivated *Vaccinium* spp. It is a close relative of *Monilinia oxycocci*, which causes cottonball disease of cranberry.

The overwintering structure of *M. vaccinii-corymbosi* is a hollow pseudosclerotium. This structure is composed of an inner and outer rind of melanized, thick-walled cells surrounding a cream-white medulla (center) made up of a dense matrix of fungal hyphae as well as host cell walls and vascular elements. The size and shape of pseudosclerotia can vary greatly.

Fig. 58. Shoot blight phase of mummy berry disease, caused by *Monilinia vaccinii-corymbosi*, on blueberry. Note the blighting of leaves and presence of grayish conidial tufts on the petioles of affected leaves. (Courtesy H. Scherm; © APS)

Fig. 59. Flower cluster blight caused by *Monilinia vaccinii-corymbosi*, the causal organism of mummy berry. The pathogen sporulates on the affected blueberry pedicels. (Courtesy H. Scherm; © APS)

If the fungus has colonized the entire ovary, the resulting pseudosclerotium is furrow ribbed or pumpkin shaped and hollow. The furrows correspond to the carpel septa and pseudosepta of the ovary. If the fungus has colonized only one or a few carpels, the pseudosclerotium may be crescent shaped or composed of several crescents attached to each other at one or both ends.

Pseudosclerotia germinate carpogenically to produce one to three (occasionally more) apothecia (Fig. 62). Apothecial initials are brown and have rhizoidal tufts near the base. Apothecia become urnulate to cupulate or discoid with age and are 3–10 mm in diameter. Asci (10–12 × 200–260 μm) are inoperculate and contain ellipsoid, hyaline ascospores that are 9–10 × 16–18 μm. Paraphyses are rounded at the tips and 2–3 μm wide.

Conidia (19–21 × 26–28 μm) are hyaline, smooth, and limoniform or subglobose. They are separated by spindle-shaped disjunctors (2 × 3–5 μm) that aid in breaking the conidial chain apart, thereby facilitating their dissemination. Microconidia (2–3 μm) are borne on germinating ascospores, conidia, and mycelium in nature and in vitro. Their function is unknown.

Disease Cycle and Epidemiology

The mummy berry disease cycle has several distinctive features: an obligatory sexual stage that leads to the production of apothecia and ascospores; a single cycle of primary infection of young, noncuticularized, vegetative tissues by ascospores; a single cycle of secondary infection of open flowers by conidia, which occurs through the pistil and ultimately leads to fruit mummification; and in general, a very tight synchronization between the phenologies of the host and the pathogen.

The pseudosclerotium is the sole oversummering and overwintering structure of *M. vaccinii-corymbosi*. Pseudosclerotia germinate carpogenically in late winter or early spring following the accumulation of sufficient chill-hours and degree-days, the former breaking their dormancy and the latter influencing the rate at which apothecia are produced. Populations of pseudosclerotia from different geographic areas and from different blueberry species or cultivars have chill-hour requirements for germination that are similar to the bud break chill-hour requirement of their blueberry hosts.

Apothecial production is favored by good contact with moist soils, but germination is reduced or prevented if pseudosclerotia are buried too deep. Temperatures of 10–16°C favor emergence of the stipe, whereas 16°C is optimal for apothecial formation. At higher temperatures (20–25°C), apothecia mature more rapidly and discharge large numbers of spores but senesce rapidly, whereas at lower temperatures (10–15°C), they last longer and in total produce more spores. Light exposure is also necessary for the stipes to mature into apothecia. In the field, apothecia may be present for about 1 month. Intermittent freezing reduces but does not eliminate the ability of apothecia to discharge viable ascospores.

Ascospore release is triggered by changes in relative humidity, rainfall, or both. The spores are windborne and may be dispersed over relatively long distances. The optimum temperature for ascospore germination is 14°C, but infection can take place at temperatures as low as 2°C. Only very young, noncuticularized tissues are susceptible to ascospore infection, with susceptibility being greatest when new shoots are approximately 5 mm long and decreasing to close to zero once shoots have expanded to more than 45 mm long. Susceptibility to infection is increased following freeze injury. Although no injury may be visible following a frost episode, there is solute leakage from host cells, which stimulates ascospore germination.

Blight symptoms develop 1–2 weeks after ascospore infection, depending on the temperature. High relative humidity favors the production of conidiophores and conidia on blighted tissue. Conidia produced on blighted shoots are disseminated to open flowers by wind, rain, and pollinators (particularly bees). The ultraviolet reflection pattern of conidia-bearing shoots changes to mimic the ultraviolet signature of flower calyxes,

Fig. 61. Ripe blueberry fruit with mummy berry disease, caused by *Monilinia vaccinii-corymbosi*, at the time of harvest. (Courtesy H. Scherm; © APS)

Fig. 60. Immature blueberry fruit infected by *Monilinia vaccinii-corymbosi*, the causal organism of mummy berry. Note the dense, white mycelial matrix in the locules of affected fruit. (Courtesy H. Scherm; © APS)

Fig. 62. Overwintered pseudosclerotium of *Monilinia vaccinii-corymbosi*, the causal organism of mummy berry, with mature and immature apothecia. (Courtesy H. Scherm; © APS)

thereby attracting pollinators that subsequently deposit the conidia on the stigmatic surfaces of open flowers during foraging. The conidia germinate in the stigmatic exudate and grow through the stylar canal, reaching the ovary within 4–7 days after inoculation. Flower susceptibility to conidial infection is greatest on the day of anthesis and decreases rapidly within 5 days after anthesis. Pollination prior to inoculation with conidia reduces but does not prevent flower infection. Once in the ovary, the fungus slowly colonizes the developing fruit, initially by forming a dense, mycelial matrix within the locules. As the fruit develops further, the fungus invades the mesocarp, wherein a fungal entostroma, the pseudosclerotium, is formed. It is at that time that the infected fruit becomes recognizable as a mummy berry.

Management

Blueberry species and cultivars differ in their relative susceptibility to mummy berry, but this has not been utilized systematically for disease management. Reduced susceptibility may be caused by one of two mechanisms, phenology-mediated disease escape or genetic resistance to primary or secondary infection. Disease escape takes place when host plants break bud or complete bloom too early or too late relative to the time when ascospores or conidia are released. For example, southern highbush blueberries in Florida (United States) and southern Georgia generally show very limited levels of infection by *M. vaccinii-corymbosi* because vegetative and flower buds break earlier than those of the native rabbiteye blueberries with which the pathogen has coevolved.

True genetic resistance can be identified in controlled experiments by conducting inoculations with the pathogen at defined host phenological stages. Results of such studies documented a significant genetic contribution to reduced susceptibility toward primary infection, secondary infection, or both. For example, resistance to the mummy berry phase in the highbush blueberry cultivar Weymouth is associated with a reduction of the growth rate of the pathogen in the stylar canal during pistil infection.

Cultural practices have been used with some success to reduce the amount of primary inoculum of *M. vaccinii-corymbosi* in blueberry plantings. Pseudosclerotia are unable to form apothecia when covered with soil to a depth of more than 2.5 cm; thus, cultivation in the alleys between the blueberry rows reduces the amount of disease pressure. Unfortunately, most pseudosclerotia are located close to the crowns of the blueberry bushes where cultivation equipment cannot be used without damaging the shallow roots of the plants. Lowbush blueberries generally are not cultivated, but fields are pruned (either mechanically or by burning) in alternate years to stimulate vegetative growth and flower bud production during the first summer and fruit production during the second year. Burning generally destroys most pseudosclerotia, whereas mechanical pruning with a flail mower is ineffective in this regard.

Early spring soil applications of lime sulfur, urea fertilizer, or certain herbicides in highbush and rabbiteye blueberry plantings may burn or desiccate the apothecia, but this approach is usually not very effective since apothecia develop over an extended period of time. Practices that ensure good soil drainage also are helpful in reducing apothecial development.

Application of fungicides currently is the most widely used and most effective management tactic against mummy berry. Active ingredients from several different fungicide classes are labeled for use against the disease, allowing their deployment in mixture or rotation schedules to prevent the development of fungicide resistance. In lowbush blueberry, where most of the losses are a result of ascospore infection, two well-timed sprays 7–10 days apart are usually adequate to provide effective disease control. In highbush and rabbiteye blueberry, three to six applications may be needed to control both ascospore and subsequent conidial infections. On cultivars and in production areas where vegetative bud break takes place before bloom, one or two prebloom applications are made to control the blight phase of the disease; this is then followed by bloom sprays to control the secondary infection of flowers. Under conditions where bloom occurs before or simultaneously with leaf bud break, only bloom applications are needed. In organic production, a *Bacillus subtilis*-based biofungicide has been used with some success to suppress secondary infection.

Selected References

Batra, L. R. 1983. *Monilinia vaccinii-corymbosi* (Sclerotiniaceae): Its biology on blueberry and comparison with related species. Mycologia 75:131-152.

Ehlenfeldt, M. K., and Stretch, A. W. 2001. Resistance to blighting by *Monilinia vaccinii-corymbosi* in diploid and polyploid *Vaccinium* species. HortScience 36:955-957.

Lehman, J. S., Igarashi, S., and Oudemans, P. V. 2007. Host resistance to *Monilinia vaccinii-corymbosi* in flowers and fruits of highbush blueberry. Plant Dis. 91:852-856.

Ngugi, H. K., Scherm, H., and Lehman, J. S. 2002. Relationships between blueberry flower age, pollination, and conidial infection by *Monilinia vaccinii-corymbosi*. Phytopathology 92:1104-1109.

Ramsdell, D. C., Nelson, J. W., and Myers, R. 1974. An epidemiological study of mummy berry disease of highbush blueberry. Phytopathology 64:222-228.

Scherm, H., Savelle, A. T., and Pusey, L. P. 2001. Interactions between chill-hours and degree-days affect carpogenic germination in *Monilinia vaccinii-corymbosi*. Phytopathology 91:77-83.

(Prepared by H. Scherm and P. D. Hildebrand)

Phomopsis Canker

Phomopsis canker has been reported on highbush blueberry in the United States in Arkansas, Illinois, Indiana, Massachusetts, Michigan, New Jersey, and North Carolina. It has also been found in Canada, Chile, and Lithuania. The disease was detected on introduced American blueberry cultivars in Romania and Scotland but apparently did not persisted or spread to commercial plantings. Unpublished reports indicate that this disease also is found in highbush blueberry plantations in northern Germany. The causal fungus is considered a quarantine pathogen in Europe. Phomopsis canker is not found on rabbiteye or lowbush blueberry, but the pathogen causes a twig blight of rabbiteye blueberry in the southeastern United States.

Symptoms

Phomopsis canker is usually first noticed by the wilting and death of canes starting in midsummer. Leaves are reddish brown and remain attached to the stem, which is referred to as flagging (Fig. 63). Cankers may form from the soil line to 1–1.5 m above the ground on 1-, 2-, and 3-year-old stems. Cankers on 1-year-old stems become evident in early to midsummer. They are narrow and elongated, brown to reddish brown, and 2–10 cm long or longer. As the canker develops, it encircles the stem and becomes somewhat flattened. All plant tissue above the canker is eventually killed. Cankers can expand into the crown, which often results in the infection and death of newly emerging stems. Cankers turn light brown to gray with age, and pycnidia form in the bleached areas.

Occasionally, the tips of succulent new shoots become infected, turn dark brown, and may take on a shepherd's crook appearance (Fig. 64). The cankers usually progress downward, become up to 30 cm long, and may encompass the whole shoot. When infected stems are cut open lengthwise, a light brown necrotic area can be seen. Often, the pith of infected stems is brown well below the cankered area (Fig. 65).

Causal Organism

Phomopsis canker is caused by the fungus *Phomopsis vaccinii* (syn. *Diaporthe vaccinii*), which is the same pathogen that causes Phomopsis twig blight and fruit rot of blueberries. The anamorphic and teleomorphic stages were described in 1931. The same fungus has been known to cause an important fruit rot of cranberries since 1916. When highbush blueberry was inoculated separately with isolates of *P. vaccinii* from blighted blueberry twigs and from the pulp of decaying cranberries, both cultures caused the blighting of shoots and girdling of branches and grew identically on artificial media.

Pycnidia are abundant on older cankers, usually on dead wood. The pycnidia (0.3–0.5 mm in diameter) are dark, ostiolate, and immersed or erumpent (Fig. 66). Both alpha and beta conidia are produced. Alpha conidia are hyaline, oval to ellipsoid, guttulate, and 2–4 × 6–11 µm. Beta conidia are long, narrow, curved, and approximately 0.75 × 12–18 µm. Beta conidia have not been induced to germinate, suggesting that they function as spermatia.

The anamorph is the stage that is usually seen on blueberry, although the teleomorph has reportedly been found on highbush blueberry in Chile. In 1940, the teleomorph was reported in culture. The perithecia in cornmeal agar cultures

Fig. 63. Highbush blueberry with stem wilting resulting from Phomopsis canker, caused by *Phomopsis vaccinii*, in the midsummer. (Courtesy A. M. C. Schilder; © APS)

Fig. 64. Young blueberry stem infected by *Phomopsis vaccinii*, the causal organism of Phomopsis canker. Note the shepherd's crook appearance. (Courtesy A. M. C. Schilder; © APS)

Fig. 65. Brown discoloration of the xylem and pith of a blueberry stem infected by *Phomopsis vaccinii* (left), the causal organism of Phomopsis canker, compared with a healthy stem (right). (Courtesy C. Garcia-Salazar; © APS)

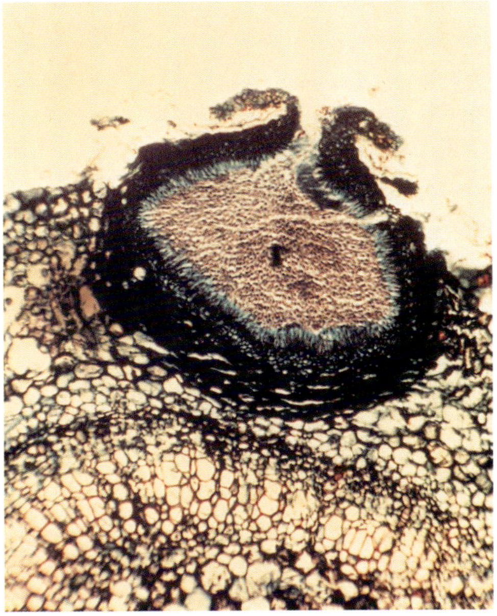

Fig. 66. Cross section of a pycnidium of *Phomopsis vaccinii*, the causal organism of Phomopsis canker. (Courtesy P. E. Parker; © APS)

were described as stromatic, and both the stroma and perithecia appeared to be identical to those of the cranberry isolate. Perithecia are elongated and thick walled, with ostioles copiously supplied with upward-directed hairs. Asci (5.8–9.6 × 32–48 µm) are oblong, fusoid, and sessile, with narrow pores. Ascospores (2.5–4.2 × 6.4–12.8 µm) are two celled, slightly constricted, and biguttulate.

Colonies on potato dextrose agar are white and zonate, with yellowish gray to brownish gray coloration around the agar plug in some strains. The mycelium is cottony or felty with a somewhat aerial, feathery margin. Mycelial growth on potato dextrose agar is maximal at 27°C, less at 22°C, and absent at 0 and 32°C.

Disease Cycle and Epidemiology

P. vaccinii overwinters as mycelium and pycnidia in cankers on stems. The etiology and epidemiology of Phomopsis canker were studied over two seasons in Michigan. As little as 3.8 mm (0.15 in.) of rain triggers the release of conidia. Conidial discharge begins at blossom bud swell and continues during each rain event until late August. Conidia are splash dispersed and germinate readily in the presence of water after 48 h at 10–27°C. Conidia can enter the plant through wounds or directly through the tips of young succulent blueberry shoots.

Plants that have been wounded mechanically or by freezing are more susceptible to infection than are undamaged plants. Stem injuries caused by mechanical harvesting probably provide infection entry points. Infection of succulent, current-season stems is more prevalent during seasons with above-average precipitation in mid- to late summer. Stem wilting and death tend to be more prevalent during hot and dry summers and following winter injury or drought stress the previous year.

Management

The disease is often introduced in the field with infected planting material. Therefore, disease-free planting material is a prerequisite for a healthy planting. Pruning to remove cankered and wilted stems, as deep into the crown as possible, helps to remove inoculum. However, cankers can extend into the crown below the soil line and result in the infection of newly emerging stems. Infected stems should be removed from the field and burned or buried to avoid further spread of the disease.

Most commercial cultivars are susceptible to Phomopsis canker. Highbush blueberry cultivars Jersey, Berkeley, Bluejay, and Bluehaven are particularly susceptible, while cultivars Bluetta, Elliott, and Rancocas are considered resistant.

In the past, benomyl plus captan as a tank mix was the standard for the control of Phomopsis canker. Spray applications every 2 weeks from bud break until late August reduced the number of wilted shoots the following year by 60%. Since benomyl is no longer available, other fungicides need to be used. Few new fungicides have been tested for control of Phomopsis canker in the past decade because of the long latent period. However, it is assumed that fungicides that are effective against Phomopsis twig blight are also effective against Phomopsis canker. The fungicides fenbuconazole, propiconazole, fosetyl-Al, pyraclostrobin, pyraclostrobin plus boscalid, chlorothalonil, and ziram are effective against Phomopsis twig blight. Dormant sprays of lime sulfur also help control the disease.

Selected References

Ames, G. K., Gergerich, R. C., Weidemann, G. J., and Patterson, C. A. 1988. First report of *Diaporthe vaccinii* on blueberry in Arkansas. Plant Dis. 72:362.

Baker, J. B., Hancock, J. F., and Ramsdell, D. C. 1995. Screening highbush blueberry cultivars for resistance to Phomopsis canker. HortScience 30:586-588.

Demaree, J. B., and Wilcox, M. S. 1947. Fungi pathogenic to blueberries in the eastern United States. Phytopathology 37:487-506.

Guerrero, C. J., and Godoy, A. I. 1989 Detection of *Phomopsis vaccinii* (Shear. Stevens and Bein) in highbush blueberry (*Vaccinium corymbosum* L.). Agric. Tec. (Santiago) 49:220-223.

Kačergius, A., Jovaišienė, Z., and Valiuskaite, A. 2004. *Phomopsis vaccinii* on *Vaccinium corymbosum* in Lithuania. Bot. Lith. 10:75-80.

Parker, P. E., and Ramsdell, D. C. 1977. Epidemiology and chemical control of Phomopsis canker of highbush blueberry. Phytopathology 67:1481-1484.

Smith, I. M., Scott, P. R., Holderness, M., McNamara, D. G., and Burger, B. V., eds. 2005. *Diaporthe vaccinii*. Quarantine Pests for Europe, 2nd ed. Data Sheets on Quarantine Pests for the European Union and for the European and Mediterranean Plant Protection Organization. CAB International, Wallingford, United Kingdom.

Teodorescu, G., Copaescu, V., and Florea, S. 1985. The behaviour of some blueberry cultivars to the main mycoses in Romania. Acta Hortic. 165:159-165.

Weingartner, D. P., and Klos, E. J. 1975. Etiology and symptomatology of canker and dieback diseases on highbush blueberries caused by *Godronia* (*Fusicoccum*) *cassandrae* and *Diaporthe* (*Phomopsis*) *vaccinii*. Phytopathology 65:105-110.

Wilcox, M. S. 1939. Phomopsis twig blight of blueberry. Phytopathology 29:136-142.

Wilcox, M. S. 1940. *Diaporthe vaccinii*, the ascigerous stage of Phomopsis, causing a twig blight of blueberry. Phytopathology 30:441-443.

Wilcox, H. J., and Falconer, M. A. 1961. New or uncommon plant pests. Plant Pathol. 10:123-124.

(Prepared by D. C. Ramsdell; Revised by A. M. C. Schilder)

Phomopsis Twig Blight and Fruit Rot

Phomopsis twig blight was one of the first diseases observed in cultivated blueberry fields. It was identified as a serious problem on blueberry in Massachusetts (United States) in 1924 and was also reported to occur in New Jersey (United States). This disease is common in most blueberry-growing regions in North America and has also been reported in South America (Chile) and Europe (Lithuania). The same pathogen causes a canker disease of blueberry and an upright dieback and fruit rot (viscid rot) of cranberry.

Twig blight has increased tremendously in prevalence and severity in the southeastern United States over the past 20 years. It can be found in all blueberry plantings in North Carolina (United States), with 100% of the bushes affected in some individual plantings. Phomopsis twig blight outbreaks are sporadic in northern growing regions, where the canker phase of the disease is more common. More than 100 blighted twigs per bush have been counted, with an estimated fruit loss of up to 75%. The fungus also causes a serious fruit rot at harvest in North Carolina, where crop losses of up to 5% have been attributed to it.

Symptoms

The primary symptom is a blighting of fruiting twigs, which results in reduced fruit production (Fig. 67). Visible symptoms may appear as early as flower bud break, when individual flower buds turn brown and die, followed by an expanding brown, necrotic area around the bud (Fig. 68). Infections can take place on fruit-bearing twigs at all stages of development, but blossoms are particularly susceptible. The lesion at first progresses rapidly, killing all buds along the stem, and may reach 20–25 cm in length. Progression of the twig blight lesion further down the stem into older wood eventually stops. Rabbiteye plants are susceptible to the pathogen and exhibit symptoms similar to those of highbush plants. These symptoms may be difficult to distinguish from cold injury and twig blights caused by other pathogens.

Reddish brown, circular to irregular lesions, 2–10 mm in diameter, may also appear on leaves in mid- to late summer (Fig. 69). Leaf lesions have a darker edge and may have a bleached center. Under humid conditions, pycnidia readily develop in the leaf lesions.

Phomopsis fruit rot takes place on ripe fruit before or after harvest. Infected berries may show white tufts of mycelium in the calyx area. They become very soft and may split under pressure and leak juice. The internal tissues turn reddish brown and mushy.

Causal Organism

The causal organism is the coelomycetous fungus *Phomopsis vaccinii* (syn. *Diaporthe vaccinii*). The sexual stage is placed in the phylum Ascomycota, order Diaporthales, and is rarely seen on blueberry.

Pycnidia are dark, ostiolate, immersed or erumpent, and nearly globose, with two types of conidia, forming white to yellowish cirrhi (Fig. 70). Alpha conidia are hyaline, one celled, ovate to fusiform, guttulate, and 2–4 × 6–11 µm (Fig. 71). Beta conidia are hyaline, one celled, filiform, straight or curved, eguttulate, and approximately 0.75 × 12–18 µm (Fig. 71). Some pycnidia only produce alpha conidia, while others produce both types. Beta conidia have not been induced to germinate, suggesting that they function as spermatia. In vitro production of alpha and beta conidia is affected by light, temperature, and culture medium. Beta conidia are usually not produced in culture after one or more transfers.

Perithecia are produced in stromata and have long, protruding necks. Asci (5.8–9.6 × 32–48 µm) are oblong, fusoid, and sessile, with narrow pores. Ascospores (2.5–4.2 × 6.4–12.8 µm) are cylindrical, two celled, and slightly constricted; each cell is typically biguttulate.

Colonies on potato dextrose agar are white and feathery to zonate, with cottony or felty mycelium. Some strains have a yellowish gray to brownish gray coloration around the agar plug. Pycnidia with creamy to yellowish spore droplets are produced in culture.

A direct tissue blot immunoassay (DTBIA) and a plate-trapped antigen enzyme-linked immunosorbent assay (PTA-ELISA) with *Phomopsis*-specific antiserum have been used to

Fig. 67. Phomopsis blossom and twig blight of blueberry, caused by *Phomopsis vaccinii*. (Courtesy A. M. C. Schilder; © APS)

Fig. 68. Expanding Phomopsis twig blight lesion originating from a blueberry flower bud, caused by *Phomopsis vaccinii*. (Courtesy P. Wharton; © APS)

Fig. 69. Phomopsis twig blight blueberry leaf lesions, caused by *Phomopsis vaccinii*. (Courtesy A. M. C. Schilder; © APS)

Fig. 70. Cirrhi of *Phomopsis vaccinii* conidia, the causal organism of Phomopsis twig blight, on an infected blueberry twig. (Courtesy P. Wharton; © APS)

detect *P. vaccinii* in plant tissues, with PTA-ELISA being more sensitive.

Disease Cycle and Epidemiology

Phomopsis twig blight is often introduced into fields with planting material. In the southeastern United States, *P. vaccinii* overwinters in blueberry twigs infected the previous season. Conidia are released during rainy periods and disseminated by rain splash and wind-driven rain from late February to early August. The greatest numbers of conidia are disseminated during the early part of the growing season from bud break through bloom. Conidia from current-season infections are disseminated from the end of May through July in North Carolina. Although conidia from current-season infections do not extend the period of spore dispersal, they do increase the total inoculum available for infection through harvest.

The open flower buds not only act as a trap for rain-dispersed conidia but also provide the fungus with an entrance into the vascular tissue of the stem. Conidia of *P. vaccinii* germinate and directly penetrate the succulent young floral parts of blueberry before and during bloom. The optimum temperature for conidium germination is 21–24°C. Once the fungus has invaded the flowers, it rapidly grows between the cortical cells of the pedicel into the intercellular spaces of the stem cortex, leading to necrosis and collapse of parenchyma cells and vascular tissue. Lesions may become visible 3–4 days after infection. Stem lesions stop expanding about 8 weeks after infection because of the formation of a layer of suberized cells in the cortex separating infected and healthy portions of the stem.

Blueberry fruit are susceptible to infection by *P. vaccinii* at all stages of development, but the infection remains latent until fruit maturity. In Michigan (United States), postharvest fruit rot tends to develop in fields with high twig blight incidence.

Management

Sanitation is an important control method since the pathogen overwinters in infected twigs. Removal of infected twigs by pruning prior to spore release in the spring decreases the amount of inoculum present at bud break. This may not be practical in large plantings but is appropriate for small plantings and home gardens. Mowing bushes after harvest (topping), as is done in southern growing areas, also reduces the amount of overwintering inoculum.

Cultivar selection is a management option, since blueberry cultivars are known to differ in their susceptibility to Phomopsis twig blight. Among northern highbush cultivars, Elliott, Bluetta, Rubel, and Bluecrop are resistant to Phomopsis twig blight, whereas Jersey, Berkeley, Bluehaven, Murphy, and Earliblue are susceptible. Southern highbush cultivar Harrison is highly susceptible and should not be grown without using a fungicidal spray program. 'Croatan' is moderately susceptible, while the cultivars Reveille, Cape Fear, Bluechip, and Wolcott are relatively resistant. Some rabbiteye cultivars are susceptible, especially Delite. Half-high cultivars Northsky and Chippewa are resistant to Phomopsis twig blight.

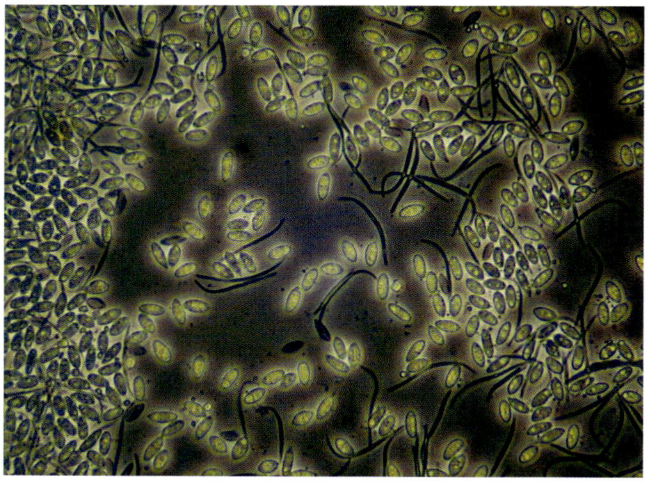

Fig. 71. Alpha conidia (fusiform, showing guttules) and beta conidia (filiform) of *Phomopsis vaccinii*, the causal organism of Phomopsis twig blight. Note that spores shown were collected from *P. vaccinii* infecting cranberry but are identical to those collected from blueberry isolates of *P. vaccinii*. (Courtesy C. Constantelos; © APS)

Protective applications of effective fungicides at 14-day intervals from bud break through full bloom control Phomopsis twig blight. Additional sprays at 14-day intervals through berry development control the fruit rot phase. Benomyl used to be the standard fungicide for control of Phomopsis twig blight but is no longer available in the United States. Other effective systemic fungicides include fenbuconazole, propiconazole, fosetyl-Al, pyraclostrobin, and pyraclostrobin plus boscalid. Chlorothalonil and ziram have good activity as protectants. Lime sulfur applied as a dormant spray before bud break also helps reduce disease pressure.

Selected References

Cline, W. O. 2000. Twig blight of blueberry. Fruit Dis. Inf. Note 10. North Carolina Cooperative Extension Service, North Carolina State University, Raleigh. Online publication.

Cline, W. O., Bloodworth, B. K., and Meister, C. W. 2002. Evaluation of fungicides to control diseases of blueberry, 2002. Fungic. Nematicide Tests 58:SMF017.

Daykin, M. E., and Milholland, R. D. 1990. Histopathology of blueberry twig blight caused by *Phomopsis vaccinii*. Phytopathology 80:736-740.

Farr, D. F., Castlebury, L. A., and Rossman, A. Y. 2002. Morphological and molecular characterization of *Phomopsis vaccinii* and additional isolates of *Phomopsis* from blueberry and cranberry in the eastern United States. Mycologia 94:494-504.

Gabler, J., Kačergius, A., and Jovaišienė, Z. 2004. Detection of *Phomopsis vaccinii* on blueberry and cranberry in Europe by direct tissue blot immunoassay and plate-trapped antigen ELISA. J. Phytopathol. 152:630-632.

Guerrero, C. J., and Godoy, A. I. 1989. Diagnosis of *Phomopsis vaccinii* (Shear. Stevens and Bein) in highbush blueberry (*Vaccinium corymbosum* L.). Agric. Tec. (Santiago) 49:220-223.

Kačergius, A., Jovaišienė, Z., and Valiuskaite, A. 2004. *Phomopsis vaccinii* on *Vaccinium corymbosum* in Lithuania. Bot. Lith. 10:75-80.

Milholland, R. D. 1982. Blueberry twig blight caused by *Phomopsis vaccinii*. Plant Dis. 66:1034-1036.

Milholland, R. D., and Daykin, M. E. 1983. Blueberry fruit rot caused by *Phomopsis vaccinii*. Plant Dis. 67:325-326.

Parker, P. E., and Ramdsell, D. C. 1977. Epidemiology and chemical control of Phomopsis canker of highbush blueberry. Phytopathology 67:1481-1484.

Polashock, J. J., and Kramer, M. 2006. Resistance of blueberry cultivars to Botryosphaeria stem blight and Phomopsis twig blight. HortScience 41:1457-1461.

Schilder, A. M. C., Gillett, J. M., and Sysak, R. W. 2000. Evaluation of fungicides for control of Phomopsis twig blight of blueberry, 1999. Fungic. Nematicide Tests 55:87-88.

Schilder, A. M. C., Gillett, J. M., and Sysak, R. W. 2001. Evaluation of fungicides for control of Phomopsis diseases in blueberries, 2000. Fungic. Nematicide Tests 56:SMF4.

Schilder, A. M. C., Gillett, J. M., and Sysak, R. W. 2003. Evaluation of fungicides for control of Phomopsis twig blight and fruit rot in blueberries, 2002. Fungic. Nematicide Tests 58:SMF037.

Smith, I. M., Scott, P. R., Holderness, M., McNamara, D. G., and Burger, B. V., eds. 2005. *Diaporthe vaccinii*. Quarantine Pests for Europe, 2nd ed. Data Sheets on Quarantine Pests for the European Union and for the European and Mediterranean Plant Protection Organization. CAB International, Wallingford, United Kingdom.

Wilcox, M. S. 1939. Phomopsis twig blight of blueberry. Phytopathology 29:136-142.

Wilcox, M. S. 1940. *Diaporthe vaccinii*, the ascigerous stages of Phomopsis, causing a twig blight of blueberry. Phytopathology 30:441-443.

(Prepared by R. D. Milholland; Revised by A. M. C. Schilder)

Phyllosticta Leaf Spot and Fruit Rot

The disease syndrome Phyllosticta leaf spot and fruit rot was first observed on cultivated bushes of *Vaccinium australe* near Beltsville, Maryland, in 1939 and has been reported on other *Vaccinium* spp., both cultivated and wild, in Georgia, North Carolina, Mississippi, and many other states in the United States. Although not considered a major pathogen in commercial production fields, a review of annual reports by state diagnostic clinics shows that *Phyllosticta* spp. are widely reported blueberry leaf spot pathogens in homeowner samples from unsprayed bushes. The *Phyllosticta* spp. that cause these diseases are potentially important as causal organisms of fruit and storage rots on highbush and rabbiteye cultivars throughout the eastern and southern United States.

Symptoms

Leaf spots are small (1–3 mm in diameter), circular, and gray in the center and are surrounded by a brown to purple margin. One to six pycnidia develop in the lesions. Further investigation and description of the symptoms are needed since much larger leaf spot lesions have been observed. These may be misidentifications or other *Phyllosticta* spp., or the size and appearance of *Phyllosticta* lesions may simply be incompletely described in the scientific literature. Fruit lesions (6–8 mm in diameter) are circular, gray to brown, slightly depressed, and hard, with several pycnidia (Fig. 72). Both leaf and fruit symptoms begin to appear in mid-May in North Carolina. The disease may appear on the fruit during the green stage or preripening period as a hard, dry rot.

Causal Organism

Phyllosticta vaccinii causes Phyllosticta leaf spot and fruit rot. Pycnidia that develop in the leaves are oval to globose, subepidermal to somewhat erumpent, ostiolate, and on average, 68–75 µm wide. Conidia (average 7.0 × 7.6 µm) are obovate and hyaline, with coarse guttules. The length-to-width ratio of the conidia is 1.4 (obovate). On potato dextrose agar, *P. vaccinii* grows slowly (0.5–1.0 mm/day) at 24°C, with optimum growth at 28°C; colonies are olive to black, with dense clusters of pycnidia.

Fig. 72. Phyllosticta fruit rot, caused by *Phyllosticta vaccinii*, on a blueberry fruit. (Courtesy R. D. Milholland; © APS)

Disease Cycle and Epidemiology

Very little is known about the life history of this fungus and the development of the disease on blueberry foliage and fruit.

Management

Spraying with fungicides immediately after harvest and at 2, 4, and 6 weeks after harvest can reduce losses from blueberry leaf spots. The highbush cultivars Jersey and Murphy are susceptible to Phyllosticta leaf spot and fruit rot.

Selected References

Cline, W. O., and Milholland, R. D. 1996. Diseases of blueberry fruit at harvest in North Carolina. Pages 219-226 in: Blueberries: A Century of Research. Food Products Press, Binghamton, NY.

Demaree, J. B., and Wilcox, M. S. 1947. Fungi pathogenic to blueberries in the eastern United States. Phytopathology 37:487-506.

Taylor, J., and Clayton, C. N. 1959. Comparative studies on Gloeosporium stem and leaf fleck and Dothichiza leaf spot of highbush blueberry. Phytopathology 49:65-67.

Weidemann, G. J., Boone, D. M., and Burdsall, H. H., Jr. 1982. Taxonomy of *Phyllosticta vaccinii* (Coelomycetes) and a new name for the true anamorph of *Botryosphaeria vaccinii* (Dothideales, Dothioraceae). Mycologia 74:59-65.

(Prepared by R. D. Milholland; Revised by W. O. Cline)

Phytophthora Root Rot

Phytophthora root rot is the most serious root disease of highbush blueberry worldwide. It is found in nearly all temperate- and warm-climate blueberry-growing regions. The disease is not as serious on rabbiteye cultivars, which are less susceptible.

Phytophthora root rot, caused by *Phytophthora cinnamomi*, was first described on blueberry from New Jersey (United States) in 1963. In 1967, the fungus was recovered from 40% of the blueberry plantings surveyed in southeastern North Carolina (United States). The disease was also reported from highbush blueberry in Arkansas in 1978 and, by 1982, was established in northwestern Arkansas, Missouri, and Oklahoma (United States). In 1996, it was reported from highbush blueberry in northwestern Italy and, in 2001, was reported to be widely distributed in blueberry plantings in Oregon (United States). The disease has also been reported in New Zealand and Australia. Because of the susceptibility of highbush blueberry and the pathogen's wide geographic distribution and broad host range, the disease is likely to become a serious problem in any growing areas where *P. cinnamomi* can survive.

Symptoms

The earliest symptoms on highbush blueberry are dark lesions on the fibrous root tissues. As the pathogen invades the root tissues, the cortex become darkened and roots eventually rot. The root systems of infected plants become reduced and younger plants can be easily pulled out of the ground. The crown may become discolored in cross section, and this advanced symptom generally leads to death of the plant. Aboveground symptoms include early leaf discoloration (yellow to red) and marginal burn, with new leaves being small and premature defoliation likely. Plants take on a twiggy appearance and are often stunted in growth with little evidence of new growth (Fig. 73). The plants are more susceptible to drought stress because of the reduced root system.

Aboveground symptoms on rabbiteye plants are usually less severe than those on highbush plants. Leaves eventually become chlorotic, and there may be some defoliation.

Fig. 73. Phytophthora root rot, caused by *Phytophthora cinnamomi*, on highbush blueberry. (Courtesy P. V. Oudemans; © APS)

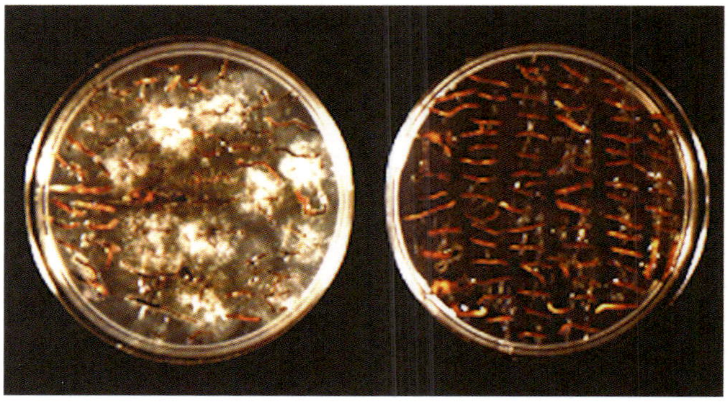

Fig. 74. Isolation of *Phytophthora cinnamomi* on a PARPH selective medium. Blueberry root samples collected from the field were washed, cut into 1-cm pieces, and pressed into the medium. After 3 days, *P. cinnamomi* colonies appeared from root sections on the left, suggesting this pathogen is the cause of Phytophthora root rot in this sample. *P. cinnamomi* colonies failed to appear on the sample on the right, suggesting this sample is free of the pathogen. (Courtesy P. V. Oudemans; © APS)

The disease often appears first in poorly drained sections of the field and in heavier soils. It can sometimes be misdiagnosed as wet feet or a mineral deficiency. For diagnosis, symptomatic root tissues can be plated onto a *Phytophthora* selective growth medium (Fig. 74) and the pathogen identified by morphology or through a molecular diagnostic procedure.

Causal Organism

The causal organism of Phytophthora root rot in blueberry is *Phytophthora cinnamomi*. This organism is classified in the kingdom Stramenopila, phylum Oomycota, class Peronosporomycetes, order Peronosporales, and family Peronosporaceae. It is not closely related to the fungi and often does not respond to many modern fungicides. The sporangia are variable in shape, often ovoid, obpyriform, ellipsoidal or elongate ellipsoidal, and nonpapillate. Sporangia proliferate either internally or externally and each produces approximately 10–20 biflagellate zoospores. Chlamydospores are the primary overwintering structure and are globose to pyriform. Chlamydospores are produced abundantly in culture. Only isolates of the A2 mating type have been described from blueberry, and therefore, the oospores with amphigynous antheridia are found in paired cultures with the A1 mating type. Oospores have only rarely been observed in single or selfed cultures and it is unlikely that they play any significant role in the disease cycle on blueberry. The most diagnostic feature of *P. cinnamomi* is the coralloid mycelium produced in culture.

Disease Cycle and Epidemiology

P. cinnamomi is a broad-host-range pathogen that can attack more than 1,000 host species, ranging from gymnosperms to angiosperms. Thus, in any particular area, a variety of distinct host species may coexist and act as a pathogen reservoir. As with many *Phytophthora* spp., *P. cinnamomi* is very dependent on water for dispersal of the self-motile, flagellated zoospores and these readily disseminate through surface water. The pathogen can also be disseminated through contaminated soil, where infected root tissues or chlamydospores may reside in a dormant state. Introduction into a blueberry planting may occur through planting of infected materials or by contaminated water supplies (i.e., flooding or irrigation with surface runoff).

Under sufficiently wet conditions, the pathogen infects and colonizes root tissues. Once an infection is established, the pathogen spreads along the row by root contact. Episodic flooding caused by excessive irrigation or precipitation also promotes pathogen spread. In saturated soils, the zoospores are attracted to the region of elongation on small, absorbing roots, where they germinate and penetrate the root just behind the root cap. In highly susceptible highbush blueberry rootlets, the fungus penetrates the epidermis and invades the vascular tissue within 24 h. Hyphae grow intracellularly in the epidermis, cortex, phloem, and xylem vessels.

Chlamydospores are the primary overwintering structure of *P. cinnamomi* and are formed in rotted roots and plant tissues. During the decomposition of the infected host tissues, chlamydospores are released into the soil, where they remain dormant until the proper conditions for germination arise. Abundant soil moisture favors root infection. Disease development is often more severe on soil types with poor drainage. Soil temperatures between 15 and 32°C favor the growth of *P. cinnamomi*.

Management

Management of Phytophthora root rot in blueberry depends on an integrated approach. First and foremost, only clean planting stock should be used. Propagation beds should not be watered excessively and should not be irrigated with surface water. Nursery plants should not be treated with an oomycete fungicide (see below) since this will mask symptom development and lead to introduction to the new plantings.

Site selection and soil preparation are also essential components for disease management. Fields should be well drained, and each management unit or irrigation zone should exhibit uniform drainage properties so that overwatering does not happen. If needed, ditches or drainage tiles can be installed to remove excessive surface water.

Irrigation should be regular and not excessive. Standing water and extended soil saturation promotes growth and reproduction of the pathogen.

Infected plants may be treated with an oomycete selective fungicide. The phenylamide fungicides provide good activity against *P. cinnamomi* when applied to soil. The phosphonate fungicides also are active against this pathogen and may be applied to soil or leaves. Caution should be exercised when using phosphorous acid salts at a pH less than 5.5 since phytotoxicity may result.

Selected References

Clark, J. R., Moore, J. N., and Draper, A. D. 1986. Inheritance of resistance to Phytophthora root rot in highbush blueberry. J. Am. Soc. Hortic. Sci. 111:106-109.

Erb, W. A., Moore, J. N., and Sterne, R. E. 1987. Response of blueberry cultivars to inoculation with *Phytophthora cinnamomi* Rands zoospores. HortScience 22:298-300.

Milholland, R. D. 1975. Pathogenicity and histopathology of *Phytophthora cinnamomi* on highbush and rabbiteye blueberry. Phytopathology 65:789-793.

Sterne, R. E. 1982. Phytophthora root rot of blueberry in Arkansas. Plant Dis. 66:604-605.

(Prepared by R. D. Milholland;
Revised by P. V. Oudemans)

Powdery Mildew

Powdery mildew can be found in most blueberry plantings throughout the United States. Damage from this disease is normally very slight. However, the disease can be very serious in some cases. Powdery mildew fungi seldom kill their host since they are obligate parasites. Instead, they utilize their nutrients, reduce photosynthesis, increase respiration and transpiration, and impair the growth of the host.

Symptoms

Symptoms of powdery mildew usually do not develop on blueberry leaves until midsummer. On some cultivars, the mycelium is conspicuous on upper leaf surfaces (Fig. 75), whereas on others, symptoms are yellowish or reddish spots and leaf puckering. The latter symptoms may be mistaken for a virus disease. On the lower leaf surface, the mycelium is barely visible as water-soaked spots in radiating, dendritic patterns (Fig. 76). White powdery colonies may be more conspicuous later in the season. Heavy infections lead to premature defoliation, which can weaken plants.

Causal Organism

Powdery mildew is caused by the fungus *Microsphaera vaccinii* (syns. *Erysiphe vaccinii*, *Microsphaera alni* var. *vaccinii*, and *Microsphaera penicillata* var. *vaccinii*), which belongs to the phylum Ascomycota, order Erysiphales. This pathogen infects plants in the family Ericaceae, including *Vaccinium*, *Rhododendron*, and *Gaylussacia* spp. It is not known whether specialized strains exist.

The mycelium is mostly superficial, consisting of numerous hyphae anchored in the epidermis of the leaf by haustoria that penetrate the epidermal cells. The mycelium produces short conidiophores from which chains of hyaline, ovoid or ellipsoid conidia (12 × 20 µm) develop. Conidia are disseminated by air currents.

At the end of the season, chasmothecia (formerly called cleistothecia) develop on powdery mildew colonies, usually on the upper leaf surface (Fig. 77). They are yellow at first and then turn brown to black as they mature. Mature chasmothecia have six to twelve dichotomously branched appendages that are 1.5–6 times as long as the diameter of the chasmothecium. The apex of each appendage branches dichotomously three or four times, with the ultimate branches not recurved. Chasmothecia contain two to eight globose to ovoid asci, each with four to eight ascospores. Ascospores (9–15 × 14–24 µm) are unicellular, ellipsoid to oblong, and hyaline.

Fig. 75. White powdery patches and reddish discoloration caused by *Microsphaera vaccinii*, the causal organism of powdery mildew, on blueberry leaves. (Courtesy W. O. Cline)

Disease Cycle and Epidemiology

M. vaccinii most likely overwinters as chasmothecia on fallen leaves on the ground. It is not known whether the fungus can also overwinter as mycelium in dormant buds. Spores are released, germinate, and cause infection when the relative humidity is fairly high but there is no free water on the plant surface. Germination begins in about 2 h, appressoria form in 6 h, and secondary hyphae form in 15 h. Conidiophores are initiated after 3 days, and mature conidia develop after 5 days. With the coming of cool weather later in the season, conidium production ceases and chasmothecia are formed.

Fig. 76. Colonies of *Microsphaera vaccinii*, the causal organism of powdery mildew, form dendritic patterns of water-soaked spots on the lower surface of a blueberry leaf. (Courtesy A. M. C. Schilder; © APS)

Fig. 77. Chasmothecia (overwintering fruiting bodies) on a colony of powdery mildew, caused by *Microsphaera vaccinii*, on a blueberry leaf. (Courtesy W. O. Cline)

Management

Control of powdery mildew is usually not needed because of the minimal damage it causes to the plant. Fungicides, such as sulfur and copper, have been shown to control powdery mildew. Modern fungicides, such as sterol inhibitors and strobilurins, are also effective against powdery mildew. Differences in cultivar susceptibility have been documented. Cultivars Pioneer, Cabot, and Wareham are highly susceptible; cultivars Jersey, Rubel, and Concord are moderately susceptible; and cultivars Berkeley, Earliblue, Ivanhoe, Stanley, Rancocas, Harding, and Katherine are resistant. Inheritance of resistance appears to be governed by multiple genes, with susceptibility being partially dominant.

Selected References

Bergman, H. F. 1939. Observations on powdery mildew on cultivated blueberries in Massachusetts in 1938. Phytopathology 29:545-546.

Demaree, J. B., and Wilcox, M. S. 1947. Fungi pathogenic to blueberries in the eastern United States. Phytopathology 37:487-506.

Moore, J. N., Bowen, H. H., and Scott, D. H. 1962. Response of highbush blueberry varieties, selections, and hybrid progenies to powdery mildew. Proc. Am. Soc. Hortic. Sci. 81:274-280.

Parmelee, J. A. 1977. The fungi of Ontario. II. Erysiphaceae (mildews). Can. J. Bot. 55:1940-1983.

Varney, E. H., and Stretch, A. W. 1966. Diseases and their control. Pages 236-279 in: Blueberry Culture. P. Eck and N. F. Childers, eds. Rutgers University Press, New Brunswick, NJ.

(Prepared by R. D. Milholland and A. W. Stretch;
Revised by A. M. C. Schilder)

Red Leaf

Red leaf is a very conspicuous disease of highbush blueberry. It is observed occasionally in Michigan (United States), especially during very cool, wet seasons. While disease incidence is generally low, as many as 20–30% of the bushes in a field may be infected. Reports of red leaf are rare on highbush blueberry in the northeastern and southeastern United States and on rabbiteye blueberry in the south. Although the pathogen attacks native *Vaccinium* spp. and cultivated cranberry in the Pacific Northwest, the disease has not been reported on highbush blueberry in this region. The pathogen systemically invades the plant and can cause significant losses when disease incidence is high. Most of the information currently available on red leaf disease comes from research conducted on lowbush blueberry.

Symptoms

In midsummer, portions of leaves turn red on one or more canes on a bush; hence the common name for the disease. Patterns vary from leaves that are completely red to irregular red patches (Fig. 78). The affected red areas pucker or cup and may be slightly thicker than the rest of the leaf. At the cellular level, the number of chloroplasts is markedly reduced and there is little chlorophyll. The red pigment is restricted to the upper epidermis and the palisade cells of the mesophyll.

The underside of the leaf beneath the red area eventually becomes covered with a white to cream-colored layer of fungal growth called a hymenium (Fig. 79). Once the pathogen has sporulated, the affected portions of leaves turn brown and dry up. Affected plants have reduced fruit set and slowly decline over a period of several years. Severely affected plants may die.

Causal Organism

Red leaf disease is caused by the fungus *Exobasidium vaccinii*, a basidiomycete pathogen in the order Exobasidiales. *E. vaccinii* is parasitic on a wide range of hosts in the family Ericaceae, including *Vaccinium*, *Rhododendron*, and *Camellia* spp. This fungus also causes white or light green spots on leaves and fruit in North Carolina (United States); however, these symptoms are quite distinct from red leaf.

Intercellular hyphae are delicate, septate, hyaline, and binucleate. The hyphae bear short, typically lobed haustoria that extend only a short distance into host cells. Hormones produced by the fungus interfere with the normal differentiation of host tissues and promote hypertrophy and hyperplasia.

The pathogen sporulates on the white hymenium (composed of basidia, basidiospores, conidia, and hyphae) found on the lower surface of infected leaves. Basidia are cylindrical, narrowly clavate, and superficial. The apex of each basidium is round and is crowned with sterigmata (usually four, but there may be two to six) (Fig. 80). The sterigmata are narrowly conical and straight or slightly curved. Basidiospores (2.0–4.5 × 9–17 μm) are mostly musiform to fusiform, slightly curved, and hyaline (Fig. 81). Initially, basidiospores are unicellular and uninucleate, but a transverse septum forms after nuclear division. Each basidiospore has a conspicuous hilar appendage. Mature basidiospores are actively discharged into the air and carried on wind currents. Some remain trapped in the tangle of hyphae in the hymenium, where they germinate through short germ tubes, one from each cell, to produce conidia.

The conidia are long, straight, narrow, and smaller (0.8–1.5 × 3.5–13.8 μm) than the basidiospores. Conidia also germinate, producing additional, smaller conidia. This growth habit is called budding but differs from the classical budding of yeasts.

Fig. 78. Red patches on leaves of highbush blueberry infected by *Exobasidium vaccinii*, the causal organism of red leaf. (Courtesy A. M. C. Schilder; © APS)

Fig. 79. Thickened, white hymenial layer of *Exobasidium vaccinii*, the causal organism of red leaf, on the lower leaf surface of highbush blueberry. (Courtesy A. M. C. Schilder; © APS)

In culture, *E. vaccinii* produces cream to yellowish white, slow-growing colonies. Hyphae are few or absent since the fungus grows by budding. The colonies are raised and cerebriform, with margins that are undulate to lobate. The sexual stage has not been observed in culture.

Disease Cycle and Epidemiology

Knowledge of the disease cycle of red leaf of highbush blueberry is incomplete. The following description comes mainly from research conducted on lowbush blueberry. Germinating basidiospores penetrate leaves directly or through stomata. Once inside host tissues, hyphae grow intercellularly. Growth is closely appressed to the cell walls of the host. Haustoria are numerous within infected cells of the leaf. These structures arise from hyphal compartments, called haustorial mother cells, that come in contact with the cell wall of the host. The haustoria are short-lobed, hyphal branches that extend into the host cell only a short distance.

A rather dense layer of hyphae grows between the leaf mesophyll cells and the lower epidermis. During the later stages of the disease, hyphae grow between the epidermal cells, pushing them apart. When the hymenium forms, the fungus ruptures the cuticle and, if stomata are present, emerges through them. Hyphae and basidia grow perpendicular to the leaf surface. It has been suggested that the swollen mesophyll cells act as a water reservoir for the formation of the hymenium. The hymenium is strictly hypophyllous.

E. vaccinii survives systemically in the host. It is not known whether the pathogen moves downward in the plant once infection has taken place. Leaves on new growth arising from previously infected stems usually develop red leaf symptoms.

Circumstantial evidence suggests that infections in highbush blueberry plantings may originate from infected wild (lowbush) blueberry stands in the surrounding landscape. Red leaf disease is favored by cool, moist conditions. The disease may be more severe in fields where excessive rates of nitrogen fertilizer have been used.

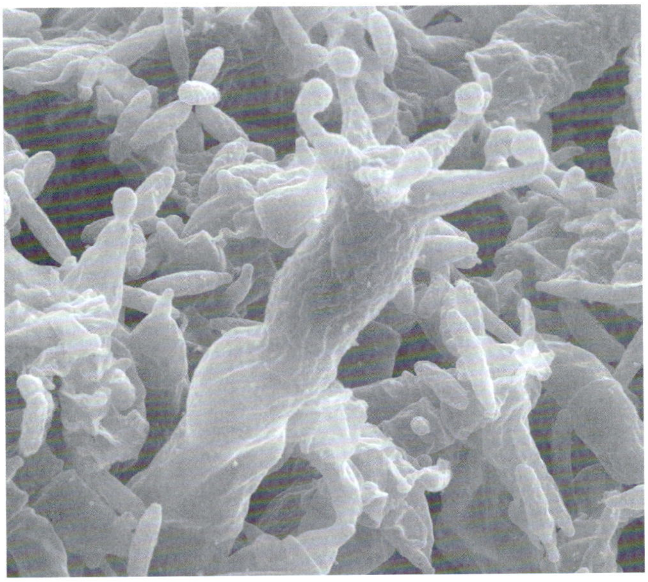

Fig. 80. Electron micrograph of a basidium of *Exobasidium vaccinii*, the causal organism of red leaf, with five immature basidiospores on sterigmata. Note the thin, elongated conidia in the hymenium surrounding the basidium. (Courtesy T. D. Miles; © APS)

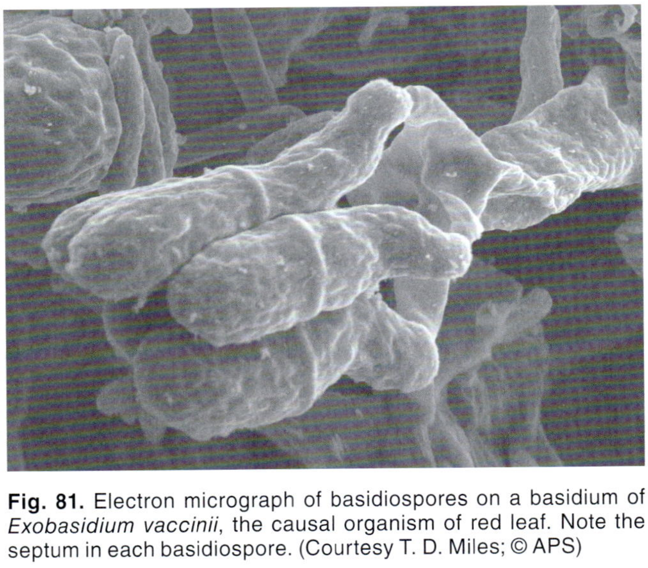

Fig. 81. Electron micrograph of basidiospores on a basidium of *Exobasidium vaccinii*, the causal organism of red leaf. Note the septum in each basidiospore. (Courtesy T. D. Miles; © APS)

Management

Because plants are systemically invaded by the pathogen, infected plants should be destroyed (burned or buried). This is an important control measure because infected bushes do not recover, and when left in the field, they serve as sources of inoculum for the further spread of the disease. While removing infected canes cannot cure infected bushes, it may help reduce the spread of the disease by removing diseased tissue prior to sporulation.

Fungicide sprays do not cure systemically infected plants, but treating healthy plants protects them from infection by basidiospores. Treatment should begin as soon as the fungus starts to sporulate (i.e., when the whitish hymenium forms). Additional applications may be needed to protect new leaf growth as long as basidiospores are being produced. Chemical control should be considered only if disease incidence is high.

Selected References

Cline, W. O. 1998. An *Exobasidium* disease of fruit and leaves of highbush blueberry. Plant Dis. 82:1064.

Hall, I. V., Forsyth, F. R., Lockhart, C. L., and Aalders, L. E. 1966. Effect of time of day, a parasitic fungus, and a genetic mutation on rate of photosynthesis in lowbush blueberry. Can. J. Bot. 44:529-533.

Hilborn, M. T., and Hyland, F. 1956. The mode of infection of lowbush blueberry by *Exobasidium vaccinii*. (Abstr.) Phytopathology 46:241.

Lockhart, C. L. 1958. Studies on red leaf disease of lowbush blueberries. Plant Dis. Rep. 42:764-767.

McGee, D. C. 1978. The severity of red leaf disease caused by *Exobasidium vaccinii* in lowbush blueberries in Maine. Plant Dis. Rep. 62:486-487.

Mims, C. W., and Nickerson, N. L. 1986. Ultrastructure of the host-pathogen relationship in red leaf disease of lowbush blueberry caused by the fungus *Exobasidium vaccinii*. Can. J. Bot. 64:1338-1343.

Mims, C. W., and Richardson, E. A. 1987. An ultrastructural study of the asexual spores of the plant pathogenic fungus *Exobasidium vaccinii*. Bot. Gaz. 148:228-234.

(Prepared by P. R. Bristow; Revised by A. M. C. Schilder)

Septoria Leaf Spot

Septoria leaf spot, caused by *Septoria albopunctata*, appears commonly on cultivated blueberries in the southeastern United States. In a field survey in Georgia (United States), Septoria leaf spot ranked as the most prevalent and most severe foliar disease, appearing in about 50% of the blueberry plantings monitored during a 2-year period. In contrast, the disease is rare in blueberry production areas in the western, northeastern, and north-central United States and Canada. The only overseas report of *S. albopunctata* is from *Vaccinium oldhamii* (natsuhaze) in Korea.

In areas where it appears, Septoria leaf spot generally is more prevalent and more severe on highbush and southern highbush cultivars than on rabbiteye cultivars. Examples of susceptible cultivars of these three species include Murphy and Bounty, Star and Bluecrisp, and Brightwell and Premier, respectively. Severe epidemics during the summer and autumn reduce the level of leaf photosynthesis and trigger premature defoliation, leading to reduced flower bud set during the autumn and lower berry yields the following year. Considerable disease-induced defoliation also has been observed in blueberry rooting beds, resulting in poor plant growth.

Infection of current-season shoots by *S. albopunctata* can lead to the formation of cankerous stem lesions. These lesions generally do not girdle the stem and are not associated with shoot blight or dieback; therefore, direct yield losses associated with the stem canker phase of the disease are probably limited.

Symptoms

Septoria leaf spot is relatively easy to diagnose based on symptoms and fungal signs. Leaf lesions are small (typically 2–4 mm in diameter) and circular, with grayish white centers and purple to brown margins (Fig. 82). Lesion expansion rates vary considerably among blueberry cultivars; for example, whereas the rabbiteye cultivars Brightwell and Premier both may become severely infected, leaf lesions on the latter are noticeably smaller than on the former.

Stem lesions are larger (5–6 mm in diameter) and slightly sunken but otherwise resemble leaf lesions in appearance (Fig. 83). Both types of lesions contain black, ovoid pycnidia of *S. albopunctata*, typically one or two but occasionally up to four or five per lesion. Under humid conditions, observation with a hand lens or a stereomicroscope reveals whitish cirrhi exuding from the ostioles of the pycnidia (Fig. 84).

Causal Organism

Septoria albopunctata was first described by Cooke on *Vaccinium arboreum* (sparkleberry) in South Carolina (United States). Only the anamorphic state of this fungus has been found to date, but molecular data suggest a taxonomic relationship with teleomorphic species in the genus *Mycosphaerella*. In a recent multigene-based phylogenetic study on *Septoria* and related genera, *S. albopunctata* clustered in a small clade distinct from the majority of other *Septoria* spp. on herbaceous and woody hosts. As such, further taxonomic work may well lead to a reclassification of the organism.

Pycnidia of *S. albopunctata* average 90 × 118 µm. Conidia are hyaline, multicelled (five to eleven), filiform (3.0–4.3 × 42–96 µm), straight or curved, and obclavate to spindle shaped.

Disease Cycle and Epidemiology

The pathogen is thought to overwinter in stem lesions and on leaf residue on the ground, but the relative importance of these two overwintering sites has not been quantified. In addition, native ericaceous plants such as *V. arboreum* and *Rhododendron minus* (dwarf rhododendron) can harbor the disease and may provide a potential source of inoculum if located in close proximity to blueberry plantings.

In untreated field plots in Georgia, foliar symptoms of Septoria leaf spot appear first between mid-April and late June, followed by a sigmoidal increase in disease severity until late September. Later in the autumn, average disease severity de-

Fig. 82. Septoria leaf spot, caused by *Septoria albopunctata*, on different southern highbush (top) and rabbiteye (bottom) blueberry cultivars. (Courtesy H. Scherm; © APS)

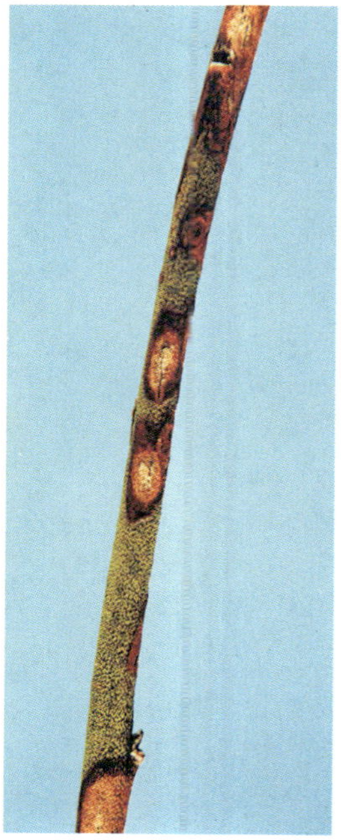

Fig. 83. Lesion from Septoria leaf spot, caused by *Septoria albopunctata*, on a highbush blueberry stem. (Courtesy R. D. Milholland; © APS)

Fig. 84. Closeup of Septoria leaf spot lesion, caused by *Septoria albopunctata*, showing four pycnidia with cirrhi on blueberry. (Courtesy P. M. Brannen; © APS)

creases because of the premature abscission of severely affected leaves. Pycnidiospores of *S. albopunctata* are splash dispersed and are available season-long. Although leaves of all ages are susceptible to infection, those in lower positions on a shoot (i.e., older leaves) and on shoots lower in the canopy generally have higher disease levels. The disease progress curve of stem lesions lags behind that of foliar symptoms. In general, disease development is favored by temperatures of 24–28°C and frequent rain or overhead irrigation.

Management

There are no cultural controls of Septoria leaf spot that could be applied in commercial blueberry production. Some rabbiteye cultivars, such as Tifblue and especially Climax, are less susceptible to Septoria leaf spot and generally do not require the application of specific management tactics against the disease. On susceptible blueberry cultivars in the southeastern United States, however, repeated fungicide applications in the summer and autumn have been shown to reduce disease severity and increase leaf retention, flower bud set, yield in the following year, or any combination of the three. At the time of this writing, several protective and systemic active ingredients are labeled for the control of Septoria leaf spot, allowing for their judicious deployment based on disease pressure, presence of other foliar diseases, and fungicide resistance management considerations. Of particular interest is the efficacy of potassium phosphite and similar phosphonate fungicides against the disease; these materials can serve as low-cost alternatives to or rotation partners with conventional fungicides.

Selected References

Brannen, P. M., Scherm, H., and Bruorton, M. D. 2002. Use of selected fungicides for control of Septoria leaf spot of blueberry, 2001. Fungic. Nematicide Tests 57:SMF46.

Brannen, P. M., Scherm, H., and Stanaland, R. D. 2005. Fungicidal control of Septoria and anthracnose leaf spots of blueberry, 2004. Fungic. Nematicide Tests 60:SMF037.

Cline, W. O. 2002. Blueberry bud set and yield following the use of fungicides for leaf spot control in North Carolina. Acta Hortic. 574:71-74.

Cooke, M. C. 1883. New American fungi. Grevillea 12:22-33.

Demaree, J. B., and Wilcox, M. S. 1947. Fungi pathogenic to blueberries in the eastern United States. Phytopathology 37:487-506.

Feau, N., Hamelin, R. C., and Bernier, L. 2006. Attributes and congruence of three molecular data sets: Inferring phylogenies among *Septoria*-related species from woody perennial plants. Mol. Phylogenet. Evol. 40:808-829.

Ojiambo, P. S., Scherm, H., and Brannen, P. M. 2006. Septoria leaf spot reduces flower bud set and yield potential of rabbiteye and southern highbush blueberries. Plant Dis. 90:51-57.

Ojiambo, P. S., Scherm, H., and Brannen, P. M. 2007. Temporal dynamics of Septoria leaf spot of blueberry and its relationship to defoliation and yield. Online. Plant Health Progress doi: 10.1094/PHP-2007-0726-05-RS

Shin, H. D., and Sameva, E. F. 2002. Taxonomic notes on the genus *Septoria* in Korea (II). Mycotaxon 83:287-300.

(Prepared by H. Scherm)

Silver Leaf

Silver leaf disease was first observed in blueberries in the year 2000 in the southern blueberry-growing region in Chile. Initially thought to be a nutritional problem, it is now considered a devastating disease; without control, the pathogen can kill the plant in 3 years and spread throughout the planting. Eight cultivars have been recognized as susceptible through field observations and plant inoculations: Bluehaven, Bluecrop, Brigitta, Duke, Draper, Elliott, Liberty, and Misty. However, additional cultivars were susceptible in artificial inoculation studies. In addition to Chile, the disease has been observed on isolated plants in the United States in North Carolina and in the Willamette Valley in Oregon.

Symptoms

The main symptoms appear during the summer, when the top surface of young leaves turns silvery and leaf color sometimes becomes purple or whitish (Fig. 85). In addition to the silvery appearance, the tips of some leaves may turn necrotic. Small areas of the epidermis are removed easily by leaves and branches brushing against each other in the wind. These areas turn dark red or brown because of sunburn of the palisade layer. Normally, the symptoms start on one or two branches, which show reduced growth and finally die in two or three seasons. In the meantime, new branches start to show symptoms until the whole bush is affected. The other distinctive symptom is the browning of the heartwood, including purple to brown rings (Fig. 86) or a concentric discoloration from the center. This browning is more accentuated at the plant crown and disappears upward. The root system does not show symptoms. Fruiting bodies (basidiocarps) develop on the surface of dead

Fig. 85. Silver gray leaf surface coloration on blueberry branches affected by silver leaf, caused by *Chondrostereum purpureum*. Leaves also turn purple or whitish. (Courtesy A. France; © APS)

Fig. 86. Purple rings in blueberry heartwood caused by the toxin movement of *Chondrostereum purpureum*, the causal organism of silver leaf. (Courtesy A. France; © APS)

or dying branches. Since this disease was first described on blueberry, the fungus has changed genetically and variations in symptoms have been detected, such as silvering and apical necrosis of leaves and necrosis of young stems, making the disease more aggressive.

Causal Organism

The causal fungus is *Chondrostereum purpureum*, which belongs to the Stereaceae family in the phylum Basidiomycota. The fungus produces basidiomata on dead wood during the autumn and winter; these are small, discoid (1–25 mm in diameter), and purple to pink, with a hairy surface and resupinate borders (Fig. 87). The hymenium contains claviform holobasidia with fusiform leptocystidia, which are about 18–30 µm long and extend above the basidia. The basidia hold four hyaline spores that are ovoid to subglobose, (2.70–)3.5(–4.56) × (5.05–)6.4(–7.94) µm, and apiculate at one end (Fig. 88).

It is suspected that this strain of *C. purpureum* is unique since the pathogen has never been reported in the main blueberry-growing areas of Chile, although the fungus has been reported in other hosts, such as apple, cherry, peach, plum, and kiwi. In addition, the original detection was at the edge of the southern blueberry-growing region in Chile, far from the area where *C. purpureum* is most prevalent in other hosts in Chile.

Fig. 87. Purple basidiocarps of *Chondrostereum purpureum*, the causal organism of silver leaf on blueberry, on the surface of affected wood. (Courtesy A. France; © APS)

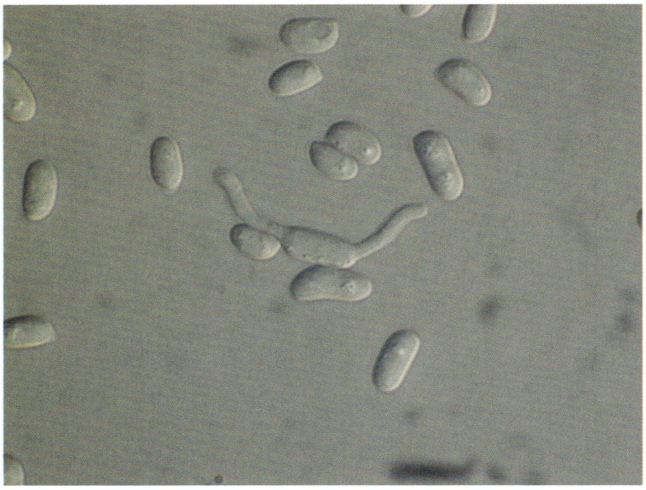

Fig. 88. Basidiospores and germination of *Chondrostereum purpureum*, the causal organism of silver leaf. (Courtesy A. France; © APS)

Disease Cycle and Epidemiology

Silver leaf disease spreads by basidiospores that are liberated from the basidia during the autumn and winter rain. Spores can be dispersed several kilometers by the wind to reach blueberry plantations. The fungus only colonizes fresh wounds on branches to reach the xylem vessels, where the spores germinate under moist conditions by producing hyphae that grow through the wood until they reach the plant crown. The fungus continues colonizing the heartwood in a concentric manner until it reaches the bark after two to four growing seasons. When the wood is dead, the fungus produces new basidiocarps and spores.

The disease can also be spread by infected cuttings, which are asymptomatic the first year but later develop symptoms and contribute to the spread of the disease after transplanting. This is probably the most common way of dissemination of this disease in Chile, where silver leaf has spread approximately 700 km in about 4 years from the site of original detection.

Management

Control of silver leaf is difficult because the pathogen is situated deep in the wood and because basidiocarps produce a large amount of inoculum on infected branches, on pruning stumps, and inside bark cracks. The first strategy is the use of healthy plants from nurseries in order to prevent disease transmission by diseased cuttings. Painting pruning wounds immediately after pruning with a fungicide- or *Trichoderma*-based product provides good protection against infection. The most susceptible cultivars (Brigitta, Liberty, Draper, and Duke) should be pruned during dry periods after harvest, usually at the end of the summer, when higher temperatures and the lack of rain prevent basidiocarp development. Symptomatic branches must be removed as soon as possible, allowing the formation of new growth and delaying the progression of symptoms. All stumps and dead material must be removed during pruning. All pruned branches must be burned. Pruning shears must be disinfected after use on diseased bushes since it is possible to get the basidiocarp tissues on the blades. The utility of removing alternate hosts is not clear for disease control since this strain appears to be specific to blueberry, but this needs to be further investigated.

Selected References

France, A., Santelices, C., Buddie, A., and Kirk, P. 2009. Silver leaf: First worldwide report of a new and harmful disease on blueberry. Acta Hortic. 810:341-344.

Latorre, B. 2004. Enfermedades de las plantas cultivadas. 6a ed. Ediciones Universidad Católica de Chile, Santiago, Chile.

Myairi, K., Fujita, K., Okuno, T., and Sawai, K. 1977. A toxic protein causative of silver-leaf disease symptoms on apple tree. Agric. Biol. Chem. 41:1897-1902.

Setliff, E. C. 2002. The wound pathogen *Chondrostereum purpureum*, its history and incidence on trees in North America. Austral. J. Bot. 50:645-651.

Tate, K. G. 1995. Silver leaf disease. Pages 30-31 in: Compendium of Stone Fruit Diseases. J. M. Ogawa, E. I. Zehr, G. W. Bird, D. F. Ritchie, K. Uriu, and J. K. Uyemoto, eds. American Phytopathological Society, St. Paul, MN.

(Prepared by A. France)

Sooty Blotch

Sooty blotch is characterized by the darkening of first- through early third-year stems of highbush blueberry (*Vaccinium corymbosum*). This condition is the result of the colonization of the waxy cuticle and epidermis by a complex of dematiaceous fungi (Fig. 89). There is some question whether sooty blotch should be considered a disease since it has been difficult to associate yield loss with symptoms. Typically, sooty blotch symptoms disappear following development of the periderm on older canes.

Symptoms

In general, green stems are stained dark green to black. Closer inspection reveals a high degree of symptom variability. Following immersion in a 10% bleach solution, symptoms are no longer visible, suggesting a very superficial colonization. Specific symptom types are described below.

General blotch is characterized by dense mycelial growth, particularly in the internodal regions (Fig. 90 left). Under a dissecting scope, the dense, filamentous, septate mycelium is dematioid, melanconiaceous, and olivaceous to black. Pycnidia have simple to complex ostioles and are either randomly or evenly distributed depending upon the portion of the stem under examination. Mycelia and pycnidia are often observed at the periphery of stomata.

Above-node blotch is seen as a dense mat of mycelium directly above the leaf and floral nodes. Atypical above-node coverage symptoms may appear as unique arrangements of punctate mycelium and fruiting bodies forming in either random or evenly distributed patterns. Evenly distributed patterns range from crescent shaped to ellipsoidal, often with fruiting bodies radiating outward from the center of the blotch in a fan-like pattern.

Fig. 89. Blueberry stems showing mild to severe sooty blotch symptoms, caused by a complex of dematiaceous fungi. (Courtesy P. V. Oudemans; © APS)

Fig. 90. General blotch symptoms on the internodal region showing dark mycelial growth (left side), as well as a region with flyspeck symptoms (right side). These sooty blotch symptoms on blueberry are caused by a complex of dematiaceous fungi. (Courtesy T. J. Waller; © APS)

Flyspeck is similar to the flyspeck observed on apples (Fig. 90 right). Dark, ovate, shiny, flattened, sclerotium-like structures are apparent. These vary in size and concentration and are often seen in association with other symptom types.

Causal Organisms

A diverse group of dematiaceous fungi are associated with sooty blotch. Various genera, including *Pseudocercospora*, *Microcyclospora*, *Hormonema*, *Discostroma*, and *Rhizosphaera*, have been associated with this condition. Many of the genera are common black yeasts in the fungal class Dothideomycetes.

Disease Cycle and Epidemiology

Although much of the disease cycle is unknown, fungi associated with this condition are presumed to overwinter on symptomatic stems as mycelium and conidia and spores. Water droplets likely disperse the fungi throughout the canopy, after which they grow in a subcuticular manner, spreading along the stem. Often the previous season's growth is the most visibly affected, suggesting that most fungal growth develops during the autumn, winter, and spring months. The degree of stem coverage can be substantial, resulting in grower concerns that yields may be reduced. There is limited evidence indicating that high levels of coverage result in reductions of yield components, such as inflorescence bud formation.

Management

Application of lime sulfur prior to symptom development has been shown to reduce the degree of fungal coverage.

Selected Reference

Batzer, J. C., Gleason, M. L., Harrington, T. C., and Tiffany, L. H. 2005. Expansion of the sooty blotch and flyspeck complex on apples based on analysis of ribosomal DNA gene sequences and morphology. Mycologia 97:1268-1286.

(Prepared by P. V. Oudemans, T. J. Waller, and J. J. Polashock)

Stem Blotch

Stem blotch was first observed in a rabbiteye blueberry planting in Georgia (United States). A *Cercospora* sp. was isolated from diseased stems of the cultivars Delite, Southland, Woodard, and Tifblue and was identified as the causal organism of stem blotch of rabbiteye blueberry in 1975. The disease was later found on the rabbiteye selection NC911, growing at the Horticultural Crops Research Station in Castle Hayne, North Carolina (United States), and has been reported on rabbiteye in Mississippi (United States). As yet, stem blotch has not been identified on other blueberry species.

Symptoms

Small, red lesions develop on succulent stems 2 months after infection. Lesions are poorly defined and slightly raised. They are 2–5 mm long during the early stages of disease development. On woody stems, symptoms appear as light to dark red discolorations; lesions do not conform to any particular shape or size but appear as a blotching of the stem (Fig. 91). Lesions encircle the stem within 1 year and may coalesce to form elongated lesions 1 cm long.

Causal Organism

Stem blotch is caused by a *Cercospora* sp. The fungus produces a light brown to olive brown, fasciculate stroma (20–25 × 20–25 µm) that emerges from a stem and upon which conidiophores and conidia are produced. Conidiophores (2–3 × 15–35 µm) are hyaline to light olive brown in mass and are

Fig. 91. Stem blotch disease of blueberry, caused by a *Cercospora* sp. (Courtesy R. D. Milholland; © APS)

rounded at the ends. Conidia (1.5–3.0 × 35–90 μm) are hyaline, obclavate-cylindrical, and straight to slightly curved, with an obconic-truncate base and subacute tip. Conidia have three to eight septa, but the septations are indistinct in many spores.

Disease Cycle and Epidemiology

The fungus penetrates through stomata and grows intercellularly within the stem cortex. Small raised lesions develop as a result of hyperplasia and hypertrophy in the cortex. Growth of the fungus causes cells to collapse. Most of the cells in the cortex become infected after 6 months, and within 1 year, the fungus completely encircles the stem and girdles it. Sporulation takes place after 6 months, when the hyphae accumulate in the substomatal cavity to form a stroma. Very little is known about how the fungus overwinters or its sources of inoculum.

Management

The use of disease-free propagation wood helps to control stem blotch. Removal of dead or infected stems reduces the inoculum level and subsequent disease development.

Selected References

Braswell, J., Spiers, J. M., and Hegwood, C. P. 2004. Establishment and Maintenance of Blueberries. Miss. State Univ. Ext. Publ. 1758.

Milholland, R. D. 1977. Cercospora stem blotch disease of rabbiteye blueberry. Phytopathology 67:816-819.

(Prepared by R. D. Milholland; Revised by W. O. Cline)

Blueberries (Lowbush)

Anthracnose

Colletotrichum gloeosporioides (syn. *Glomerella cingulata*) causes a blossom blight in lowbush blueberry similar in appearance to that caused by *Botrytis cinerea* (see Botrytis Blossom Blight below). Serious outbreaks of the blossom blight phase have not been reported. Rot develops in mature lowbush berries as in highbush fruit, with acervuli producing orange spores in droplets or tendrils on the fruit. Rot generally develops as a sunken, soft area near the calyx or blossom end of the fruit. In other aspects, the disease is presumed to be similar to anthracnose in highbush and rabbiteye blueberries (see Anthracnose Fruit Rot under Blueberries (Highbush and Rabbiteye) above).

The pathogen is widespread and infects, on average, 3% of the fruit grown in Maine (United States). Because almost all lowbush fruit is frozen at harvest, before the appearance of significant decay, the disease has historically been of little concern. However, recent fresh marketing of wild blueberries could increase the attention given to this fruit rot as a postharvest problem.

Biennial pruning by field burning, which destroys stems and scorches leaf litter and fallen berries, reduces the incidence of fruit infection about half as much as does pruning by flail mowing.

Selected Reference

Lambert, D. H. 1990. Postharvest fungi of lowbush blueberry fruit. Plant Dis. 74:285-287.

(Prepared by D. H. Lambert; Revised by S. L. Annis)

Botrytis Blossom Blight

Botrytis blossom blight is usually present at low levels in most lowbush blueberry fields but can develop into a consistent problem in areas where damp weather, coastal fogs, or frost are common during bloom. The disease is sometimes confused with Monilinia blight (primary stage of mummy berry), which also produces patches of blighted blossoms associated with particular clones. The berries may become infected, but the disease incidence is usually low and not considered a significant factor in the production or processing of the lowbush blueberry crop that is primarily frozen for processing. However, infected berries in whole packs destined for the fresh market can be a problem.

Symptoms

Botrytis blossom blight most commonly appears as a blight of open corollas that turn brown and calyxes that turn purple (Fig. 92). Brown lesions may develop on leaves touched by infected blossoms. Under favorable conditions, twig tips and stems are also infected and turn brown (Fig. 93). Because clones vary in susceptibility and developmental stage, severe outbreaks are often limited to distinct clones and appear as light brown patches in the field.

Botrytis blossom blight is easily differentiated from other problems by the whiskery appearance of black conidiophores on dead blossoms (Fig. 94). Blossoms infected by *Monilinia vaccinii-corymbosi* (see Mummy Berry under Blueberries (Highbush and Rabbiteye) above), in contrast, are usually killed before they are fully open and the affected leaves are destroyed. Nonpollinated blossoms infected by *Botrytis cinerea* develop a purple calyx and often shed their corollas.

Fig. 92. Individual blueberry corollas and calyxes affected by Botrytis blossom blight, caused by *Botrytis cinerea*. (Courtesy P. D. Hildebrand of the Department of Agriculture and Agri-Food, Government of Canada; © Her Majesty the Queen in Right of Canada, as represented by the Minister of Agriculture and Agri-Food Canada)

Fig. 93. Blueberry blossom clusters and stems affected by Botrytis blossom blight, caused by *Botrytis cinerea*, following a prolonged period of wetness. (Courtesy P. D. Hildebrand of the Department of Agriculture and Agri-Food, Government of Canada; © Her Majesty the Queen in Right of Canada, as represented by the Minister of Agriculture and Agri-Food Canada)

Causal Organism

Botrytis blossom blight is caused by the facultative parasite *Botrytis cinerea*. The teleomorph of *Botrytis cinerea* (syn. *Botryotinia fuckeliana*) has not been reported from lowbush blueberry. (For more details on this pathogen, see Botrytis Blight and Fruit Rot under Blueberries (Highbush and Rabbiteye) above.)

Disease Cycle and Epidemiology

B. cinerea overwinters in diseased fruit, stems, and leaves as sclerotia and mycelium. Leaves shed in the autumn of the vegetative (sprout) year can be a source of inoculum but numerous weed hosts, such as sheep sorrel, wild strawberry, and wild raspberry, commonly found within fields and along border areas are also an important inoculum source. Abundant sporulation of the pathogen can often be found on dead tissues of these weed hosts before the disease appears on blueberry blossoms.

Conidia are unable to infect blossoms at early stages of development, but infection increases substantially at the prebloom stage and appears readily at the full bloom and senescent stages. Green berries are relatively resistant to infection unless contacted by an infected blossom. Blossom infection is favored by wet periods with moderate temperatures. At 20°C, high levels of infection require only 10 h of wetness. At lower temperatures, wet periods of longer duration are required for high levels of infection. Under marginal conditions for disease development, infected blossoms turn brown and drop prematurely, frequently contributing to unexplained yield losses. However, if wet periods persist for 2 or more days and temperatures are favorable for infection and disease development, the fungus rapidly colonizes the blossom and spreads to the ovary and the peduncle of the blossom cluster and may progress into the stem, causing these tissues to turn brown. Under these conditions, affected blossoms within clusters tend to cling together and remain attached to the stems. Also, under persistent wet conditions, healthy blossoms that have fallen may become lodged on leaves and stems, where they become infected and act as a nutrient base from which the pathogen can infect these tissues, causing further damage to the crop. Sporulation of the fungus takes place within a few days after infection. The fungus may sporulate prolifically on blossoms that have fallen to the ground or on those that have become lodged, especially within dense canopies. Severe outbreaks typically take place in coastal areas that experience prolonged periods of rain or fog that favor repeated cycles of sporulation and infection.

Fig. 94. Conidiophores of *Botrytis cinerea*, the causal organism of Botrytis blossom blight, on dead blueberry blossoms. (Courtesy P. D. Hildebrand of the Department of Agriculture and Agri-Food, Government of Canada; © Her Majesty the Queen in Right of Canada, as represented by the Minister of Agriculture and Agri-Food Canada)

Management

Botrytis blossom blight may be controlled with applications of older fungicides, such as ferbam, captan, or benzimidazoles. Newer and more effective fungicides include fludioxonil, boscalid, fenhexamid, and pyraclostrobin. Fungicides should be applied before forecasted wet weather beginning at midbloom and thereafter on a 7- to 10-day schedule according to need based on forecasted wet weather. Strains of *B. cinerea* tolerant to benzimidazoles do not currently predominate but have resulted in control failures. Resistant strains can appear with relatively little selection pressure. Heavy fertilization should be avoided because it contributes to a dense canopy, which favors disease buildup. Burn pruning does not significantly reduce disease severity during the following harvest year.

Selected References

Hildebrand, P. D., McRae, K. B., and Lu, X. 2001. Factors affecting flower infection and disease severity of lowbush blueberry by *Botrytis cinerea*. Can. J. Plant Pathol. 23:364-370.

Johnson, S. B. 1979. Etiology of Botrytis blossom and stem blight of lowbush blueberries. M.S. thesis. University of Maine, Orono.

Lambert, D. H. 1990. Postharvest fungi of lowbush blueberry fruit. Plant Dis. 74:285-287.

Lambert, D. H., Wright, W. A., and Frey, M. A. 1988. Evaluation of fungicides for control of Botrytis blight on lowbush blueberry, 1987. Fungic. Nematic. Tests 43:94.

Pelletier, E. N., and Hilborn, M. T. 1954. Blossom and twig blight of lowbush blueberries. Maine Agric. Exp. Stn. Bull. 529.

Smith, B. J. 1998. Botrytis blossom blight of southern blueberries: Cultivar susceptibility and effect of chemical treatments. Plant Dis. 82:924-927.

(Prepared by D. H. Lambert; Revised by P. D. Hildebrand)

Brown Leaf Spot

Brown leaf spot was recognized as a disease of lowbush blueberry before 1931. It has been reported in Maine (United States) and eastern Canada, where it is universally distributed; however, there is serious loss only in particular years and areas.

Symptoms

Medium brown spots 1–4 mm in diameter develop on the upper and lower surfaces of leaves. Larger lesions are often angular and discrete (Fig. 95). On the leaves of some blueberry clones, dead areas may be surrounded by discolored zones. The round, determinate shape and distinct, light-colored center typical of highbush Septoria leaf spot lesions are usually absent in the lowbush lesions. Pycnidia in dead areas become more visible after the leaves are incubated in a moist chamber. Severely affected leaves drop prematurely, resulting in substantial defoliation.

Causal Organism

Brown leaf spot is associated with a *Septoria* sp., but the pathogen has not been formally described. Conidia produced on potato dextrose agar are long and uniformly slender (1.5 × 60–90 µm), with three to five inconspicuous septa. Differences in symptoms, spore morphology, and geographic distribution suggest that this species is distinct from the highbush pathogen *S. albopunctata*. Inoculations of highbush blueberry with isolates from lowbush blueberry have not produced symptoms.

Fig. 95. Discrete lesions of brown leaf spot, caused by a *Septoria* sp., on a lowbush blueberry leaf. (Courtesy D. Lambert; © APS)

Disease Cycle and Epidemiology

The distribution of brown leaf spot is widespread but varies considerably from year to year. The disease is more severe in locations with coarse, dry soils. Disease severity decreases with nitrogen fertilization.

Management

The brown leaf spot pathogen is sensitive to some fungicides, but treatment to control only this disease would not be economical.

Selected References

Lockhart, C. L. 1961. Diseases and their control. Pages 31-35 in: The Blueberry in the Atlantic Provinces. Can. Dep. Agric. Publ. 754.

Markin, F. L. 1931. Progress of investigations: Blueberry diseases. Maine Agric. Exp. Stn. Bull. 360:204-206.

(Prepared by D. H. Lambert; Revised by P. D. Hildebrand)

Exobasidium Fruit and Leaf Spot

Exobasidium fruit and leaf spot develops sporadically in lowbush and highbush blueberries and on a few other *Vaccinium* spp., but yield losses usually are not severe.

Symptoms

Symptoms first appear as small, scattered to confluent, chlorotic spots on leaves. Within 2–3 weeks, the spots expand to form flat, white to pale yellow, circular lesions up to 5–7 mm in diameter. These symptoms are nearly identical to those on highbush blueberry (see Fig. 39). The lower surface of the lesions eventually becomes covered with a thin, powdery, "white bloom" consisting of the hymenium of the fungus similar to that of red leaf disease (see Red Leaf below). As the lesions age, they thicken slightly, with the upper surface becoming reddish or brown in the center and sometimes having faint targetlike concentric rings. The lower surface may become overgrown by other fungi. Heavily infected leaves shed prematurely, but damage to the plants is negligible. Symptoms of infection on berries appear as small, circular, firm spots that fail to ripen normally and remain green or pink (Fig. 96). Affected tissues in ripe fruit extend 2–3 mm into the berry.

Causal Organism

The *Exobasidium* sp. causing fruit and leaf spot is similar to, but different from, *Exobasidium vaccinii*, which is associated with red leaf disease (see Red Leaf below). The hymenium of the fungus consists of cylindrical to clavate, basidia with usually four sterigmata. Basidiospores are 2–2.5 × 8–12 µm, hyaline, banana shaped, straight or slightly curved, and somewhat hooked at the base. Conidia are cylindrical to ellipsoid and approximately 1.2 × 6.5–8 µm. The fungus grows slowly on potato dextrose agar and has a restricted, yeastlike, pale salmon growth habit with a raised, moist, wrinkled surface.

Disease Cycle and Epidemiology

Little is known about the epidemiology of this fungus. Symptoms on leaves appear in June, indicating that infection takes place when leaves are young, but symptoms do not appear on later flushes of growth, suggesting that spores are available for infection only during late spring and early summer. Infec-

Fig. 96. Symptoms of Exobasidium fruit spot, caused by an *Exobasidium* sp., on lowbush blueberries. (Courtesy P. D. Hildebrand of the Department of Agriculture and Agri-Food, Government of Canada; © Her Majesty the Queen in Right of Canada, as represented by the Minister of Agriculture and Agri-Food Canada)

tion on leaves of lowbush blueberry in a controlled environment takes place readily at 20°C and with a 48-h wet period.

Management

No attempt is made to control this disease, which appears only sporadically. Affected berries destined for the fresh market usually can be graded out if the infection incidence is not too high.

Selected References

Brewer, M. T., Turner, A. N., Brannen, P. M., Cline, W. O., and Richardson, E. A. 2014. *Exobasidium maculosum*, a new species causing leaf and fruit spots on blueberry in the southeastern USA and its relationship with other *Exobasidium* spp. parasitic to blueberry and cranberry. Mycologia 106(3):415-423.

Cline, W. O. 1998. An Exobasidium disease of fruit and leaves of highbush blueberry. Plant Dis. 82:1064.

Nickerson, N. L., and Vander Kloet, S. P. 1997. Exobasidium leaf spot of lowbush blueberry. Can. J. Plant Pathol. 19:66-68.

(Prepared by P. D. Hildebrand)

Fusicoccum (Godronia) Canker

Fusicoccum (Godronia) canker is not normally a disease of concern on lowbush blueberry. However, in 2002, a widespread epidemic occurred in Canada in Nova Scotia and to a lesser extent in New Brunswick and Prince Edward Island. The disease appeared in 91% of the fruiting fields in Nova Scotia, and the incidence of affected cropping stems averaged 15% and was as high as 38%. Because the lesions usually girdled the stems, these data can be regarded as crop loss figures. Interestingly, the disease has failed to develop since 2002, although an occasional infected stem can be found, indicating that the pathogen and disease persist in fields at very low levels.

Symptoms

Lesions of Godronia canker on lowbush blueberry are similar in appearance to those appearing on highbush blueberry, but smaller. Lesions first appear on 1-year-old fruiting stems in late May or early June and result in tip dieback. Affected stems may appear in patches or randomly throughout a field. Lesions usually appear about midway up an infected stem and are almost always centered about a leaf bud. They are dark orange-brown, surrounded by a purple border, and may be up to 2 cm long. In time, the centers of lesions become tan. Black pycnidia develop in the center of the lesion and tend to encircle the leaf bud (Fig. 97). Leaf and flower buds above the lesion are prevented from developing and die, causing tip dieback. Stem tissues below the lesion are not affected and leaf shoots begin to grow vigorously (Fig. 98) so, by early July, a severely infected field will appear to have recovered but will have a reduced yield. Lesions that are more slow to develop may not result in tip dieback but the stems remain unthrifty with reddish leaves and small berries.

Causal Organism

Godronia canker is caused by the ascomycete fungus *Godronia cassandrae*. Apothecia of this organism can occasionally be found on old, dead stems, but their role in the disease is not known. The anamorph stage of *G. cassandrae* (syn. *Fusicoccum putrefaciens*) on blueberry can be readily identified from the pycnidia and conidia (Fig. 99) developing in lesions. The organism is similar to the one causing canker on highbush

Fig. 97. Fusicoccum (Godronia) canker lesion, caused by *Godronia cassandrae*, with pycnidia on a lowbush blueberry stem. (Courtesy P. D. Hildebrand of the Department of Agriculture and Agri-Food, Government of Canada; © Her Majesty the Queen in Right of Canada, as represented by the Minister of Agriculture and Agri-Food Canada)

Fig. 98. Fusicoccum (Godronia) canker, caused by *Godronia cassandrae*, resulting in tip dieback and vigorous growth below the lesion on a blueberry stem. (Courtesy P. D. Hildebrand of the Department of Agriculture and Agri-Food, Government of Canada; © Her Majesty the Queen in Right of Canada, as represented by the Minister of Agriculture and Agri-Food Canada)

Fig. 99. Conidia of *Godronia cassandrae*, the synonym of the causal organism of Godronia canker, from lowbush blueberry. (Courtesy P. D. Hildebrand of the Department of Agriculture and Agri-Food, Government of Canada; © Her Majesty the Queen in Right of Canada, as represented by the Minister of Agriculture and Agri-Food Canada)

blueberry. (For more details on this pathogen, see Fusicoccum (Godronia) Canker under Blueberries (Highbush and Rabbiteye) above.)

Disease Cycle and Epidemiology

The epidemiology of this pathogen is not well understood in lowbush blueberry. It can usually be found at low levels in unmanaged areas along the edges of fields or in ditch banks on 2- to 3-year-old stems. However, the reason for its sudden widespread occurrence in 2002 and the return to low levels of infection in subsequent years is not explained. Mature conidia of the fungus can be found in pycnidia from late May to early November in Nova Scotia and are released in response to rainfall. Conidia germination of lowbush blueberry isolates on water agar takes place maximally at 20–25°C within 30 h, after 50 h at 10°C, and not at all at 5°C. The appearance of lesions usually midway up the stem suggests that infection takes place at the tip of developing sprout stems during the previous year from early to mid-July when they are about half grown.

Management

Burn pruning effectively controls Godronia canker in lowbush blueberry. In the 2002 epidemic, fields that were burned during the previous year had low disease levels compared with those that were mowed. Other specific control measures are not applied since this disease typically is not a problem.

Selected References

Hildebrand, P. D., Renderos, W. E., Davies, R. J., and McRae, K. B. 2003. Godronia canker of lowbush blueberry. Can. J. Plant Pathol. 25:425.

Lockhart, C. L. 1975. Effect of temperature on the development of *Godronia cassandrae* f. *vaccinii* cankers on lowbush blueberry. Can. Plant Dis. Surv. 55:29-30.

Lockhart, C. L., and Delbridge, R. W. 1972. Occurrence and pathogenicity of *Godronia cassandrae* f. *vaccinii* on lowbush blueberry in Nova Scotia. Can. Plant Dis. Surv. 52:119-121.

(Prepared by P. D. Hildebrand)

Gloeosporium Stem and Leaf Spot

Gloeosporium minus causes small red lesions on the stems and small red flecks or spots on the leaves of both highbush and lowbush blueberries. If the number of leaf spots is significant and infection takes place as the leaves are developing, then the leaves may be malformed and puckered. Spots may become necrotic during the year after infection. The bark of infected shoots may become roughened. The fungus readily invades tissue killed by other pathogens or stresses.

G. minus also causes fruit rot and is one of the more common fungi isolated from both lowbush and highbush blueberries. Before the pathogen was identified, the disease was known as blossom end rot in New Jersey (United States). (For more details on this disease and the causal organism, see Gloeosporium Leaf Spot and Stem Canker under Blueberries (Highbush and Rabbiteye) above.)

Selected References

Lambert, D. H. 1990. Postharvest fungi of lowbush blueberry fruit. Plant Dis. 74:285-287.

Taylor, J. 1958. Stem canker and related blueberry diseases. N. C. State Tech. Bull. 132.

(Prepared by F. L. Caruso; Revised by S. L. Annis)

Leaf Rust

Leaf rust was once considered a relatively minor disease, but since the early 2000s, more severe outbreaks have been appearing, resulting in premature defoliation in most lowbush blueberry production areas. If defoliation of sprout stems happens in early September, the formation of flower buds in the autumn and yields in the subsequent year are reduced. Leaf rust also affects other *Vaccinium* spp. and related plants (see Leaf Rust under Blueberries (Highbush and Rabbiteye) above and Rust under Cranberries below).

Symptoms

Early symptoms of leaf rust may be detected with a hand lens in late July and appear as water-soaked lesions on the undersides of leaves. Symptoms become more pronounced in August and September, when these lesions penetrate to the upper surface and turn reddish to brown (Fig. 100). Symptom development varies among blueberry clones with some showing only flecking while others show well-defined lesions up to 2 mm in diameter. Yellow pustules (the uredinia of the fungus) develop on the undersides of the spots (Fig. 101). Diseased leaves are shed prematurely.

Fig. 100. Lowbush blueberry leaves affected by leaf rust, caused by *Thekopsora minima*. (Courtesy N. L. Nickerson; © APS)

Causal Organism

Leaf rust of lowbush blueberry is caused by the heteroecious leaf rust fungus *Thekopsora minima*, with pycnia and aecia (Fig. 102) on *Tsuga* spp. (hemlock) and uredinia and telia on *Vaccinium* spp. and other Ericaceae members. Previously in North America, leaf rust on blueberry was considered to be caused by a species complex (a group of several biologically distinct fungi) known as *Pucciniastrum vaccinii* (syn. *P. myrtilli*), consisting of a western form with conical aecia that resemble uredinia and an eastern form with cylindrical aecia. *T. minima* is now regarded as the eastern form and *Naohidemyces vaccinii* is regarded as the western form.

Aecia of *T. minima* on hemlock are hypophyllous, subepidermal in origin, cylindrical with fragile peridial cells; the outer walls of the cells are thin and smooth and the inner walls are verrucose. Aceciospores are yellow, subglobose, and 15–18 × 18–26 µm. Uredinia are hypophyllous, dome shaped, and scattered or clustered on the leaf spots; they are bullate and dehiscent by a central pore and have thick-walled ostiolar cells. Urediniospores are yellow to orange, ellipsoid to oblong, evenly echinulate, and 18–32 × 20–24 µm. Teliospores form intraepidermally and consist of two to eight cells (18–32 × 20–35 µm) adhering laterally in a surface view, with a germ pore on each cell close to the junction of the cells in the center of the sporeballs.

Disease Cycle and Epidemiology

The fungus overwinters as teliospores on diseased leaves of its ericaceous hosts. In the spring, the teliospores germinate to produce basidiospores that infect developing needles of its alternate host hemlock. Inconspicuous pycnia form on the needles, followed by yellow aecia from which aeciospores are released and infect blueberry leaves. On blueberries, leaf rust is most commonly found within the range of hemlock, but sporadic outbreaks have also been reported in regions far from hemlock, suggesting that the urediniospores may travel long distances by wind. Factors that influence disease development on blueberry have not been studied, but the optimum temperature for spore germination of *T. minima* on agar is 21.5°C, the optimum temperature for symptom development on *Rhododendron* spp. is 23°C, and the optimum temperature for uredinial production is 19.5°C.

Fig. 101. Uredinia and urediniospores on the underside of blueberry leaves infected with *Thekopsora minima*, the causal organism of leaf rust. (Courtesy P. D. Hildebrand of the Department of Agriculture and Agri-Food, Government of Canada; © Her Majesty the Queen in Right of Canada, as represented by the Minister of Agriculture and Agri-Food Canada)

Fig. 102. Aecia of the blueberry leaf rust fungus, *Thekopsora minima*, on eastern hemlock. (Courtesy P. D. Hildebrand of the Department of Agriculture and Agri-Food, Government of Canada; © Her Majesty the Queen in Right of Canada, as represented by the Minister of Agriculture and Agri-Food Canada)

Management

In lowbush blueberry, the effects of leaf rust can be reduced by an application of fungicide (e.g., chlorothalonil and prothioconazole) in sprout fields during the early stages of outbreaks in late July or early August. Fungicide application delays disease development and subsequent defoliation, allowing adequate time for flower bud formation in the autumn.

Selected References

Arthur, J. C. 1934. Manual of the Rusts in United States and Canada. Purdue Research Foundation, Lafayette, IN.

Demaree, J. B., and Wilcox, M. S. 1947. Fungi pathogenic to blueberries in the eastern United States. Phytopathology 37:487-506.

Hiratsuka, Y. 1965. The identification of *Uraecium holwayi* on hemlock as the aecial stage of *Pucciniastrum vaccinii* in western North America. Can. J. Bot. 43:475-478.

Markin, F. L. 1931. Notes on blueberry diseases in Maine. Plant Dis. Rep. 15:11-14.

Pfister, S. E., Halik, S., and Bergdahl, D. R. 2004. Effect of temperature on *Thekopsora minima* urediniospores and uredinia. Plant Dis. 88:359-362.

Sato, S., Katsuya, K., and Hiratsuka, Y. 1993. Morphology, taxonomy and nomenclature of *Tsuga*-Ericaceae rusts. Trans. Mycol. Soc. Jpn. 34:47-62.

Ziller, W. G. 1974. The Tree Rusts of Western Canada. Can. For. Serv. Publ. 1329. Department of the Environment, Victoria, British Columbia, Canada.

(Prepared by N. L. Nickerson; Revised by P. D. Hildebrand)

Other Fungi Causing Fruit Rot

The majority of the harvest of lowbush blueberries is frozen, but there is a small and growing fresh-market industry. Some of the fungi that commonly rot lowbush blueberry fruit (e.g., *Botrytis cinerea* and *Colletotrichum gloeosporioides*) have been mentioned in previous sections. Other fungi found rotting lowbush blueberry fruit include *Alternaria* spp., *Pestalotia vaccinii*, *Penicillium* spp., *Phomopsis* spp., and *Trichoderma* spp. *Alternaria* infection produces greenish gray mycelium with chains of spores on the berries. *Pestalotia vaccinii* produces wet, black masses of spores that dry and harden over time. Berries infected with *Phomopsis* spp. produce raised dark pycnidia with gray to white wet spore masses. Berries infected with *Trichoderma* spp. produce dense cottony cushions of white mycelium on the surface of the berry. Different *Penicillium* spp. have been found growing in lowbush blueberry fruit; they all have greenish blue spore masses with mycelium produced on the surface. The percentage of lowbush blueberry fruit with fruit rot typically ranges up to 6.5% depending upon the field.

Selected Reference

Lambert, D. H. 1990. Postharvest fungi of lowbush blueberry fruit. Plant Dis. 74:285-287.

(Prepared by S. L. Annis)

Fig. 103. Stems with attached red-brown leaves resulting from Phomopsis canker, caused by *Phomopsis vaccinii*, on lowbush blueberry. (Courtesy S. L. Annis; © APS)

Phomopsis Canker

Phomopsis vaccinii (syn. *Diaporthe vaccinii*) causes a stem disease that kills individual or clusters of stems in both sprout and crop year fields. Phomopsis canker is widely spread throughout lowbush blueberry fields but in low abundance. Stems die rapidly in mid- to late summer, typically after dry conditions. The stems are red-brown from the rhizome upward and usually have a visible lesion at the base of the stem just above the soil level. The leaves turn red-brown and die but remain attached to the stems (Fig. 103). While a highly visible disease, it is of minor importance and does not warrant chemical control. (For more detail on the disease, see Phomopsis Canker under Blueberries (Highbush and Rabbiteye) above.)

(Prepared by S. L. Annis)

Powdery Mildew

Powdery mildew is universally present in lowbush blueberry fields and contributes to leaf reddening and early abscission. Because the onset of infection is usually late in the season, however, it has not been an important disease. (For more detail on the disease, see Powdery Mildew under Blueberries (Highbush and Rabbiteye) above.)

Symptoms

The appearance of powdery mildew differs considerably among clones but varies less within individual clones. Symptoms range from open, red, polygonal rings in resistant plants (Fig. 104) to solid red spots with or without mycelium to leaves covered in white mycelium. Leaves sometimes are also distorted and red-green with heavier infections (Fig. 105). Rings are easily confused with those caused by *Blueberry red ringspot virus*. Infection is more common on the upper surface of lower leaves, particularly in areas with dense stem production. Abundant chasmothecia form on leaves with mycelium. Infected leaves abscise prematurely.

Causal Organism

Blueberry powdery mildew is caused by *Microsphaera vaccinii* (syns. *M. penicillata* var. *vaccinii*, *M. alni* var. *vaccinii*, and *Erysiphe vaccinii*). Its host range has not been determined. (For more details on this pathogen, see Powdery Mildew under Blueberries (Highbush and Rabbiteye) above.)

Fig. 104. Red polygonal rings caused by the powdery mildew fungus, *Microsphaera vaccinii*, on a resistant lowbush blueberry leaf. (Courtesy D. Lambert; © APS)

Fig. 105. Distorted leaves of a lowbush blueberry clone covered with mycelium and conidiophores of the powdery mildew fungus, *Microsphaera vaccinii*. (Courtesy D. Lambert; © APS)

Disease Cycle and Epidemiology

The extent to which chasmothecia or infected buds serve as sources of overwintering inoculum has not been determined for lowbush blueberry. Postharvest burning does not reduce powdery mildew incidence in subsequent years, which suggests either that most of the inoculum survives in the more protected leaf litter or that dispersal from adjacent fields is substantial. Conidia are produced abundantly on some clones but not others. Powdery mildew may be evident in July on more susceptible clones, but substantial infection does not appear until harvest or thereafter in most years. The disease may be more severe if plants are deficient in phosphorus.

Management

Powdery mildew may be controlled with appropriate fungicides but usually the expense is not justified because of the late appearance of the disease. Lowbush blueberry clones differ in their susceptibility to the pathogen.

Selected References

Lambert, D. H. 1990. Effects of pruning method on the incidence of mummy berry and other lowbush blueberry diseases. Plant Dis. 74:199-201.

Trevett, M. F., Hilborn, M. T., and Durgin, R. E. 1969. An apparent relationship between powdery mildew and nutrient element balance in lowbush blueberry leaves. Maine Agric. Exp. Stn. Res. Life Sci. 17:10-14.

(Prepared by D. H. Lambert; Revised by S. L. Annis)

Red Leaf

Red leaf disease appears on lowbush blueberry and occasionally on highbush blueberry. The host range of the causal organism is not known, and it is possible that the same organism also attacks other *Vaccinium* spp. Red leaf is a systemic disease that weakens blueberry plants and reduces fruit production, but in most fields, disease incidence is so low that the effect on yield is negligible. The incidence of diseased stems has been reported to be as high as 30% in some fields, but more commonly, the incidence is less than 5%. Fields that have been in production for many years tend to have a higher incidence of disease, but this increases only by 0.1% per year.

Symptoms

Symptoms of red leaf do not appear in the year of infection and are relatively inconspicuous on first-year shoots; it may be several years before diseased plants become noticeable. Once infection is well established, however, symptoms usually appear on new growth every year because the pathogen is perennial and systemic in the shoots and rhizomes.

Symptoms are apparent as soon as new growth begins in the spring but are most conspicuous on second-year and older shoots during June and early July. Diseased plants are easily recognized by the presence of shoots with bright red leaves (Fig. 106). These shoots typically grow in conspicuous clusters or patches scattered throughout the field (Fig. 107). The red leaves may be slightly enlarged but do not show the extreme hypertrophy characteristic of many diseases caused by *Exobasidium* spp.

Eventually, the lower surface of the red leaves becomes covered with a dense, white, feltlike deposit consisting of the hymenium (spore-bearing layer) of the fungus. This deposit distinguishes diseased plants from those showing natural bronzing. Shoots on diseased plants growing in shaded areas may have leaves that are pale pink, yellowish, or even pale green instead of bright red, but the presence of the white fungal growth on the underside of the leaves is diagnostic.

After the fungus has sporulated, the affected leaves wither and are shed prematurely, starting with the lowermost leaves and proceeding to the tip of the shoot. Later in the season, a new flush of axillary shoots with small, red leaves may arise from axillary buds on infected shoots. Diseased plants often fail to flower and usually do not produce mature fruit; over a period of several years, they weaken and eventually may die. Ripe fruit from infected stems have little flavor.

Causal Organism

Red leaf is caused by a species of the basidiomycete fungus *Exobasidium*. Currently, the pathogen is classified as *E. vaccinii*, but recent research has shown that it does not fit the modern concept of *E. vaccinii*. Further studies are needed to elucidate the taxonomic status of the red leaf causal organism.

The slender, intercellular hyphae of the fungus form a netlike mycelium that becomes systemic and perennial in blueberry shoots and rhizomes. The fungus sporulates in late June and early July on the lower surface of the leaves of systemically infected plants. The basidia, which have four to ten sterigmata, are often difficult to find in squash mounts because they usually collapse after the spores are released. The hyaline, banana-shaped (musiform) basidiospores (2.0–3.5 × 9.5–15.0 μm) are at first unicellular but usually develop one to three transverse septa during germination. On the hymenium, basidiospores often germinate to produce secondary spores, but in culture, they usually develop germ tubes and hyphae (Fig. 108). Hyaline, needle-shaped conidia (0.8–1.5 × 3.5–13.8 μm), which develop from secondary spores or from existing conidia, are often abundant on the hymenium; they are usually unicellular but may develop one transverse septum.

Disease Cycle and Epidemiology

The pathogen overwinters as a mycelium in systemically infected blueberry plants. It is present in dormant buds and active apical meristems; spreads within young, growing shoots and rhizomes; and sporulates on the underside of leaves. A second flush of diseased axillary shoots may emerge later in the season.

Germinating basidiospores can infect healthy plants by the direct penetration of young stems. It is not known whether leaf infections can also occur. Conidia from cultures of the fungus can infect plants under controlled conditions, which suggests that conidia produced on the hymenium are capable of infecting

Fig. 106. Second-year stems of lowbush blueberry affected by red leaf, caused by an *Exobasidium* sp. Note the white fungal growth on the underside of the diseased leaves. (Courtesy P. D. Hildebrand of the Department of Agriculture and Agri-Food, Government of Canada; © Her Majesty the Queen in Right of Canada, as represented by the Minister of Agriculture and Agri-Food Canada)

Fig. 107. Red leaf, caused by an *Exobasidium* sp., in a lowbush blueberry field. (Courtesy N. L. Nickerson; © APS)

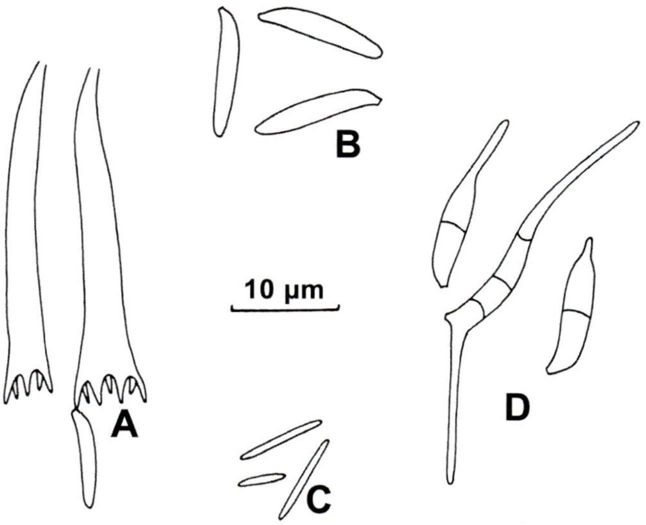

Fig. 108. An *Exobasidium* sp., causal organism of red leaf, basidia (**A**), basidiospores (**B**), conidia (**C**), and germinating basidiospores on water agar (**D**). (Courtesy N. L. Nickerson; © APS)

plants in the field. Greenhouse-grown plants inoculated with cultures of the fungus became infected through stem wounds, but it is not known whether this mode of infection takes place in the field. The spread of red leaf by spores is thought to occur only when rainy periods coincide with the period of spore release in late June or early July.

Management

In most lowbush blueberry fields, no attempt is made to control red leaf because disease incidence is very low and the disease does not appear to spread. Some fungicides inhibit the pathogen under experimental conditions, but none have been shown to control the disease once the fungus becomes systemic in the rhizome. Rhizome-seated infections are not controlled by burn pruning lowbush blueberry, although the practice may destroy new infections that have not yet become systemic. Studies in Maine (United States) and Nova Scotia (Canada) showed that the incidence of red leaf in fields pruned by mowing was similar to the incidence in fields pruned by burning.

Eradication of diseased plants with 2,4-D (2,4-dichlorophenoxyacetic acid) or glyphosate effectively controls red leaf but may damage the surrounding healthy plants. Lowbush blueberry clones inoculated with the pathogen in experimental tests differed in susceptibility, and it is likely that host resistance also plays a role in limiting the spread of the disease in genetically diverse natural plantings.

Selected References

Hilborn, M. T., and Hyland, F. 1956. The mode of infection of lowbush blueberry by *Exobasidium vaccinii.* (Abstr.) Phytopathology 46:241.

Hildebrand, P. D., Nickerson, N. L., McRae, K. B., and Lu, X. 2000. Incidence and impact of red leaf disease caused by *Exobasidium vaccinii* in lowbush blueberry fields in Nova Scotia. Can. J. Plant Pathol. 22:364-367.

Lambert, D. H. 1990. Effects of pruning method on the incidence of mummy berry and other lowbush blueberry diseases. Plant Dis. 74:199-201.

Lockhart, C. L. 1958. Studies on red leaf disease of lowbush blueberries. Plant Dis. Rep. 42:764-767.

McGee, D. C. 1978. The severity of red leaf disease caused by *Exobasidium vaccinii* in lowbush blueberries in Maine. Plant Dis. Rep. 62:486-487.

Mims, C. W., and Nickerson, N. L. 1986. Ultrastructure of the host-pathogen relationship in red leaf disease of lowbush blueberry caused by the fungus *Exobasidium vaccinii.* Can. J. Bot. 64:1338-1343.

Mims, C. W., and Richardson, E. A. 1987. An ultrastructural study of the asexual spores of the plant pathogenic fungus *Exobasidium vaccinii*. Bot. Gaz. 148:228-234.

Nickerson, N. L., and MacNeill, B. H. 1987. Studies on the spread of red leaf disease, caused by *Exobasidium vaccinii*, in lowbush blueberries. Can. J. Plant Pathol. 9:307-310.

(Prepared by N. L. Nickerson; Revised by P. D. Hildebrand)

Septoria Leaf Spot and Stem Canker

Septoria leaf spot and stem canker is caused by a *Septoria* sp. that has not yet been formally described. This species can cause extensive spotting on leaves and stems and is associated with premature leaf and fruit drop. The disease occurs annually to varying degrees in most lowbush blueberry fields throughout eastern Canada and Maine (United States). It is on both *Vaccinium angustifolium* and *V. myrtilloides*, the two host species that typically comprise commercial stands of lowbush blueberry. Based on disease symptoms, it appears that this *Septoria* sp. is likely distinct from the one associated with brown leaf spot (see Brown Leaf Spot above).

Symptoms

In sprout fields, leaf lesions appear on the lower leaves of stems, while in fruiting fields, leaf lesions appear throughout the canopy. The lesions first appear in mid- to late June as minute water-soaked spots on the leaf undersurface and are visible only with magnification. The spots are translucent with transmitted light. With time, the spots increase in number and size and become slightly raised. The symptoms may resemble edema, which also produces slightly raised water-soaked spots. The center of a Septoria leaf spot may become dark brown and slightly sunken, while the outer edge remains water-soaked and slightly raised. Tissues between the lesions may become brown and symptoms may be more pronounced along the leaf veins. The leaf undersurface may have many lesions of different ages. Later in July, the spots begin to coalesce, producing slightly sunken, necrotic spots sometimes bound by leaf veins or with irregular borders and variable in size up to approximately 5 mm in diameter. The lesions penetrate to the upper surface and appear as red-purple spots with irregular borders. Symptom intensity and color of spots varies among clones. Leaves deep in the canopy of sprout and fruiting fields may turn yellow instead of red-purple and appear peppered with the dark green water-soaked spots. In sprout fields, leaves on the lower half of the stems are more severely affected and may drop (Fig. 109). In the upper canopy of fruiting fields, the reddish purple discoloration continues to intensify into late July and early August, when the leaves of affected clones take on a reddish brown discoloration (Fig. 110). Clones may become completely defoliated by harvest time. Green berries may be infected occasionally and show small, raised water-soaked spots. By mid-August, it is frequently difficult to discern leaf spots caused by this *Septoria* sp. from those caused by leaf rust, powdery mildew, or other minor leaf-spotting pathogens.

Lesions also appear on fruiting stems and are easy to identify in the spring before the foliage develops. They initially appear as red-purple spots variable in size (0.5–10 mm long) and are confined to the lower half of the stem. They expand with time and the centers turn brown and become slightly sunken (Fig. 111). Most lesions have a red-purple border. A few dark brown to black pycnidia develop in the lesions. Over time, the lesions may coalesce, causing much of the lower half of the stems to be affected, but stems usually are not killed. As the season progresses, the lesions may turn from brown to tan and the pycnidia also lose their dark color and become less visible when they no longer produce conidia. The stem lesions result from infections from the previous spring during the sprout phase of growth but do not become visible until mid- to late September, when they appear as faint tan or reddish blemishes. The lesions then intensify in color and expand during April and May of the following year.

The stem lesions appear in the internodes, in contrast to lesions caused by *Godronia cassandrae* that are almost always centered on leaf nodes and kill stem tissues distally from the point of infection (see Fusicoccum (Godronia) Canker above).

Causal Organism

This *Septoria* sp. has not been formally described. Conidia from plant material are hyaline, straight or curved, filiform (obclavate when immature) with up to three to five inconspicuous septa, and variable in length, 1.5–2 × 20–110 µm (average 58 µm) (Fig. 112). Few pycnidia are produced in stem lesions compared with the many on overwintered leaves. Pycnidia in stem lesions are frequently surrounded by a halo of tan epidermal tissue that has become separated from subepidermal tissue during pycnidium development. The organism can be isolated from current-season leaves once leaf symptoms become pronounced. On potato dextrose agar, initial growth appears cream colored and yeastlike and is composed of masses of conidia growing in clusters along hyphae. Colonies also have sparse, white, aerial mycelium. Colonies are slow growing, remain compact, become slightly sunken in the agar, and eventually become covered with charcoal to black mycelium.

Although the conidia of this *Septoria* sp. are similar in size to the species associated with brown leaf spot, leaf symptoms are different, suggesting that the two species are distinct (see Brown Leaf Spot above). Unlike *Septoria albopunctata*, which infects highbush blueberry and produces pycnidia in the lesions on current-season leaves (see Septoria Leaf Spot under Blueberries (Highbush and Rabbiteye) above), the *Septoria* sp. that infects lowbush blueberry typically does not produce mature pycnidia in large numbers on infected leaves until the spring of the following year. However, occasionally it can be induced to produce mature pycnidia on current-season infected leaves by incubating them for 7–10 days in a moist chamber, especially if they are coinfected with other pathogens.

Disease Cycle and Epidemiology

The disease cycle of this *Septoria* sp. is best understood within the context of the production cycle of lowbush blueberry (Fig. 113). Blueberry fields are managed on a biennial production cycle in which fields are pruned to ground level

Fig. 109. Leaf spotting, discoloration, and defoliation caused by a *Septoria* sp., the causal organism of Septoria leaf spot, on lower portions of lowbush blueberry sprout stems. (Courtesy P. D. Hildebrand of the Department of Agriculture and Agri-Food, Government of Canada; © Her Majesty the Queen in Right of Canada, as represented by the Minister of Agriculture and Agri-Food Canada)

Fig. 110. Symptoms of leaf spotting caused by a *Septoria* sp., the causal organism of Septoria leaf spot, on the upper surface of blueberry leaves from fruiting stems, ranging from individual spots to large necrotic areas where spots have coalesced (left to right). Lesion size, number, and color vary on clones and with age of infection. (Courtesy P. D. Hildebrand of the Department of Agriculture and Agri-Food, Government of Canada; © Her Majesty the Queen in Right of Canada, as represented by the Minister of Agriculture and Agri-Food Canada)

Fig. 111. Lesions with pycnidia caused by a *Septoria* sp., the causal organism of Septoria stem canker, on a fruiting stem of lowbush blueberry. (Courtesy P. D. Hildebrand of the Department of Agriculture and Agri-Food, Government of Canada; © Her Majesty the Queen in Right of Canada, as represented by the Minister of Agriculture and Agri-Food Canada)

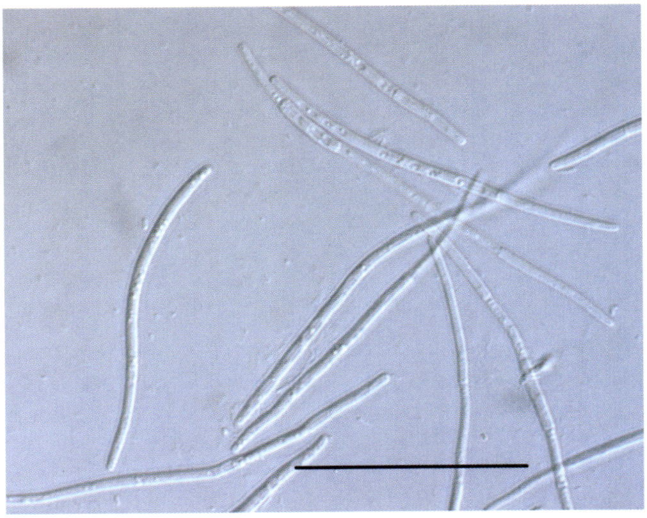

Fig. 112. Conidia of a *Septoria* sp., the causal organism of Septoria leaf spot and stem canker, from lowbush blueberry. Scale bar = 50 µm. (Courtesy P. D. Hildebrand of the Department of Agriculture and Agri-Food, Government of Canada; © Her Majesty the Queen in Right of Canada, as represented by the Minister of Agriculture and Agri-Food Canada)

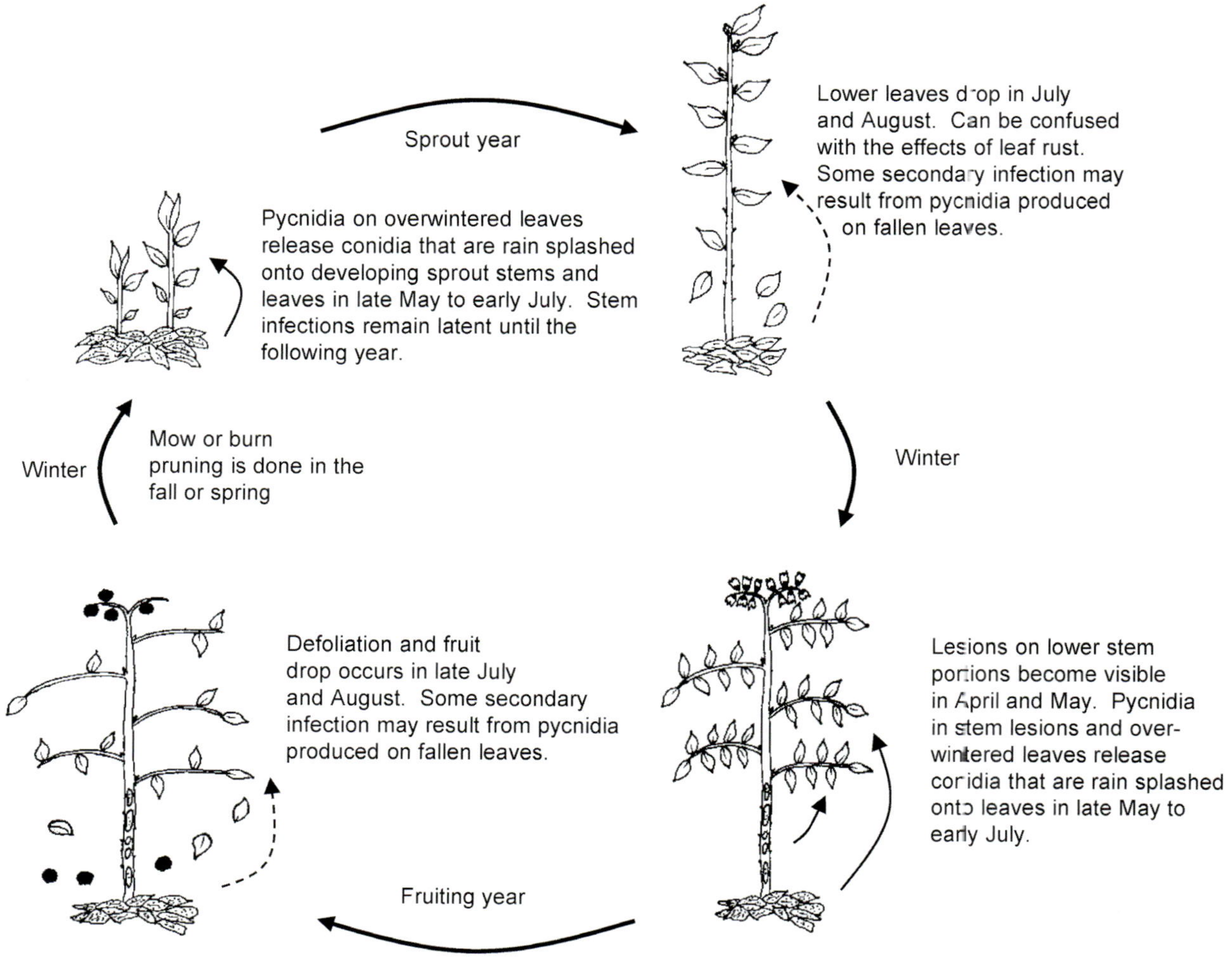

Fig. 113. Disease cycle of a *Septoria* sp., the causal organism of Septoria leaf spot and stem canker, on lowbush blueberry. (Courtesy P. D. Hildebrand of the Department of Agriculture and Agri-Food, Government of Canada; © Her Majesty the Queen in Right of Canada, as represented by the Minister of Agriculture and Agri-Food Canada)

by mowing or burning in one year to encourage the regrowth of sturdy sprout stems with fruit buds, followed by flowering and fruit production in the second year. In pruned fields, new sprout stems begin to emerge during early June and continue to lengthen apically, producing new leaves through July and early August. In the spring, pycnidia in overwintered leaves develop through April and May and release conidia most intensively over a 3- to 4-week period beginning in late May in response to rain. Only those stem portions and leaves that have expanded during this time are exposed to inoculum and eventually show symptoms. The characteristic minute water-soaked spots begin to appear on the leaves in late June, intensify in number, and slowly expand with time. By mid- to late July, the leaves begin to drop if the infection is severe. Although stem tissues that have been exposed to inoculum also become infected, lesions do not become pronounced until the spring.

In the spring of the fruiting year, pycnidial development in overwintered leaves and conidial release follows the same seasonal pattern that occurs in sprout fields, with most of it happening during the bloom period. In addition, the stem infections that occurred during the previous year develop into the characteristic lesions on the lower half of the stems that were exposed to inoculum. Stem symptoms intensify through April and May, and pycnidia are produced in the lesions. Release of conidia from pycnidia in the stems and leaves happens concurrently, but pycnidia in the leaves represent a far greater inoculum source. By mid- to late July, severely diseased leaves begin to drop and, under dry conditions, defoliation and berry drop may become severe.

Although secondary inoculum may be produced on infected current-season leaves, it is considered less important than the primary inoculum from the overwintered leaves. Secondary inoculum may develop in late July and August on infected fallen leaves, particularly if they are infected with other leaf pathogens.

Management

Burn pruning after the fruiting phase of production can reduce the incidence of this disease, but if conditions are highly favorable for infection during the subsequent sprout phase, then the number of leaf and stem infections may still be high enough to give substantial inoculum levels for the following fruiting year. In order for burning to reliably reduce the incidence of this disease, it must be done intensively and evenly to destroy overwintering inoculum in the leaf litter. Fungicide applications during the conidial release period can effectively protect plants from infection, but the decision to apply control measures can be difficult to make. The incidence level of stem lesions early in the fruiting year may help to gauge inoculum potential and guide in deciding whether to apply fungicides. Controlling the disease in conjunction with Botrytis blight during the bloom period of the fruiting year can be beneficial, but attempting to control the disease in the sprout phase likely would not be economical.

Selected References

Hilborn, M. T. 1955. Blueberry diseases and their control. Pages 27-32 in: Producing Blueberries in Maine. Maine Agric. Exp. Stn. Bull. 479.

Hildebrand, P. D., Renderos, W. E., and Filmore, S. A. E. 2010. Severity of Septoria leaf spot and stem canker and leaf rust in lowbush blueberry fields pruned by mowing or burning. Can. Plant Dis. Surv. 90:155-157.

Lockhart, C. L. 1961. Diseases and their control. Pages 31-35 in: The Blueberry in the Atlantic Provinces. Can. Dep. Agric. Publ. 754.

Markin, F. L. 1931. Progress of investigations: Blueberry diseases. Maine Agric. Exp. Stn. Bull. 360:204-206.

(Prepared by P. D. Hildebrand and W. E. Renderos)

Valdensinia Leaf Spot

Valdensinia leaf spot was first observed in commercial lowbush blueberry fields in Nova Scotia (Canada) in 1997, although the disease was previously known to occur on blueberry and numerous non-Ericaceous hosts growing in the wild in Canada, the United States, and other parts of the world. Since then, it has been reported in blueberry fields in the other Atlantic Provinces and Quebec (Canada) and in Maine (United States). Outbreaks are associated with periods of prolonged wet weather and there may be substantial yield losses resulting from premature defoliation.

Symptoms

Symptoms on blueberry leaves appear as circular, brown lesions, sometimes with a purple-red border, that are variable in size up to 1 cm in diameter (Fig. 114). Necrotic tissues within lesions may appear concentrically zonate and a single, unusually large conidium may be visible in the center of each lesion. Valdensinia leaf spot may be confused with other minor leaf-spotting pathogens, but the presence of the large conidium in the center of a lesion is characteristic for this disease. A single lesion frequently results in leaf abscission while the leaf is still green. Leaf drop may be rapid and severe when infections are during June, but the shedding of newly infected leaves occurs more slowly in late July and August. Defoliated cropping stems recover by producing new leaf shoots. Sprout stems produce new vegetative shoots in leaf axils where flower buds normally appear, resulting in reduced yields during the subsequent year. Developing fruit may also become infected, showing circular, sunken brown lesions that eventually encompass the whole berry, which eventually drops.

Fig. 114. Lowbush blueberry leaves infected with *Valdensia heterodoxa*, the causal organism of Valdensinia leaf spot. (Courtesy P. D. Hildebrand of the Department of Agriculture and Agri-Food, Government of Canada; © Her Majesty the Queen in Right of Canada, as represented by the Minister of Agriculture and Agri-Food Canada)

Causal Organism

Valdensinia leaf spot is caused by *Valdensia heterodoxa*, the synonym of the ascomycete (Sclerotiniaceae) fungus *Valdensinia heterodoxa*. *Valdensia heterodoxa* has two types of conidia. The primary conidia (aleuriospores) are large (450–600 μm), multicelled, hyaline to brownish, and star shaped with a whorl of usually four and sometimes three or five appendages (Fig. 115). The secondary conidia (phialospores, *Gliocladium*-like) are found only in artificial culture and are small (2.0–2.5 μm), hyaline, and produced in a gloeoid mass. Upon maturation, the primary conidia have a unique dispersal mechanism whereby the appendages that are initially curved upward are levelled and eventually lowered against the substratum. Hygroscopic action then propels the conidia into the air. The main body of the conidia is sticky, which enables it to adhere to leaf surfaces. If a conidium does not encounter and impact the lower surface of a leaf, it may impact the upper leaf surface on its downward trajectory. Multiple germ tubes grow from a conidium and rapidly colonize leaf tissues. New conidia are subsequently formed on colonized leaves, providing inoculum for the further spread of the disease. The midveins and sometimes the lateral veins of colonized leaves may become sclerotized, which enables the fungus to overwinter; then in the spring, conidia are formed on the sclerotia (Fig. 116). If leaves become skeletonized and disintegrate during the autumn and winter, it is very difficult to find sclerotia in the spring. Apothecia are rarely found but are presumably also a source of inoculum in the spring. They are stipitate, disk shaped, 1.5–3.5 mm in diameter, and up to 7 mm high and form on sclerotized leaf veins of host plants. Asci are inoperculate, cylindrical to slightly clavate, thin walled

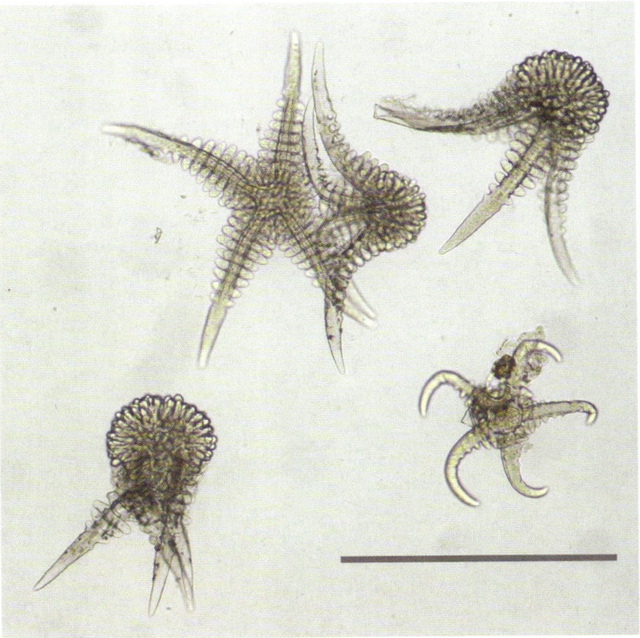

Fig. 115. Multicelled conidia of *Valdensia heterodoxa*, the causal organism of Valdensinia leaf spot, at various stages of development; appendages develop first, followed by the conidium head. Scale bar = 400 μm. (Courtesy P. D. Hildebrand of the Department of Agriculture and Agri-Food, Government of Canada; © Her Majesty the Queen in Right of Canada, as represented by the Minister of Agriculture and Agri-Food Canada)

tapering at the base, and 6–10 × 97–110 µm. Ascospores are hyaline, narrowly ellipsoid with an occasional broadly ellipsoid spore present, and initially aseptate, becoming symmetrical or asymmetrical with one to two septa with age and the second septum asymmetrically placed.

Disease Cycle and Epidemiology

Valdensinia leaf spot is often noticed first in moist, shaded areas of fields during early June as small foci of about 10 cm in diameter encompassing a few defoliated stems, but foci can also develop in nonshaded areas. With prolonged wet weather, the foci can expand rapidly in size with new foci developing around the original ones and eventually coalescing to affect large areas of the field. Field-to-field spread of the disease is believed to happen mainly by mechanical transmission. Young infected leaves are colonized rapidly, shed, become soft and, under wet conditions, easily stick to footwear and machinery passing through diseased areas.

Conidia are formed on infected leaves at 10–25°C, with an optimum temperature of 15–20°C. None are produced at 5 and 30°C. Conidia infect leaves rapidly within 6–8 h of wetness at temperatures of 15–25°C, with an optimum temperature of 20°C. Infection is moderate at 10°C (12 h of wetness), slow at 5°C (24 h of wetness), and does not occur at 30°C. Sclerotized leaves remain viable for at least 2 years. Overfertilization of fields can dramatically increase conidium production and the incidence and size of sclerotia in infected leaves.

Management

Lowbush blueberry is typically pruned either mechanically or by burning in alternate years to stimulate vegetative growth and flower bud production during the first summer and fruit production during the second year.

Good control of Valdensinia leaf spot can be achieved by the intensive burn pruning of affected fields to consume infected leaf debris. If the burn prune operation is not intensive or uniform enough, sclerotized leaves survive and provide inoculum for the following growing season. Mechanical pruning with a flail mower is ineffective in this regard.

Several fungicides (chlorothalonil, pyraclostrobin, boscalid, and fluazinam) are labelled against this disease in Canada. One or two applications of fungicide made during early disease development can slow the disease sufficiently to maintain adequate foliage for normal fruit development in the fruiting year or flower bud development in the sprout year. However, if wet weather persists, fungicide efficacy is greatly reduced. Work activities in affected fields should be planned to avoid mechanical transmission to healthy fields and contaminated equipment should be washed to remove diseased leaves.

Fig. 116. Conidia of *Valdensia heterodoxa*, the causal organism of Valdensinia leaf spot, on the sclerotized midvein of an overwintered blueberry leaf. (Courtesy P. D. Hildebrand of the Department of Agriculture and Agri-Food, Government of Canada; © Her Majesty the Queen in Right of Canada, as represented by the Minister of Agriculture and Agri-Food Canada)

Selected References

Hildebrand, P. D., and Renderos, W. E. 2007. Valdensinia leaf spot (*Valdensinia heterodoxa*) of commercial lowbush blueberry in Atlantic Canada: An emerging new threat. (Abstr.) Can. J. Plant Pathol. 29:90.

Norvell, L. L., and Redhead, S. A. 1994. *Valdensinia heterodoxa* (Sclerotiniaceae) in the United States. Can. J. For. Res. 24:1981-1983.

Peyronel, B. 1923. [On an unusual polyphagous parasite: *Valdensia heterodoxa* n. gen. et n. sp.]. Stn. Sper. Agrar. Ital. 56:521-538. [In Italian.]

Redhead, S. A. 1974. Epistolae mycologicae IV. *Valdensinia heterodoxa* Peyr. (Sclerotiniaceae). Syesis 7:235-238.

Redhead, S. A., and Perrin, P. W. 1972. *Asterobolus*: A new parasitic hyphomycete with a novel dispersal mechanism. Can. J. Bot. 50:409-412.

Redhead, S. A., and Perrin, P. W. 1972. *Asterobolus*: A synonym of *Valdensia*. Can. J. Bot. 50:2083-2084.

Vogelgsang, S., and Shamoun, S. F. 2002. Growth, sporulation, and conidia discharge of *Valdensinia heterodoxa*, a foliar pathogen of salal, as influenced by temperature and photoperiod *in vitro*. Mycol. Res. 106:480-490.

(Prepared by P. D. Hildebrand and W. E. Renderos)

Witches'-Broom

Witches'-broom is a relatively minor disease of highbush and lowbush blueberries and other *Vaccinium* spp. in North America, Europe, Siberia, and Japan. Although heavily infected plants produce no fruit, disease incidence is usually so low that crop losses are negligible. However, high levels of infection can be found in blueberry plants located near fir (*Abies*) trees, the alternate host of the rust fungus that causes witches'-broom.

Symptoms

Diseased blueberry plants have broomlike masses of swollen, spongy shoots with shortened internodes and leaves of reduced size (Fig. 117). The brooms usually begin to develop during the year following infection and then persist for many years, producing infected new growth each spring. Young stems on the broom are initially reddish or yellow, but as the season progresses, they become brown and shiny, then dull, and eventually dry and cracked. Heavily infected plants produce no fruit.

Causal Organism

The heteroecious rust fungus *Pucciniastrum goeppertianum*, the causal organism of witches'-broom, produces spermagonia and aecia on several species of fir and telia on blueberry and other *Vaccinium* spp. In some cases, spermagonia are absent.

The cylindrical to somewhat flattened aecia develop in yellowish lesions on the lower surface of fir needles; they mature on the current year's needles in late summer or autumn and, in western North America, also on the previous year's needles in the spring and early summer. Aeciospores (10–18 × 18–30 µm) are broadly ellipsoid with reddish orange contents and a colorless, finely warted wall. No uredinia are produced.

The infection in blueberry and other *Vaccinium* spp. is systemic. Teliospores develop in a single, closely appressed layer in the epidermal cells inside telia that form a continuous, reddish brown layer around the swollen stems. The teliospores are oblong or ellipsoid (10–14 × 13–30 µm), are usually divided into four cells by vertical septa, and have golden brown walls that are thickened at the apex.

Fig. 117. Lowbush blueberry with young (**A**) and old (**B**) witches'-broom, caused by *Pucciniastrum goeppertianum*. (Courtesy S. L. Annis; © APS)

Disease Cycle and Epidemiology

In eastern North America, aeciospores produced on balsam fir needles in the summer are carried by wind to blueberry plants, where they germinate on the leaves or stems. In western North America, aecia also mature on the previous year's fir needles in the spring and early summer. The fungus penetrates the epidermis of blueberry or other *Vaccinium* spp. and eventually becomes systemic and perennial in the blueberry host, stimulating the production of excessive lateral buds that develop into the characteristic broomlike masses of swollen shoots with shortened internodes and reduced leaves.

The incubation period before the appearance of symptoms is approximately 1 year; symptoms are usually first observed in the spring. No uredinia are produced, but telia develop in late summer and autumn in a layer on the swollen stems of the brooms, where they overwinter. During the following spring, teliospores germinate, producing basidiospores that are windborne to fir trees, where they infect the current year's needles. Because the brooms are perennial and produce new growth each spring, they may continue to serve as sources of inoculum for many years in highbush blueberry. Basidiospore release and infection of fir trees coincides with the period of rapid elongation and maximum susceptibility of the fir shoots. Flecking may be observed on fir needles within 2 weeks and mature aecia within 5 weeks of infection.

Host alternation is believed to be essential for this rust because there is no uredinial stage. However, some investigators have observed heavily infected *Vaccinium* spp. in areas remote from fir and have suggested that basidiospores may be able to infect blueberry as well as fir. Experimental evidence for this is lacking.

Management

Because the pathogen is perennial and systemic in blueberry crowns and rhizomes, burning and other pruning methods do not eliminate witches'-broom. The best control strategy is to eradicate the alternate host (fir trees) within 460 m of the blueberry plants. This may not be practical, however, in areas such as eastern North America, where balsam fir is often abundant near blueberry fields and is grown commercially for Christmas tree production. Eradication of diseased blueberry plants with a recommended herbicide over time can decrease the disease incidence in an affected field.

Little is known about clonal resistance to witches'-broom. Some hybrids of highbush and lowbush blueberries have proved highly susceptible in eastern Canada. The cultivar Rancocas appears to be one of the least susceptible highbush cultivars and might serve as a good parent in future crosses with lowbush clones.

Selected References

Arthur, J. C. 1934. Manual of the Rusts in United States and Canada. Purdue Research Foundation, Lafayette, IN.

Demaree, J. B., and Wilcox, M. S. 1947. Fungi pathogenic to blueberries in the eastern United States. Phytopathology 37:487-506.

Hilborn, M. T. 1955. Blueberry diseases and their control. Pages 27-32 in: Producing Blueberries in Maine. Maine Agric. Exp. Stn. Bull. 479.

Hiratsuka, Y., McArthur, L. E., and Emond, F. J. 1967. A distinction between *Pucciniastrum goeppertianum* and *P. epilobii* with clarification of status of *Peridermium holwayi* and *P. ornamentale*. Can. J. Bot. 45:1913-1915.

Markin, F. L. 1931. Progress of investigations: Blueberry diseases. Maine Agric. Exp. Stn. Bull. 360:204-206.

Markin, F. L. 1943. Blueberry diseases in Maine. Maine Agric. Exp. Stn. Bull. 419:394-433.

Van Sickle, G. A. 1973. A survey of production losses due to witches' broom of blueberry in the Maritime provinces. Plant Dis. Rep. 57:608-611.

Van Sickle, G. A. 1975. Basidiospore production and infection of balsam fir by a needle rust, *Pucciniastrum goeppertianum*. Can. J. Bot. 53:8-17.

Van Sickle, G. A. 1977. Seasonal periodicity in the discharge of *Pucciniastrum goeppertianum* basidiospores. Can. J. Bot. 55:745-751.

(Prepared by N. L. Nickerson and F. L. Caruso;
Revised by S. L. Annis)

Diseases Caused by Viruses

Blueberries

Blueberry latent spherical virus

As part of a study to detect viruses in blueberries at Iwate University, Japan, 'Blueray' tissue was mechanically inoculated onto *Chenopodium quinoa* indicator plants. Plants developed chlorosis indicative of virus presence. After virus purification and genome characterization, it was determined that a new virus, named *Blueberry latent spherical virus* (BLSV) may infect blueberry.

Symptoms

As implied by its name, BLSV does not cause visual symptoms in many cultivars, including Berkeley, Bluecrop, Bluejay, Collins, Coville, Herbert, Jersey, Pemberton, and Rubel. When the virus was transmitted to *Nicotiana benthamiana*, it caused a mild, transient mosaic, while *Luffa cylindrica* developed chlorotic spots on systemically infected leaves. *Gomphrena globosa*, *Ipomoea tricolor*, *Nicotiana clevelandii*, *Nicotiana occidentalis*, and *Nicotiana tabacum* are BLSV hosts but no symptoms have been observed on these plants.

Causal Agent

BLSV is a member of the family *Secoviridae*, most closely related to *Peach rosette mosaic virus*, a member of subgroup C of the genus *Nepovirus*. A reverse transcription-polymerase chain reaction detection assay was developed and used to determine the presence of the virus in field-grown plants as part of the study that led to the discovery of BLSV. Although the virus is symptomless in single infections, it is unknown whether it acts synergistically with other viruses to cause symptoms.

Disease Cycle

The virus is readily mechanically transmissible to several herbaceous indicators but not to blueberry seedlings, even when inoculating with purified virus. The virus is graft transmissible from blueberry to blueberry. Other members of the genus *Nepovirus* are readily transmitted by nematodes in the genera *Xiphinema* and *Longidorus* but no nematode transmission experiments have been conducted with BLSV. The distribution and population structure of the virus is unknown since no surveys have been conducted to determine its geographic range.

Management

BLSV does not cause symptoms in blueberry and, as such, there are no management recommendations.

Selected Reference

Isogai, M., Tatuto, N., Ujiie, C., Watanabe, M., and Yoshikawa, N. 2012. Identification and characterization of blueberry latent spherical virus, a new member of subgroup C in the genus *Nepovirus*. Arch. Virol. 157:297-303.

(Prepared by I. E. Tzanetakis and R. R. Martin)

Blueberry latent virus

During the process of identifying an agent associated with a new blueberry disorder, double-stranded RNA (dsRNA) was isolated from several plants. The dsRNA was not associated with the particular disorder and it was later determined that it is present in the majority of northern highbush, southern highbush, and rabbiteye cultivars. The dsRNA belongs to a new virus species named *Blueberry latent virus* (BBLV).

Symptoms

As the name implies, no symptoms have been associated with the presence of BBLV. Its role in mixed infections with other viruses has not been studied but such an interaction is highly unlikely because of the nature of the virus.

Causal Agent

BBLV is a member of a new family of dsRNA viruses and closely related to viruses isolated from tomato, rose, rhododendron, and broad bean. The virus genome is unique among plant viruses since it does not appear to code for a movement protein, indicating that movement within the plant is probably done only through cell division. Several isolates of BBLV have been sequenced. Genome diversity never exceeded 0.5%, indicating a very strict purification pressure or recent emergence of the virus in blueberry. Virus RNA has been detected in the palisade and spongy mesophyll as well as in the vascular bundles. A high rate of seed transmission, reaching 100%, has been demonstrated for three crosses that included northern highbush, southern highbush, and rabbiteye cultivars. A reverse transcription-polymerase chain reaction assay as well as antibodies against one of the virus proteins have been developed, and the virus has been detected in more than 50% of the samples collected from all major blueberry-producing areas of the United States and about 30% of the samples tested in Japan.

Management

BBLV does not cause symptoms in blueberry and, as such, there are no management recommendations.

Selected References

Isogai, M., Nakamura, T., Ishii, K., Watanabe, M., Yamagishi, N., and Yoshikawa, N. 2011. Histochemical detection of Blueberry latent virus in highbush blueberry plant. J. Gen. Plant Pathol. 77:304-306.

Martin, R. R., Zhou, J., and Tzanetakis, I. E. 2011. Blueberry latent virus: An amalgam of *Partitiviridae* and *Totiviridae*. Virus Res. 155:175-180.

(Prepared by I. E. Tzanetakis and R. R. Martin)

Blueberry virus A

Leaf yellowing on highbush blueberry 'Spartan' prompted an investigation as to whether a virus was the causal agent of the disorder. After double-stranded RNA extraction from symptomatic material, a single band of 17 kilobases was identified, indicative of virus infection. Shotgun cloning and sequencing revealed the presence of a novel virus designated as *Blueberry virus A* (BVA).

Symptoms

Several symptomatic and asymptomatic 'Spartan' plants and a small number of symptomless 'Bluecrop', 'Coville', and

'Sierra' individual plants were tested for the presence of the virus. The majority of the plants were found to be infected, but curiously, the percentage of infection in symptomless material was higher than that in the symptomatic plants. Grafting of virus-free 'Blueray', 'Sierra', and 'Spartan' scions onto infected 'Spartan' rootstocks was successful but no symptoms were observed during 2 years of observation. These observations indicate that the virus is symptomless in single infections.

Causal Agent

BVA is a member of the family *Closteroviridae* and most closely related to members of the genus *Closterovirus*. A reverse transcription-polymerase chain reaction detection assay has been developed and used successfully to determine the presence of the virus in elite germplasm. The virus is present in Japan and North America, where it was associated with severe decline, presumably because of BVA interactions with other viruses infecting the material.

Disease Cycle

It has not been possible to transmit BVA using leaf sap from infected bushes to inoculate herbaceous hosts or blueberry seedlings. BVA taxonomy is still in flux, but its similarity with members of the genus *Closterovirus* prompted the testing of the cotton aphid (*Aphis gossypii*) as a potential vector. All transmission attempts with this aphid were unsuccessful. Knowledge of virus epidemiology, including distribution and population structure, is minimal, as is our understanding of the virus' potential to alter symptoms in mixed infections with any of the other viruses reported in this compendium.

Management

There are currently no management recommendations for BVA.

Selected References

Isogai, M., Muramatu, S., Watanabe, M., and Yoshikawa, N. 2013. Complete nucleotide sequence and latency of a novel blueberry-infecting closterovirus. J. Gen. Plant Pathol. 79:123-127.

Martin, R. R., Polashock, J. J., and Tzanetakis, I. E. 2012. New and emerging viruses of blueberry and cranberry. Viruses 4:2831-2852.

(Prepared by I. E. Tzanetakis and R. R. Martin)

Fruit Drop

Blueberry fruit drop disease was first observed in British Columbia, Canada, in the late 1990s, and similar symptoms have since been observed in the United States in Washington, Oregon, and New York. The fruit drop symptoms have been most obvious in the cultivar Bluecrop, in which nearly 100% of the fruit drops from the bush when the berries are 3–5 mm in diameter. Symptoms of poor fruit set in other cultivars may or may not be related to the fruit drop disease.

Symptoms

Plants affected with fruit drop show a reddening of the young leaves early in the season and a candy striping of the corolla (Fig. 118) similar to what is seen with *Blueberry shoestring virus*. After flowering, the young leaves lose the red coloration and the bushes appear normal. Approximately 3 weeks prior to ripening, when the fruit is about 3–5 mm in diameter, the fruit turns blue and drops from the bush. At harvest, affected plants bear no blue fruit (Fig. 119). Prior to harvest, affected bushes can be seen from the edge of the field since the branches, not having any fruit, are upright and the bushes appear taller than neighboring bushes that are laden with fruit (Fig. 120). The affected bushes grow more vigorously than the fruiting bushes, presumably because they are allocating fewer resources to fruit production and more to vegetative growth.

Causal Agent

The viral double-stranded RNA purified from diseased plants that was cloned and sequenced initially was found to be widespread in blueberry and not related to the disease, since it was present in multiple cultivars of northern and southern highbush and rabbiteye blueberries, including symptomless plants of 'Bluecrop' (see *Blueberry latent virus* above). Subsequently, a DNA virus was amplified from symptomatic but not from asymptomatic 'Bluecrop' plants using rolling circle amplification, which amplifies circular DNA molecules. This virus has been cloned and sequenced and found to be related to the caulimoviruses, *Dahlia mosaic virus* and *Cauliflower mosaic virus*, but codes for a single open reading frame and likely represents a new genus in the family *Caulimoviridae*. Detection primers were developed and the presence of the amplified viral fragment was highly correlated with symptoms in 'Bluecrop', the virus was present in 20 of 20 symptomatic plants tested and only in 1 of 20 asymptomatic plants in the same section of an affected field. The presence of the virus has been confirmed in symptomatic plants in British Columbia and Washington, but

Fig. 118. Early-season symptoms of blueberry fruit drop, caused by an unknown virus, showing reddening of the new leaves and corollas. (Courtesy R. R. Martin)

Fig. 119. Flower cluster on a 'Bluecrop' blueberry bush with fruit drop, caused by an unknown virus, just prior to harvest. (Courtesy R. R. Martin)

Fig. 120. View of 'Bluecrop' blueberry plants from the edge of a field. Branches of fruit drop-affected bushes (rear center), caused by an unknown virus, are much more upright than the healthy bushes that are laden with fruit (foreground). (Courtesy R. R. Martin)

testing has not been carried out in other regions. The virus was detected in several symptomless plants of 'Duke' and 'Liberty' adjacent to a field of 'Bluecrop' with a high number of infected plants. The susceptibility and impact of this virus in other cultivars are unknown.

Disease Cycle

It has not been possible to transmit the causal agent of blueberry fruit drop using leaf or flower sap from infected bushes to inoculate herbaceous hosts or to graft transmit from blueberry to herbaceous hosts. The disease can be graft transmitted from blueberry to blueberry. In some case, the disease has spread rapidly within affected fields, with a radial pattern of spread. In other fields, the disease appears to spread quite slowly with only a few newly affected bushes per year. The disease has spread to adjacent fields, but no vector has yet been identified.

Management

There are currently no management recommendations for fruit drop disease.

Selected References

Diaz-Lara, A., Keller, K. E., and Martin, R. R. 2014. A new virus isolated from blueberry exhibiting a complete fruit drop symptom. (Abstr.) Phytopathology 104(Suppl. 3):S3.33.

Martin, R. R., Zhou, J., and Tzanetakis, I. E. 2011. Blueberry latent virus: An amalgam of *Partitiviridae* and *Totiviridae*. Virus Res. 155:175-180.

(Prepared by R. R. Martin)

Leaf Mottle

Leaf mottle, a pollen-borne viral disease caused by *Blueberry leaf mottle virus* (BLMV), has been reported in highbush blueberries in Michigan (United States) and half-high blueberry clones in New Brunswick (Canada). A serologically distinct strain of BLMV, also known as *Grapevine Bulgarian latent virus* (GBLV), was found in a single grapevine in New York (United States). GBLV has also been detected in grapevines in Bulgaria, Hungary, and Portugal. This strain is not known to attack blueberry. Leaf mottle is considered economically important only in Michigan. The virus has been serologically confirmed in 25 commercial fields, and visual diagnosis suggests that it probably is present in another 100 Michigan plantations.

Symptoms

Leaf mottle is most severe on bushes of the highbush blueberry cultivar Rubel, which may decline and die over a period of 5–10 years. Bushes that have been infected for a few years have dead stems, and a small amount of regrowth may be found near the crown area (Fig. 121). Leaves are malformed and mottled and may be straplike. Translucent, roughly circular spots may be visible when a leaf is held up to light.

On 'Jersey', symptoms are much milder and stem dieback is not prevalent. Leaves are smaller than normal and pale yellowish green and may be straplike (Fig. 122). The terminal leaves arising in the crown area are rosetted, appearing to be piled on top of each other as a result of the shortened internodes of shoots (Fig. 122). Older, chronically infected plants are stunted and produce little or no crop (Fig. 123). Symptoms on 'Blueray' are similar to those on 'Jersey'. The disease has been diagnosed in only a few 'Blueray' bushes in one field, and the damage to this cultivar did not appear very severe.

Causal Agent

Although BLMV is a member of the nepovirus group, it does not appear to have a nematode vector but spreads by infected pollen. BLMV virus particles are isometric and 28–30 nm in diameter. BLMV has three components with sedimentation coefficients of 53S (top), 120S (middle), and 128S (bottom). The middle and bottom components contain single-stranded RNA with molecular masses of 2.15 and 2.35 × 10^6 Da, respectively. The protein coat subunit has a molecular mass of 54,000 Da.

BLMV is sap transmissible to a few herbaceous hosts. *Chenopodium quinoa* develops circular, chlorotic lesions (2–3 mm in diameter) on inoculated leaves, followed by terminal dieback in 7–10 days. BLMV is seedborne in up to 20% of mechanically inoculated *Chenopodium quinoa* plants. *Nicotiana clevelandii* also displays pinpoint-sized local lesions on systemically infected terminal leaves 2–3 weeks after inoculation. *Chenopodium amaranticolor* and *Nicotiana tabacum* cv. Xanthi also become symptomatic 7–10 days after mechanical inoculation.

BLMV is highly immunogenic and enzyme-linked immunosorbent assay (ELISA) readily detects BLMV in infected blueberry pollen as well as leaf, blossom, and fruit bud tissues. Commercial ELISA kits are available for testing bushes for BLMV.

Disease Cycle and Epidemiology

BLMV particles are present on the surface of and inside pollen grains from infected blueberry bushes. Honeybees spread infected pollen from bush to bush and, in the process, infect healthy bushes. The latent period between initial infection and the onset of symptoms is approximately 3–4 years. Wild

Fig. 121. Severe dieback of stems and leaf regrowth on 'Rubel' highbush blueberry infected by *Blueberry leaf mottle virus*, the causal agent of leaf mottle. (Courtesy D. C. Ramsdell)

lowbush and highbush blueberry plants have been found to be infected with BLMV at distances of up to 100 m into wooded areas adjacent to a commercial planting containing infected bushes. This suggests that wild blueberry plants can serve as a reservoir of BLMV. BLMV is seedborne in 1.5% of blueberry seedlings from an infected bush.

In controlled experiments conducted in cages with small, pot-grown bushes of cultivars Jersey, Rubel, Spartan, Elliott, Patriot, Blueray, Bluecrop, and Bluejay, honeybees transmitted BLMV from an infected source plant. Honeybees carry virus-laden pollen in the pollen baskets on their legs and on the hairs on their bodies. Patterns of disease spread in the field are likely determined mostly by bee foraging behavior. Bees also carry diseased pollen back to the hive. Indirect evidence suggests that bees can pick up contaminated pollen from other bees in the hive. Experiments with marked bees in blueberry plantations have shown that bees forage on bushes up to 1,600 m from the hive. Furthermore, honeybees visit other hives up to 600 m away. BLMV has been shown to survive in infected pollen in the hive and remain infectious for up to 10 days.

Fig. 122. Pale yellowish green discoloration, rosetting, and narrowing of 'Jersey' highbush blueberry leaves infected by *Blueberry leaf mottle virus*, the causal agent of leaf mottle. (Courtesy A. M. C. Schilder; © APS)

Fig. 123. Declining 'Jersey' blueberry bush with pale green leaves infected by *Blueberry leaf mottle virus*, the causal agent of leaf mottle. (Courtesy A. M. C. Schilder; © APS)

Management

The fact that honeybees are essential in pollinating and setting a profitable crop in highbush blueberry greatly complicates the control of leaf mottle. However, the first approach is prevention, accomplished by planting virus-tested clean stock. In fields that contain diseased bushes, all infected plants must be identified and removed. Bushes should be killed with an appropriate brush killer or herbicide to prevent regrowth and flowering. Because of the 3- to 4-year latent period between infection and the appearance of symptoms, visual diagnosis is not always sufficient; thus individual plants must be tested by ELISA. Dormant fruit buds, blossoms, and terminal leaves are reliable tissues for testing. However, since the virus is unequally distributed in the bush, six or seven samples should be taken from different parts of the bush to make a composite sample for testing.

Beehives should be placed as far as possible from infection foci in a field or in neighboring fields. Hives should not be moved from one field to another, from early-flowering to late-flowering cultivars, or to blueberries farther north as the season progresses.

Selected References

Anonymous. 1997. Blueberry leaf mottle nepovirus. In: Quarantine Pests for Europe, 2nd ed. I. M. Smith, D. G. McNamara, P. R. Scott, and M. Holderness, eds. EPPO and CABI, Wallingford, United Kingdom.

Boylan-Pett, W., Ramsdell, D. C., Hoopingarner, R. A., and Hancock, J. F. 1991. Honeybee foraging behavior, in-hive survival of infectious pollen-borne blueberry leaf mottle virus and transmission of the virus in highbush blueberry. Phytopathology 81:1407-1412.

Childress, A. M., and Ramsdell, D. C. 1986. Lack of evidence for a nematode vector of blueberry leaf mottle virus (BBLMV). Acta Hortic. 186:87-91.

Childress, A. M., and Ramsdell, D. C. 1986. Detection of blueberry leaf mottle virus in highbush blueberry pollen and seed. Phytopathology 76:1333-1337.

Childress, A. M., and Ramsdell, D. C. 1987. Bee-mediated transmission of blueberry leaf mottle virus via infected pollen in highbush blueberry. Phytopathology 77:167-172.

Jaswal, A. S. 1990. Occurrence of blueberry leaf mottle, blueberry shoestring, tomato ringspot and tobacco ringspot virus in eleven halfhigh blueberry clones grown in New Brunswick, Canada. Can. Plant Dis. Surv. 70:113-117.

Ramsdell, D. C. 1987. Blueberry leaf mottle. Pages 112-114 in: Virus Diseases of Small Fruits. R. H. Converse, ed. U. S. Dep. Agric. Agric. Handb. 631. U.S. Government Printing Office, Washington, DC.

Ramsdell, D. C., and Stace-Smith, R. 1979. Blueberry leaf mottle, a new disease of highbush blueberry. Acta Hortic. 95:37-47.

Ramsdell, D. C., and Stace-Smith, R. 1981. Physical and chemical properties of the particles and ribonucleic acid of blueberry leaf mottle virus. Phytopathology 71:468-472.

Ramsdell, D. C., and Stace-Smith, R. 1983. Blueberry leaf mottle virus. Descriptions of Plant Viruses, No. 267. Commonwealth Mycological Institute and Association of Applied Biologists, Kew, Surrey, England.

Uyemoto, J. K., Taschenberg, E. F., and Hummer, D. K. 1977. Isolation and identification of a strain of grapevine Bulgarian latent virus in Concord grapevine in New York State. Plant Dis. Rep. 61:949-953.

(Prepared by D. C. Ramsdell; Revised by A. M. C. Schilder)

Mosaic

Mosaic disease was reported for the first time in Michigan (United States) in the 1950s in highbush blueberry plantings and has seen been observed in North America in the United States in Arkansas, Indiana, New Jersey, New York, Oregon, and Washington and in British Columbia, Canada, with little within-field spread observed. Since the turn of the century, this disease has been observed in many blueberry-growing areas around the world, including New Zealand, Europe, South Africa, Argentina, and Chile, and appears to be much more common in northern areas of North America, where previously it spread little. Mosaic affects highbush cultivars Bluecrop, Blueray, Brigitta, Burlington, Berkeley, Herbert, Pioneer, Rubel, Cabot, Concord, Earliblue, Jersey, Stanley, Toro, and likely others and occasionally *Vaccinium pallidum* (a lowbush dryland blueberry). It has not been observed on rabbiteye blueberry but this may be because the presumed virus vector is not present in areas where rabbiteye is grown. Fruit on diseased bushes ripen late and are of poor quality. A yield reduction of 15% caused by mosaic was reported for the cultivar Bluecrop.

Fig. 124. 'Legacy' highbush blueberry with leaf symptoms of mosaic, potentially caused by Blueberry mosaic associated virus. (Courtesy R. R. Martin; Reproduced from Martin et al., 2012; © 2012 by the authors; licensee MDPI, Basel, Switzerland, Creative Commons Attribution license 3.0)

Fig. 125. 'Bluecrop' highbush blueberry with leaf symptoms of mosaic, potentially caused by Blueberry mosaic associated virus. (Courtesy R. R. Martin)

Symptoms

Symptoms on foliage include mild to brilliant mottle and mosaic patterns of chrome yellow, yellow, and yellow-green (Fig. 124). Occasionally, leaves display areas of pink coloration (Fig. 125). Symptomatic shoots may develop erratically throughout the bush. Mosaic symptoms are not produced every year and presumably depend on specific environmental factors, such as heat, which affects gene silencing. In experiments in which the cultivars Herbert, Stanley, and Burlington were grafted at bud break with mosaic-infected buds, symptoms developed 27, 31, and 51 days after grafting, respectively.

Causal Agent

The causal agent of blueberry mosaic is not transmissible by mechanical means or by dodder to herbaceous plants; however, it is graft transmissible. A new virus, Blueberry mosaic associated virus (BlMaV), closely related to members of the genus *Ophiovirus*, has been found in all symptomatic plants collected from several areas in North America and Europe but Koch's postulates have yet to be completed.

Disease Cycle and Epidemiology

Until recently mosaic has spread slowly from bush to bush in the field; however, since about 2000, this disease has become widespread and, in some cases, affects a large percentage of plants in a field. The vector is unknown but all other ophioviruses identified to date are transmitted by soilborne fungi in the genus *Olpidium*. Diseased planting stocks are responsible for the introduction of mosaic into commercial plantings.

Management

There are reliable detection protocols developed to detect diverse isolates of the virus from both North America and Europe. Use of virus-tested plants to establish new plantings is the best way to control mosaic until a vector identity is confirmed. Infected plants should be removed from the field.

Selected References

Martin, R. R., Polashock, J. J., and Tzanetakis, I. E. 2012. New and emerging viruses of blueberry and cranberry. Viruses 4:2831-2853.

Ramsdell, D. C., and Stretch, A. W. 1987. Blueberry mosaic. Pages 119-120 in: Virus Diseases of Small Fruits. R. H. Converse, ed., U. S. Dep. Agric. Agric. Handb. 631. U.S. Government Printing Office, Washington, DC.

Thekke-Veetil, T., Ho, T., Keller, K. E., Martin, R. R., and Tzanetakis, I. E. 2014. A new ophiovirus is associated with blueberry mosaic disease. Virus Res. 189:92-96.

Thekke-Veetil, T., Polashock, J. J., Marn, M. V., Plesko, I. M., Schilder, A. C., Keller, K. E., Martin, R. R., and Tzanetakis, I. E. 2015. Population structure of blueberry mosaic associated virus: Evidence of reassortment in geographically distinct isolates. Virus Res. 201:79-84.

Varney, E. H. 1957. Mosaic and shoestring, virus diseases of cultivated blueberry in New Jersey. Phytopathology 47:307-309.

(Prepared by D. C. Ramsdell;
Revised by R. R. Martin and I. E. Tzanetakis)

Necrotic Ring Blotch

Since 2006, a new disease has been observed on southern highbush blueberry (*Vaccinium corymbosum* interspecific hybrids) in several states in the southeastern United States. Current evidence indicates that these symptoms are caused by a new virus, *Blueberry necrotic ring blotch virus* (BNRBV). Diagnostic enzyme-linked immunosorbent assay and polymerase chain reaction (PCR) tests failed to detect any of the previously described blueberry viruses, and no fungal or bacterial pathogens were associated with the symptoms. After cloning and sequencing the double-stranded RNA (dsRNA) isolated from symptomatic tissue, the virus was identified. The initial distribution of necrotic ring blotch was limited to very sporadic reports and observations in southeastern Georgia in 2006 and 2007, but by 2008, the disease was found in multiple locations throughout the major blueberry production counties of Georgia. Necrotic ring blotch has since been observed in Florida, Mississippi, South Carolina, and North Carolina. Premature defoliation occurs on severely affected plants. To date, the disease has not been observed in the native, more widely grown rabbiteye blueberry (*Vaccinium virgatum*) cultivars.

Symptoms

Symptoms of necrotic ring blotch are relatively easy to recognize during the early stages, but as lesions coalesce, leaf symptoms can be confused with those of fungal leaf spots. Initial symptoms of necrotic ring blotch are observed as irregularly shaped, circular spots or blotches with green centers on the top of leaves that penetrate to the bottom surface of leaves (Fig. 126). A symptom variation, often observed on the same plant, is a greasy or oily spot, which greatly resembles a chemical burn injury. These symptoms are distinct from those caused by *Blueberry red ringspot virus*, which produces ring spots that are more reddish, have thinner margins, are less blotchy in appearance, and generally appear on the upper leaf surface and stems. However, a distinct red-lesion variant strain of BNRBV has been observed in Florida, although rings are generally concentric and can spread to cover the entire leaf (Fig. 127). With necrotic ring blotch, symptoms are more prevalent on older leaves (i.e., the lower and inner leaves of plants), but they can eventually inundate a bush throughout. The outer ring colors can initially vary from dark brown to purplish black, but as mentioned, they often coalesce to form solid brown to black necrotic spots or blotches that are similar to those of Septoria leaf spot or other fungal leaf spots. However, Septoria leaf spot never has green centers, and fungal structures (pycnidia and cirrhi) are observed in spots caused by *Septoria* spp. Plants with severe symptoms of either necrotic ring blotch or Septoria leaf spot can defoliate prematurely.

Causal Agent

The presence of dsRNA in symptomatic leaf tissue indicated the presence of virus(es), which was confirmed by cloning and sequencing the dsRNA. Diagnostic primers were developed for detection by reverse transcription-polymerase chain reaction (RT-PCR). All of more than 50 symptomatic plants from North Carolina, Georgia, and Florida tested positive in the RT-PCR assay, and more than 2,500 asymptomatic plants from nurseries and production fields tested negative. This perfect correlation strongly suggests that BNRBV is the causal agent associated with the symptoms. The red-lesion variant, BRNBV-RL, was also detected with the detection primers. Sequence analysis revealed that BNRBV has an approximately 14-kb genome divided into four positive-sense RNA segments, has replicase and movement domains that are conserved across single-stranded RNA viruses from the families *Virgaviridae* and *Bromoviridae* and the genus *Cilevirus*. BNRBV contains two helicase (HEL) domains, which is unique among viruses, located on RNA1 and RNA2. Both HEL domains have conserved motifs and represent two distinct clades, suggesting that the virus may be the result of recombination events in plants with coinfections of multiple viruses. Based on genome organization and sequence divergence, BNRBV likely represents a new genus of plant viruses.

Disease Cycle and Epidemiology

There is currently very limited information about the epidemiology of necrotic ring blotch, including its means of transmission or spread, the potential involvement of vectors, and the role of alternate hosts or other sources of inoculum. Preliminary data indicate that the disease is not readily transmitted by vegetative propagation. Because of their association with diseased plants, one or more arthropod vectors are suspected. Research also indicates that the disease is significantly aggregated within and across rows, and in some cases, pronounced edge effects are observed, all indicative of a slow-moving vector. There is also evidence that BNRBV is a local-lesion virus, with limited systemic movement.

Management

No definitive control measures are currently known. However, there is strong evidence for host resistance or tolerance, since some cultivars are often found to harbor extensive symptoms (i.e., Star, O'Neal, and FL 86-19), whereas others on the same farm or field tend to be symptomless. Although spread through vegetative propagation appears to be limited, suspect plants should not be utilized for propagation through hardwood or softwood cuttings. Furthermore, tissue-cultured plants may help to reduce the level of initial introduction of this disease.

Fig. 126. Blueberry leaves with the initial symptoms of necrotic ringblotch, caused by *Blueberry necrotic ringblotch virus*. The leaves have irregularly shaped, circular, dark brown to purplish black blotches with green centers on the top surface. (Courtesy P. Harmon; © APS)

Fig. 127. Blueberry leaves with symptoms of a red-lesion variant strain *Blueberry necrotic ringblotch virus*, the causal agent of necrotic ringblotch. The leaves have red, concentric rings that expanded across the leaves, some have green centers. (Courtesy P. Harmon; © APS)

If an arthropod vector is identified, vector management with insecticides may also be recommended.

Selected References

Cantu-Iris, M., Harmon, P. F., Londoño, A., and Polston, J. E. 2013. A variant of blueberry necrotic ring blotch virus associated with red lesions in blueberry. Arch. Virol. 158:2197-2200.

Martin, R. R., Polashock, J. J., and Tzanetakis, I. E. 2012. New and emerging viruses of blueberry and cranberry. Viruses 4:2831-2852.

Quito-Avila, D. F., Brannen, P. M., Cline, W. O., Harmon, P. F., and Martin, R. R. 2013. Genetic characterization of Blueberry necrotic ring blotch virus, a novel RNA virus with unique genetic features. J. Gen. Virol. 94:1426-1434.

(Prepared by P. M. Brannen, H. Scherm, and R. R. Martin)

Necrotic Ringspot

Necrotic ringspot is a nematode-vectored disease of highbush blueberry caused by *Tobacco ringspot virus* (TRSV). The disease was first discovered in New Jersey (United States) and has been reported in the United States in Arkansas, Connecticut, Illinois, Michigan, and Oregon and in New Brunswick (Canada) and Chile. Necrotic ringspot causes a slow, steady decline in bush productivity in susceptible cultivars, including Collins, Pemberton, Rubel, and Stanley. TRSV has also been detected in cultivars Bluecrop and Elliott. In 1978, a strain of TRSV was reported to kill stems of the cultivar Jersey, which had earlier been reported as resistant. In 1990, TRSV was reported in half-high blueberry clones in New Brunswick. Necrotic ringspot has not been reported to appear in lowbush or rabbiteye blueberries.

Fig. 128. Necrotic and chlorotic lesions and downward cupping of blueberry leaves with necrotic ringspot, caused by *Tobacco ringspot virus*. (Courtesy K. Cox; © APS)

Fig. 129. 'Stanley' highbush blueberry with greatly reduced leaf size and leaf rosetting from necrotic ringspot, caused by *Tobacco ringspot virus*. (Courtesy D. C. Ramsdell)

Symptoms

Symptoms vary with the cultivar. On most cultivars, affected leaves are misshapen, crinkled, or cupped downward and have small necrotic spots 2–3 mm in diameter (Fig. 128). The necrotic spots may fall out, creating a shot hole effect. Small twigs may also show small necrotic areas. On cultivars Stanley and Concord, however, leaves do not show necrotic spots but they are extremely small and rosetted because of the shortening of the internodes (Fig. 129). On 'Jersey', one strain of TRSV also causes a severe stem dieback, and there are small amounts of regrowth near the crown area. Infected bushes of all cultivars show a steady decline in productivity and may die eventually (Fig. 130).

Causal Agent

TRSV is a member of the nepovirus group, family *Secoviridae*. It is a spherical virus that is 28 nm in diameter. The genome consists of two single-stranded, positive-sense RNAs with molecular masses of 1.4 and 2.4 × 10^6 Da. The protein coat is composed of a single subunit with a molecular mass of 55,000 Da. The virus particles sediment in sucrose density gradients as three similarly sized components: empty protein shells and two nucleoproteins containing different amounts of RNA. The top, middle, and bottom components of TRSV have sedimentation coefficients of 53, 91, and 126S, respectively.

The virus is a good immunogen. Virus-specific antiserum is used to detect TRSV in infected blueberry tissues by enzyme-linked immunosorbent assay (ELISA). Commercial ELISA kits are available for testing blueberry plants for TRSV.

TRSV appears in a wide range of herbaceous and woody hosts. The virus is readily sap transmitted to herbaceous indicator plants as follows: grind 1–2 g of young symptomatic leaves in 0.05 M phosphate buffer, pH 7.0, containing 1.0% (v/v) nicotine alkaloid, and rub the sap onto Carborundum-dusted leaves. *Chenopodium quinoa* develops necrotic primary local lesions and apical dieback within 7–10 days. In 'National Pickling' cucumber (*Cucumis sativus*), chlorotic local lesions (2–4 mm in

Fig. 130. Declining blueberry bushes caused by infection by *Tobacco ringspot virus*, the causal agent of necrotic ringspot. (Courtesy A. M. C. Schilder; © APS)

diameter) develop on inoculated cotyledons and are followed by systemic chlorosis or chlorotic and necrotic lesions on new leaves. *Nicotiana tabacum* cv. Burley shows necrotic ring spots on inoculated leaves and systemic mosaic symptoms on new leaves. Ouchterlony gel double-diffusion tests with TRSV antiserum can confirm the presence of TRSV in herbaceous hosts.

Disease Cycle and Epidemiology

Necrotic ringspot is spread by the dagger nematode, *Xiphinema americanum* sensu lato. TRSV is not transmitted as efficiently by *X. americanum* as is *Tomato ringspot virus*. The virus is acquired within 24 h and is transmitted by both adult and larval stages of the nematode. TRSV particles adhere to the region of the stylet extension and anterior esophagus of the nematode. *X. americanum* can transmit the virus for about 10 weeks after initial acquisition. In blueberry fields, infection foci are roughly circular and may widen at a rate of about 1 m per year.

TRSV is endemic to eastern North America. Certain weeds, such as common dandelion (*Taraxacum officinale*), common plantain (*Plantago major*), and common chickweed (*Stellaria media*), are alternate hosts of TRSV and may serve as reservoirs for acquisition of the virus by dagger nematodes. The weed host range of TRSV is very similar to that of *Tomato ringspot virus*, with one major exception: TRSV infects *Plantago major* rather than *Plantago lanceolata* (narrow-leaved plantain), whereas the reverse is true for *Tomato ringspot virus*.

TRSV may be introduced into a field via infected planting stock, usually by growers who propagate their own stock or buy uncertified stock.

Management

The best control of necrotic ringspot is prevention. Sites should be inspected for the presence of the dagger nematode before planting. If the nematode is present, soil fumigation should be considered (e.g., with 1,3-dichloropropene and chloropicrin), especially if TRSV has been detected in a previous planting. Only certified, virus-tested clean stock should be used in new plantings.

If infection is found in an established field, control is more complicated. The infected bushes must be identified by inspection and ELISA, and all infected bushes should be removed, including the crown and major roots. Removing nonsymptomatic bushes immediately adjacent to the infected area is also recommended. The soil should be fumigated prior to replanting with certified virus-tested plants.

Selected References

Converse, R. H., and Ramsdell, D. C. 1982. Occurrence of tomato and tobacco ringspot viruses and of dagger and other nematodes associated with cultivated highbush blueberries in Oregon. Plant Dis. 66:710-712.

Harrison, B. D., and Murant, A. F. 1977. Nepovirus group. Descriptions of Plant Viruses, No. 185. Commonwealth Mycological Institute and Association of Applied Biologists, Kew, Surrey, England.

Jaswal, A. S. 1990. Occurrence of blueberry leaf mottle, blueberry shoestring, tomato ringspot and tobacco ringspot viruses in eleven halfhigh blueberry clones grown in New Brunswick, Canada. Can. Plant Dis. Surv. 70:113-117.

Lister, R. M., Rainiere, L. C., and Varney, E. H. 1963. Relationships of viruses associated with ringspot diseases of blueberry. Phytopathology 53:1031-1035.

Medina, C., Matus, J. T., Zuniga, M., San-Martin, C., and Arce-Johnson, P. 2002. Occurrence and distribution of viruses in commercial plantings of *Rubus*, *Ribes* and *Vaccinium* species in Chile. Cienc. Invest. Agrar. 33:23-28.

Ramsdell, D. C. 1978. A strain of tobacco ringspot virus associated with a decline disease of Jersey highbush blueberry. Plant Dis. Rep. 62:1047-1051.

Ramsdell, D. C. 1987. Necrotic ringspot of blueberry. Pages 114-116 in: Virus Diseases of Small Fruits. R. H. Converse, ed. U. S. Dep. Agric. Agric. Handb. 631. U.S. Government Printing Office, Washington, DC.

Varney, E. H., and Rainiere, L. C. 1960. Necrotic ringspot, a new virus disease of cultivated blueberry. (Abstr.) Phytopathology 50:241.

Wang, S., Gergerich, R. C., Wickizer, S. L., and Kim, K. S. 2002. Localization of transmissible and nontransmissible viruses in the vector nematode *Xiphinema americanum*. Phytopathology 92:646-653.

(Prepared by A. M. C. Schilder and D. C. Ramsdell)

Peach Rosette Mosaic

Peach rosette mosaic virus (PRMV) is very rare in blueberry and has been reported in this crop only in southeastern Michigan (United States) and southeastern Ontario (Canada). The disease has been observed on highbush blueberry in an experimental planting in which bushes were planted in infested soil at a vineyard site. No information is available on the economic importance of the virus in blueberry, but in grape and peach, PRMV causes a serious disease. Several countries that import blueberry plants consider PRMV a quarantine pathogen and require testing to confirm that blueberry planting material is free from this virus.

Symptoms

To date, peach rosette mosaic has been recognized only in the highbush blueberry cultivars Jersey and Berkeley. In 'Jersey', affected leaves become strap shaped (Fig. 131), distorted into a crescent shape (Fig. 132), or both. In 'Berkeley', leaves develop a spoon shape and are much smaller than normal (Fig. 133). Symptomatic leaves are not equally distributed on infected bushes. No necrotic or chlorotic lesions have been observed on leaves. These symptoms are unique to diseases induced by *Tobacco ringspot virus* and *Tomato ringspot virus*, respectively (see Necrotic Ringspot above and Tomato Ringspot below). No twig or fruit symptoms have been observed with PRMV infection.

Fig. 131. 'Jersey' highbush blueberry with the strap-leaf symptom of peach rosette mosaic, caused by *Peach rosette mosaic virus*. (Courtesy D. C. Ramsdell)

Causal Agent

PRMV is a member of the nepovirus group, family *Secoviridae*. It is a single-stranded, positive-sense RNA virus with isometric particles, 28 nm in diameter, that are angular in profile. There are three types of isometric particles, with sedimentation coefficients of 52S (top), 115S (middle), and 135S (bottom). The middle and bottom components contain two single-stranded RNAs with molecular masses of 2.2 and 2.5 × 10^6 Da, respectively (as determined by gel electrophoresis under denaturing conditions). The coat protein is a single polypeptide with a molecular mass of 55,000 Da.

PRMV can be transmitted by mechanical inoculation and grafting but not by contact between plants. This virus is also transmitted by seed.

PRMV is a moderately good immunogen. Gel double-diffusion tests may be used to identify PRMV in sap from infected *Chenopodium quinoa*. Enzyme-linked immunosorbent assay is used to detect PRMV in blueberry leaf tissues.

PRMV is readily detected by sap inoculations to *C. quinoa* and *C. amaranticolor*. On *C. quinoa*, faint chlorotic local lesions appear on inoculated leaves within 4–10 days and are followed by systemic leaf epinasty and abscission. Noninoculated terminal leaves are mottled and twisted. The terminal growing point usually dies within 2 weeks. On *C. amaranticolor*, faint chlorotic local lesions may or may not develop on inoculated leaves; noninoculated terminal leaves show mottling and the terminal growing point dies after 2 weeks.

Fig. 132. 'Jersey' highbush blueberry with leaf distortion symptoms of peach rosette mosaic, caused by *Peach rosette mosaic virus*. (Courtesy D. C. Ramsdell)

Fig. 133. 'Berkeley' highbush blueberry with leaf distortion symptoms of peach rosette mosaic, caused by *Peach rosette mosaic virus*. (Courtesy D. C. Ramsdell)

Disease Cycle and Epidemiology

There is no direct information concerning the spread of PRMV in highbush blueberry; however, considerable information is available concerning its spread in grapevines. In vineyards in Michigan and peach orchards in Ontario, PRMV is spread by the nematodes *Xiphinema americanum* sensu lato and *Longidorus diadecturus*, respectively. One or more infection foci, roughly circular in shape, are evident in diseased plantings. These may expand at a rate of about 1 m per year in all directions, which is a typical rate of spread for a nepovirus in a woody perennial. The virus infects several common weeds, including Carolina horse nettle (*Solanum carolinense*), common dandelion (*Taraxacum officinale*), and curly dock (*Rumex crispus*). PRMV is seedborne in these weeds and in grape.

Management

Because peach rosette mosaic is so rare, control measures have not been developed for blueberry. However, care must be taken when planting blueberry in former vineyard or peach orchard sites that previously were found to be infected with PRMV. The recommendations concerning the control of PRMV that have been developed for grape should also prove useful for blueberry. In vineyards, long-term fallowing of soils has not been an effective means of control because the dagger nematode (*X. americanum*) survives and remains viruliferous on the remaining grape roots. Thus, as many roots as possible should be removed from infested sites, and the soil should be thoroughly fumigated. Before new plantings are established, nematode surveys should be done. Only virus-tested certified planting stock should be planted. (See Necrotic Ringspot above for more detailed information on the control of nematode-vectored virus diseases.)

Selected References

Allen, W. R., Van Schagen, J. G., and Eveleigh, E. S. 1982. Transmission of peach rosette mosaic virus to peach, grape, and cucumber by *Longidorus diadecturus* obtained from diseased orchards in Ontario. Can. J. Plant Pathol. 4:16-18.

Dias, H. F. 1975. Peach rosette mosaic virus. Descriptions of Plant Viruses, No. 150. Commonwealth Mycological Institute and Association of Applied Biologists. Kew, Surrey, England.

Ramsdell, D. C. 1987. Peach rosette mosaic virus in blueberry. Page 116 in: Virus Diseases of Small Fruits. R. H. Converse, ed. U. S. Dep. Agric. Agric. Handb. 631. U.S. Government Printing Office, Washington, DC.

Ramsdell, D. C., and Gillett, J. M. 1981. Peach rosette mosaic virus in highbush blueberry. Plant Dis. 65:757-758.

Ramsdell, D. C., and Myers, R. L. 1974. Peach rosette mosaic virus, symptomatology and nematodes associated with grapevine 'degeneration' in Michigan. Phytopathology 64:1174-1178.

Ramsdell, D. C., and Myers, R. L. 1978. Epidemiology of peach rosette mosaic virus in a Concord grape vineyard. Phytopathology 68:447-450.

(Prepared by D. C. Ramsdell; Revised by A. M. C. Schilder)

Red Ringspot

Red ringspot disease was first described in highbush blueberry in New Jersey (United States) as ringspot in 1950. The causal agent of the disease was shown to be graft transmissible in 1954 and was, therefore, assumed to be a virus or virus complex. The disease, now called red ringspot, was confirmed to be caused by a virus, designated *Blueberry red ringspot virus* (BRRSV). The method of natural spread of BRRSV is unknown. The disease continues to be a problem in New Jersey, where it is actively spreading, and plantings in Arkansas

(United States) also have been found to have a high incidence of BRRSV infection. The disease has also been reported in the United States in Michigan, Connecticut, Massachusetts, New York, North Carolina, and Oregon, but at very low incidence. It is not known to spread in Michigan, suggesting that the vector of the virus may not be present there. Impact in New Jersey has been difficult to document and many cultivars appear to yield a normal crop when infected. One report from Michigan found that infected 'Blueray' plants yielded smaller berries and exhibited an overall 25% crop loss.

The cultivars that commonly show symptoms of red ringspot are Blueray, Bluetta, Burlington, Cabot, Coville, Darrow, Earliblue, and Rubel. The cultivar Jersey is apparently immune, and the cultivar Bluecrop exhibits excellent field resistance.

Red ringspot of blueberry is very similar to cranberry ringspot disease. Electron microscopy of cranberry leaf tissues with ringspot disease has revealed viral inclusion bodies that are identical to those in BRRSV-infected blueberry leaves. It was recently shown that the red ringspot diseases of blueberry and cranberry are caused by the same virus.

Symptoms

One-year-old or older stems often develop red ring spots or red blotches that are roughly circular in outline and 4–6 mm in diameter. Reddish brown circular spots 3–5 mm in diameter appear on older leaves in mid- to late summer (Fig. 134). These spots are most prominent on the upper surface of the leaf but, in some cultivars, can be visible on the underside of the leaf as well. Younger leaves usually remain symptomless. The powdery mildew fungus (*Microsphaera vaccinii*) can cause similar symptoms, but the fungus-induced leaf spots are prominent on both sides of the leaves and tend not to be circular. The virus-induced red ring spots can coalesce, causing the leaves to appear prematurely red during late summer and early autumn. Circular, light-colored areas of blotching about 2–3 mm in diameter may develop on affected fruit of some cultivars (Fig. 135). Similar symptoms have been seen in rabbiteye blueberry stems and fruit. The cultivar Ozarkblue produces misshapen fruit when infected with BRRSV (Fig. 136). The cultivar Bluetta may exhibit a disorder on leaves only that resembles red ringspot, but this disorder is probably genetic in nature.

Fig. 134. Red ringspot symptoms, caused by *Blueberry red ringspot virus*, on 'Duke' blueberry leaves. (Courtesy J. J. Polashock)

Causal Agent

BRRSV is a member of the genus *Soymovirus* in the family *Caulimoviridae* of reverse-transcribing, DNA-containing viruses. Virus particles are spherical and 42–46 nm in diameter. The viral genomic nucleic acid is a circular double-stranded DNA of 8,303 base pairs. The coat protein subunit is a single polypeptide with a molecular mass of approximately 44,000 Da. BRRSV differs from other members of the family *Caulimoviridae* in that there are two components in cesium chloride gradients, with buoyant densities of 1.3 and 1.4 g/cm^3. Typical caulimovirus inclusion bodies are found in ultrathin sections made through leaf spots.

BRRSV has not been shown to be sap transmissible. Enzyme-linked immunosorbent assay (ELISA) reagents have been developed and were shown to detect BRRSV in symptomatic leaves but not in symptomless leaves from an infected bush; however, the BRRSV ELISA test has not been robust and is not currently recommended. The virus can more reliably be detected using polymerase chain reaction with BRRSV-specific primers.

Disease Cycle and Epidemiology

Natural spread of red ringspot is apparent in New Jersey but not in Michigan. Circumstantial evidence in New Jersey suggests

Fig. 135. Fruit symptoms of red ringspot, caused by *Blueberry red ringspot virus*, on a 'Duke' blueberry. (Courtesy J. J. Polashock)

Fig. 136. Misshapen fruit as a result of red ringspot, caused by *Blueberry red ringspot virus*, on 'Ozarkblue' blueberries. (Courtesy W. O. Cline)

that the vector may be a mealybug (*Dysmicoccus* sp.). Mealybugs are common inhabitants of highbush blueberry plantings in New Jersey and are rarely, if ever, found in Michigan plantings. The blueberry aphid, *Illinoia pepperi*, is abundant in blueberry bushes in both states. Attempts to transmit BRRSV have not been reproducible with mealybugs or with the blueberry aphid.

Management

Only virus-tested certified stock should be planted. Diseased plants in should be rogued and destroyed. Since natural autumn coloration and other disorders, such as powdery mildew, can be mistaken for BRRSV infection, plants should be tested before roguing.

Selected References

Gillett, J. M., and Ramsdell, D. C. 1988. Blueberry red ringspot virus. Descriptions of Plant Viruses, No. 327. Commonwealth Mycological Institute and Association of Applied Biologists, Kew, Surrey, England.

Glasheen, B. M., Polashock, J. J., Lawrence, D. M., Gillett, J. M., Ramsdell, D. C., Vorsa, N., and Hillman, B. I. 2002. Cloning, sequencing, and promoter identification of Blueberry red ringspot virus, a member of the family *Caulimoviridae* with similarities to the "Soybean chlorotic mottle-like" genus. Arch. Virol. 147:2169-2186.

Hull, R., Geering, A., Harper, G., Lockhart, B. E., and Schoelz, J. E. 2005. Family *Caulimoviridae*. Pages 385-396 in: Virus Taxonomy: Eighth Report of the International Committee on the Taxonomy of Viruses. Elsevier Academic Press, San Diego, CA.

Hutchinson, M. T., and Varney, E. H. 1954. Ringspot: A virus disease of cultivated blueberry. Plant Dis. Rep. 38:260-263.

Kim, K. S., Ramsdell, D. C., Gillett, J. M., and Fulton, J. P. 1981. Virions and ultrastructural changes associated with blueberry red ringspot disease. Phytopathology 71:673-678.

Polashock, J. J., Ehlenfeldt, M. K., and Crouch, J. A. 2009. Molecular detection and discrimination of *Blueberry red ringspot virus* strains causing disease in cultivated blueberry and cranberry. Plant Dis. 93:727-733.

Ramsdell, D. C., Kim, K. S., and Fulton, J. P. 1987. Red ringspot of blueberry. Pages 121-123 in: Virus Diseases of Small Fruits. R. H. Converse, ed. U. S. Dep. Agric. Agric. Handb. 631. U.S. Government Printing Office, Washington, DC.

(Prepared by J. J. Polashock and B. I. Hillman)

Scorch

Blueberry scorch, caused by *Blueberry scorch virus* (BlScV), is becoming one of the most pervasive diseases of highbush blueberry. The disease was first described as sheep pen hill disease (SPHD) in New Jersey (United States) in 1970, but it was not at that time shown to be a virus-associated disease. Blueberry scorch was reported as a virus-associated disease in 1980 in a planting of 'Berkeley' highbush blueberry near Puyallup, Washington (United States). SPHD and blueberry scorch were later shown to be caused by different strains of BlScV than those described from Washington. The disease has since been observed in other highbush blueberry cultivars on the west coast of North America in Oregon (United States), Washington, and British Columbia (Canada). On the east coast of North America, the disease has been reported in the United States in New Jersey, Connecticut, and Massachusetts. In 2009, BlScV was detected in field-grown blueberry plants in Michigan, the largest blueberry-producing state in the United States. The disease also has been reported in Italy and is likely to be in other parts of Europe. The virus has infected all cultivars of highbush blueberry that have been tested in grafting experiments. BlScV has also been found in asymptomatic cranberry plants and black huckleberry. Scorch has not been reported in lowbush or rabbiteye blueberry.

Symptoms

Symptoms vary depending on the strain of the virus as well as the plant cultivar. The symptoms of scorch range from the complete necrosis of flowers (Fig. 137) and leaves, which eventually leads to death of the bush in some cultivars, to symptomless infection that does not cause significant yield loss in other cultivars. From experiments done in the Pacific Northwest in North America, the cultivars Atlantic, Berkeley, Bluejay, Blueray, Dixi, Herbert, Lateblue, Pemberton, Rubel, Spartan, and Weymouth were shown to exhibit a flower and leaf necrosis combined with marginal leaf chlorosis. Cultivars Olympia and Stanley exhibited only the marginal leaf chlorosis (Fig. 138), while cultivars Bluecrop, Concord, Jersey, and Washington were symptomless.

Blighting appears just before the corolla opens. Blighted flowers are initially brown but bleach to gray over time. On the west coast of North America, the scorched blossoms often are retained throughout the summer. In cultivars that show the severe blossom blight symptom, the twigs die back 4–10 cm. On the east coast, symptoms are similar, but tend to be exhibited on more cultivars; only 'Jersey' was found to be asymptomatic. An oak-leaf line pattern often appears on the leaves in early autumn (Fig. 139) and blossoms are not retained. Such a pattern is often difficult to detect and may be evident on only a few

Fig. 137. 'Elliott' highbush blueberry showing blighted blossoms resulting from blueberry scorch, caused by *Blueberry scorch virus*. (Courtesy J. J. Polashock)

Fig. 138. Blueberry leaves with chlorotic margins resulting from blueberry scorch, caused by *Blueberry scorch virus*. (Courtesy P. R. Bristow; © APS)

leaves. Infected plants generally decline each year and eventually produce little or no fruit.

There is a latent period of 1–2 years between infection and the expression of symptoms. Symptoms usually appear on one or a few branches and progressively spread to affect the entire bush in subsequent years. The productivity of individual plants declines as more of the plant becomes affected. Affected bushes show symptoms every year, but severity can vary considerably from year to year. This latter property may be problematic from the control perspective, since a grower may be reluctant to remove an infected plant that bears very poorly one year and better the next.

Symptoms caused by BlScV can be confused with injury caused by spring frost or with mummy berry blossom blight or Botrytis blossom blight. Because no distinctive symptoms may be associated with this virus, it is best to test bushes by enzyme-linked immunosorbent assay or other positive virus identification means to confirm the presence or absence of the virus.

Causal Agent

BlScV is a member of the genus *Carlavirus* of the family *Betaflexiviridae*. The virions of BlScV are flexuous, rod-shaped particles with a monopartite, positive-sense, single-stranded RNA genome of approximately 8,500 nucleotides and a coat protein with a molecular mass of approximately 35.2 kDa, with some variation in these numbers depending on virus strain. The virus is related to other members of the carlavirus group, most closely to *Lily symptomless virus* and *Potato virus S*. BlScV has been transmitted mechanically and by several aphid species to the experimental herbaceous hosts *Chenopodium quinoa* and *Chenopodium amaranticolor*. Several strains of BlScV have been sequenced in their entirety, and variation among strains can be considerable.

Disease Cycle and Epidemiology

BlScV is transmitted by aphids experimentally and in the field. Experiments done in western North America demonstrated that the blueberry aphids *Ericaphis fimbriata* and *Illinoia pepperi* were capable of virus transmission. An associated study in which trap plants ('Berkeley') were placed among infected plants in commercial fields at 2-week intervals during the growing season (mid-April to mid-October) showed that transmission takes place between mid-May and mid-August. Studies in New Jersey have shown that any of several aphid species may transmit the virus with differing efficiencies and that *Ericaphis fimbriata* was the most efficient. Transmission of the virus is associated with colonization of the trap plants by aphids. Like other viruses of this type, transmission by aphids is thought to be nonpersistent, with rapid acquisition of the virus from infected plants and retention in the aphid for only a brief time. Studies with *Ericaphis fimbriata* as the vector and *Chenopodium quinoa* as the experimental host plant demonstrated that the virus can be acquired from an infected plant within 5 min and transmitted immediately to an uninfected plant. Retention time within aphids has not been measured.

Fig. 139. Highbush blueberry leaf in the autumn with an oak-leaf line pattern symptom resulting from scorch, caused by *Blueberry scorch virus*. (Courtesy J. J. Polashock)

The virus spreads quickly in the field in a radial pattern from a point source. In the absence of disease control measures, the number of infected bushes in a field approximately doubles each year until about 50% of the bushes are infected; the rate subsequently declines, but eventually all the bushes become infected.

Management

The single most important component of an effective control program for BlScV is initiating a planting with virus-free stock. Because BlScV is symptomless in several cultivars, it is crucial that virus-tested plants be used when a field is established or enlarged. Certification programs are especially important in curbing the spread of this virus within growing areas and across state or international borders.

If plants in a field are infected with BlScV, an aggressive aphid control program combined with removal of diseased bushes should be used to eliminate the virus from a field. Tolerant cultivars can be a potential drawback to this strategy. Since tolerant bushes are symptomless, growers are not likely to remove them but rather fill in vacant spaces with more tolerant plants. Unless all of the infected plants are removed, tolerant plants remaining in the field serve as a continuous source of inoculum. In such cases, a continuous aphid control program is necessary to prevent the virus from spreading to other cultivars that do develop symptoms.

Field-to-field spread is limited to rather short distances (less than 1 km), but the virus moves readily to neighboring fields that are separated by 5–10 m. This pattern of movement to nearby farms is consistent with the virus spreading primarily via an aphid vector in a nonpersistent fashion. Mechanical harvesters may hasten the spread of BlScV along rows by physically moving aphids from plant to plant, but given the difficulty of experimental mechanical transmission of the virus, it is unlikely that this is an important direct mechanism of spread.

Selected References

Bernardy, M. G., Dubeau, C. R., Braun, A., Harlton, C. E., Bunckle, A., Wegener, L. A., Lowery, D. T., and French, C. J. 2005. Molecular characterization and phylogenetic analysis of two distinct strains of *Blueberry scorch virus* from western Canada. Can. J. Plant Pathol. 27:581-591.

Bristow, P. R., Martin, R. R., and Windom, G. E. 2000. Transmission, field spread, cultivar response, and impact on yield in highbush blueberry infected with *Blueberry scorch virus*. Phytopathology 90:474-479.

Cavileer, T. D., Halpern, B. T., Lawrence, D. M., Podleckis, E. V., Martin, R. R., and Hillman, B. I. 1994. Nucleotide sequence of the carlavirus associated with blueberry scorch and similar diseases. J. Gen. Virol. 75:711-720.

Ciuffo, M., Pettiti, D., Gallo, S., Masenga, V., and Turina, M. 2005. First report of *Blueberry scorch virus* in Europe. Plant Pathol. 54:565.

Lawrence, D. L., and Hillman, B. I. 1994. Synthesis of infectious transcripts of blueberry scorch carlavirus *in vitro*. J. Gen. Virol. 75:2509-2512.

Martin, R. R., and Bristow, P. R. 1988. A carlavirus associated with blueberry scorch disease. Phytopathology 78:1636-1640.

Wegener, L. A., Punja, Z. K., and Martin, R. R. 2004. First report of *Blueberry scorch virus* in cranberry in Canada and the United States. Plant Dis. 88:427.

Wegener, L. A., Martin, R. R., Bernardy, M. G., MacDonald, L, and Punja, Z. K. 2006. Epidemiology and identification of strains of

Blueberry scorch virus on highbush blueberry in British Columbia, Canada. Can. J. Plant Pathol. 28:250-262.

(Prepared by J. J. Polashock and B. I. Hillman)

Shock

Blueberry shock disease was first observed in Washington (United States) in 1987 and initially was confused with blueberry scorch, caused by *Blueberry scorch virus* (BlScV). However, in the case of shock, affected plants produced a second flush of leaves after flowering and the plants appeared normal by late summer, except for the lack of fruit. In subsequent years, plants flowered and fruited normally and did not exhibit any symptoms. BlShV has been identified throughout the Pacific Northwest (Oregon and Washington [United States] and British Columbia [Canada]) and recently in Michigan (United States) (2009) and Nova Scotia (Canada) (2008).

Symptoms

Blueberry shock virus (BlShV) can cause a complete flower and leaf necrosis of highbush blueberry during the bloom period. The symptoms caused by BlShV are similar to those caused by BlScV. Affected young leaves wilt, with blackened veins and black streaks down the petiole, or they blight to an orange color. Blighted blossoms look identical to those blighted by BlScV. Blighted blossoms and leaves drop, so that by early summer, the affected plants may be completely defoliated (Fig. 140). In the early stages of leaf necrosis, the leaves may have irregular spots (Fig. 141), and at mid- to late bloom, the affected leaves can turn deep purple in some cultivars. As the season progresses, the plants appear to recover since a second flush of leaves is produced. By late summer, these plants appear normal, except that they produce little or no fruit and, prior to harvest, may appear more vigorous than adjacent healthy plants (Fig. 142). The severe symptoms only develop once on a given bush or part of a bush. In some cultivars, bushes may exhibit a red leaf spot symptom the year following the severe shock reaction. In 2010 and 2011, under cool and wet conditions during April and May, multiple cultivars exhibited a red leaf symptom on plants that had been in the recovered phase the previous year. It is not clear whether these environmental conditions are responsible for this red leaf symptom in recovered plants.

Leaves that do not blight are often slightly chlorotic, with thin red ring spots. The red ring spots are always visible on both the upper and lower surfaces of the leaves, in contrast to the ring spots caused by *Blueberry red ringspot virus*, which may be visible only on the upper surface of leaves of some cultivars.

In young plants, the entire bush usually develops symptoms the year after infection. However, in larger bushes often one or two canes show symptoms the year after infection, and the following year, those canes will appear normal and the rest of the bush will exhibit symptoms. Thus, it is common for bushes in the field to exhibit symptoms for 2 years and, in some cases, even 3 years. However, in such cases, any one part of the bush only develops the severe shock reaction once.

BlShV has been detected in all blueberry cultivars tested, including Berkeley, Bluecrop, Bluehaven, Bluejay, Blueray, Bluetta, Collins, Darrow, Dixi, Duke, Earliblue, Elizabeth, Elliott, Herbert, Jersey, Lateblue, Meader, Northland, Patriot, Pemberton, Spartan, Stanley, Toro, Weymouth, and 1613A. The symptom reaction of some cultivars is unknown because the infected plants tested were from commercial fields, and some of the plants may have recovered before the testing. 'Berkeley' plants dually infected by BlShV and BlScV may exhibit a red leaf spot symptom (Fig. 143).

Causal Agent

BlShV is a member of the ilarvirus group, a group of single-stranded, positive-sense RNA viruses with tripartite genomes. The virions are quasi-isometric spheres 26–29 nm in diameter. The coat protein subunit has a molecular mass of 27,300 Da,

Fig. 140. Highbush blueberry planting with early summer defoliation and blossom blighting and drop resulting from shock, caused by *Blueberry shock virus*. (Courtesy P. R. Bristow; © APS)

Fig. 141. Early stages of leaf necrosis in 'Bluegold' highbush blueberry resulting from shock, caused by *Blueberry shock virus*. (Courtesy R. R. Martin)

Fig. 142. 'Duke' blueberry showing symptoms of blueberry shock disease (right), caused by *Blueberry shock virus*, compared with a healthy plant (left). (Courtesy R. R. Martin)

and the four RNAs have molecular masses of 1.03, 0.84, 0.57, and 0.30 × 10^6 Da. Polyclonal antiserum prepared against this virus gave a weak cross-reaction with *Prunus necrotic ringspot virus* and *Apple mosaic virus* but did not react with 10 other ilarviruses.

BlShV can be transmitted mechanically to *Nicotiana clevelandii* from blossoms of infected bushes ground in 0.05 M phosphate buffer, pH 7.2. It has not been possible to transmit the virus mechanically from other blueberry tissues. The virus can be transmitted to a number of *Nicotiana* spp. and can be maintained in *Nicotiana* plants in a greenhouse at temperatures below 22°C but not at higher temperatures. Pollen collected from infected bushes and stored at –80°C can serve as an excellent source of inoculum for experimental purposes.

Disease Cycle and Epidemiology

Trap plants ('Berkeley') placed in infected blueberry fields at weekly intervals during the growing season (mid-April through September) became infected only during the bloom period. In transmission electron microscopy using gold-labeled protein A in combination with antibodies to the virus, spherical virus particles have been visualized on the surface of and inside pollen grains from infected blueberry flowers but not pollen grains from flowers of healthy plants. This suggests that the virus is pollen-borne, which is consistent with other members of the ilarvirus group. Furthermore, BlShV is readily transmitted to trap plants ('Berkeley') when they are caged with infected field plants and honeybees but only occasionally transmitted when honeybees are excluded. It is not known whether insects are involved in the horizontal transmission of BlShV, as has been suggested for *Tobacco streak virus*. The symptomless infected plants serve as a source of inoculum for the continued spread of the virus. If left uncontrolled, the disease spreads throughout a field but, eventually, the entire field recovers and a good crop is produced in well-managed fields.

Depending on the cultivar, the virus spreads rapidly in the field in a radial pattern from an infection focus. The number of infected plants doubles every year during the early stages of disease spread within a field. In some cultivars, such as Berkeley, Bluetta, Bluegold, Duke, Earliblue, Liberty, Draper, Pemberton, and Reka, the virus spreads very quickly through a field, while in other cultivars, such as Bluecrop and Blueray, the virus spreads more slowly. BlShV has not been detected in native vegetation surrounding blueberry fields infected with the virus nor in weeds within infected fields.

Management

The introduction of BlShV into new areas is controlled primarily through the use of certified virus-tested planting stocks.

Fig. 143. Highbush blueberry leaves with a red leaf spot symptom resulting from double infection by *Blueberry scorch virus* and *Blueberry shock virus*. (Courtesy P. R. Bristow; © APS)

Care must be taken not to use planting stock from a field that is infected with BlShV but has recovered and, therefore, does not show symptoms of the disease.

There is interest in the Pacific Northwest to produce plants that have been preinoculated with BlShV so that growers would start with plants that have already gone through the recovery phase. Experiments are underway to propagate from infected mother plants to determine whether the progeny would be "protected" and not develop symptoms in grower's fields. This presents the potential problem of careful tracking of plant material to ensure the "protected" plants are not distributed outside of the area where the disease is present.

Since the virus is widespread in the Pacific Northwest, eradication is not recommended when the disease first is observed in a field. The current recommendation is to let the disease run its course, since a reoccurrence of symptoms has not been observed in fields infected for up to 15 years. In cultivars within which the virus spreads quickly, an entire 10- to 20-ha field may be infected over 3–5 years, whereas in cultivars within which the disease spreads slowly, it takes more than 10 years for bushes throughout the field to become infected.

Current-season infections cannot be detected by enzyme-linked immunosorbent assay (ELISA) until the flower buds begin to swell the following spring. To avoid false-negative results, a sample for an ELISA test should consist of four or five buds or leaves from several locations on the bush. The virus can also be detected by reverse transcription-polymerase chain reaction, although again not during the year of infection.

Selected References

Bristow, P. R., and Martin, R. R. 1999. Transmission and the role of honeybees in field spread of blueberry shock ilarvirus, a pollen-borne virus of highbush blueberry. Phytopathology 89:124-130.

MacDonald, S. G., Martin, R. R., and Bristow, P. R. 1991. Characterization of an ilarvirus associated with a necrotic shock reaction in blueberry. Phytopathology 81:210-214.

Martin, R. R., Bristow, P. R., and Wegener, L. A. 2006. Scorch and shock: Emerging virus diseases of highbush blueberry and other *Vaccinium* species. Acta Hortic. 715:463-467.

(Prepared by R. R. Martin)

Shoestring

Shoestring, an aphid-vectored disease caused by *Blueberry shoestring virus* (BSSV), is the most widespread virus disease of highbush blueberry. It was first reported in New Jersey in 1957 and was later found in Michigan, New York, North Carolina, and Washington (United States). Shoestring also is found in lowbush blueberry in Nova Scotia, Canada, but has not been observed in rabbiteye blueberry. The disease has not been reported from other parts of the world.

In 1981, an estimated 145,000 bushes on 4,000 ha were infected in Michigan, which resulted in bush and crop losses of more than $3 million. By 2008, the disease was still widespread in this state, particularly in old 'Jersey' fields. Yield reductions of up to 25% have been documented on infected bushes. In addition, a high percent of the fruit from infected bushes is reddish purple at maturity rather than the desired blue color, which lowers the fruit grade.

Shoestring has been observed in the highbush blueberry cultivars Blueray, Burlington, Coville, Earliblue, Elliott, Jersey, June, Rancocas, Rubel, and Weymouth. Cultivars Jersey, Elliott, and Blueray are particularly susceptible. In a survey of 26 counties in the Upper and Lower Peninsula of Michigan, BSSV was found in 6–13% of wild lowbush blueberry clones and 12–50% of highbush blueberry bushes tested in forested areas.

Symptoms

There is a latent period of 4 years between infection and the expression of symptoms in the field. Several different symptoms are associated with shoestring. The most prominent and reliable symptom is elongated reddish streaks about 3–20 mm long on current-year and 1-year-old stems, especially on the side directly exposed to the sun (Fig. 144). During the blossom period, flowers on infected bushes exhibit pinkish to reddish corollas or streaks on the corollas. (Note that in the cultivar Blueray, which has naturally reddish corollas, this trait may be mistaken for a symptom of shoestring.)

Infected leaves are elongated or straplike (Fig. 145); hence the name shoestring. This symptom may be apparent on many of the leaves on the bush or, in some cases, on just a few leaf clumps in the crown area, low in the bush. Some leaves may show a red oak-leaf pattern on the leaf blade (Fig. 146). Some leaves may be crescent shaped and partially or entirely reddened (Fig. 146).

Infected bushes bear less fruit than healthy bushes. In addition, the fruit does not ripen normally and remains reddish purple instead of turning blue (Fig. 147). On actively growing bushes that have been infected for many years, the terminal third of the stems may be crooked. Infected bushes slowly decline over time and may be stunted (Fig. 148).

Fig. 144. Red streaks on current-year stems of 'Jersey' highbush blueberry resulting from shoestring, caused by *Blueberry shoestring virus*. (Courtesy M. Longstroth)

Fig. 145. Narrow, straplike leaves are a symptom resulting from shoestring, caused by *Blueberry shoestring virus*, in highbush blueberry. (Courtesy M. Longstroth)

Causal Agent

BSSV is a member of the sobemovirus group since it shares physicochemical properties described for other members of this group. Unlike other sobemoviruses, however, BSSV has an aphid vector. BSSV is an isometric virus 28 nm in diameter. The genome is monopartite (composed of single-stranded, positive-sense RNA), with a molecular mass of 1.45×10^6 Da. The complete genome is 4,200 nucleotides long. The coat protein is a single polypeptide with a molecular mass of 30,000 Da. Virions contain 20% nucleic acid and 80% protein.

Fig. 146. Red oak-leaf pattern on leaves and curved or crescent-shaped leaves of 'Jersey' highbush blueberry infected by *Blueberry shoestring virus*, the causal agent of shoestring. (Courtesy M. Longstroth)

Fig. 147. 'Jersey' highbush blueberry with reddish purple fruit, a symptom resulting from shoestring, caused by *Blueberry shoestring virus*. (Courtesy D. C. Ramsdell)

BSSV is sap transmissible, with difficulty, to small succulent blueberry seedlings but only when purified virus is mechanically inoculated at a high concentration (0.1–0.5 mg/ml). The virus is not mechanically transmissible to herbaceous hosts; several attempts to mechanically inoculate 44 herbaceous indicator plants with purified virus have failed.

BSSV is serologically unrelated to other viruses. In agar double-diffusion tests, the virus gave no reaction against antisera to 33 viruses with isometric particles, including members of the bromovirus, comovirus, cucumovirus, ilarvirus, nepovirus, tombusvirus, and tymovirus groups and several unclassified viruses. Antiserum to BSSV did not react with any of 24 viruses belonging to these groups. The enzyme-linked immunosorbent assay (ELISA) readily detects BSSV in infected leaf, blossom, and fruit bud tissues. Commercial ELISA kits are available for testing potentially infected bushes.

Fig. 148. Stunting of 'Jersey' highbush blueberry (on the right) caused by infection with *Blueberry shoestring virus*, the causal agent of shoestring. (Courtesy R. Isaacs; © APS)

The virus is found in the mesophyll, epidermis, meristem, xylem, phloem, and palisade parenchyma; within cells, it is found in cytoplasm, nuclei, and cell vacuoles. When examined in a transmission electron microscope, epidermal cells of symptomatic blueberry leaf tissue have viruslike particles (VLPs) in small crystalline arrays and as individual particles. In palisade parenchyma cells, VLPs were found in small cytoplasmic vacuoles and were also associated with chloroplast membranes. VLPs also were found in xylem parenchyma, where they were confined to small, clear areas at the periphery of the cells but not in mature xylem vessels or leaf phloem tissue. In root tissues, VLPs were present in varying amounts in almost all xylem parenchyma cells examined, usually in greater numbers than in leaf xylem parenchyma.

Disease Cycle and Epidemiology

BSSV is vectored by the blueberry aphid, *Illinoia pepperi*, in a persistent circulatory manner. The aphid is monoecious and overwinters as eggs on blueberry bushes. In Michigan, the first aphid population peak is in mid- to late May. The aphids colonize the terminal leaves and succulent shoot tips, especially in the crown area of the bush. The vast majority of the aphids are apterous (wingless) until the late summer and autumn, when a low percentage become alate (winged).

Epidemiological studies have shown that field transmission of BSSV begins when the first aphid population emerges in the spring and ends just before leaf drop in the autumn (Fig. 149). The disease tends to spread down the row from bush to bush. In most plantings, bushes touch each other within the row (0.9–1.2 m apart) but not between rows (3 m apart). Aphids move from infected bushes to adjacent healthy ones as they feed.

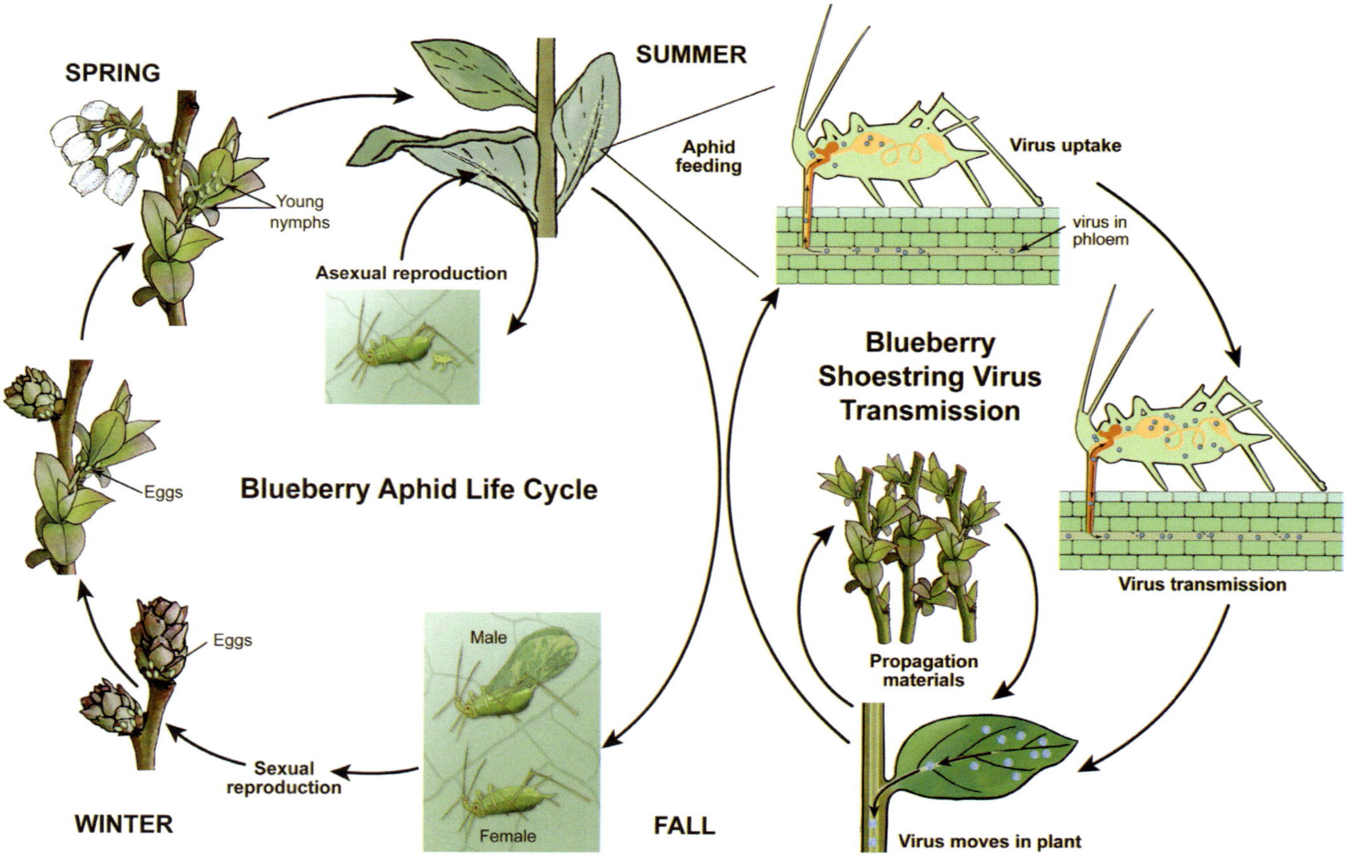

Fig. 149. Life cycle of the blueberry aphid, *Illinoia pepperi*, and transmission of *Blueberry shoestring virus*, the causal agent of shoestring, in blueberry. (Courtesy M. Cameron; Reproduced, by permission of Michigan State University, from Isaacs et al. 2008)

Aphids are transported longer distances by over-the-row mechanical harvesters. Such movement was tracked in a study in which a bush was exposed to rubidium and aphids were captured at regular intervals down the row, immediately following the harvester, and tested. Aphids with rubidium were detected up to 67 bushes from the point source.

Management

Because of the long latent period before symptoms become obvious, roguing infected bushes is not a commercially feasible practice. However, when plantings are established from infected cuttings, symptoms may be visible sooner and roguing would be advised. Well-timed applications of effective aphicides greatly impede further spread of BSSV within the field and are the principal control strategy. Insecticide timing is important. The first application should be made during late May and early June, when aphids are first evident on terminals in the crown area. Spraying with ground equipment is more effective than by airplane. In custom mechanical harvesting, the harvester should be thoroughly washed with water to remove aphids before it enters another field.

Virus-tested planting stock should always be planted as a preventative measure. The only highbush cultivar that shows field resistance to BSSV is Bluecrop; the cultivar Atlantic also shows good field resistance but is not widely planted. Based on artificial inoculation, the southern highbush cultivars Georgia Gem and Misty, and the rabbiteye cultivar Brightwell are resistant, whereas the rabbiteye cultivar Climax is susceptible.

Selected References

Acquaah, T., Ramsdell, D. C., and Hancock, J. F. 1995. Resistance to blueberry shoestring virus in southern highbush and rabbiteye cultivars. HortScience 30:1459-1460.

Hancock, J. F., Morimoto, K. M., Pritts, M. P., and Ramsdell, D. C. 1986. Blueberry shoestring virus in natural populations of highbush and lowbush blueberry. HortScience 21:1059-1060.

Hartmann, J. X., Bath, J. E., and Hooper, G. R. 1973. Electron microscopy of viruslike particles from shoestring-diseased highbush blueberry, *Vaccinium corymbosum* L. Phytopathology 63:432-436.

Isaacs, R., Schilder, A., Miles, T., and Longstroth, M. 2008. Michigan Blueberry Facts: Blueberry Aphid & Blueberry Shoestring Virus. Mich. State Univ. Ext. Bull. E-3050.

Lesney, M. S., Ramsdell, D. C., and Sun, M. 1978. Etiology of blueberry shoestring disease and some properties of the causal virus. Phytopathology 68:295-300.

Morimoto, K. M., and Ramsdell, D. C. 1985. Aphid vector population dynamics and movement relative to field transmission of blueberry shoestring virus. Phytopathology 75:1217-1222.

Morimoto, K. M., Ramsdell, D. C., Gillett, J. M., and Chaney, W. G. 1985. Acquisition and transmission of blueberry shoestring virus by its aphid vector *Illinoia pepperi*. Phytopathology 75:709-712.

Ramsdell, D. C. 1979. Physical and chemical properties of blueberry shoestring virus. Phytopathology 69:1087-1091.

Ramsdell, D. C. 1979. Blueberry shoestring virus. Descriptions of Plant Viruses, No. 204. Commonwealth Mycological Institute and Association of Applied Biologists, Kew, Surrey, England.

Ramsdell, D. C. 1979. Blueberry shoestring. Pages 103-105 in: Virus Diseases of Small Fruits. R. H. Converse, ed. U. S. Dep. Agric. Agric. Handb. 631. U.S. Government Printing Office, Washington, DC.

Terhune, B. T., Ramsdell, D. C., Klomparens, K. L., and Hancock, J. F. 1991. Quantification of blueberry shoestring virus RNA and antigen in its aphid vector, *Illinoia pepperi*, during acquisition, retention, and transmission. Phytopathology 81:1096-1102.

Varney, E. H. 1957. Mosaic and shoestring, virus diseases of cultivated blueberry in New Jersey. Phytopathology 47:307-309.

(Prepared by D. C. Ramsdell;
Revised by A. M. C. Schilder and J. J. Polashock)

Tomato Ringspot

Tomato ringspot virus (ToRSV) causes a nematode-vectored disease, tomato ringspot, of highbush blueberry in New York, Oregon, Pennsylvania, and Washington in the United States but has not been reported elsewhere in the world on this crop or on other *Vaccinium* spp. ToRSV can cause a serious and, in some cultivars, lethal disease in blueberry.

Symptoms

The occurrence and intensity of tomato ringspot symptoms vary among blueberry cultivars. Infected leaves are often cupped and malformed, with circular spots 2–5 mm in diameter. Young leaves may be strap shaped and mottled (Fig. 150) and confused with leaves infected with *Blueberry shoestring virus*. Necrotic brown spots 2–5 mm in diameter may appear on infected canes (Fig. 151). Infection is rarely uniform within

Fig. 150. Young leaves of 'Earliblue' highbush blueberry infected with *Tomato ringspot virus*, the causal agent of tomato ringspot. Note the strap-leaf symptom and mottling. (Courtesy R. H. Converse)

Fig. 151. Shoots of 'Earliblue' highbush blueberry infected with *Tomato ringspot virus*, the causal agent of tomato ringspot. Note the chlorotic or necrotic leaf lesions and necrotic stem lesions. (Courtesy R. H. Converse)

a plant: some shoots may be defoliated, others may have necrotic leaves, and still others may appear normal (Figs. 152 and 153). Fruit production and quality are reduced (in some cases to zero) in infected plants. Diseased plants may be defoliated by midharvest, and infected plants may eventually die, often after a severe winter.

Symptoms of ToRSV are similar to those of necrotic ringspot, which is caused by *Tobacco ringspot virus*. The two viruses are readily distinguished by enzyme-linked immunosorbent assay (ELISA) of sap from infected blueberry leaves and are also distinguishable by nucleic acid probes.

Causal Agent

Like the other members of the genus *Nepovirus*, ToRSV has a bipartite, single-stranded RNA genome. Both RNA species are required for infection. ToRSV particles are isometric and 28 nm in diameter. The coat protein shell is composed of 60 subunits, each with a molecular mass of 55,000 Da. Monoclonal and polyclonal antibodies against this coat protein are commercially available and can be used to detect the virus in suspect blueberry plants, usually by ELISA. Young leaves from infected blueberry plants give the highest ELISA readings. Complementary DNA and RNA probes prepared against this virus are also commercially available for detecting its presence in infected plants.

ToRSV has a broad herbaceous host range. It naturally infects many wild and cultivated species of woody and herbaceous plants in 35 families, including many common weeds, such as chickweed (*Stellaria media*), dandelion (*Taraxacum officinale*), and narrow-leaved or buckhorn plantain (*Plantago lanceolata*), and fruit crops, such as apple, grape, peach, and raspberry.

ToRSV can be sap inoculated as follows: grind 1–2 g of young, symptomatic leaves in 0.05 M phosphate buffer, pH 7.0, containing 1% (v/v) nicotine alkaloid, and rub the sap on Carborundum-dusted leaves. *Chenopodium quinoa* develops necrotic primary local lesions and apical dieback within 7–10 days. On 'National Pickling' cucumber (*Cucumis sativus*), chlorotic local lesions 2–5 mm in diameter appear on inoculated cotyledons and are followed by systemic necrosis and chlorotic or necrotic lesions on new leaves.

Disease Cycle and Epidemiology

ToRSV is transmitted in the soil from roots of infected plants to healthy roots by dagger nematodes (*Xiphinema* spp.). The virus is retained for several months on specific attachment sites on the lining of the alimentary canal of the vector nematode. Transmission appears to depend on this retention. The speciation of the dagger nematodes that transmit ToRSV is in flux. *Xiphinema americanum* sensu lato includes the currently known nematode vectors of this virus.

Nematode-inoculated hosts often remain free of the symptoms caused by ToRSV, even when sap-inoculated plants of the same species exhibit characteristic foliar symptoms. ToRSV is seedborne in many of its weed and crop hosts. Plants infected through seed rarely exhibit symptoms but they can act as foci of virus acquisition by dagger nematodes that feed on their roots. Blueberry plants that are tolerant of ToRSV or that were only recently infected can also serve as foci of infection in a field.

ToRSV appears to spread slowly in blueberry fields in Oregon. Roughly oval patches of plants that have been killed or weakened by the virus develop over the course of several years.

Management

An integrated approach is required to manage ToRSV. Before planting, the soil should be tested for the presence of dagger nematodes, and if present, the field should be fumigated with a registered nematicide according to approved local methods. The timing of soil sampling is important since nematodes migrate vertically in the soil profile during the course of a season. Nematode populations are known to vary seasonally

Fig. 152. 'Earliblue' highbush blueberry infected with *Tomato ringspot virus*, the causal agent of tomato ringspot, but exhibiting uneven symptom distribution. Some shoots are defoliated; some have necrotic, distorted leaves; and some have apparently normal leaves. (Courtesy R. H. Converse)

Fig. 153. Shoots of 'Blueray' highbush blueberry infected with *Tomato ringspot virus*, the causal agent of tomato ringspot. Note the interveinal chlorosis, narrow leaves, puckered leaves, and some apparently healthy leaves. The virus is unevenly distributed and most leaf samples from this bush tested negative for *Tomato ringspot virus* by enzyme-linked immunosorbent assay. (Courtesy R. R. Martin)

in Oregon, with the lowest levels in late summer and the highest levels in early spring. The use of certified planting stock that is known to be free of ToRSV is essential for the control of ToRSV. No direct studies with blueberry are available, but studies in red raspberry fields in Oregon have shown that post-planting application of nematicides to a planting infected with ToRSV is unlikely to reduce the rate of spread of the virus. Efforts should be made to minimize soil movement in fields where ToRSV is known to be found. Grass alleyways between the rows greatly reduce the movement of soil by machinery and reduce the rate of spread of ToRSV. When using machinery in multiple fields, always start with the healthiest fields first and finish with fields where ToRSV may be a problem. Weeds that are hosts of ToRSV (see above) should be controlled in the planting to eliminate virus reservoirs for vector nematodes. However, a weed cover crop, even with those that are hosts of ToRSV, is better than having bare soil that can be moved by equipment traveling through the field.

Tolerant blueberry cultivars should be selected (none are known to be immune, although 'Bluecrop' has rarely been observed to be infected in the field, even under extreme disease pressure), and highly susceptible cultivars should be avoided in situations in which ToRSV infection is likely. A thorough evaluation of the degree of tolerance of the important highbush blueberry cultivars to infection by ToRSV via nematode vectors has not been carried out. The cultivars Atlantic, Bluecrop, Dixi, Earliblue, N51G, Olympia, Pemberton, and Stanley have been reported in the literature to be infected in the field in Oregon and/or Washington. Observations in a 40-year-old blueberry field in Oregon indicated that Berkeley, Earliblue, Pemberton, and Stanley were the most susceptible cultivars and that the cultivars Atlantic, Blueray, Dixi, and Jersey appeared to be infected much more slowly than the other four susceptible cultivars. In three separate field situations in Oregon, the cultivar Bluecrop has not become infected in fields with high ToRSV disease pressure.

Where ToRSV has been identified in an established blueberry planting, not only affected plants but also the adjacent symptomless plants (which may already be infected) should be rogued. If a field is removed because of ToRSV, or for spot treatments of rogued areas, nematicide treatment, weed-free fallow for 24 months, or rotation with a crop that is a nonhost of ToRSV can be very effective in controlling ToRSV. In the latter case, the virus is eliminated from the system, while the nematode survives.

Selected References

Converse, R. H., and Ramsdell, D. C. 1982. Occurrence of tomato and tobacco ringspot viruses and of dagger and other nematodes associated with cultivated highbush blueberries in Oregon. Plant Dis. 66:710-712.

Forer, L. B., and Stouffer, R. F. 1982. *Xiphinema* spp. associated with tomato ringspot virus infection of Pennsylvania fruit crops. Plant Dis. 66:735-736.

Johnson, F. 1972. Tomato ringspot virus of blueberries in Washington. (Abstr.) Phytopathology 62:1104.

Pinkerton, J. N., and Martin, R. R. 2005. Management of *Tomato ringspot virus* in red raspberry with crop rotation. Int. J. Fruit Sci. 5:55-67.

Stace-Smith, R. 1984. Tomato ringspot virus. Descriptions of Plant Viruses, No. 290. Commonwealth Mycological Institute and Association of Applied Biologists, Kew, Surrey, England.

(Prepared by R. H. Converse; Revised by R. R. Martin)

Diseases Caused by Bacteria

Blueberries

Bacterial Canker

Bacterial canker has been a major stem disease on susceptible highbush blueberry in the western portions of Oregon and Washington (United States) and British Columbia (Canada). The disease has not been reported in any other production region in North America, but it has been reported in southern Chile, and a similar, if not identical, disease called bacterial blight has been reported in several blueberry plantings in southern Australia (Tasmania).

Symptoms

Symptoms first appear in January or early February in the Northern Hemisphere as water-soaking on 1-year-old stems. These areas rapidly develop into reddish brown to black, irregularly shaped cankers. Cankers have definite margins and extend from a few millimeters long to the entire length of the stem (Fig. 154). Buds within the canker are killed. Cankers usually surround the stem and sometimes girdle it, killing buds and new growth above that point. This shoot-tip dieback is the most common symptom observed on young plants in nurseries. In other cases, the vascular tissues are not damaged and the buds above the canker grow normally. Leaves turn orange to tan if infection takes place after vegetative buds have produced leaves (Fig. 155). The disease also appears on cuttings in propagation beds. Necrotic leaf spots can also develop during wet springs.

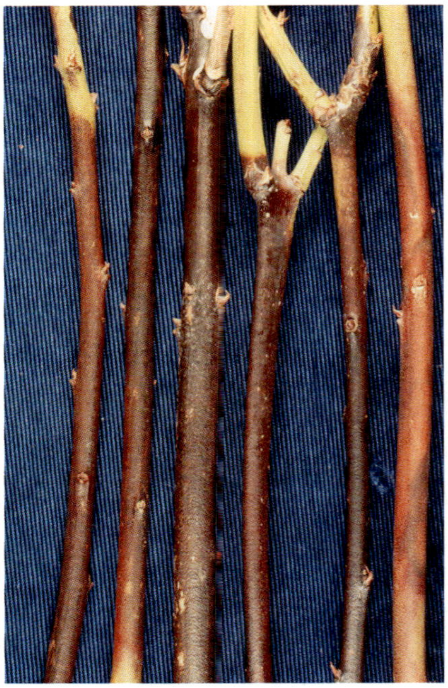

Fig. 154. One-year-old branches of highbush blueberry 'Bluetta' with bacterial canker, caused by *Pseudomonas syringae*. (Courtesy P. R. Bristow; © APS)

Causal Organism

Bacterial canker is caused by *Pseudomonas syringae*, a gram-negative, non-spore-forming bacterium. Isolates from blueberry and from lilac (*Syringa* spp.) infect both hosts in cross-inoculation trials. Disease incidence and severity, however, are always greatest with isolates from the same host. The blueberry isolate does not infect *Vaccinium* spp. native to the Pacific Northwest, other ericaceous plants, or sweet or sour cherry. In serological tests, isolates from lilac and blueberry react only with their own antiserum.

P. syringae is negative for dihydrolase and oxidase, and it does not pit polypectate gel or convert nitrates to nitrites. The bacterium produces the fluorescent pigment fluorescein, the toxin syringomycin, and the polysaccharide levan when grown on a sucrose-based medium. It utilizes lactate, and most isolates are positive for the tobacco hypersensitivity reaction.

Disease Cycle and Epidemiology

P. syringae survives and multiplies in buds and on the bark as an epiphyte; it may also survive epiphytically on weeds and grasses. Survival in soil is considered to be poor, although this topic has received little attention.

The presence of the disease on blueberry depends on the strain of the bacterium. Pathogenicity may depend on the ability of the bacterium to nucleate ice because disease incidence and severity in the spring are often greater when there has been frost injury the previous autumn. The presence of ice-nucleating *P. syringae* on the plant has been shown in other host–pathogen systems to raise the temperature above that at which ice crystals would normally form in plant tissues. Not all isolates nucleate ice, and ice nucleation does not happen unless the population of ice-nucleating bacteria is relatively high. Wounds caused by ice crystals allow the pathogen to enter and damage plant cells. Cold weather (temperatures below 10°C) and moisture favor development of the disease.

Once inside the stem, the bacterium appears to be only locally systemic. Only 1-year-old stems are attacked. Apparently, latent infections do not develop once stems have passed the second growing year. Bacterial canker can be particularly severe on young plants in new plantings because a high proportion of the wood is at the susceptible age.

P. syringae can be disseminated by wind, rain, or insects or in infested propagation wood or infested nursery stock. Equipment such as pruning tools and mechanical harvesters may carry the pathogen from plant to plant in addition to causing entry wounds.

Management

Two applications of Bordeaux mixture or fixed copper fungicide/bactericide in the autumn (early October and early November in the Northern Hemisphere) reduce the number of infected stems per plant the following spring. Other spray schedules have not been tested. These sprays probably reduce the population of the pathogen on stems. Treatments with copper may also protect wounds caused by freeze injury by reducing the population of ice-nucleating bacteria. In British Columbia, a spring application of fixed copper is recommended in addition to the autumn sprays. Bordeaux mixture is a good bactericide that is not easily washed off by rain. Adding a spreader–sticker to spray suspensions lengthens the period of control under the cool, wet conditions of the Pacific Northwest. To minimize the risk of injuring tender new growth, spring applications should be made when conditions favor rapid drying. Bacteria resistant to streptomycin and copper-based bactericides have been found in nurseries in Oregon, on many crops throughout the production area in Oregon's Willamette Valley, and in blueberry production fields in British Columbia.

Diseased wood should be pruned out and destroyed before the onset of autumn rains. Late-summer applications of nitrogen fertilizer should be avoided because it prolongs succulent growth and delays hardening off of the stems, which in turn can lead to winter injury.

Resistant cultivars should be planted if the local environment favors disease development. Highbush cultivars Elliott, June, Rancocas, and Weymouth are resistant to bacterial canker; cultivars Atlantic, Bluejay, Blueray, Chandler, Coville, Darrow, Evelyn, Jersey, Patriot, Scammell, and N51G (sometimes called Eberhardt) are susceptible. The reactions of cultivars Bluecrop, Bluetta, Burlington, Duke, Pioneer, Rubel, and Spartan are intermediate between those of the susceptible and the resistant cultivars. This classification is based on field observations of cultivar reactions and not on controlled inoculation tests.

Fig. 155. Flagging of discolored leaves on branches of highbush blueberry 'Patriot' girdled by *Pseudomonas syringae*, the causal organism of bacterial canker. (Courtesy P. R. Bristow; © APS)

Selected References

Canfield, M. L., Baca, S., and Moore, L. W. 1986. Isolation of *Pseudomonas syringae* from 40 cultivars of diseased woody plants with tip dieback in Pacific Northwest nurseries. Plant Dis. 70:647-650.

Guerrero, C. J., and Lobos, A. W. 1989. Determination of *Pseudomonas syringae* on highbush blueberry (*Vaccinium corymbosum* L.) in southern Chile. Agric. Tec. (Santiago) 49:224-227.

Lelliot, R. A., Billings, E., and Hayward, A. C. 1966. A determinative scheme for the fluorescent plant pathogenic pseudomonads. J. Appl. Bacteriol. 29:470-489.

Lindow, S. E. 1983. The role of bacterial ice nucleation in frost injury to plants. Annu. Rev. Phytopathol. 21:363-384.

Scheck, H. J., Pscheidt, J. W., and Moore, L. W. 1996. Copper and streptomycin resistance in strains of *Pseudomonas syringae* from Pacific Northwest nurseries. Plant Dis. 80:1034-1039.

Stace-Smith, R., Wooley, P. H., and Vaughan, E. K. 1953. A new disease of cultivated blueberry. (Abstr.) Phytopathology 43:589.

Vaughan, E. K. 1956. A strain of *Pseudomonas syringae* pathogenic on cultivated blueberry. (Abstr.) Phytopathology 46:640.

Vaughan, E. K., and Boller, C. A. 1954. Experiments on control of bacterial canker of blueberry. Plant Dis. Rep. 38:867-868.

Wilson, S. J., and Sampson, P. J. 1984. Bacterial blight of blueberry. Australas. Plant Pathol. 13:57.

(Prepared by P. R. Bristow and L. W. Moore;
Revised by J. W. Pscheidt)

Bacterial Leaf Scorch

Bacterial leaf scorch of blueberry, caused by the bacterium *Xylella fastidiosa*, was identified as a major disease of some southern highbush blueberry cultivars in 2006. It has also been identified in rabbiteye blueberry but, to date, has not caused the acute problems seen in southern highbush blueberry. Losses on susceptible southern highbush cultivars can be severe. Bacterial leaf scorch is most prevalent in the southeastern United States, where southern highbush blueberry production is also prevalent, especially in Georgia and Florida.

Symptoms

The initial symptom of bacterial leaf scorch is a marginal leaf scorch or burn (Fig. 156) similar to that observed with extreme drought, fertilizer salt burn, or root rots. Sometimes, the scorched leaf area is bordered by a darker band between the healthy and scorched tissue, and the remaining green tissue may take on an oak-leaf pattern within the leaf. Leaf symptoms can be uniformly distributed throughout the plant, but in the early stages, scorching may be limited to individual stems or one side of the plant. Spring growth of bacterial leaf scorch-infected bushes is characterized by very-small-diameter twigs. Eventually, leaves abscise, and young twigs or stems may take on a characteristic yellow, skeleton-like appearance that is sometimes referred to as yellow stem (Fig. 157). After leaf drop, the bush eventually dies; mortality generally takes place within 1–2 years. Often, where a bush has died from bacterial leaf scorch, a neighboring bush shows symptoms the following year.

Fig. 156. Bacterial leaf scorch symptoms (late summer) observed on blueberry leaves infected with *Xylella fastidiosa*. Note the marginal leaf burn surrounded by a dark line of demarcation between green and dead tissues. (Courtesy P. M. Brannen; © APS)

Fig. 157. Complete defoliation of a *Xylella fastidiosa*-infected blueberry plant. The yellow stems are distinctive and characteristic of bacterial leaf scorch. (Courtesy P. M. Brannen; © APS)

The plant can drop virtually all leaves and yet the stems and root system appear sound, and vascular discoloration generally is not observed. Stem dieback (necrosis) does not generally appear in the early stages of disease development but may be observed on twigs, stems, and canes after leaf drop.

Causal Organism

Xylella fastidiosa is a genetically diverse, fastidious, gram-negative, xylem-limited, plant-pathogenic bacterial species. While causing disease symptoms on numerous monocotyledonous and dicotyledonous plant species, it can also infect host plants without causing disease symptoms; among these are various grasses and herbaceous weeds that are found throughout the southeastern blueberry production region. Native blueberries also likely harbor the bacterium. Therefore, a bacterial reservoir is presumed to be readily available for new infections and epidemic initiation within new plantings.

The *X. fastidiosa* strains found in blueberry are distinctly different from those of *X. fastidiosa* subsp. *piercei*, and strains that cause bacterial leaf scorch of blueberry are primarily recombinant *X. fastidiosa* subsp. *multiplex* strains. As in other *Xylella*-incited diseases, it is assumed that the bacterium blocks xylem vessels, thereby preventing water and nutrient flow throughout the plant. This bacterium only survives in plant xylem or within its insect vectors, sharpshooter leafhoppers (Cicadellidae) and spittlebugs (Cercopidae); bacteria attach to the insect foregut, and while insects ingest xylem fluid during the feeding process, they also egest fluid that contains the bacterium, resulting in infection and spread. In general, *Xylella*-incited diseases are more prevalent in warmer environments, which promote winter survival of the insect vectors as well as the bacterium in its host plants. The rapid onset of bacterial leaf scorch in the southeastern United States may be partially attributable to recently warmer winters that may have aided the survival of vectors and the bacterium.

The bacterium cannot be easily observed with a light microscope, so confirmation is only possible through enzyme-linked immunosorbent assay or polymerase chain reaction laboratory techniques. In addition, bacterial isolation on various specialized media, followed by the above-mentioned techniques, is further recommended for disease confirmation.

Disease Cycle and Epidemiology

The epidemiology of bacterial leaf scorch has not been studied extensively. The disease cycle of this bacterium in grape, peach, and plum is well known and, currently, is presumed to be similar in blueberry. Infected weeds and grassy hosts probably serve as key reservoirs and overwintering sites of the bacterium. In the spring and early summer, insect vectors (sharpshooters and other xylem-feeding insects such as spittlebugs) transmit the bacterium through feeding on infected plant tissues and subsequent feeding on healthy plants. In other systems, the glassy-winged sharpshooter, *Homalodisca vitripennis*, is a very important vector. This insect is abundant in south Georgia and Florida, where it is known to be the major vector of *X. fastidiosa* in peach and European wine grapes; it is also the most prevalent vector found in southern highbush blueberry production sites. Although nonvector means of spread have been hypothesized, there is yet no clear evidence that bacterial leaf scorch can be transmitted through propagation, root grafting, or pruning, and insects are considered to be the primary means of bacterial plant-to-plant movement, as in other systems.

Upon infection, the bacterium multiplies and moves throughout the plant xylem, which becomes clogged through a combination of tyloses, gumming, and bacterial exudates. Once the xylem stops functioning properly, symptoms develop and eventually the plant dies. Plant death can be relatively rapid but, in general, does not occur until the second season after symptoms develop.

Management

Identification of *Xylella fastidiosa*-infected plants is possible in the field, and once such plants are identified, they should be immediately removed from the field and destroyed. It is essential that new plants not be propagated from *Xylella fastidiosa*-infected plants. Currently, there is no testing program for propagation; however, propagators should never take cuttings from plants with visual disease symptoms.

Field resistance has been observed in some southern highbush blueberry cultivars, and greenhouse studies in Florida have further confirmed resistance. The cultivar FL 86-19, also known as V1, is particularly susceptible to infection and disease development by *X. fastidiosa*. Compared with other southern highbush or rabbiteye cultivars, FL 86-19 quickly develops symptoms and high bacterial titers after artificial inoculation, which correlates well with observed susceptibility in the field. Cultivars O'Neal and Star are also infected and develop symptoms readily. Cultivars V5 and Emerald have apparent field resistance to this bacterium. This indicates that breeding may be utilized to develop cultivars that are resistant to *X. fastidiosa*. However, some varieties have shown differential resistance to various *X. fastidiosa* strains, making selection of resistant cultivars difficult.

Currently, there are no chemical controls against the bacterium. However, it may be possible to slow disease progression by vector suppression, killing the insects that transmit the bacterium. Application of soil-applied neonicotinoid insecticides (imidacloprid or thiamethoxam) is tentatively recommended as plants begin their initial flush of growth in the spring. Neonicotinoid materials work best when there is sufficient soil moisture and active growth. If leafhoppers are present, vector suppression by soil-applied neonicotinoid materials can be enhanced by alternating with foliar applications of pyrethroid insecticides and with organophosphates such as phosmet.

Selected References

Alderz, W. C., and Hopkins, D. L. 1979. Natural infectivity of two sharpshooter vectors of Pierce's disease of grape in Florida. J. Econ. Entomol. 72:916-919.

Brannen, P. M., Krewer, G., Boland, B., Horton, D., and Chang, C. J. 2008. Bacterial Leaf Scorch of Blueberry (C 922). UGA College of Agricultural and Environmental Sciences, Athens, Georgia.

Chang, C. J., and Walker, T. J. 1988. Bacterial leaf scorch of northern red oak: Isolation, cultivation, and pathogenicity of a xylem-limited bacterium. Plant Dis. 72:730-733.

Chang, C., Brannen, P., Krewer, G., Boland, R., and Donaldson, R. 2007. Bacterial leaf scorch of blueberry: A new disease caused by *Xylella fastidiosa*. (Abstr.) Phytopathology 97:S20.

Chang, C. J., Donaldson, R., Brannen, P. M., Krewer, G., and Boland, R. 2009. Bacterial leaf scorch, a new blueberry disease caused by *Xylella fastidiosa*. HortScience 44:413-417.

Davis, M. J., Purcell, A. H., and Thomson, S. V. 1980. Isolation media for the Pierce's disease bacterium. Phytopathology 70:425-429.

Hopkins, D. L. 1983. Gram-negative, xylem-limited bacteria in plant disease. Phytopathology 73:347-350.

Hopkins, D. L. 1989. *Xylella fastidiosa*: Xylem-limited bacterial pathogen of plants. Annu. Rev. Phytopathol. 27:272-290.

Nissen, L., Denny, T., Brannen, P., and Chang, C. 2009. *Xylella fastidiosa* strains causing bacterial leaf scorch of blueberry in Georgia are genetically distinct from those causing Pierce's disease of grape. (Abstr.) Phytopathology 99:S93.

Wells, J. M., Raju, B. C., Hung, H. Y., Weisburg, W. G., Mandelco-Paul, L., and Brenner, D. J. 1987. *Xylella fastidiosa* gen. nov., sp. nov.: Gram-negative, xylem-limited, fastidious plant bacteria related to *Xanthomonas* spp. Int. J. Syst. Bacteriol. 37:136-143.

(Prepared by P. M. Brannen, C. J. Chang, and D. L. Horton)

Bacterial Leaf Spot

Bacterial leaf spot of blueberry (*Vaccinium corymbosum*) caused by *Burkholderia andropogonis* was first reported in Australia in 1986. Sporadic outbreaks occurred on young leaves of hardwood cuttings and potted young blueberry plants in New Jersey (United States) in the early 1990s. In many cases, severe outbreaks (Fig. 158) led to reduced plant vigor that resulted in poor establishment and the development of cuttings.

Symptoms

Typical leaf symptoms are irregular-shaped, reddish brown necrotic lesions (Fig. 159). However, symptoms also have been reported as brown to black lesions with a tan center and a yellow halo surrounding the edges. In young lesions, water soaking can be observed along lesion edges. Older lesions become dry and brittle, which may lead to a shot hole appearance. In severe cases, lesions coalesce to form larger lesions but are delimited by leaf veins.

Causal Organism

Bacterial leaf spot is caused by *Burkholderia andropogonis*. This pathogen was previously known as *Pseudomonas andropogonis* and is now reclassified as a member of the genus *Burkholderia* within the beta subdivision of proteobacteria. The species is known to cause leaf spot on a broad range of hosts, including maize, sorghum, common vetch, clover, red kidney bean, velvet bean, chickpea, bougainvillea, statice, carnation, and carob. In severe cases, spotting can extend to other tissues, including stems and buds. Multiple blueberry cultivars are susceptible to the pathogen, with some cultivars, such as Berkeley, Duke, and Herbert, showing greater susceptibility than others. Greenhouse studies indicate that cranberry leaves inoculated with the bacterium also develop disease symptoms, although

Fig. 158. Symptoms of bacterial leaf spot, caused by *Burkholderia andropogonis*, in a blueberry propagation bed. (Courtesy P. V. Oudemans; © APS)

Fig. 159. Bacterial leaf spot symptoms, caused by *Burkholderia andropogonis*, on highbush blueberry. (Courtesy P. V. Oudemans; © APS)

this disease has not been reported in cranberries in the field. Different strains of the bacterium, including those isolated from blueberry, vary in their host range and virulence.

B. andropogonis is gram negative and rod shaped. The bacterium produces circular, glistening white colonies with entire margins that appear after 48–72 h of growth at 30°C on nutrient-rich media, such as yeast extract–dextrose–calcium carbonate (YDC) agar.

Disease Cycle and Epidemiology

Little is known about the disease cycle and epidemiology. Epiphytic growth likely occurs under hot and humid conditions. Bacteria are spread by overhead irrigation or wind-driven rain and rain splash, which also enhance infection and symptom development. In contrast, mechanical damage or frost do not seem to play as important a role. The disease is prevalent during early spring; young succulent leaves, especially those produced on hardwood cuttings used for propagation, are most susceptible.

Management

There are no registered pesticides for the control of bacterial leaf spot. Good cultural practices that reduce the level of pathogen inoculum include proper sanitation and removal of severely infected tissue. Changes in management practices that minimize infection, e.g., minimizing irrigation to avoid long periods of leaf wetness and reducing direct impact from sprinklers, are the best methods for control. A reduction in the rate of infection and disease progression can allow plants to recover and outgrow disease symptoms. After transplanting to field nurseries or commercial plantings, incidence of the disease is greatly diminished.

Selected References

Kobayashi, D. Y., Stretch, A. W., and Oudemans, P. V. 1995. A bacterial leaf spot of highbush blueberry hardwood cuttings caused by *Pseudomonas andropogonis*. Plant Dis. 79:839-842.

Moffet, M. L., Hayward, A. C., and Fahy, P. C. 1986. Five new hosts of *Pseudomonas andropogonis* occurring in eastern Australia: Host range and characterization of isolates. Plant Pathol. 35:34-43.

(Prepared by D. Y. Kobayashi and P. V. Oudemans)

Crown Gall

Crown gall of highbush and rabbiteye blueberries is caused by a number of soilborne bacteria that enter wounds in the host at or below the soil line and induce the plant to form galls. Crown gall is occasionally a problem in propagation beds and new plantings but may also be found in production fields on mature plants. This disease is infrequent in incidence overall but may be severe in fields, where it is found when cultural practices encourage disease development.

Symptoms

Galls are most common at the base of canes or on major roots (Fig. 160) and are rare on smaller lateral roots. Occasionally, galls form on branches higher in the bush, especially in fields where there has been flooding or when pruning has been done under wet conditions.

Young galls are cream colored to light brown; they turn dark brown to black and become rough and hard as they age (Fig. 161). In the Pacific Northwest, galled areas on canes may appear elongated because of the convergence of several smaller galls growing together. Small galls may sometimes appear as roughened areas under the bark of younger stems. Galls vary in size, with a few becoming large (5–16 cm in diameter). New galls sometimes form by breaking through the cortex of existing galls. No bud proliferation is associated with crown galls.

Infected plants may be stunted or weak compared with healthy ones. When older plants (2–5 years old) are affected, the foliage discolors prematurely in the summer. The foliage on an affected branch or plant initially takes on a reddish hue and then becomes yellowish brown as the disease progresses. Very young plants may be killed outright.

Causal Organisms

Crown gall is caused by soilborne bacteria of the family Rhizobiaceae. The causal organism of the disease in blueberry-growing regions in the eastern United States was identified as *Agrobacterium rubi*, which was initially isolated from galls on floricanes of *Rubus* spp. In the Pacific Northwest, *A. tumefaciens* has been shown to cause galling of blueberry, while in Argentina, both *A. tumefaciens* and *A. rubi* have been isolated from blueberry galls. The genes that confer bacterial pathogenicity are carried on a large extrachromosomal piece of DNA, which is referred to as the tumor-inducing (Ti) plasmid. This

Fig. 160. Galls on the lower stem and roots of an infected 'Duke' highbush blueberry plant. Crown gall is caused by soilborne bacteria of the family Rhizobiaceae. (Courtesy M. Serdani; © APS)

Fig. 161. Typical older, darkened, erumpent galls on young stems of blueberry 'Draper'. Crown gall is caused by soilborne bacteria of the family Rhizobiaceae. (Courtesy M. Serdani; © APS)

plasmid can be lost from pathogenic strains or transferred from pathogenic to nonpathogenic strains of the bacterium. Traditionally, pathogenic isolates have been named *A. tumefaciens* and those that lack the Ti plasmid were known as *A. radiobacter*. Additionally, strains of agrobacteria that possess a root-inducing (Ri) plasmid, which stimulates the proliferation of many fine roots (hairy root disease), were known as *A. rhizogenes*. More recently, it has been suggested that *A. tumefaciens* and *A. rhizogenes* do not differ significantly from species in the genus *Rhizobium*, and the names *Rhizobium radiobacter* and *R. rhizogenes*, respectively, have been proposed. *A. rubi* was renamed *R. rubi*, and may contain a Ti plasmid or be nonpathogenic. Under the rules of bacterial nomenclature, either the older nomenclature or the newer names may be used, and here the traditional names will continue to be used to prevent confusion, with the understanding that agrobacteria are a genetically variable group and that pathogenicity is determined by plasmid-borne genes. Nonpathogenic isolates (those that do not possess the Ti or Ri plasmid) are often associated with *A. tumefaciens* in galls. With few exceptions, isolations from uninfected plant tissues and soil yield only nonpathogenic *A. radiobacter*.

The term "biovar" (abbreviated bv.) is used to describe groupings of bacterial strains within a species. Three biovars have been identified for *Agrobacterium* spp.; however, because biovar 3 is isolated almost exclusively from grapevine galls, it was elevated to species status (*A. vitis*). Not enough isolates of *Agrobacterium* spp. from blueberry have been examined to definitively identify which biovars are common on this host, but biovar 2 strains consistently have been isolated from galls of blueberry grown in several geographic regions. Biovars 1 and 2, as well as *A. rubi*, were isolated from nursery and field plants in Argentina, but only the few biovar 1 isolates tested produced galls on blueberry. Isolates of *A. tumefaciens* from other crops, including apple, dahlia, hop, and peach, did not induce gall formation in blueberry. Similarly, isolates from blueberry were unable to infect apple and peach. However, isolates obtained from blueberry were pathogenic when assayed on tomato seedlings and young blueberry plants. Recovery of pathogenic isolates from blueberry in Oregon (United States) has been difficult. Mutation to a nonpathogenic state after infection may be one reason for the preponderance of nonpathogenic isolates from these blueberry galls. These isolates may just have a different host range than isolates from other sources, or they may be host specific.

A. tumefaciens is rod shaped (0.6–1.0 × 1.5–3.0 µm), aerobic, and gram negative; does not produce endospores; and is motile by means of flagella circumferentially arranged near one end of the cell. Colonies on potato dextrose agar are usually white to cream colored, convex, and glossy, with entire margins.

Disease Cycle and Epidemiology

Although the bacteria apparently survive in soil, isolating tumorigenic strains from soil has been very difficult. In the host, the pathogen requires a wound to initiate infection. Wounds may result from natural causes, such as lateral root formation or bud scale and leaf scars, or from mechanical causes, such as pruning, cultivation, harvesting, insect feeding, or frost damage. Wounds generally remain susceptible for 2–4 days under the warm conditions of summer and can remain susceptible for weeks at temperatures below 13°C. During this period, infection is also influenced by soil moisture and pathogen populations. The disease tends to be less of a problem in acid soils.

Crown gall development is initiated when the pathogen enters a wound, attaches to a plant cell, and inserts a segment of DNA (called T-DNA) from the Ti plasmid (carried by the pathogen) into a chromosome of the healthy blueberry cell. Expression of T-DNA genes results in the overproduction of plant hormones. The elevated synthesis of growth hormones, which is controlled by the bacterial genes, stimulates the plant cells to divide and enlarge abnormally, forming a gall, which harbors the pathogen intercellularly. Gall development is usually accompanied by the synthesis of opines, which are unique amino acids produced in plant tissues infected by *A. tumefaciens*.

Small galls appear within 2–4 weeks after infection when temperatures are 20°C or above. Symptom development is delayed by temperatures below 15°C, and latent infections may not show signs of gall formation until higher temperatures return during the second growing season. Infection is inhibited at temperatures above 32°C.

Once the T-DNA is incorporated into the genome of the host, galls may continue to develop in the absence of the bacterium. This probably explains why new galls develop after the older ones die during the winter. New infections from bacteria surviving in the plant may also cause new galls. The motile bacteria move in the water films surrounding soil particles or are carried along with water as it percolates through the soil. The bacteria can spread over longer distances in irrigation water or floodwater, in soil adhering to equipment, or on plant material. Severe incidence of crown gall in blueberry occurred in one field in Oregon after the grower pruned plants in the rain.

Management

All blueberry cultivars are apparently susceptible to crown gall. Thus, the most effective means of control is to establish plantings in uninfested soil with pathogen-free planting stock. Fields where the previous crop was infected should be avoided for 2 or 3 years or more. A nonhost crop, such as grasses (grains or pasture) or vegetables, should be grown during this period to significantly reduce the population of the pathogen.

Soil fumigation has generally been ineffective in eliminating *A. tumefaciens* from field soil. Using high rates of chloropicrin to sterilize the planting medium before using it in propagation beds does eliminate the pathogen.

Good cultural practices, including sanitation, minimize the risk of introducing the pathogen into disease-free plantings. All nursery stock should be carefully inspected, and all plants with crown gall symptoms should be removed from the site and discarded. This practice is especially important in nurseries to avoid the contamination of healthy plants. Bare-root stock is easier to inspect than stock grown in soil in plastic bags or pots. Plants grown in containers should be inspected when they are planted.

The adoption of management practices that minimize wounding of plant tissues is beneficial in controlling crown gall. Soil insects, such as root weevils, should be controlled to reduce root wounding. Practices such as crown grinding to narrow the base of large, mature plants in order to accommodate mechanical harvesters should be avoided.

Use of an uninfested planting medium in propagation beds is critical. Pruning equipment should be sterilized when propagation wood is collected. Pruners may be sterilized with a 10–20% solution of commercial bleach, a 0.5% solution of potassium permanganate, or shellac thinner (70% ethyl alcohol).

The biological control agent *A. radiobacter* strain K1026 (a plasmid transfer-deficient mutant of strain *A. radiobacter* K-84) has shown promise for controlling crown gall on a number of crops, although its efficacy on blueberry is unknown. Biological control by *A. radiobacter* strain K1026 is strictly preventive in nature and does not cure infections that have already occurred. A paintable crown gall control product containing 2,4-xylenol and meta-cresol may be used on very young galls to prevent further development, but the effects may be transitory and the product can injure some plants. Since there is no sure cure for infected plants, excluding *A. tumefaciens* from a field by strict sanitation remains the most reliable means of control.

Selected References

Alippi, A. M., Lopez, A. C., and Balatti, P. A. 2012. Diversity among agrobacteria isolated from diseased plants of blueberry (*Vaccinium corymbosum*) in Argentina. Eur. J. Plant Pathol. 134:415-430.

Belanger, C., Canfield, M. L., Moore, L. W., and Dion, P. 1995. Genetic analysis of nonpathogenic *Agrobacterium tumefaciens* mutants arising in crown gall tumors. J. Bacteriol. 177:3752-3757.
Canfield, M. L., Putnam, M. L., White, T. J., and Moore, L. W. 1995. Isolation of *Agrobacterium tumefaciens* from blueberry (*Vaccinium corymbosum*). (Abstr.) Phytopathology 85:1194.
Demaree, J. B., and Smith, N. R. 1952. Blueberry galls caused by a strain of *Agrobacterium tumefaciens*. Phytopathology 42:88-90.
Otten, L., Burr, T., and Szegedi, E. 2008. *Agrobacterium*: A disease-causing bacterium. Pages 2-46 in: *Agrobacterium*: From Biology to Biotechnology. T. Tzfira and V. Citovsky, eds. Springer Science + Business Media, New York.
Young, J. M., Kuykendall, L. D., Martínez-Romero, E., Kerr, A., and Sawad, H. 2001. A revision of *Rhizobium* Frank 1889, with an emended description of the genus, and the inclusion of all species of *Agrobacterium* Conn 1942 and *Allorhizobium undicola* de Lajudie et al. 1998 as new combinations: *Rhizobium radiobacter*, *R. rhizogenes*, *R. rubi*, *R. undicola* and *R. vitis*. Int. J. Syst. Evol. Microbiol. 51:1689-1695.

(Prepared by P. R. Bristow and L. W. Moore;
Revised by M. L. Putnam)

Stunt

Blueberry stunt was first reported in New Jersey in 1942. The pathogen is vectored by leafhoppers and, in addition to New Jersey (United States), has been reported in eastern Canada and in Maine, Massachusetts, New Hampshire, New York, Michigan, North Carolina, Pennsylvania, Maryland, Virginia, and Arkansas (United States). Early literature suggests that all cultivars of highbush blueberry are susceptible, but there has been no comprehensive screening for susceptibility of more recently released cultivars. Rabbiteye blueberry (*Vaccinium ashei*), lowbush blueberry (*V. angustifolium*), *V. stamineum*, and *V. myrtilloides* are also reported to be naturally infected. The disease organism has been graft transmitted to *V. altomontanum*, *V. amoenum*, and *V. elliottii*.

Symptomatic bushes are dwarfed and decline over time compared with healthy bushes. Although flowers can appear normal, fruit set and berry size decrease following infection, and eventually, affected plants may fail to produce fruit.

Symptoms

On highbush blueberry, overall dwarfing of the bush is the primary symptom; hence the name stunt. Affected leaves are typically cupped slightly downward and may be rounded at the tip. Leaves may have chlorotic margins, and chlorosis may be found between lateral veins that retain their normal green color (Fig. 162). Later in the season, the older leaves sometimes develop interveinal red streaks (Fig. 162). Symptoms often appear first at the base of the plant on new growth. During late summer to early autumn, the chlorotic areas on the leaves or the entire bush may turn a brilliant red, while normal plants are still green (Fig. 163). Stem internodes are shortened, and branches appear excessively bushy because of the growth of normally dormant buds. Symptoms on rabbiteye blueberry are similar.

Causal Organism

Transmission electron microscopy of ultrathin sections of symptomatic blueberry tissues revealed the presence of a phytoplasma (formerly described as a mycoplasma-like organism or MLO) in the phloem of the diseased tissues (Fig. 164). The causal organism, in regions where it has been determined, is primarily '*Candidatus* Phytoplasma asteris', subgroup 16SrI-E. Other species and strains have also been described from some regions. Phytoplasmas are specialized bacteria that lack a cell wall and have a small genome (680–1,600 kilobases). Cells are pleomorphic with a diameter of 160–700 nm. DNA of the pathogen can be detected in nucleic acid extracts from affected plants using polymerase chain reaction and universal primers for phytoplasmas.

The causal organism is easily graft transmitted. Plants grafted at bud break may develop symptoms within several months. Dodder (*Cuscuta campestris* and *Cuscuta subinclusa*) has been used successfully to transmit the causal organism to periwinkle (*Catharanthus roseus*).

Disease Cycle and Epidemiology

Stunt phytoplasma is spread by the sharp-nosed leafhopper *Scaphytopius magdalensis*. Controlled transmission experiments suggest that two additional species, *Scaphytopius acutus* and *Scaphytopius frontalis*, are also vectors. Studies of the three leafhopper species showed that they reach two distinct population peaks during the growing season in Michigan. The first starts at the cessation of bloom and ends in early July; the second begins by mid-August and ends by mid-October. Epidemiological studies in which small, healthy trap plants were placed in the field under stunt-diseased bushes showed that the

Fig. 162. Highbush blueberry 'Duke' with leaf symptoms of stunt disease, caused by a phytoplasma. (Courtesy J. J. Polashock)

Fig. 163. Highbush blueberry with autumn (October) symptoms of stunt, caused by a phytoplasma. Note that nearly the entire bush has red leaves. The plant immediately adjacent is a healthy plant of the same cultivar. (Courtesy J. J. Polashock)

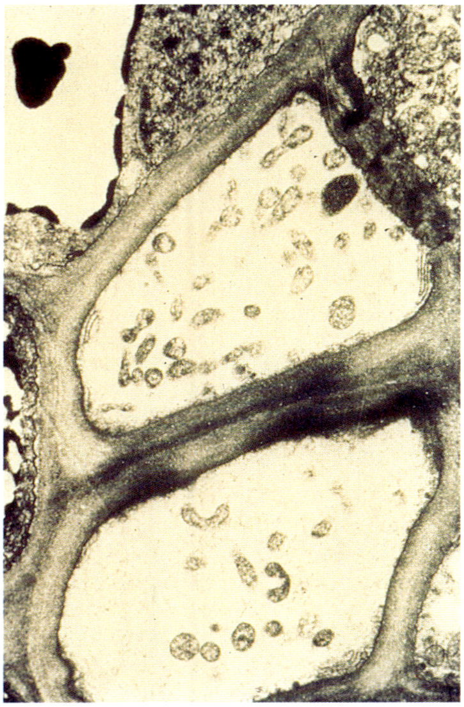

Fig. 164. Stunt phytoplasma in phloem sieve tubes of highbush blueberry (×64,800). (Courtesy D. C. Ramsdell)

infection periods coincide with leafhopper activity. *Scaphytopius* spp. have been trapped from pin cherry, black cherry, chokecherry, dewberry, wild raspberry, and blackberry plants in and adjacent to a commercial highbush blueberry plantation.

Management

Targeted sprays to control leafhopper populations, rouging of infected plants, and planting of clean nursery stock is essential for controlling spread of the disease. In warmer climates, such as in North Carolina and Arkansas (United States), however, disease control may be more difficult to achieve because additional generations of leafhoppers may occur each year. The incidence of this disease appears to be increasing, perhaps because of the trend toward using targeted, rather than broad spectrum, insecticides that are less effective at controlling leafhopper populations.

Selected References

Bagadia, P. G., Polashock, J., Bottner-Parker, K. D., Zhao, Y., Davis, R. E., and Lee, I. M. 2013. Characterization and molecular differentiation of 16SrI-E and 16SrIX-E phytoplasmas associated with blueberry stunt disease in New Jersey. Mol. Cell. Probes 27:90-97.

Chen, T. A. 1971. Mycoplasmalike organisms in sieve tube elements of plants infected with blueberry stunt and cranberry false blossom. Phytopathology 61:233-236.

Dale, J. L., and Mainland, C. M. 1981. Stunt disease in rabbiteye blueberry. HortScience 16:313-314.

Hutchinson, M. T. 1955. An ecological study of the leafhopper vectors of blueberry stunt. J. Econ. Entomol. 48:1-8.

Lee, I.-M., Hammond, R. W., Davis., R. E., and Gunderson, D. E. 1993. Universal amplification and analysis of pathogen 16S rDNA for classification and identification of mycoplasmalike organisms. Phytopathology 83:834-842.

Maeso Tozzi, D. C., Ramsdell, D. C., Taboada, O., Lee, I.-M., and Davis, R. E. 1993. Epidemiological studies on the stunt disease of highbush blueberry. Ann. Appl. Biol. 123:579-599.

Maramorosch, K. 1955. Transmission of blueberry stunt virus by *Scaphytopius magdalensis*. J. Econ. Entomol. 48:106.

Ramsdell, D. C., and Stretch, A. W. 1987. Blueberry stunt. Pages 106-108 in: Virus Diseases of Small Fruits. R. H. Converse, ed. U. S. Dep. Agric. Agric. Handb. 631. U.S. Government Printing Office, Washington, DC.

Tomlinson, W. E., Jr., Marucci, P. E., and Doehlert, C. A. 1950. Leafhopper transmission of blueberry stunt disease. J. Econ. Entomol. 43:658-662.

Wilcox, R. B. 1942. Blueberry stunt, a virus disease. Plant Dis. Rep. 26:211-213.

(Prepared by D. C. Ramsdell; Revised by J. J. Polashock)

Disease Caused by an Alga

Blueberries

Algal Stem Blotch

Algal stem blotch is only known to appear on blueberry in Florida (United States). In individual fields, the disease can be widespread and very damaging. The southern highbush cultivar Misty is particularly susceptible.

Symptoms

Initial symptoms on green stems are small reddish blotches that expand and form irregular patches. Lesions may eventually girdle canes. The centers of old lesions become bleached to gray. Under humid conditions, orange velvety tufts or mats develop on the lesions (Figs. 165 and 166). These tufts may initially appear green. Other conspicuous symptoms are pale yellow leaves on stunted canes (Fig. 167). Cane dieback may also occur, but it is not clear whether this is a direct symptom or is caused by secondary pathogens, such as *Botryosphaeria* spp., that invade damaged stems.

Causal Organism

Algal stem blotch is caused by a parasitic green alga, *Cephaleuros virescens*, which is a member of the Trentepohliales, an order of aerophilic filamentous green algae in the phylum Chlorophyta. This pathogen causes leaf, fruit, and stem spots

Fig. 165. Orange tufts of algal growth are visible on lesions of algal stem blotch, caused by *Cephaleuros virescens*, on a blueberry cane. (Courtesy P. Harmon; © APS)

on a wide range of hosts in the tropics and subtropics, including citrus, mango, palms, and magnolia. Along the U.S. Gulf Coast, *C. virescens* has been recorded on 287 plant species and cultivars, including 80 that incite stem infections. *C. virescens* has also been reported to cause cane lesions and stem cracking on blackberry in Arkansas and Louisiana (United States).

The thallus of the alga consists of threadlike, branched filaments that grow in a weblike fashion underneath the cuticle. It is not known whether the pathogen also colonizes the epidermis of blueberry canes. Cell and tissue death can occur beneath the thallus, causing the bark to crack. The orange tufts observed are hairlike stalks protruding through the cuticle. The orange coloration is caused by the presence of the pigments astaxanthin and beta-carotene in algal cells. The stalks may be sterile or fertile, bearing multiple ovoid sporangia (24 × 30 µm) on swollen terminal cells. Sporangia release biflagellate zoospores, which are the infective propagules (Fig. 168).

The pathogen is also capable of sexual reproduction by the release of biflagellate gametes from flask-shaped gametangia. *C. virescens* is homothallic. The gametes fuse in pairs to form dwarf sporophytes, which in turn produce microsporangia. The microsporangia release quadriflagellate zoospores whose role has not been elucidated.

Although *C. virescens* can be grown on artificial media, zoospores have never been produced in culture nor has pathogenicity been demonstrated successfully.

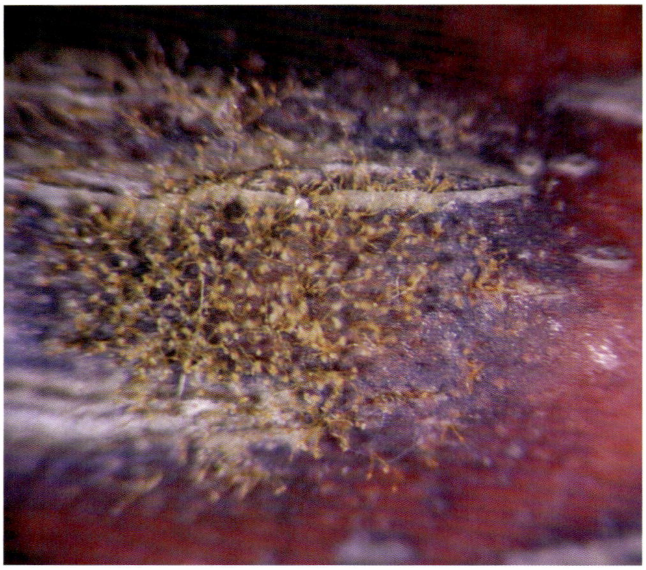

Fig. 166. Sporangiophores of *Cephaleuros virescens*, the causal organism of algal stem blotch, on a lesion on a blueberry cane. (Courtesy P. Harmon; © APS)

Fig. 167. Blueberry canes infected by *Cephaleuros virescens*, the causal organism of algal stem blotch, are stunted with pale yellow leaves. (Courtesy P. Harmon; © APS)

Disease Cycle and Epidemiology

Since little is known about the behavior of *C. virescens* on blueberry, most of the following information is based on reports on other hosts. *C. virescens* mainly colonizes the cane epidermis and survives on infected stems and in plant debris for some time. Warm weather and frequent rains favor disease development. Under humid conditions in early to midsummer, sporangia are produced on cane lesions. Rain splash, wind-driven rain, or wind dislodge sporangia from diseased host tissue. In the presence of free water, motile zoospores are released from sporangia and spread by rain splash to susceptible tissues. The zoospores penetrate the host cuticle and cause new infections. Because of prolific sporulation, the disease can be very aggressive. Environmental stress and foliar attack by mites, insects, or other diseases may predispose plants to infection by *C. virescens*.

Management

Algal stem blotch can be avoided by starting new plantings with disease-free planting stock. Inoculum can be removed by pruning out and destroying infected canes. Reducing plant stress by proper fertilization, irrigation, and management of foliar pests and diseases may reduce disease pressure. Practices that improve air circulation, such as pruning, removing weeds, and wider plant spacing, also discourage disease development. Planting highly susceptible cultivars in disease-prone areas should be avoided. Regular application of copper-based fungicides can be used as a protective measure.

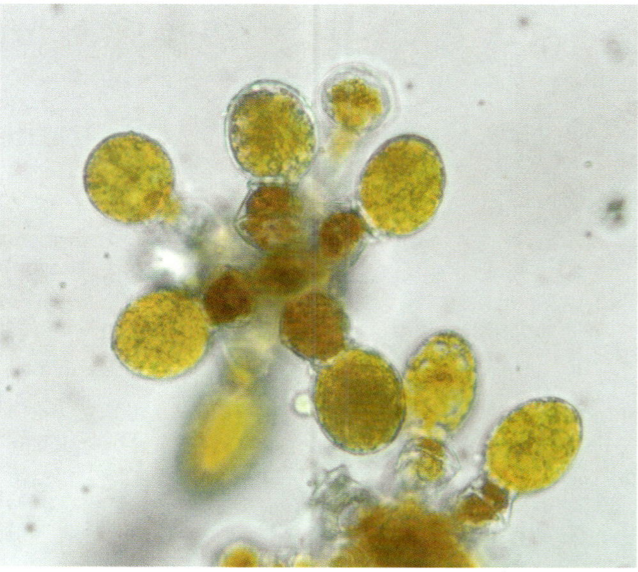

Fig. 168. *Cephaleuros virescens* zoosporangia with delimited zoospores. *Cephaleuros virescens* is the causal organism of algal stem blotch. (Courtesy G. E. Holcomb; Reproduced from Elliott, M. L., Broschat, T. K., Ushida, J. Y., and Simone, G. W. 2004. Compendium of Ornamental Palm Diseases and Disorders. American Phytopathological Society, St. Paul, MN; © APS)

Selected References

Holcomb, G. E., Vann, S. R., and Buckley, J. B. 1998. First report of *Cephaleuros virescens* in Arkansas and its occurrence on cultivated blackberry in Arkansas and Louisiana. Plant Dis. 82:263.

Joubert, J. J., and Rijkenberg, F. H. J. 1971. Parasitic green algae. Annu. Rev. Phytopathol. 9:45-64.

Lim, T. K. 1994. Algal leaf spot. Page 41 in: Compendium of Tropical Fruit Diseases. R. C. Ploetz, G. A. Zentmyer, W. T. Nishijima, K. G. Rohrbach, and H. D. Ohr, eds. American Phytopathological Society, St. Paul, MN.

Marlatt, R. B., and Alfieri, S. A., Jr. 1981. Hosts of a parasitic alga, *Cephaleuros* Kunze, in Florida. Plant Dis. 65:520-522.

Nelson, S. C. 2008. *Cephaleuros*: The plant-parasitic green algae. Plant Disease of the Week, No. 5. Online publication. University of Hawaii at Manoa, Hilo, HI.

(Prepared by A. M. C. Schilder)

Disease Caused by a Parasitic Plant

Blueberries and Cranberries

Dodder

Dodder is an obligate parasite that invades the vascular tissue of host plants to obtain carbohydrates and water. These parasitic vines are grouped into the genus *Cuscuta*, which is sometimes placed in its own family, Cuscutaceae, or more commonly, as one of approximately 50 genera in the family Convolvulaceae. Taxonomy is very difficult for this genus, relying on the careful examination of specific floral parts. Recent research has utilized plastid and nuclear DNA to delimit species. More than 150 species within three subgenera have been described in Yuncker's well-recognized monograph. The genus is distributed worldwide, although the majority of species are found in the Americas from Canada to Chile. Although some species have a very narrow host range, the genus *Cuscuta* has a wide host range, including dicotyledonous plants such as forage legumes, tomato, potato, carrot, sugar beet, citrus, numerous ornamental species, and many weeds and wild plants, as well as monocots such as onions and asparagus.

Although many species are capable of infecting both blueberry and cranberry, only two species have been documented to cause economically important infection. In New Jersey (United States), both crops are infected by *C. compacta*. This species has also been reported to attack blueberry in North Carolina (United States). In Massachusetts and Wisconsin (United States), cranberry is typically infected by *C. gronovii*, commonly called swamp dodder (Fig. 169). Recently, *C. cephalanthi* has been reported as infecting commercial cranberry vines in Massachusetts, but its economic importance and distribution remains to be determined. Research is ongoing to survey the range of dodder species present in Massachusetts. Management of dodder is of great concern for the Massachusetts cranberry industry and of relatively less importance for cranberry in Wisconsin and New Jersey. Dodder can be problematic for blueberry in North Carolina, New Jersey, and Michigan (United States).

Symptoms

Dodder seedlings produce haustoria that penetrate the cuticle and vascular tissues of the host plant, which may include various weed species as well as cranberry or blueberry plants. Heavily infected plants are weakened and stunted, and leaves may turn red. Yields are reduced because the parasite deprives the host of the carbohydrates and nutrients needed to produce new plant material. In addition, extensive dodder growth prevents sunlight from reaching the leaves of the host and reduces photosynthesis.

Causal Organisms

Dodder reproduces from seeds that are light tan to dark brown (Fig. 170). Seeds are contained in capsules in groups of one to four. Germination of dodder seed is controlled by dormancy related to a hard seed coat. Dormancy can be broken by scarification. Because seed may remain viable for as long as 30 years, it is difficult to reduce the soil seed bank. Preliminary work in Massachusetts indicates that 8–25% of the most recently produced seed may germinate the following year.

The dodder seedling lacks a radicle, but the emerging end of the embryonic axis produces a bulbous swelling that absorbs water from the soil. Dodder lacks true roots. Germination is followed by the emergence of a slender arched (or hooked) seedling. After unbending, the seedling begins to rotate counterclockwise in a circular motion (circumnutate). When it comes in contact with a vertical structure, the seedling begins to wrap

Fig. 169. Cranberry infected with swamp dodder (*Cuscuta gronovii*). (Courtesy H. A. Sandler; © APS)

Fig. 170. Dodder seeds (*Cuscuta* sp.). (Courtesy T. A. Bewick)

around that structure (Fig. 171). The thigmotropic response only occurs if red light is present. Once the dodder seedling has produced several coils around its host plant, haustoria form and the infection process begins.

Cuscuta spp. produce an intertwining mass of leafless stems that are usually yellow or orange, but green, red, and whitish stems have been observed (Fig. 172). Seedlings do contain a small amount of chlorophyll, but they must make a successful attachment to a susceptible host within days of emergence to survive. Once haustoria penetrate the host, the portion of the dodder stem in contact with the soil dies. Small white flowers are produced midsummer, and hard-coated seeds are produced (early autumn) in capsules that grow in clusters (Fig. 173). Dodder can also propagate vegetatively by means of stem pieces, which when removed from the parent plant rapidly form haustoria and infect new host plants.

Disease Cycle and Epidemiology

Dodder seedlings cannot infect a host unless they are able to form several coils around the stem. The haustorium develops by first affixing and flattening itself against the host. A wedge-shaped protrusion penetrates the epidermis cells. Once inside the plant, the surface cells of the haustorium produce hyphae. The literature may refer to these hyphae as search hyphae, although most agree that they do not actively seek the host's conducting tissues but find the cells accidentally as the hyphae grow. The hyphae, which can grow intercellularly or intracellularly, are very rich in cytoplasm and cells have very large nuclei. Hyphae penetrate the cells by using enzymes to degrade the cell walls of the host. Older search hyphae are surrounded by two walls, one of parasite origin and one of host origin. When a search hypha comes in contact with a host sieve tube element, the hypha differentiates a number of fingerlike projections called contact hyphae. This second type of hypha inserts itself between companion cells and sieve tube elements. There is little evidence of penetration of host phloem cells by the contact hyphae.

Once the contact hyphae are in position, they induce the host phloem to release nutrients into the intercellular spaces, and the hyphae then absorb the released nutrients. This process creates an extremely strong physiologic sink in the host plant, as indicated by studies done with radioactively labeled assimilates. The sink created by dodder is often stronger than all other sinks in the host, including developing fruits and seeds. Experimental evidence indicates that assimilates flow only from the haustoria to developing dodder stems and buds. This parallels the situation in other angiosperms, in which assimilates flow from photosynthetically mature leaves to immature leaves and buds. This parallel strongly suggests that dodder haustoria can be viewed as modified mature leaves. Others contend that haustoria are modified roots; evidence exists both for and against this position.

Because dodder lacks leaves and hence has a very low transpiration rate, very little water and few minerals move from the host xylem to the dodder. Detailed experiments have indicated that most materials reach dodder stems through the phloem of the host.

Management

Prevention of infestation is key for dodder management. Dodder seed capsules are buoyant. Since cranberry growers use water for many reasons, including harvest, an excellent dispersal mechanism is in place for the parasite. In addition, observations suggest that dodder seed can be spread through sprinkler irrigation systems, so this method of dispersal is of concern for blueberry and cranberry growers alike. All efforts should be made to move water from clean uninfested areas to infested areas. Many cranberry growers separate dodder seed capsules from fruit during the harvesting process and attempt further removal of the dodder capsules when abscised cranberry leaves are removed in a postharvest flood.

Early emerging weeds should be controlled. Dodder species that infect blueberry and cranberry are also able to infect many weed species, including some grasses. Dodder often infects early emerging weeds during early spring and then subsequently infect the crop plants as succulent tissues are produced. Control practices that suppress the growth of weeds, especially those that emerge early, can help reduce the spread of dodder.

Fig. 171. Dodder seedling (*Cuscuta* sp.) wrapped on a weed and cranberry stem in a cranberry bed. (Courtesy H. A. Sandler; © APS)

Fig. 172. Dodder (*Cuscuta* sp.) growing on blueberry stems. (Courtesy P. V. Oudemans; © APS)

Fig. 173. Dodder (*Cuscuta* sp.) seed capsules on dormant blueberry stems. (Courtesy P. M. Brannen; © APS)

Edges and warm areas of production areas should be scouted for newly emerging seedlings, and these should be removed as soon as possible. A single dodder plant can spread 4.5–7.5 m in 1 year. Physically removing dodder strands and infested crop plants can eliminate small areas of dodder infestation. Hand removal is easiest when dodder is in the early stages of infestation. Certainly, dodder should be removed before it produces flowers and sets seed. Dodder plants that have been removed must be relocated far away from the production site or burned.

Herbicides can be used for the control of dodder, but registrations vary from state to state. Preemergence herbicides are typically applied prior to bud break of the crop, but specific recommendations should be given by local experts. For both blueberry and cranberry systems, germination occurs in early spring. Research has shown that a low level of germination may occur throughout the growing season, making postemergence control necessary as well. Postemergence dodder control can be achieved with glyphosate (*N*-(phosphonomethyl)glycine), but care must be taken not to contact and injure the host. Although much of the dodder is killed when the herbicide is translocated through the phloem, haustoria often survive and may limit control. A biological herbicide composed of *Alternaria destruens* was commercially available, but field tests in cranberry could not demonstrate adequate efficacy. Work is currently underway to examine the pathogenicity of a *Colletotrichum* sp. Handheld flame torches could also be used to damage and kill dodder stems, especially when used before seed set.

Additional cultural practices can be interspersed into an integrated management plan. For both crops, the areas adjacent to irrigation reservoirs can be scouted for dodder infestations and any dodder should be physically removed or burned. For cranberry, application of uniform layers of sand at least 2.5 cm deep can suppress dodder seed germination in cranberry. Since sand is typically applied every 3–5 years, this method cannot provide control on an annual basis. Short (24–48 h) spring floods (typically mid-May) can also suppress dodder infestations. Summer floods (24–72 h) can also control dodder but cause crop loss.

Selected References

Dawson, J. H., Musselman, L. J., Wolswinkel, P., and Dorr, I. 1994. Biology and control of *Cuscuta*. Rev. Weed Sci. 6:265-317.

Gaertner, E. E. 1950. Studies of seed germination, seed identification, and host relationships in dodder, *Cuscuta* spp. Cornell Univ. Agric. Exp. Stn. Mem. 294.

Ghantous, K. M., Stefanovic, S., and Sandler, H. A. 2012. Initial investigations into dodder species variation in southeastern Massachusetts. Proc. Northeast. Weed Sci. Soc. 66:60.

Kuijt, J. 1977. Haustoria of phanerogamic parasites. Annu. Rev. Phytopathol. 15:91-118.

Monaco, T. J., and Mainland, C. M. 1981. *Cuscuta compacta* on blueberries in North Carolina. Haustorium Parasit. Plants Newsl. 7:1. www.odu.edu/~lmusselm/haustorium/pdf/haustorium07.pdf

Parker, C., and Riches, C. R. 1993. Parasitic weeds of the world: Biology and control. CAB International, Oxon, United Kingdom.

Sandler, H. A. 2010. Managing *Cuscuta gronovii* (swamp dodder) in cranberry requires an integrated approach. Sustainability 2:660-683. www.mdpi.com/2071-1050/2/2/660/

Sandler, H. A., Else, M. J., and Sutherland, M. 1997. Application of sand for inhibition of swamp dodder (*Cuscuta gronovii*) seedling emergence and survival on cranberry (*Vaccinium macrocarpon*) bogs. Weed Technol. 11:318-323.

Stefanovic, S., Kuzmina, M., and Costea, M. 2007. Delimitation of major lineages within *Cuscuta* subgenus *Grammica* (Convolvulaceae) using plastid and nuclear DNA sequences. Am. J. Bot. 94:568-589.

Yunker, T. G. 1932. The genus *Cuscuta*. Mem. Torrey Bot. Club 18.

Yunker, T. G. 1965. *Cuscuta* (monograph). North American Flora, Ser. II, Vol. 4. New York Botanical Garden, New York.

(Prepared by T. A. Bewick; Revised by H. A. Sandler)

Diseases Caused by Fungi

Cranberries

Berry Speckle

Berry speckle, formerly Botryosphaeria fruit rot, appears in all major cranberry-growing regions of the United States and in Nova Scotia, Canada. The disease does not cause serious economic loss, but the causal fungus, *Phyllosticta elongata*, is easily confused with *Phyllosticta vaccinii*, which causes the more serious disease, early rot. Because berry speckle adversely affects the appearance of berries, it lowers the quality of fruit grown for the fresh market. Infected fruit generally do not decay before harvest but may rot after harvest and cause significant economic loss if fruit is stored unrefrigerated.

Symptoms

Fruit infections may appear as small, light-colored lesions 1–2 mm in diameter on the fruit epidermis (Fig. 174). Lesions remain light in color as the surrounding epidermis begins to redden, or they may appear as faint red ring spots on green fruit. In storage, infected fruit may develop a soft, watery decay that resembles the field decay symptoms of early rot. Infected leaves have no visible symptoms.

Causal Organism

Berry speckle disease is caused by the fungus *Phyllosticta elongata* (syn. *Guignardia vaccinii*). Older literature may refer to the causal organism as *Botryosphaeria vaccinii*. *P. elongata* can be distinguished from the related fungus *P. vaccinii*, which causes early rot, on the basis of their characteristics in culture and the morphology of their respective conidia and pycnidia.

In culture on potato dextrose agar at 24°C, *P. elongata* grows moderately fast (2–3 mm/day) and ranges in color from blue-gray to green-gray, whereas *P. vaccinii* grows relatively slowly

Fig. 174. Berry speckle, caused by *Phyllosticta elongata*, on cranberry. (Courtesy G. J. Weidemann; © APS)

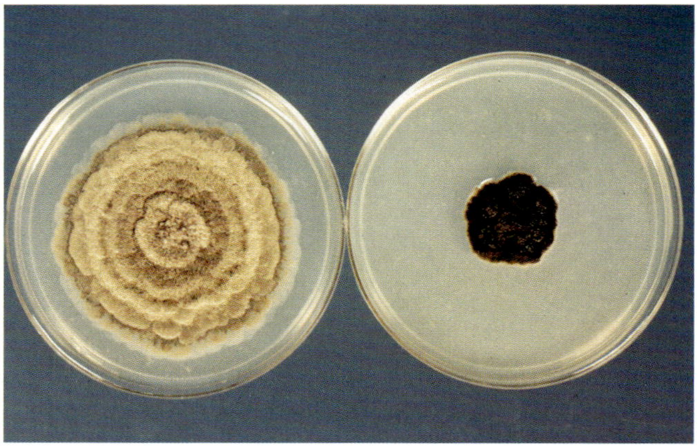

Fig. 175. Cultures of *Phyllosticta elongata* (left), the causal organism of berry speckle, and *Phyllosticta vaccinii* (right), the causal organism of early rot, growing on potato dextrose agar. (Courtesy D. M. Boone; © APS)

(0.4–1 mm/day) and ranges in color from olive to black (Fig. 175). Conidia of *P. elongata* are more elongate than those of *P. vaccinii* and possess a longer taillike appendage (mean of 30 µm versus mean of 13 µm). Pycnidia of *P. elongata* are somewhat smaller, have larger cells, and have a more poorly differentiated ostiole than those of *P. vaccinii*.

Perithecia of *G. vaccinii* (80–150 µm in diameter) are black and ostiolate. Asci (10–16 × 45–70 µm) have eight spores and are clavate and bitunicate. Ascospores (4–5 × 10–16 µm) are single celled, hyaline, and ellipsoidal to subrhomboidal. Pycnidia (80–120 µm in diameter) are single, hypophyllous, black, and ostiolate. Pycnidium walls are one to four cells thick (5–15 µm) but somewhat thicker at the apex. Conidiogenous cells (1–4 × 4–14 µm) are cylindrical. Conidia (5–8 × 9–14 µm) are single celled, hyaline, and obovate to oblong. Conidia have a mucilaginous apical appendage that is 4–128 µm long; however, most are 20–40 µm long.

Disease Cycle and Epidemiology

P. elongata (*G. vaccinii*) overwinters as pycnidia and perithecia in fallen leaves and possibly in desiccated fruit. In the spring, ascospores and conidia are disseminated to developing leaves and fruit, where they cause latent infections. Infections become more numerous as the season progresses, particularly on fruit. Infected leaves remain symptomless, although several researchers have suggested that infected leaves may drop prematurely. Infected fruit may develop a superficial speckling, but decay of fruit in the field is rare. No fruiting structures develop in infected host tissues until leaves and fruit senesce and drop in the autumn.

Management

Specific control measures generally are not required for this disease and, therefore, none have been developed. Routine fungicide applications for fruit rots probably are sufficient to minimize fruit speckle and may be beneficial in some years for fruit produced for the fresh market. Decay of fruit in storage is minimal at normally recommended storage temperatures below 5°C. The removal and/or destruction of plant debris that accumulates around the perimeter of beds during flood-harvesting operations may reduce the amount of overwintering inoculum. Delayed dormant applications of Bordeaux fungicide (8-8-100) prior to leaf development may be helpful.

Selected References

Jeffers, S. N. 1991. Seasonal incidence of fungi in symptomless cranberry leaves and fruit treated with fungicides during bloom. Phytopathology 81:636-644.

Weidemann, G. J., and Boone, D. M. 1983. Incidence and pathogenicity of *Phyllosticta vaccinii* and *Botryosphaeria vaccinii* on cranberry. Plant Dis. 67:1090-1093.

Weidemann, G. J., and Boone, D. M. 1984. Development of latent infections on cranberry leaves inoculated with *Botryosphaeria vaccinii*. Phytopathology 74:1041-1043.

Weidemann, G. J., Boone, D. M., and Burdsall, H. H., Jr. 1982. Taxonomy of *Phyllosticta vaccinii* (Coelomycetes) and a new name for the true anamorph of *Botryosphaeria vaccinii* (Dothideales, Dothioraceae). Mycologia 74:59-65.

(Prepared by G. J. Weidemann;
Revised by F. L. Caruso and J. J. Polashock)

Bitter Rot

Bitter rot is a fungal disease of cranberry fruit in both *Vaccinium macrocarpon* and *V. oxycoccos*. It was originally considered one of the most important types of rot in cranberry. This disease has been reported from all of the cranberry-producing areas in the United States and Canada. Bitter rot outbreaks tend to be sporadic and the causal organism may be the predominant fruit rot species one year and nearly absent in subsequent years.

Symptoms

Bitter rot appears as a typical cranberry fruit rot (Fig. 176) and is indistinguishable from most other fruit rots. Symptoms begin to appear in August and increase through to the time of harvest. It is necessary to confirm the species by culturing and morphological identification. Other methods, such as DNA array hybridization, can be used to identify the pathogen in cultured isolates, spores, or uncultured berry tissue.

Causal Organisms

Glomerella cingulata (syn. *Colletotrichum gloeosporioides*) is the primary causal organism of bitter rot. The species is classified in the Glomerellaceae, a monotypic family in the class Sordariomycetes, phylum Ascomycota. Conidia are produced in acervuli formed on leaves, stems, and dead tissues. The fungus also produces conidia abundantly in culture. The conidia are flesh colored in mass but hyaline when viewed singly. Conidia (4.5–6 × 12–18 µm) are oblong-elliptical or sometimes slightly smaller at one end (Fig. 177A). The acervuli initially are subepidermal but become erumpent when mature.

Perithecia are common for certain isolates in vitro but are often sterile. Isolates can be divided as those that form perithecia and conidia and others that are strictly conidial. The

Fig. 176. Symptoms of bitter rot (fruit rot; field rot), caused by *Glomerella cingulata*, in a cultivated cranberry field. (Courtesy P. V. Oudemans; © APS)

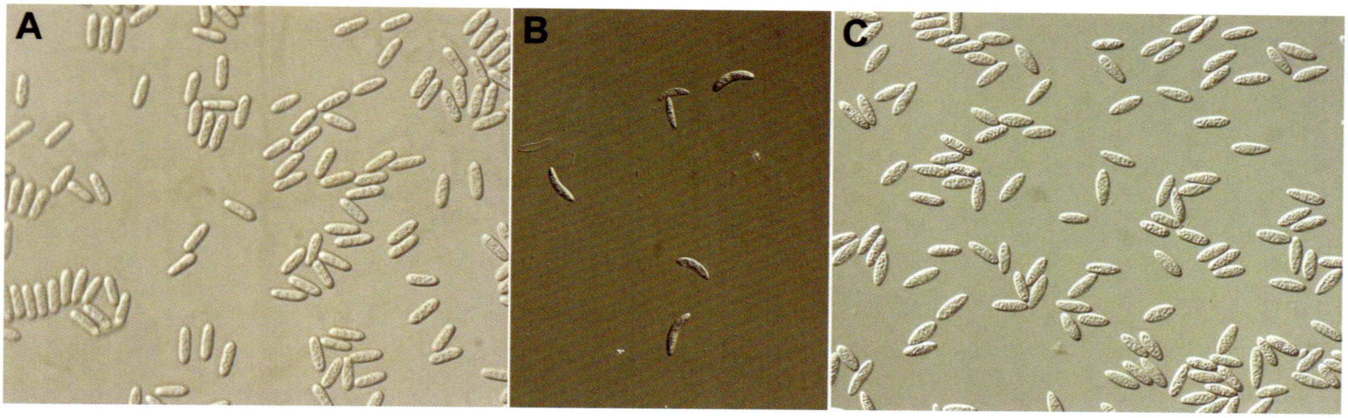

Fig. 177. Conidia and ascospores of the bitter rot fungi. **A,** *Glomerella cingulata* conidia; **B,** *Glomerella cingulata* ascospores; and **C,** *Colletotrichum fioriniae* conidia. Note the differences in conidial morphology between the two species. (Courtesy P. V. Oudemans; © APS)

perithecia are membranous and subglobose or somewhat pear shaped. Asci (10–12 × 60–70 µm) are clavate and sessile or short stipitate. Ascospores (5–7.5 × 9–18 µm) (Fig. 177B) are irregularly biseriate, oblong-elliptical, and occasionally slightly inequilateral or curved. They change from hyaline to pale greenish yellow when mature.

Colletotrichum fioriniae is also reported as a cause of bitter rot and may be more common in northern growing areas. The former species name for this pathogen is *Colletotrichum acutatum* (syn. *Glomerella acutata*). The species is classified in the Glomerellaceae, a monotypic family in the subclass Hypocreomycetidae, class Sordariomycetes, phylum Ascomycota. *C. fioriniae* produces fusiform, one-celled, hyaline conidia averaging approximately 3 × 14 µm (Fig. 177C). The conidia are produced on conidiophores in acervuli just below the epidermis of the infected tissue and emerge as orange- to salmon-colored, sticky masses. On a berry, the fungus causes the fruit to shrivel and become raisinlike, with acervuli forming over the fruit surface. In culture, the fungus is often chromogenic when grown on standard growth media. Isolates from cranberry are typically deep red on the underside of the colony, although some are nonchromogenic. The upper surface of the colony is a light-colored, fluffy mycelium with black or dark gray rays or sections. The sexual stage of this species has not been observed on cranberry.

Disease Cycle and Epidemiology

C. gloeosporioides and *C. fioriniae* overwinter on infected living and dead leaves, stems, and occasionally, rotted fruit. Conidia are most likely the principal inoculum source, although perithecia and ascospores of *G. cingulata* have been detected in field-collected materials. Perithecia have been observed on pedicels produced during the previous season. Conidia that cause infection in new growth are produced in acervuli on the overwintering sites and are dispersed by water droplets. Conidial production is probably correlated with weather and with cultural practices used in a particular cranberry bed.

Fruit are infected during the early stages of development, and after a certain point in fruit development, new infections may occur but do not lead to field rot. Following infection, the fungus remains latent until the fruit begins to mature. As the fruit mature, the fungus resumes growth and colonizes the fruit tissues, causing the internal tissues to rot but the skin remains intact. Most of the rot develops before harvest and early in the storage of fresh fruit.

The amount of bitter rot varies from season to season. In some years *G. cingulata* is the predominant pathogen present in cranberry beds; however, in subsequent years, it may be much reduced. *C. fioriniae* is more common in some growing regions and may be an increasing problem.

Management

Fungicides such as chlorothalonil, ferbam, maneb, mancozeb, and azoxystrobin have all shown activity in the control of bitter rot. The fungicide fenbuconazole is ineffective against these pathogens. Bitter rot management should follow recommended fruit rot control methods. Fungicide application should begin during early bloom and extend 2–3 weeks past petal fall. Protection of the newly formed fruit is most important.

Cultural practices that encourage sound fruit, such as good water management and proper fertilization, aid in reducing bitter rot losses.

Selected References

Oudemans, P. V., Caruso, F. L., and Stretch, A. W. 1998. Cranberry fruit rot in the Northeast: A complex disease. Plant Dis. 82:1176-1184.

Robideau, G. P., Caruso, F., Oudemans, P. V., McManus, P. S., Renaud, M. A., Auclair, M. E., DeVerna, J., and Lévesque, C. A. 2008. Detection of cranberry fruit rot using DNA array hybridization. Can. J. Plant Pathol. 30:226-240.

Shear, C. L., Stevens, N. E., and Bain, H. F. 1931. Fungous diseases of the cultivated cranberry. U. S. Dep. Agric. Tech. Bull. 258.

Stiles, C. M., and Oudemans, P. V. 1999. Distributions of cranberry fruit-rotting fungi in New Jersey and evidence for host nonspecific resistance. Phytopathology 89:218-225.

Zuckerman, B. M. 1958. Relative importance of cranberry rot fungi during the storage and harvest season in Massachusetts, 1956–1957. Plant Dis. Rep. 42:1214-1221.

(Prepared by A. W. Stretch; Revised by P. V. Oudemans)

Black Rot

Black rot is a postharvest disease of cranberry that appears in all cranberry-growing regions of North America. The disease was first reported in 1920 and currently is prevalent on water-harvested fruit in Massachusetts, New Jersey, and Wisconsin (United States). Black rot is not a serious problem in Washington and Oregon (United States), where water-harvested fruit is quick frozen for processing. The disease has also been reported as a minor rot of stored fruit in Michigan (United States). Fruit that is dry harvested, especially fruit that has insect injury, may also develop black rot.

Symptoms

Symptoms usually develop on fruit within the first 2 months of storage at 2–4°C. They range from a pale gray-brown mottle to the more typical uniform jet black discoloration (Fig. 178). The rot is relatively firm and dry during the early stages of decay, and berries gradually become dried and shriveled. Pycnidia of the pathogen often develop in the epidermis.

Causal Organisms

Three species of anamorphic fungi are known to cause black rot: *Allantophomopsis cytisporea*, *Allantophomopsis lycopodina*, and *Strasseria geniculata* (syn. *S. oxycocci*). *Allantophomopsis cytisporea* and *Allantophomopsis lycopodina* have also been referred to as light and dark strains, respectively, of *Ceuthospora lunata* and *Apostrasseria lunata*. *Ceuthospora lunata* and *Apostrasseria lunata* are now recognized synonyms of *Allantophomopsis cytisporea*. The synonym *Phacidium lunatum* (Phacidiaceae, Helotiales) on leaves of *Gaultheria procumbens* was reported for *Allantophomopsis cytisporea*. Teleomorphs for *Allantophomopsis lycopodina* and *S. geniculata* are not known.

On V8 agar, *Allantophomopsis cytisporea* produces a relatively flat, pale to medium gray to creamy white colony that covers a 9-cm-diameter plate in 7 days at room temperature (Fig. 179). A pink or yellow diffusible pigment usually is produced. Colonies of *Allantophomopsis lycopodina* (Fig. 179) differ from those of *A. cytisporea* primarily in the production of a darker gray-green mycelium and the lack of a diffusible pigment. Colonies of *S. geniculata* (Fig. 179) resemble those of *Allantophomopsis cytisporea* but are paler in color, exhibit irregularly petaloid growth, and grow more slowly.

The three black rot fungi produce hyaline, unicellular, allantoid to lunate conidia (Fig. 180). The conidia of *Allantophomopsis cytisporea* and *Allantophomopsis lycopodina* have a mucoid, cone-shaped apical appendage and an inconspicuous, irregularly shaped, mucoid basal appendage. The conidia of *Allantophomopsis cytisporea* (Fig. 180B) are uninucleate and 2–3.5 × 7–10 µm (mean of 2.9 × 8.3 µm); those of *Allantophomopsis lycopodina* (Fig. 180A) are binucleate and 2–3.5 × 8–15 µm (mean of 2.8 × 11.2 µm). The conidia of *S. geniculata* (Fig. 180C) are 2–3.5 × 7–14 µm (mean of 2.8 × 10.6 µm) and have a cellular, tubular basal appendage up to 22 µm long and a mucoid apical appendage.

Fig. 178. Cranberry fruit with black rot symptoms (top right and bottom), caused by *Allantophomopsis cytisporea*, *A. lycopodina*, or *Strasseria geniculata*, and a healthy fruit (top left). (Courtesy L. M. Carris; © APS)

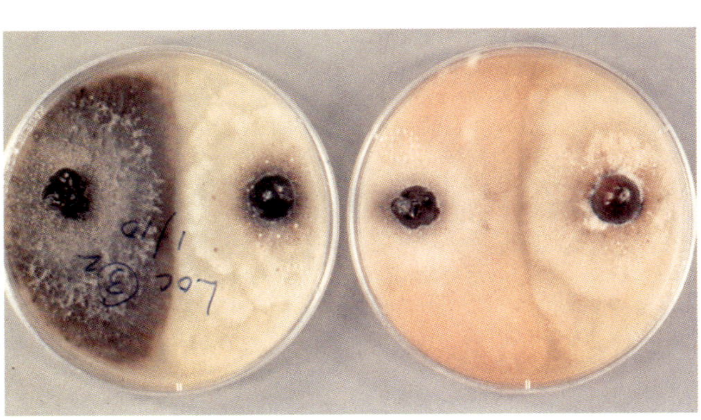

Fig. 179. Two-week-old colonies of cranberry black rot fungi on V8 agar (left to right): *Allantophomopsis lycopodina*, *Strasseria geniculata*, *A. cytisporea*, and *S. geniculata*. (Courtesy L. M. Carris; © APS)

Disease Cycle and Epidemiology

The black rot fungi probably overwinter as mycelium and immature pycnidia in necrotic cranberry leaves and rotted fruit in the duff layer of cranberry beds. Pycnidia then develop in the spring and summer. Mature pycnidia are present by harvest in the autumn. Conidia are released from pycnidia in the presence of water. Harvest water can contain numerous conidia of the black rot fungi, and this probably represents the main source of inoculum for fruit infection. The pathogens infect the fruit through wounds caused during harvest, and disease incidence is directly proportional to the length of time fruit remain in the floodwater during harvest. Conidia may remain viable on cranberry fruit for up to 6 weeks and can germinate and infect fruit that is wounded during postharvest handling.

Leaves may become infected during or after harvest from conidia that adhere to the leaf cuticle. The black rot fungi are isolated infrequently from symptomless leaf and fruit tissues.

Black rot also occurs in dry-harvested fruit. The timing and mode of infection in dry-harvested bogs are not known precisely, but they may be related to rainfall and the duration of wet periods.

Allantophomopsis lycopodina is an aerial pathogen of cultivated lingonberry (*Vaccinium vitis-idaea*) when plants are water-stressed. The pathogen causes dark brown lesions on leaves and pedicels, and branch tip dieback. Berries forming from necrotic pedicels may become shriveled and blackened.

Allantophomopsis lycopodina has been associated with seedling mortality of western white pine (*Pinus monticola*) and Douglas-fir (*Pseudotsuga menziesii*) in British Columbia (Canada) and has been isolated from leaf litter of rhododendron in the Netherlands (as *Guignardia philoprina*). A strain of *Allantophomopsis lycopodina* isolated from a dead beech branch in Japan produces a secondary metabolite, allantopyrone A, shown to be cytotoxic to human leukemic cells in vitro. Another strain isolated from a piece of rotting deciduous wood in France produces an antifungal antibiotic, allantofuranone.

Management

Fungicides have not been effective in managing black rot. Minimizing damage to the fruit and the length of time that fruit remains in water during harvest reduces disease incidence. Rapid drying or freezing of fruit after harvest may help limit black rot development. If possible, fruit destined for the fresh market should be dry harvested. Storing fruit for processing at temperatures slightly above 0°C greatly reduces the incidence of black rot.

Selected References

Carris, L. M. 1990. Cranberry black rot fungi: *Allantophomopsis cytisporea* and *Allantophomopsis lycopodina*. Can. J. Bot. 68:2283-2291.

DiCosmo, F., Nag Raj, T. R., and Kendrick, W. B. 1983. Prodomus for a revision of the Phacidiaceae and related anamorphs. Can. J. Bot. 61:31-44.

Hopkins, J., and Funk, A. 1987. *Apostrasseria lunata* associated with seedling mortality of western white pine and Douglas-fir in British Columbia. Can. Plant Dis. Surv. 67:23.

Jeffers, S. N. 1991. Effects of fungicides applied during bloom on yield, yield components, and storage rots of cranberry. Plant Dis. 75:244-250.

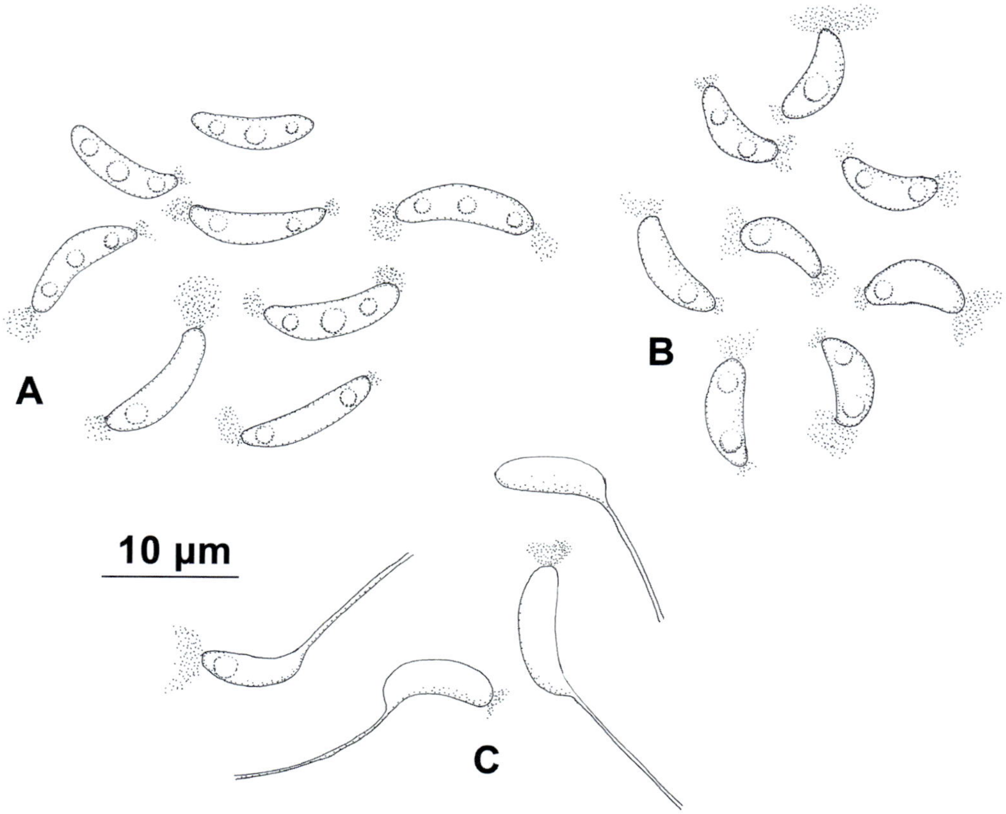

Fig. 180. Conidia of black rot fungi: **A,** *Allantophomopsis lycopodina*; **B,** *A. cytisporea*; and **C,** *Strasseria geniculata*. (Courtesy L. M. Carris; © APS)

Okane, I., Nakagiri, A., and Ito, T. 2001. Identity of *Guignardia* sp. inhabiting ericaceous plants. Can. J. Bot. 79:101-109.

Olatinwo, R. O., Hanson, E. J., and Schilder, A. M. C. 2003. A first assessment of the cranberry fruit rot complex in Michigan. Plant Dis. 87:550-556.

Oudemans, P. V., Caruso, F. L., and Stretch, A. W. 1998. Cranberry fruit rot in the Northeast: A complex disease. Plant Dis. 82:1176-1184.

Putnam, M. L. 2005. *Allantophomopsis lycopodina*—A new aerial pathogen of lingonberry (*Vaccinium vitis-idaea*). Plant Pathol. 54:248.

Schüffler, A., Kautz, D., Liermann, J. C., Opatz, T., and Anke, T. 2009. Allantofuranone, a new antifungal antibiotic from *Allantophomopsis lycopodina* IBWF58B-05A. J. Antibiot. 62:119-121.

Schwarz, M. R., and Boone, D. M. 1983. Black rot of cranberry caused by *Strasseria oxycocci*. Plant Dis. 67:31-32.

Schwarz, M. R., and Boone, D. M. 1985. Effect of wounding on incidence of black rot of cranberry in Wisconsin. Plant Dis. 69:225-227.

Shiono, Y., Yokoi, M., Koseki, T., Murayama, T., Aburai, N., and Kimura, K. 2010. Allantopyrone A, a new α-pyrone metabolite with potent cytotoxicity from an endophyte fungus, *Allantophomopsis lycopodina* KS-97. J. Antibiot. 63:251-253.

Stretch, A. W., and Ceponis, M. J. 1983. Influence of water immersion time and storage period on black rot development in cold-stored, water-harvested cranberries. Plant Dis. 67:21-23.

(Prepared by L. M. Carris)

Black Spot

In the Pacific Northwest, particularly Washington and Oregon (United States), red leaf spot (caused by *Exobasidium rostrupii*) causes serious injury at times. Much of the injury, however, is caused by the associated disease black spot. The black spot fungus infects plants through red leaf spot lesions. Berries are also attacked, but disease incidence on fruit is not high enough to be economically important. Although red leaf spot appears in all cranberry-growing areas, black spot is rarely observed elsewhere, and rarely warrants control measures.

Symptoms

Numerous black lesions, usually centered around leaf scars, form on runners and uprights (Fig. 181). Lesions generally girdle the stem, although they may be confined to one side of the stem. Girdling is followed by defoliation and death of the portion of the upright or runner beyond the lesion; this usually happens in August or September, when it is too late for lateral branches to set a mixed bud. Hence, damage to the next year's crop is probably greater than damage in the year the disease appears.

Fig. 181. Red leaf spots in cranberry colonized by the secondary invader *Mycosphaerella nigromaculans*, the causal organism of black spot. (Courtesy P. R. Bristow; © APS)

Initially, lesions and then dead stems become thickly covered with minute black fruiting bodies of the pathogen (Fig. 182). Red leaf spot lesions on leaves and berries also become blackened. Lesions on berries are black and sunken and vary in size from tiny spots to large lesions covering more than half the berry. The lesions that develop in the field before harvest do not lead to rotting in storage. The healthy portion of spotted berries remains firm unless invaded by other fungi.

Causal Organism

Black spot is caused by the ascomycetous fungus *Mycosphaerella nigromaculans*. Its asexual stage, the synonym *Ramularia nigromaculans*, is known in culture but has not been reported in the field. Immature perithecia form in the autumn and are thickly scattered in black stem and berry lesions. Mature perithecia (60–150 µm in diameter) are black, subglobose, and embedded in the host tissue; they have thin, membranous walls and small, papillate ostioles. The sessile, slightly curved asci (3–4.5 × 23–35 µm) arise from a cushionlike mass of hyaline tissue at the base of the perithecium. There are no paraphyses. Each ascus has eight hyaline, elliptical ascospores (1.8–4 × 7.5–10 µm). A constricted septum divides each ascospore into two nearly equal cells (the upper cell is slightly wider than the lower).

In culture, ascospores germinate to give rise to the asexual stage. The pathogen grows slowly on cornmeal agar; the colony is initially gray and then becomes dark brown. The dense hyphae are mostly embedded in the agar. Branched, erect, yellowish brown conidiophores arise from the submerged hyphae. The unicellular, globose to elliptical conidia (2–3.5 × 2–6 µm) are formed in chains and, when mature, are the same color as the conidiophores.

Disease Cycle and Epidemiology

Knowledge of the disease cycle of black spot is incomplete. The pathogen overwinters as immature perithecia in stem lesions. Although immature perithecia are abundant in sunken berry lesions in the autumn, no mature ones are found in the spring, probably because other fungi invade the infected berries. Perithecia on diseased vines mature in the spring. It is not known whether ascospores are the primary inoculum, when or where primary infection takes place, or what role, if any, the asexual stage plays in the disease cycle. Possibly, ascospores germinate and the fungus grows saprophytically in the duff layer or on the bark of older runners, producing conidia that infect the host later in the summer when red leaf spot lesions are present on developing berries. Another hypothesis is that the perithecia mature over a long period of time and that ascospores from the later-maturing ones infect the host during the summer.

Fig. 182. Close-up of the secondary invader *Mycosphaerella nigromaculans*, the causal organism of black spot, colonizing cranberry stem lesions and leaf spots caused by *Exobasidium rostrupii*. (Courtesy P. R. Bristow; © APS)

Black spot, like red leaf spot, is influenced greatly by weather conditions. Cloudy, rainy, or misty weather promotes disease development. The disease is most prevalent in shaded or sheltered areas of the bog.

Management

Cutting back surrounding trees and brush to improve air circulation should help to control black spot. Repeated applications of Bordeaux mixture have been observed to control black spot. If indeed *M. nigromaculans* requires red leaf spot lesions before it can attack the cranberry plant, then controlling red leaf spot disease should also control black spot.

Selected References

Bain, H. F. 1926. Cranberry disease investigations on the Pacific Coast. U. S. Dep. Agric. Bull. 1434.

Shear, C. L., Stevens, N. E., and Bain, H. F. 1931. Fungous diseases of the cultivated cranberry. U. S. Dep. Agric. Tech. Bull. 258.

(Prepared by P. R. Bristow; Revised by F. L. Caruso)

Blotch Rot

Blotch rot is a fungal disease of cranberry fruit in both *Vaccinium macrocarpon* and *V. oxycoccos*. It has been reported from all of the cranberry-producing areas in the United States and Canada and is an economically important component of the fruit rot complex. The causal organism, *Physalospora vaccinii*, has been reported from native and cultivated *Vaccinium* populations of North America and Europe. Blotch rot originally was named to distinguish it from other cranberry fruit rots. However, it is very difficult to differentiate from other fruit rot diseases based on symptoms.

Symptoms

Blotch rot is similar to most other fruit rots by visual examination (see Fig. 176 under Bitter Rot). Positive identification generally requires isolation of the causal organism from affected fruit or through molecular methods, such as DNA array hybridization. Blotch rot may progress from the blossom end; the developing lesion is approximately circular, flattened or sunken, and slightly discolored compared with the surrounding healthy tissue. Lesions eventually develop a translucent epidermis and have softened pericarp tissues. In later stages of rot development, small, dark blotches may appear on the epidermis. On healthy leaves, the fungus produces dark characteristic appressoria that can be observed with a hand lens. The fungus sporulates via perithecia on dead leaves and is rarely seen sporulating on the fruit.

Causal Organism

The causal organism *Physalospora vaccinii* is in the class Sordariomycetes, phylum Ascomycota. This species has some unresolved taxonomic issues. Two distinct morphological types can be separated by colony color (dark versus light) as well as by perithecial and ascospore morphology. The two strains can also be separated using molecular analysis; however, they are still classified as one species. The frequency of isolation of the two strains varies among growing regions; the light strain tends to be rare in New Jersey and Massachusetts but common in Wisconsin (United States). The morphology of the dark strain best fits the species description of *P. vaccinii*. In culture, the ascospores of the dark strain are broadly obovoid with blunt ends and 12–18 × 26–33 µm, whereas those of the white strain have pointed ends and are 16–19 × 44–50 µm. The ascospores are biseriate or irregularly uniseriate, hyaline until almost mature, and pale yellowish brown when mature. The surface of spores of both strains has a punctate ornamentation. In vitro, perithecia are produced most abundantly on cornmeal agar. Perithecia of the dark strain produced in vitro are slightly smaller in diameter (316 versus 332 µm) but greater in height (520 versus 479 µm) than those of the white strain.

Fig. 183. *Physalospora vaccinii*, the causal organism of blotch rot, sporulating (perithecia) on a detached cranberry leaf. (Courtesy P. V. Oudemans; © APS)

Fig. 184. Appressoria and ascospores of *Physalospora vaccinii* on a cranberry leaf surface. Appressoria can remain intact on the leaf surface of a least 2 years. (Courtesy P. V. Oudemans; © APS)

On cranberry leaves, the perithecia of *P. vaccinii* (300–400 µm in diameter) are hypophyllous, scattered, depressed globose to pyriform, and subepidermal. Beaks of perithecia are erumpent, and black, nonseptate spines (8–9 × 50–70 µm) are at the base. The walls of the perithecium are membranous, consisting of a single layer of cells. Asci (30–48 × 136–180 µm) have eight spores and are clavate and short stipitate. The paraphyses (5–8 × 200–300 µm) are stout and septate and occasionally branch near the ends. No anamorph has been observed for this species.

Disease Cycle and Epidemiology

P. vaccinii is commonly isolated from cranberry leaves as well as from rotted and sound fruit. It is occasionally isolated from the pedicels and stems of growing plants. The fungus sporulates on dead or dying leaves but rarely on fruit (Fig. 183). The ascospores are forcibly ejected from the perithecia and germinate with a single broad germ tube. An obvious dark, lobate appressorium is formed at the end of the germ tube (Fig. 184). It is able to penetrate leaves through stomata by means of infection hyphae that grow from appressoria. Once established on leaves, the pathogen is thought to become quiescent, invading and colonizing the leaves only after they begin to senesce. These latent infections on leaves are likely the primary method of survival from season to season. Since sporulation does not begin until the leaves senesce, infections may remain quiescent for 1–2 years, which is the normal survival time of cranberry leaves. Perithecia form on the upper surface of dying leaves and ascospores are ejected into the canopy (Fig. 183). Spore production begins in early June but continues through the growing season, reaching a peak in late summer to early autumn. Fruit may be infected throughout the season; however, fruit infected during the bloom period or shortly after fruit set are most likely to be rotted. Leaf infections (Fig. 185) continue to increase throughout the growing season.

Vines used as planting material invariably contain latent infections. Thus, the use of field-produced vines for planting new acreage also introduces the fungus.

Management

Fungicides such as chlorothalonil, ferbam, maneb, mancozeb, fenbuconazole, and azoxystrobin have all shown activity in the control of blotch rot. Blotch rot management should follow recommended fruit rot control methods. Fungicide application should begin during early bloom and extend 2–3 weeks past petal fall. Protection of the newly formed fruit is most important.

Cultural practices that encourage sound fruit, such as good water management, proper fertilization, and adequate pruning, also aid in reducing blotch rot losses. Excessive use of nitrogen fertilizer that leads to heavy vine growth increases disease incidence.

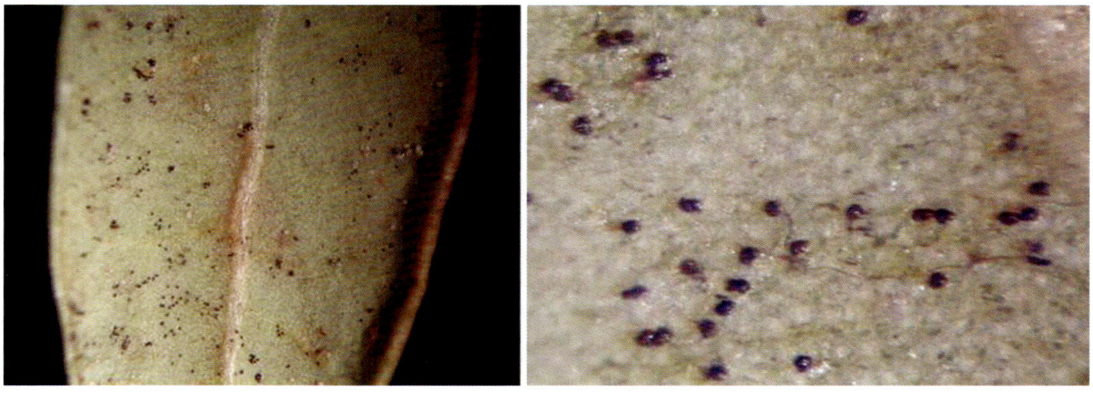

Fig. 185. Appressoria of *Physalospora vaccinii* on the underside of a cranberry leaf (left) and magnified for detail (right). (Courtesy P. V. Oudemans; © APS)

Selected References

Brown, K. J. 1982. *Physalospora vaccinii* and its effects on cranberries in Wisconsin. M.S. thesis. University of Wisconsin, Madison.

Friend, R. J. 1968. Incidence and pathogenicity of fungi found on cranberry in Wisconsin. Ph.D. thesis. University of Wisconsin, Madison.

Jeffers, S. N. 1991. Seasonal incidence of fungi in symptomless cranberry leaves and fruit treated with fungicides during bloom. Phytopathology 81:636-644.

McManus, P. S., Caldwell, R. W., Voland, R. P., Best, V. M., and Clayton, M. K. 2003. Evaluation of sampling strategies for determining incidence of cranberry fruit rot and fruit rot fungi. Plant Dis. 87:585-590.

Olatinwo, R. O., Hanson, E. J., and Schilder, A. M. C. 2003. A first assessment of the cranberry fruit rot complex in Michigan. Plant Dis. 87:550-556.

Oudemans, P. V., Caruso, F. L., and Stretch, A. W. 1998. Cranberry fruit rot in the Northeast: A complex disease. Plant Dis. 82:1176-1184.

Robideau, G. P., Caruso, F., Oudemans, P. V., McManus, P. S., Renaud, M. A., Auclair, M. E., DeVerna, J., and Lévesque, C. A. 2008. Detection of cranberry fruit rot using DNA array hybridization. Can. J. Plant Pathol. 30:226-240.

Shear, C. L., Stevens, N. E., and Bain, H. F. 1931. Fungous diseases of the cultivated cranberry. U. S. Dep. Agric. Tech. Bull. 258.

Stiles, C. M., and Oudemans, P. V. 1999. Distribution of cranberry fruit-rotting fungi in New Jersey and evidence for nonspecific host resistance. Phytopathology 89:218-225.

(Prepared by D. M. Boone; Revised by P. V. Oudemans)

Cladosporium Leaf Spot

Cladosporium leaf spot is uncommon but may cause significant losses when it does occur. The pathogen, *Cladosporium oxycocci*, causes small, brownish lesions on the previous season's leaves. Cladosporium leaf spots are distinguished from Protoventuria leaf spots and Pyrenobotrys leaf spots by their distinct gray to white centers (Fig. 186). During the course of the growing season, infected leaves become chlorotic and may fall prematurely. The fungus can be observed on the upper surface of the leaves. The fungus also causes small, red lesions, sometimes with light-colored centers, on berries. These lesions are most noticeable before the fruit develops full coloration. Injury to the berries is superficial.

Conidiophores of *C. oxycocci* are hypophyllous, simple, septate, flexuous, yellow-brown, and erect or spreading. They are arranged in small tufts arising from a sclerotioid base. Conidia (one to three per conidiophore) are acrogenous, yellow-brown, cylindrical to clavate, continuous to uniseptate, and 3–4 × 15–24 µm.

Fig. 186. Cranberry leaves with well-defined lesions with light gray centers, characteristic of Cladosporium leaf spot, caused by *Cladosporium oxycocci*. (Courtesy D. M. Boone; © APS)

Control measures are not warranted for this disease.

Selected References

Shear, C. L. 1907. Cranberry diseases. U. S. Dep. Agric. Bull. 110.

Shear, C. L., Stevens, N. E., and Bain, H. F. 1931. Fungous diseases of the cultivated cranberry. U. S. Dep. Agric. Tech. Bull. 258.

(Prepared by F. L. Caruso)

Cottonball

Cottonball causes shoot blight and fruit rot of cranberry (*Vaccinium macrocarpon*) and small cranberry (*V. oxycoccos*) in northern temperate climates worldwide. On cultivated cranberry, the disease is sporadic but economically important in British Columbia and Quebec (Canada) and especially in Wisconsin (United States), where incidence of the fruit rot phase of cottonball can exceed 50%. Low incidences of cottonball have been reported in Massachusetts, Oregon, and Washington (United States) and in Nova Scotia (Canada). Cottonball is a chronic problem in some older peat-based cranberry beds but is less common in newer sand-based plantings. The fruit rot phase of cottonball makes fruit unmarketable; thereby reducing yields and increasing the costs associated with sorting berries prior to processing or packaging. The shoot blight (also called tip blight) phase of cottonball usually does not cause direct losses of economic significance. Cottonball is sometimes referred to as hard rot because diseased berries typically remain firm and transform into hardened pseudosclerotia, unlike berries affected by other fruit rot pathogens.

Symptoms

Shoot blight symptoms first appear about 3–4 weeks after bud break, at the time that flower buds are separating from the shoot axis. Diseased shoot tips are tan, with necrosis extending down the stem and into the leaves (Fig. 187). The tan discoloration on the leaves is usually U- or V-shaped and centered along the base of the midvein. Unopened flowers also may become tan and necrotic. By early bloom, the fungus forms white ectostromata, often on crooked-over shoot tips or flower peduncles. The conidia-bearing ectostromata are present throughout bloom. Diseased shoots are interspersed with, and greatly outnumbered by, healthy shoots, making shoot blight inconspicuous in the field. Shoot blight is patchy in distribution and is often more prevalent in low, wet areas such as along center ditches in the cranberry bed. After bloom, diseased shoots become brittle and break from the healthy portion of the vine, leaving little evidence of the shoot blight phase of cottonball.

Infected immature berries show no symptoms of cottonball. However, white, cottonlike mycelia of *Monilinia oxycocci* colonize one or more locules of developing berries (Fig. 188). Later in the summer when healthy berries turn red, berries with cottonball remain yellowish green, although the fruit surface corresponding to uninfected locules may turn red. By the time of harvest, infected berries turn dark tan with brown necrotic striations that correspond to carpel walls of infected locules (Figs. 189 and 190). Infected berries remain firm, unless they become colonized by other pathogens or saprophytes. Eventually, the cottonball pathogen colonizes the pericarp of diseased berries, forming black, hard pseudosclerotia within the dried berry skins. Pseudosclerotia, also referred to as mummies, sometimes resemble dried, shrunken cranberries. In other cases, pseudosclerotia are hollow or consist only of the hardened carpel walls.

Causal Organism

Cottonball is caused by the ascomycetous fungus *Monilinia oxycocci* (syn. *Sclerotinia oxycocci*). The genus *Monilinia* was segregated from the genus *Sclerotinia* by Honey, who further subdivided the genus into the sections Disjunctoriae (having disjunctors between mature catenulate conidia) and Junctoriae (lacking disjunctors). *M. oxycocci* is in the section Disjunctoriae, whose members have similar life cycles and restricted host ranges.

Pseudosclerotia germinate to form one to many dark reddish brown cupulate apothecia with discs 3–12 mm in diameter and stipes 10–70 mm long (Fig. 191). Each ascus (8–12 × 150–190 μm)

Fig. 189. Mature cranberry fruit with cottonball, caused by *Monilinia oxycocci*. (Courtesy P. S. McManus; © APS)

Fig. 187. Cranberry shoot infected by *Monilinia oxycocci*, the causal organism of cottonball. Note the V-shaped brown discoloration at the leaf base, the crooking of the shoot tip, and the white mantle of the conidia. (Courtesy P. S. McManus; © APS)

Fig. 190. Cross section of mature cranberry fruit colonized by *Monilinia oxycocci*, the causal organism of cottonball. Four healthy berry halves at lower left. (Courtesy P. S. McManus; © APS)

Fig. 188. Immature cranberry fruit colonized by *Monilinia oxycocci*, the causal organism of cottonball. A healthy berry is in the center. (Courtesy P. S. McManus; © APS)

Fig. 191. Apothecia of *Monilinia oxycocci*, the causal organism of cottonball on cranberry, germinating from a pseudosclerotium. (Courtesy P. S. McManus; © APS)

contains four large (6–12 × 12–22 µm) and four small (4–6 × 10–14 µm) ascospores. Conidia of *M. oxycocci* (16–23 × 25–28 µm) are hyaline, limoniform to subglobose, and produced in branched or unbranched moniliform chains. Disjunctors that separate conidia in a chain are produced in planta but may be absent in culture. Microconidia may be produced from ascospores, conidia, or hyphae, but their function is not known.

Growth of *M. oxycocci* in vitro is slow and varies greatly among isolates, depending on culture medium, temperature, and light. In general, vegetative growth and stroma production are optimal on potato dextrose agar at 24°C in the dark, with 4°C the minimum temperature and 32°C the maximum temperature for vegetative growth. At 20–24°C, colony diameter after 16 days is approximately 50 mm. Conidium production is optimal on V8 juice agar at 20°C in the dark.

Disease Cycle and Epidemiology

M. oxycocci overwinters in pseudosclerotia at or near the soil surface or, less commonly, attached to the plant. Pseudosclerotia are highly resistant structures and probably remain viable for several years. However, pseudosclerotia produced in the past 1–3 years are probably the most important, since they are only lightly buried in leaf debris, sand, or both. Pseudosclerotia that are in moist soil, moss, or leaf debris germinate to form apothecia in early spring. Apothecia start to discharge ascospores at about the time that cranberry shoot growth resumes (bud break). Ascospores are discharged for a 25- to 35-day period, continuing until just before bloom. Maximum discharge occurs daily over a 10- to 14-day period and peaks when the majority of shoots show 1–3 cm of new growth. The availability of airborne ascospores peaks at 1600 to 1800 hours. The distance over which ascospores are disseminated by wind is not known, although the patchy distribution of shoot blight symptoms in the field suggests a steep dispersal gradient. The environmental conditions that lead to the infection of actively growing shoots by ascospores and subsequent shoot blight symptoms are unknown, although disease development appears to be favored by prolonged periods of wetness and cool to moderate weather conditions.

Shoot blight symptoms and white ectostromata that bear conidia first appear just prior to bloom. Conidia are disseminated during a 10- to 14-day period that coincides with bloom. Dispersal of conidia is correlated with warm conditions, low relative humidity, and wind speed, and peaks at 1000 to 1800 hours. In analogous diseases of ericaceous hosts (e.g., mummy berry of blueberry), insects have been implicated in delivering conidia to flowers as they pollinate. Although this may also happen with cottonball of cranberry, it has not been documented. Conidia germinate on the stigmata of open flowers, after which germ tubes grow through the stylar canal and reach ovaries by 72 h after inoculation. Germination on stigmata is enhanced by prior pollination and the absence of free water. Although conidia germinate on other flower parts, infection is successful only through the inoculation of stigmata. Ascospores are capable of infecting open flowers, but this probably rarely happens, because ascospores are depleted prior to bloom.

Following successful pollination and infection of flowers, *M. oxycocci* grows and nearly fills the locules of developing fruit. The fungus colonizes the pericarp, and by harvest, hardened pseudosclerotia form in up to 50% of the berries with cottonball. Pseudosclerotia do not develop later, and diseased berries without pseudosclerotia that remain in the bed following harvest decompose by the following spring. Pseudosclerotia float and may be disseminated within and among beds by water.

Management

Cottonball is managed by cultural methods and application of fungicides. Because diseased fruit and some pseudosclerotia float, most potential inoculum is removed during harvest. Reflooding beds and removing floating debris after harvest further reduces the amount of inoculum that remains in the beds. Because germination of pseudosclerotia is favored by moisture, beds should be graded to provide adequate soil drainage. Sprinkler heads should be adjusted to prevent prolonged soil saturation. Dense moss should be controlled, since it also provides a moist environment conducive to germination of pseudothecia. In the field, there is no consistent trend in the relative resistance of various cultivars. Likewise, under controlled conditions in a greenhouse, flowers of the cultivars Ben Lear, Pilgrim, Searles, and Stevens were equally susceptible to infection. Regardless of cultivar, cottonball incidence is generally higher in older beds than in younger beds, possibly because inoculum accumulates slowly over several years.

Fungicides are applied to control cottonball according to host phenology, with up to two applications at bud break to prevent the infection of succulent shoots by ascospores and up to two applications during bloom to prevent the infection of flowers by conidia. Under low disease pressure, adequate control is achieved by spraying during bloom only. The sterol demethylation inhibitor (DMI) fungicides fenbuconazole and propiconazole are effective for controlling both the shoot blight and fruit rot phases of cottonball. Other fungicide classes and bacterial biological controls have generally been less effective than the DMI fungicides. However, the long-term and often exclusive use of DMI fungicides to control cottonball increases the risk of fungicide resistance in populations of *M. oxycocci*. Therefore, cottonball management recommendations include minimizing the number of DMI fungicide sprays and alternating DMI fungicides with unrelated fungicides.

Selected References

Jeffers, S. N. 1991. Managing cranberry cottonball caused by *Monilinia oxycocci* with fungicides. Plant Dis. 75:502-506.

McManus, P. S., Best, V. M., Voland, R. P., and Leininger, B. L. 1999. Sensitivity of *Monilinia oxycocci* to fenbuconazole and propiconazole in vitro and control of cranberry cottonball in the field. Plant Dis. 83:445-450.

McManus, P. S., Best, V. M., and Voland, R. P. 1999. Infection of cranberry flowers by *Monilinia oxycocci* and evaluation of cultivars for resistance to cottonball. Phytopathology 89:1127-1130.

Sanderson, P. G., and Jeffers, S. N. 1992. Cranberry cottonball: Dispersal periods of primary and secondary inocula of *Monilinia oxycocci*, host susceptibility, and disease development. Phytopathology 82:384-392.

Sanderson, P. G., and Jeffers, S. N. 2001. Vegetative growth and conidium production by *Monilinia oxycocci* in vitro. Mycologia 93:9-16.

(Prepared by S. N. Jeffers and P. G. Sanderson;
Revised by P. S. McManus)

Early Rot

Early rot was one of the first cranberry fruit rot diseases reported in the late nineteenth century. Halsted described a scald of cranberries in 1889 that was very similar to early rot as it is known today. It is a very destructive disease of cranberry fruit in the eastern cranberry-growing areas of North America and has been increasing in importance in Wisconsin (United States). This pathogen can cause considerable losses in the field. Symptoms are seen on flowers, leaves, and stems, making it one of the most easily identified types of fruit rot. The leaf drop phase of the disease can be severe on young beds and it can severely delay the establishment phase of new plantings.

Symptoms

On developing cranberry fruit, early rot first appears as a small, dark-colored spot. Under favorable conditions, this spot enlarges rapidly, usually in a concentric manner, until the whole berry becomes soft. Frequently, the spot exhibits several dark, concentric rings resembling a dark wood grain (Fig. 192). Abundant pycnidia also develop on the fruit surface.

Infected flowers become blighted and covered with pycnidia. Flower symptoms are sometimes mistaken for blast, a condition caused by excessive heat. Leaves may develop irregular, reddish brown spots with pycnidia on the upper surface and infected leaves ultimately abscise (Fig. 193).

Causal Organism

Early rot is caused by *Phyllosticta vaccinii*. Until recently, *P. vaccinii* was thought to have a sexual stage, described in the literature as synonym *Guignardia vaccinii*, but the perethecial stage has been found to be that of the recently described fungus *P. elongata*. No sexual stage for *P. vaccinii* has been found.

P. vaccinii grows slowly in culture. The colonies are raised, with irregular margins. The mycelium is olive green and turns black with age. The black, closely aggregated pycnidia form dense, stromalike clusters. Conidia are released in slimy, cream-colored or white masses.

On leaves, pycnidia are 85–160 µm in diameter and are epiphyllous, single, black, globose to subglobose, and somewhat erumpent. The apex is thickened, prominent, and somewhat clypeus-like, with a well-differentiated, circular ostiole. The conidiogenous cells (1–4 × 3–13 µm) are cylindrical and blastic. Conidia are single celled, hyaline to faintly greenish yellow, and obovate; they contain coarse oil droplets and are 7–9 × 8–13 µm, with a gelatinous apical appendage 3–70 µm long (Fig. 194). Brown appressoria (4–7 × 8–13 µm) are often formed immediately adjacent to the spore wall.

Disease Cycle and Epidemiology

Because there was no distinction between *P. vaccinii* and *P. elongata* until 1982, there is confusion concerning the disease cycle and epidemiology of early rot as presented in earlier literature. However, *P. elongata* rarely causes fruit rot. *P. vaccinii* was likely the primary species described during the late nineteenth century. One major difference between the two is that *P. elongata* is frequently isolated from ripe fruit at harvest, whereas most fruit infected by *P. vaccinii* has rotted by September and is not harvested. Also, only a small proportion of sound fruit contain latent infections of *P. vaccinii* at harvest, whereas *P. elongata* is common in sound fruit. *P. vaccinii* can overwinter in leaves, stems, and stolons of infected plants. These symptomless infections give rise to pycnidia that develop on leaves and stems soon after the winter flood is removed. The conidia are produced and released from pycnidia in late spring and throughout the summer. Conidia are also produced on blasted flowers, blighted leaves, stem cankers, and rotted berries. Young leaves are infected as they develop, and berries are generally infected soon after pollination. Attempts to infect mature or nearly mature fruit in the laboratory have been unsuccessful. In experimental plots, *P. vaccinii* can infect 100% of the fruit.

The disease is typically most severe on young beds (1–5 years old), where it causes leaf drop as well as fruit rot. The pathogen is easily transmitted to newly planted beds on planting stock. It also is easily transmitted between growing regions by transporting infected planting stock, and it can survive as symptomless infections until conditions become conducive.

Management

Fungicides such as chlorothalonil, ferbam, maneb, mancozeb, fenbuconazole and azoxystrobin have all shown activity in the control of early rot. Fungicide sprays at 5–10% bloom, after midbloom, and once or twice thereafter at 10-day intervals effectively control both blast of flowers and young fruit and early rot of mature fruit. For new plantings, applications may be required throughout the season as long as new leaves are developing. Cranberries that have been treated with fungicides during

Fig. 192. Symptoms of early rot, caused by *Phyllosticta vaccinii*, on immature cranberry fruit. Note the abundance of pycnidia forming on the fruit surface. Also, the dark and light brown pattern on the fruit is characteristic. (Courtesy P. V. Oudemans; © APS)

Fig. 193. Symptoms of early rot, caused by *Phyllosticta vaccinii*, causing leaf drop. Note the abundant pycnidia on the leaves and some stem cankers. (Courtesy P. V. Oudemans; © APS)

Fig. 194. Spores of *Phyllosticta vaccinii*, the causal organism of early rot. Note the spore shape and the presence of "tails." Spores isolated from plant material can be used to confirm the presence of the pathogen on cranberry. (Courtesy P. V. Oudemans; © APS)

the growing season and are sound at harvest seldom develop early rot in storage.

Selected References

Halsted, B. D. 1889. Some fungus diseases of the cranberry. N. J. Agric. Exp. Stn. Bull. 64.

McManus, P. S. 1998. First report of early rot of cranberry caused by *Phyllosticta vaccinii* in Wisconsin. Plant Dis. 82:350.

Oudemans, P. V., Caruso, F. L., and Stretch, A. W. 1998. Cranberry fruit rot in the Northeast: A complex disease. Plant Dis. 82:1176-1184.

Stiles, C. M., and Oudemans, P. V. 1999. Distributions of cranberry fruit-rotting fungi in New Jersey and evidence for nonspecific host resistance. Phytopathology 89:218-225.

Weidemann, G. J. 1980. Reassessment of *Phyllosticta* on cranberry. Ph.D. thesis. University of Wisconsin, Madison.

Weidemann, G. J., and Boone, D. M. 1983. Incidence and pathogenicity of *Phyllosticta vaccinii* and *Botryosphaeria vaccinii* on cranberry. Plant Dis. 67:1090-1093.

Weidemann, G. J., Boone, D. M., and Burdsall, H. H., Jr. 1982. Taxonomy of *Phyllosticta vaccinii* (Coelomycetes) and a new name for the true anamorph of *Botryosphaeria vaccinii* (Dothideales, Dothioraceae). Mycologia 74:59-65.

(Prepared by D. M. Boone; Revised by P. V. Oudemans)

End Rot

In 1931, C. L. Shear considered end rot, caused by *Godronia cassandrae*, to be "the most important storage rot of the cranberry." Today, end rot is still considered an important disease of cranberry, and although it appears in all cranberry-growing areas in North America, its relative impact varies considerably over time and geography. Although primarily a storage disease that develops after the fruit has been harvested, it can also manifest itself as a leaf spot and twig blight. The twig blight phase is found largely in the Pacific Northwest, particularly in Oregon (United States) and British Columbia (Canada), but it is not economically significant. The leaf spot phase of end rot is sporadic in occurrence and, at worst, results in minor defoliation.

Symptoms

End rot first manifests itself as a soft, watery rot starting at either end of the berry but, more commonly, at the blossom end; the soft portions of the berry are clearly delineated from the sound portion (Fig. 195). As the rot progresses, the whole berry becomes soft and elastic to the touch. These berries, often called poppers, are distended by gas produced during the rotting process and may burst from the increased pressure. The fruits eventually collapse, shrink, and become yellowish or brownish.

The upper surface of leaves is also susceptible to infection. Leaf spots are reddish brown at first, turn tan or gray in the center, and develop a black border as they enlarge. Dark pycnidia develop within the central area. The surrounding tissue outside the black border often turns brilliant red or dark red and may cover a major portion of the leaf. Affected leaves may drop.

The fungus tends to colonize the outer senescent bark tissue of the vines but does not usually cause injury. Occasionally, the fungus invades young twigs, causing a minor twig blight that is not easily distinguishable from other twig blights.

Causal Organism

End rot is caused by the fungus *Godronia cassandrae* f. *vaccinii* (syn. *Fusicoccum putrefaciens*). The ascogenous stage has been found on stems and mummified fruit but is rarely seen. Apothecia are 0.5–0.7 mm in diameter on stems and 0.5–1.8 mm in diameter on mummified fruit. They are sessile or nearly so, depressed, urceolate, and tawny brown. Asci (5–10 × 58–117 μm) are cylindrical. Ascospores (4–7 × 50–70 μm) are hyaline, filiform, and nearly straight and have four to seven septa (mostly six to seven septa) when mature. Paraphyses are filiform, numerous, and longer than the asci; they often branch toward the apex.

Pycnidia are subglobose to pyriform, rather thick walled, more or less roughened, and tawny to tawny brownish; embedded at first, they become erumpent or subsuperficial when mature. Pycnidia are sessile or subsessile; simple or irregularly chambered forms are 160–400 μm in diameter, and larger chambered forms are 400–450 μm in diameter (Fig. 196). Conidia (pycnospores) are elliptic to fusiform, hyaline or faintly yellowish in mass, aseptate or pseudoseptate, and 2–3 × 8–18 μm (usually 2.5 × 10–12 μm) (Fig. 197). Conidiophores (2 × 25–36 μm) are simple or branched and cylindrical or somewhat tapered where the conidia are attached.

Cultures on potato dextrose agar appear white at the margins and pale lemon yellow and yellow-chrome to orange-citrine at the center. Only the pycnidial stage has been produced in culture. No physiologic specialization is known.

Disease Cycle and Epidemiology

The fungus overwinters in senescent bark on the vines and in dead leaves and rotted fruit in underlying debris. It is also readily isolated from apparently healthy vines throughout the year. Conidia sporulate throughout the year, with maximum sporulation from April through August. Although *G. cassandrae* f. *vaccinii* can be isolated from new shoots as early as June and from the blossom end of healthy-appearing berries soon after fruit set, evidence suggests that damage or senescence is usually required before infection can occur. The fungus may enter

Fig. 195. Cranberries with symptoms of end rot, caused by *Godronia cassandrae* f. *vaccinii*. (Courtesy F. L. Caruso; © APS)

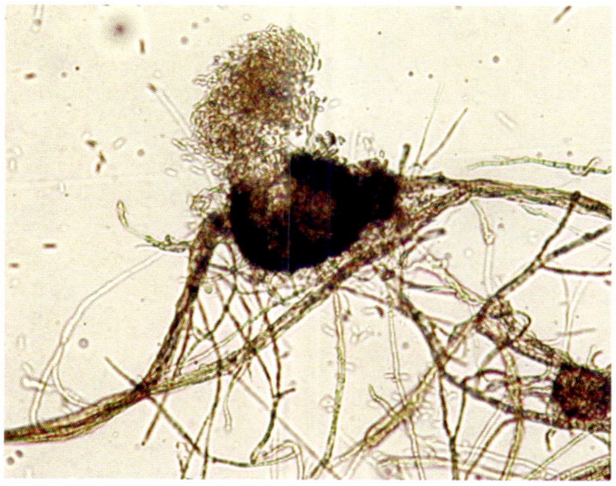

Fig. 196. Pycnidia of *Godronia cassandrae* f. *vaccinii*, the causal organism of end rot, releasing spores. (Courtesy J. S. Mika)

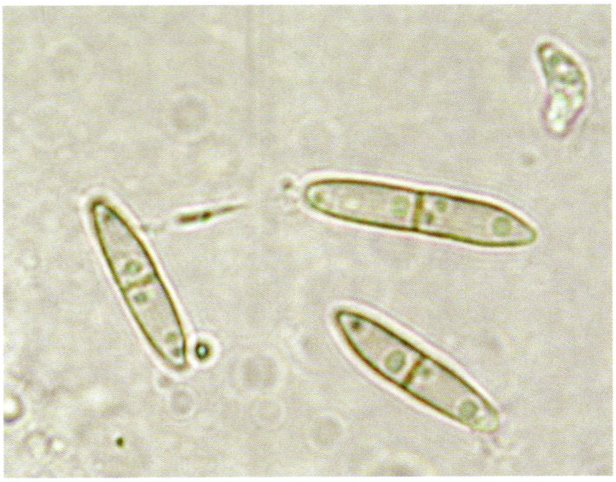

Fig. 197. Pycnospores of *Godronia cassandrae* f. *vaccinii*, the causal organism of end rot. (Courtesy J. S. Mika)

harvested fruits through pedicel scars, persistent senescent sepals, or wounds. The raking process used in flooded beds increases the amount of end rot compared with that in dry-harvested beds, which indicates that some inoculation takes place at harvest time and that spores of the fungus may be carried by the floodwater.

Relatively low temperatures (optimum 20°C) favor the growth of the fungus, but some growth occurs even at near-freezing temperatures. Conidia germinate at temperatures as low as 8°C, but 20°C is optimal. This preference for lower temperatures is probably a major factor in the relative importance of *Godronia* spp. as major storage rot fungi in Michigan and Wisconsin (United States). Even within Michigan, there are significant differences in *Godronia* populations between northern and southern regions.

Management

Because end rot is largely a storage rot, control is usually necessary only if the berries are to be used for the fresh market. Fungicides that control end rot are registered in some states but are not available everywhere cranberry is grown. Fungicides used to control other storage rots also control end rot. Whatever fungicide is used, three applications are usually required; the most critical times are between late blossom and early berry development.

Certain conditions predispose berries to the development of end rot. Excessive use of nitrogen fertilizer during the growing season, bruising of berries during harvesting, prolonged periods of continuous flooding and high temperatures, and high humidity and poor aeration during storage have all been implicated.

End rot development can be delayed somewhat by refrigeration of the berries. However, the fungus continues to develop slowly even at temperatures near freezing.

Although cultivar differences in susceptibility to the end rot fungus have been reported, these differences vary from one growing region to another. Reports from Wisconsin indicate that the cultivars Beckwith, Bergman, Early Black, Franklin, Howes, McFarlin, Pilgrim, Stevens, and Wilcox are relatively resistant, while the cultivars Ben Lear, Crowley, and Searles are susceptible. In the Pacific Northwest, both Stevens and Bergman may be quite susceptible, depending on the conditions at harvest.

Selected References

Bain, H. F. 1926. Cranberry disease investigations on the Pacific Coast. U. S. Dep. Agric. Bull. 1434.

Bristow, P. R., and Windom, G. E. 1985. The impact of machine-harvesting and fungicides on rot and physiological breakdown occurring in cold-stored cranberries. (Abstr.) Phytopathology 75:1285.

Caruso, F. L. 1989. Relationship of temperature to the fungi involved in cranberry fruit rot. (Abstr.) Phytopathology 79:1154-1155.

Groves, J. W. 1965. The genus *Godronia*. Can. J. Bot. 43:1195-1276.

Jeffers, S. N. 1991. Effects of fungicides applied during bloom on yield, yield components, and storage rots of cranberry. Plant Dis. 75:244-250.

Jeffers, S. N. 1991. Seasonal incidence of fungi in symptomless cranberry leaves and fruit treated with fungicides during bloom. Phytopathology 81:636-644.

Olatinwo, R. O., Schilder, A. M. C., and Kravchenko, A. N. 2004. Incidence and causes of postharvest fruit rot in stored Michigan cranberries. Plant Dis. 88:1277-1282.

Shear, C. L., Stevens, N. E., and Bains, H. F. 1931. Fungous diseases of the cultivated cranberry. U. S. Dep. Agric. Tech. Bull. 258.

(Prepared by H. S. Pepin and D. M. Boone;
Revised by J. S. Mika)

Fairy Ring

Fairy ring disease (Figs. 198 and 199) is frequently encountered on mature beds in the eastern cranberry-growing region of North America, particularly Massachusetts and New Jersey (United States). It has not been reported from other regions. The disease causes an opening of the canopy in the cranberry bed, resulting in numerous negative effects. It was traditionally a disease that developed on older cranberry beds and was considered to have minimal impact on yield. However, the development

Fig. 198. A typical fairy ring developing on 'Ben Lear' cranberry in New Jersey. Note that the leading edge of the ring has characteristic dieback. In the inner portion of the ring, there is excessive weed growth. (Courtesy P. V. Oudemans; © APS)

Fig. 199. A close-up of the leading edge of a fairy ring in 'Ben Lear' cranberry. (Courtesy P. V. Oudemans; © APS)

of this disease in established, highly productive cultivars has raised significant concerns among growers.

There is much confusion regarding the causal organism of fairy ring. In 1931, the causal organism was reported as *Psilocybe agrariella*; however, this has not been confirmed. Research conducted since suggests another pathogen, a *Helicobasidium* sp. (syn. *Thanatophytum* sp.), as the primary cause.

Symptoms

The first indication of fairy ring is a small area of dead or weak vines. Unlike other dieback diseases in cranberry beds, which are patchy or irregular in distribution, the area of fairy ring expands outward in all directions at a rate of 30–55 cm per growing season. Fairy rings are often initiated at ditches, in which case, the area of dead vines resembles a semicircle.

Vines affected by fairy ring turn yellow to rust-colored in a very short time and die. If berries remain attached to the upright, they wither as they would during drought stress. The upright usually dies from the tip downward. Because the symptoms of fairy ring are easily confused with those of upright dieback, fungal isolations from affected uprights are sometimes necessary to determine whether the causal organism of upright dieback is present.

When the dead area reaches a diameter of 2 m or more, the middle often partially revives with healthy cranberry plants. The result is an outwardly expanding ring of dieback of the cranberry vines.

Fairy ring is a soilborne disease causing infected runners (stolons) and roots to appear stunted and black. The cranberry canopy is easily detached from the soil, similar to pulling up a carpet. Research has shown that the runners become infected with a pathogen, resulting in the characteristic dieback. Remote sensing is an accurate way to diagnose fairy ring presence and spread (Fig. 200).

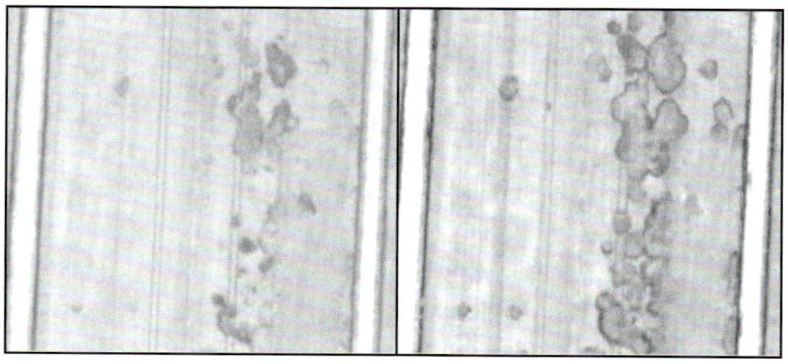

Fig. 200. Fairy rings seen in aerial photographs showing the spread of fairy ring disease in cranberry from 2006 to 2008. Images were captured by the QuickBird satellite operated by DigitalGlobe Inc. Image on the left was collected on July 18, 2006. Image on the right was collected on June 27, 2008. Note the expansion of the affected area. (Courtesy P. V. Oudemans; © APS)

Causal Organisms

Psilocybe agrariella var. *vaccinii* was described as the causal pathogen of fairy ring disease in the early 1900s. This mushroom-forming basidiomycete is rarely observed fruiting in cranberry beds, making its role as the sole causal organism questionable. Digging up cranberry sod and keeping it moist and shaded has been reported to stimulate fruiting body production. Fruiting bodies have been observed along the sides of the hole where the sod was removed and on the banks of ditches. Several studies have implicated various other soilborne fungi as the causal organism from fairy rings in New Jersey. *Phialophora* spp., binucleate *Rhizoctonia* spp., a *Neofabraea* sp., a *Papulaspora* sp., *Phomopsis vaccinii*, and *Glomerella cingulata* have all been isolated from symptomatic vines.

A dark brown mycelial network was observed growing along vegetative shoots of symptomatic cranberry vines (Fig. 201 left). The fungus was found to produce infection pads and was present on vines of all the rings sampled ($n = 22$). The fungus could not be isolated directly from symptomatic vines; however, isolates were obtained from the surrounding hyphal network. Observations made of the runner directly below the infection pads revealed necrotic tissues and evidence of fungal penetration. The fungus was binucleate with distinct right-angled branching patterns (Fig. 201 right). Preliminary inoculation studies showed that the pathogen could reinfect cranberry and initial symptoms were concordant with those observed from field-collected material. Based on sequencing of the ribosomal internal transcribed spacer region, isolates of this fungus are closely allied with a *Helicobasidium* sp. (syn. *Thanatophytum* sp.) that is the causal organism of violet root rot in other crops.

Disease Cycle and Epidemiology

Fairy ring develops as an expanding ring, killing vines at a rate of approximately 50 cm per year. New rings form on individual cranberry beds at a rate of one to three rings per year, and the disease can spread to new beds where it was not present during the previous growing season. The internal portion of the ring never recovers to full yield potential, and weeds colonized the affected areas. The affected areas also show greater levels of fruit rot and this may lead to increased seed deposition. Germination of the cranberry seeds leads to establishment of

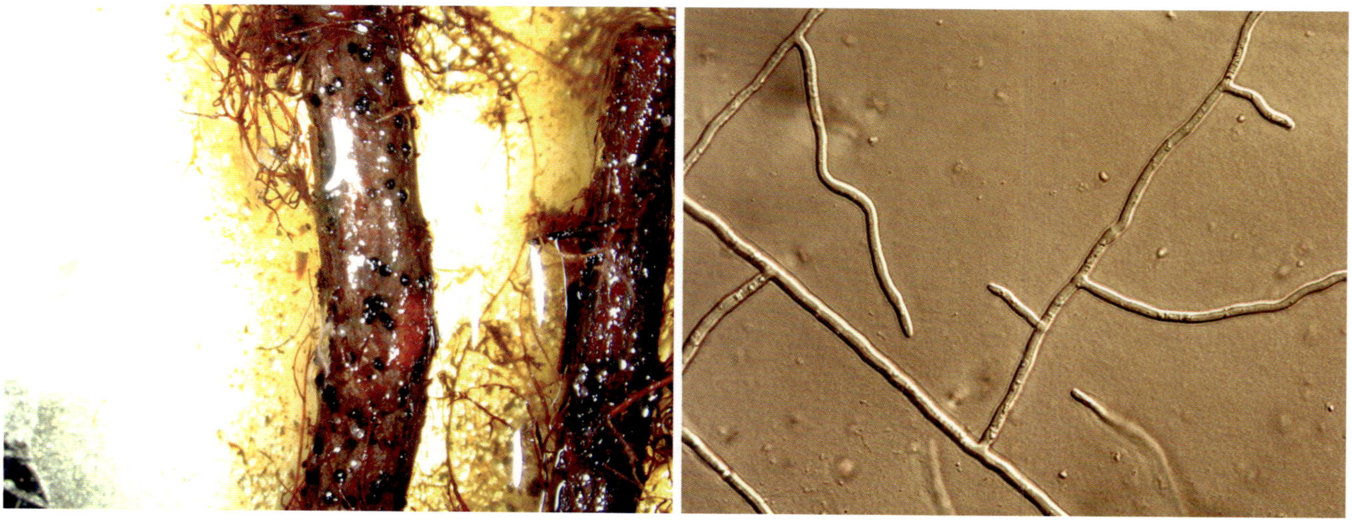

Fig. 201. Diagnosis of fairy ring disease: cranberry runners with roots and black infection pads (left) and characteristic mycelium in culture showing right-angled branching (right). (Courtesy P. V. Oudemans; © APS)

unselected cranberry genotypes with lower yield potential. The combined effect of ring expansion, vine death, and growth of vigorous, low-bearing genotypes causes the lifespan of cranberry beds to be significantly reduced. Thus, the impact of fairy ring to cranberry cultivation is multifold.

There are no commercially satisfactory methods for control of fairy ring, and it continues to spread in both Massachusetts and New Jersey. There are also no data to explain the mechanism for movement within or between cranberry beds; however, the impact of fairy ring is greatest on high-yielding genetically uniform cultivars. Significant cranberry acreage is currently under renovation with new hybrid cultivars and there is a perceived threat because of the potential susceptibility of these cultivars to the disease.

Fairy ring causes a 50–70% yield loss in the area affected. This loss can be attributed to lower productivity of the vines as well as to increased fruit rot. As rings enlarge, the area in the center of the rings recover, and the most serious loss is along the leading edge of the ring. Another form of loss is an increase in the genetic diversity of the bed. Since cranberry plantings can remain productive for several decades, maintenance of genetic uniformity is critical, especially for new hybrid cultivars.

Management

Fairy ring is a difficult disease to control. Shear reported that disease management techniques, such as ditching around rings, would stop expansion. Alternatively, drenching the affected crop with a 5% copper sulfate solution at a rate of 40.7 L/m^2 was used to stop the advance of the disease. Likewise, Zuckerman reported that ferbam (ferric dimethyl-dithiocarbamate) applied at a rate of 0.34 kg of a.i. per square meter gave excellent results when applied as a fall drench in 40.7 L/m^2. Treatment with ferbam at the recommended rates caused significant overgrowth of the vines possibly because of the stimulation of soil microflora. Azoxystrobin or fenbuconazole applied as a drench provide moderate levels of control. Recovered areas inside the perimeter of the rings should be replanted to maintain the planted genotype.

Selected References

Chang, L.-P. 1989. Fungi associated with fairy ring and dieback of cranberry (*Vaccinium macrocarpon*): Histology and pathogenicity. M.S. thesis. Rutgers University, New Brunswick, NJ.

Hlubik, W. T. 1988. A species of *Phialophora* associated with fairy ring and dieback on cranberry. M.S. thesis. Rutgers University, New Brunswick, NJ.

Oudemans, P. V., Constantelos, C., Wasilwa, L., Polashock, J., Caruso, F. V., and Carris, L. M. 2002. Identification of the causal agent of fairy ring disease on cranberry. (Abstr.) Phytopathology 92(Suppl.):S61-S62.

Oudemans, P. V., Polashock, J. J., and Vinyard, B. T. 2008. Fairy ring disease of cranberry: Assessment of crop losses and impact on cultivar genotype. Plant Dis. 92:616-622.

Shear, C. L., Stevens, N. E., and Bain, H. F. 1931. Fungous diseases of the cultivated cranberry. U. S. Dep. Agric. Tech. Bull. 258.

Zuckerman, B. M., Rochefort, K. J., and Rounsville, G. B. 1968. Control of fairy ring disease of the cultivated cranberry. Plant Dis. Rep. 52:87-88.

(Prepared by F. L. Caruso; Revised by P. V. Oudemans)

Phytophthora Root and Runner Rot

Phytophthora root and runner rot is the most important disease known to affect the stolons and roots of cranberry. The disease was formally described in 1986–1987 in Massachusetts (United States) and is now known to appear in many other cranberry-growing regions of North America. Phytophthora root and runner rot is most extensive in New Jersey (United States) and Massachusetts, and in beds where the disease is severe, crop loss can be significant.

Like many diseases caused by *Phytophthora* spp., Phytophthora root and runner rot of cranberry appears most often in beds that exhibit areas of poor drainage or experience prolonged periods of soil saturation or standing water during the growing season. It is likely that symptoms of the disease have been attributed to anaerobiosis and root asphyxiation prior to identification of the pathogen.

Symptoms

The most obvious symptom of *Phytophthora cinnamomi* infection is the development of bare patches with dead or dying vines around the outside edge (Fig. 202). These areas devoid of vine growth range in size from less than 1 m^2 to large portions of a bed; most affected areas fall between these extremes. In the early stages of disease development and at the perimeter of affected areas, vines are weak and unthrifty and may be brittle. Vines affected by Phytophthora root and runner rot decline from the root system up. The root system is weakened and reduced in mass, which causes secondary symptoms, such as stunting of the upright shoots, reduced drought tolerance, and lack of vigor. The leaves tend to be small and few in number and turn red prematurely in the autumn. Discolored areas (brown to black) appear beneath the periderm or bark on the runners. Because of the lack of fibrous roots, affected vines are easily pulled from the soil (Fig. 203).

Fig. 202. Injury from Phytophthora root and runner rot, caused by *Phytophthora cinnamomi*, in a cranberry bed. (Courtesy E. Saalau Rojas; © APS)

Fig. 203. Cranberry vines infected with *Phytophthora cinnamomi*, the causal organism of Phytophthora root and runner rot, can be easily pulled from the soil. (Courtesy P. V. Oudemans; © APS)

Other *Phytophthora* spp. are less pathogenic and may cause reductions in vigor and, under conditions of water stress, a vine dieback may occur (Fig. 204). These symptoms tend to develop in small discrete patches and do not resemble *P. cinnamomi* infections. The vines typically die from the base of the plant upward and the roots systems are reduced in volume.

Phytophthora spp. may produce chronic as well as acute infections in cranberry beds, and symptoms of chronic infection may not be easy to spot initially. Affected areas appear green and normal from a distance, but vines are stunted and vigor and yield are reduced. They may die rapidly if subjected to a stress, such as drought. Because cranberry plants spread by runner growth, poor root development in the runners also slows coverage of a bed by vines. As a result, vine density in beds with chronic Phytophthora root and runner rot may be uneven, with many open spots colonized by weeds. Rushes (*Juncus* spp.) are common invaders in *Phytophthora*-infected areas and are indicators of poor drainage.

Research has shown that compared with the occasional acute infection, chronic infections may result in greater yield loss over time because they tend to be more widespread. Chronic infections are common, and detection methods such as remote sensing are very useful.

Symptoms of Phytophthora root and runner rot are similar to those of related root disorders, such as damage by root-feeding insects and nematodes, other fungal diseases such as fairy ring, or simply wet feet. Because management of each of these problems differs, it is extremely important to get an accurate diagnosis before undertaking any control measures.

Identification is accomplished in various ways. Field identification based on symptomology is the most rapid method and, with experience, can be relatively accurate. It is not definitive, however, and should be followed up with a laboratory test, especially in new locations where the pathogen has not been seen previously or when there is any doubt. The pathogen can be isolated in the laboratory from symptomatic root or runner tissue by using a selective growth medium (PARPH) that contains defined antibiotics (Fig. 205). Few organisms besides *Phytophthora* spp. are capable of growing on PARPH medium. A trained plant pathologist or diagnostician generally makes the positive identification by microscopic observation.

Causal Organisms

Phytophthora spp. are classified in the kingdom Stramenopila, phylum Oomycota, class Peronosporomycetes, order Peronosporales, and family Peronosporaceae. Several *Phytophthora* spp. have been isolated from cranberry tissues, and all have been proved to be pathogenic to some extent. In general, *Phytophthora* spp. grow relatively slowly in culture, and isolation has been more successful since the development of selective media such as PARPH.

Hyphae of *Phytophthora* spp. are hyaline and coenocytic but develop cross walls with age. Under low-power magnification (×10), hyphae in culture appear coarse and branch at wide angles. Sporangia are rare on solid agar media for all species isolated from cranberry. However, they usually form readily when pieces of agar with mycelium or infected plant tissue are placed in water or another liquid (e.g., broth media, soil extract solution, or pond or lake water). Only homothallic species produce oospores in single-strain culture, but successful formation of oospores in vitro depends on the species and growth medium.

P. cinnamomi is the primary causal organism of Phytophthora root and runner rot in Massachusetts, New Jersey, and Oregon (United States). This species is a much more virulent pathogen than any of the other oomycetes recovered thus far and is the most important causal organism of the disease. *P. cinnamomi* produces characteristic coralloid hyphae that often have botryose swellings. Sporangia are broadly ellipsoidal and nonpapillate and proliferate internally. This species is heterothallic, so it does not produce oospores in single-strain culture; however, it does produce spherical chlamydospores. Cardinal temperatures for growth in culture are 4–7°C minimum, 28°C optimum, and 33–36°C maximum.

A second species was identified as *Phytophthora* taxon cranberry. Sporangia are nonpapillate and ellipsoid-limoniform. This species is homothallic with plerotic oospores and paragynous antheridia. Oospores are increasingly granulated when colonies were grown above 15°C, with approximately 80% abortion in isolates grown at 25°C.

A variety of additional morphological types have been isolated from cranberry roots in British Columbia and Quebec (Canada) and in Massachusetts, New Jersey, Oregon, Washington, and Wisconsin (United States). Most of these species exhibit low to moderate pathogenicity. Isolates described as *P. cryptogea* may cause stunting, but the impact under commercial production systems has not been quantified. Sporangia of all the morphological types appear relatively similar. They are produced on simple or irregularly branched sporangiophores, are usually ovoid to obpyriform, and proliferate internally. *P. cryptogea* and three of the undetermined morphological types are heterothallic and do not produce oospores in single-strain culture; however, when oospores are produced in paired cultures, antheridia are amphigynous. *P. megasperma*

Fig. 204. Dispersed vine dieback in a cranberry bed with Phytophthora root and runner rot, caused by *Phytophthora* taxon cranberry. (Courtesy P. V. Oudemans; © APS)

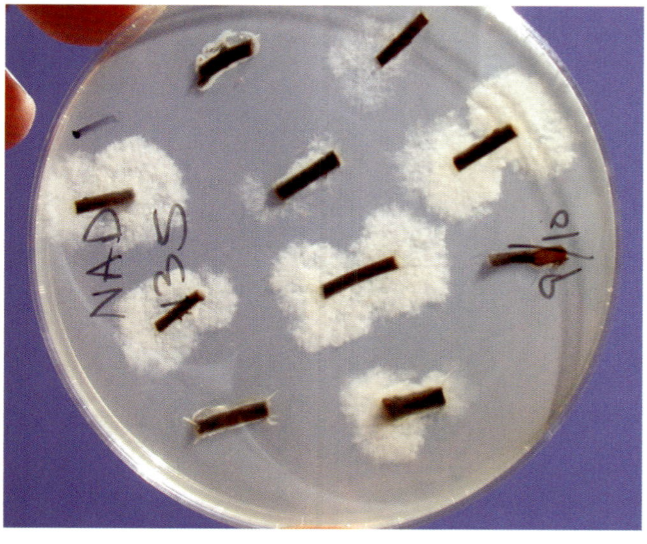

Fig. 205. Isolation from symptomatic cranberry root or runner tissue of a *Phytophthora* sp. using selective medium. (Courtesy P. V. Oudemans; © APS)

and the other undetermined morphological type are homothallic, producing ample oospores with paragynous antheridia on appropriate media.

Disease Cycle and Epidemiology

Zoospores of *Phytophthora* spp. are carried in irrigation water from surface sources (but not from well water, which is typically lacking these propagules) and are regularly introduced into cranberry beds. In Wisconsin, *Phytophthora* spp. can be recovered in the spring and late autumn from water in both flooding canals and drainage ditches. In New Jersey, *P. cinnamomi* was recovered from surface water throughout most of the growing season, while *Phytophthora* taxon cranberry was recovered only during the cooler months when water temperatures were below 18°C. Symptoms of infection develop in only a small percentage of the exposed acreage, however, because of the excellent drainage and low pH (3.0–4.5) of soils in most cranberry beds. Development of a *Phytophthora* infection is associated with poor drainage. In water-saturated soil, *Phytophthora* spp. can reproduce and spread quickly. Infections in roots and runners produce sporangia that release swimming zoospores that carry infection to other vines, and these cycles of infection continue as long as the soil is wet and temperatures are favorable. *P. cinnamomi* can survive low temperatures but grows and reproduces best at higher temperatures. Zoospores are detected in irrigation systems when surface water temperatures exceed 15°C. Zoospores are carried to cranberry vines in irrigation water. Infection requires prolonged soil wetness from flooding or poor drainage as well as soil temperatures above approximately 10°C. Under these conditions, zoospores that reach susceptible vines germinate to produce a mycelium that colonizes and kills roots and runners. Sporangia form on the mycelium and release more zoospores. Zoospores are attracted to nearby uninfected vines by chemical signals and swim toward them, spreading the infection. When the soil dries out or temperatures drop, chlamydospores (resting spores) form in rotted root and runner tissues. Chlamydospores are resistant to drying and freezing and allow *Phytophthora* spp. to overwinter in cranberry beds. When soils are once again warm and wet, chlamydospores germinate to form sporangia, beginning the infection cycle again.

Management

Controlling Phytophthora root and runner rot requires the integration of several practices. Drainage is the most important soil property determining the degree of damage caused by *Phytophthora* spp., thus water management is most critical, in particular, eliminating the causes of soil saturation. Uniform drainage allows soil water content to be managed to a level at which infections are minimized. Problems arise in areas where soil drainage is highly variable and where it is not possible to sufficiently irrigate well-drained areas without overirrigating poorly drained areas. Uniform drainage should be a key consideration in establishing new beds. Puddles or standing water are the first places *Phytophthora* infections occur. Drainage methods that remove standing water, such as ditches or underdrains, are very useful in controlling Phytophthora root and runner rot.

Irrigation uniformity is also crucial because overwatering can increase the chance of infection. Repeated cycles of wetting and drying, especially extreme cycles, are conducive to Phytophthora root and runner rot. Irrigation timing should be scheduled to consistently maintain soil moisture near the optimum level rather than leaving long intervals between irrigation events. Soil pH is of questionable value for the control of *Phytophthora* spp. Since cranberry is an acid-loving plant, it can tolerate relatively low pH levels. However, the use of sulfur to reduce pH in areas with symptoms of Phytophthora root and runner rot can lead to additional damage if the soils are poorly drained. This practice should, therefore, be used with extreme caution.

New beds should be constructed with the aim of providing uniform drainage. Irrigation systems should be carefully laid out so that sprinkler heads are placed according to design specifications. For irregular beds, sprinkler heads placed closer than specified in the design should be outfitted with reduced-volume nozzles. Unevenly drained beds result in the creation of zones with different irrigation requirements and become very difficult to manage. Before planting, the grower should examine the bed for low points or poorly drained areas that may be vulnerable to the development of Phytophthora root and runner rot. Implementing a drainage plan during new bed construction avoids problems down the road. Sanitation practices to prevent introduction of the pathogen are generally recommended in cranberry culture, even though the pathogen is already widely distributed. Caution should be exercised when moving plant material or soil between watersheds with different *Phytophthora* spp.

Most of the commercial cranberry cultivars are susceptible to Phytophthora root and runner rot. The disease has been diagnosed in all cultivars grown commercially in Massachusetts, New Jersey, and Wisconsin. However, preliminary results from artificial inoculation of greenhouse-grown plants suggest that a numbered selection, #35, is more resistant to *P. cinnamomi* than any other cultivar tested.

Fungicides can be used to protect living plants from Phytophthora root and runner rot. Once appropriate cultural practices have been implemented, treatment with fungicides can aid in the establishment of new vines. Both phenylamide fungicides as well as phosphonate fungicides work well against *P. cinnamomi. Phytophthora* taxon cranberry is inherently insensitive to phenylamide fungicides. Both types of fungicides perform best when used to protect plants following improvements to drainage or correction of overwatering.

Selected References

Caruso, F. L., and Wilcox, W. F. 1990. *Phytophthora cinnamomi* as a cause of root rot and dieback of cranberry in Massachusetts. Plant Dis. 74:664-667.

Erwin, D. C., and Ribeiro, O. K. 1996. Phytophthora Diseases Worldwide. American Phytopathological Society, St. Paul, MN.

Jeffers, S. N. 1988. *Phytophthora* species associated with a cranberry decline syndrome in Wisconsin. (Abstr.) Phytopathology 78:1572.

Oudemans, P. V. 1999. *Phytophthora* species associated with cranberry root rot and surface irrigation water in New Jersey. Plant Dis. 83:251-258.

Polashock, J. J., Vaiciunas, J., and Oudemans, P. V. 2005. Identification of a new *Phytophthora* species causing root and runner rot of cranberry in New Jersey. Phytopathology 95:1237-1243.

Pozdnyakova, L., Oudemans, P. V., Hughes, M. G., and Gimenez, D. 2002. Spatial detection and quantification of Phytophthora root rot effects on cranberry yield. Comput. Electron. Agric. 37:57-70.

(Prepared by S. N. Jeffers and F. L. Caruso;
Revised by P. V. Oudemans)

Powdery Mildew

Although powdery mildew has been observed occasionally in cranberry beds, it appears primarily on plants grown in the greenhouse and is not economically significant in commercial cranberry production. The disease has been reported to appear periodically on greenhouse-grown plants in Massachusetts, New Jersey, Washington, and Wisconsin (United States).

Symptoms

The first signs of powdery mildew are small patches of diffuse white mycelium, with or without powdery spores, on leaves and stems of the succulent, apical portion of growing shoots.

Both upper and lower leaf surfaces are colonized, but signs usually are more obvious on the upper surface.

As the disease progresses, white mycelium and spores completely cover the apical portion of the affected shoots (Fig. 206), and the fungus colonizes leaves and nonwoody stems of all ages. Shoot tips and young, newly opened leaves that are heavily infected become distorted and necrotic. On plants where the disease is well established and unmanaged, uprights may be stunted and leaves turn red and drop prematurely. Eventually, conspicuous, dark brown cleistothecia are produced, primarily on lower leaf surfaces and occasionally on stems.

Causal Organism

Powdery mildew of cranberry is caused by *Microsphaera vaccinii* (syns. *M. penicillata* var. *vaccinii*, *M. alni* var. *vaccinii*, and *Erysiphe vaccinii*). This species is thought to be unique to North America and attacks various genera in the family Ericaceae. *M. vaccinii* also causes powdery mildew of blueberry and other *Vaccinium* spp., but preliminary evidence suggests that isolates exhibit some degree of host specialization. For example, *M. vaccinii* from cranberry infects small cranberry (*V. oxycoccos*) and lingonberry (*V. vitis-idaea*) but does not infect highbush blueberry or highbush–lowbush hybrids of blueberry. In Europe, powdery mildew of *Vaccinium* spp. is caused by *Podosphaera* spp.

Isolates of *M. vaccinii* from cranberry produce cylindrical to ellipsoidal conidia in short chains that fragment easily. Conidia are hyaline and 12–18 × 26–36 µm. Cleistothecia are globose and 72–144 µm in diameter (mean of 105 µm); they are yellow to golden brown when immature and dark brown (nearly black) when mature. Cleistothecium appendages provide key taxonomic characteristics. Appendages on cranberry isolates are hyaline except at the base, nonseptate, and relatively long (two to six times the cleistothecium diameter). The number of appendages per cleistothecium ranges from four to twenty-one (mean of nine). Appendage tips branch dichotomously three or four times, with the ultimate branch tips usually recurved; most descriptions of *M. vaccinii* on other hosts indicate that ultimate branch tips are straight. Cleistothecia have three to eight asci, and asci have four or eight hyaline ascospores.

Disease Cycle and Epidemiology

The disease cycle and epidemiology of powdery mildew on cranberry have not been studied. Why the disease appears in cranberry-growing regions on plants in the greenhouse but not on plants in the field is not known. Perhaps the wetland environment in which many cranberries grow is unfavorable for powdery mildew development. In the greenhouse, conidia appear to be the only means of infection. Conidia germinate and infect leaves and nonwoody stems on all aerial parts of the plant but do not infect flowers or berries. Succulent, young shoots and shoot tips appear to be most susceptible to infection. Stresses on plant growth, such as nutrient deficiency or drought, seem to increase susceptibility and hasten disease development. The pathogen proliferates under the normal ranges of greenhouse temperature (18–27°C), relative humidity (30–90%), and photoperiod (12–16 h of light).

Fig. 206. Cranberry shoot with the succulent apical portion covered with powdery mildew, caused by *Microsphaera vaccinii* (left), and a healthy shoot (right). (Courtesy S. N. Jeffers; © APS)

Management

Host resistance probably is not a satisfactory method of powdery mildew control. Six cultivars inoculated in the greenhouse (Ben Lear, Crowley, McFarlin, Pilgrim, Searles, and Stevens) all became diseased and exhibited no obvious differences in susceptibility. Powdery mildew can be managed effectively in the greenhouse with routine applications of sterol demethylation inhibitor or strobilurin fungicides and sulfur. Currently, however, no fungicide is registered to control powdery mildew on cranberry.

Selected References

Braun, U. 1984. A short survey of the genus *Microsphaera* in North America. Nova Hedwigia 39:211-243.

Jeffers, S. N., Drilias, M. J., and Sanderson, P. G. 1990. Powdery mildew of cranberry: A first report. (Abstr.) Phytopathology 80:973.

Parmelee, J. A. 1977. The fungi of Ontario. II. Erysiphaceae (mildews). Can. J. Bot. 55:1940-1983.

(Prepared by S. N. Jeffers; Revised by F. L. Caruso)

Protoventuria Early Leaf Spot

Protoventuria early leaf spot develops in all the major cranberry-growing regions but appears to be more common in Oregon and Washington (United States). Disease incidence may be high in localized areas. Whether the disease has any appreciable effect on plant vigor or yield is not known.

Symptoms

Infected second-year leaves develop distinct reddened circles or rings on the upper surface as early as March in the Pacific Northwest (Fig. 207). Close examination of leaf spots reveals the presence of dark, radiating, subcuticular hyphae (Fig. 208). The reddening is transient; by midsummer, leaf spots appear darkened because of the proliferation of subcuticular hyphae. Dark pseudothecia covered with prominent spines are produced superficially on leaf spots, usually after the leaf tissue has become senescent.

Fig. 207. Protoventuria early leaf spot, caused by *Protoventuria barriae*, on 'Stevens' cranberry. (Courtesy L. M. Carris; © APS)

Causal Organism

Protoventuria early leaf spot is caused by the fungus *Protoventuria barriae*. The globose to ovoid pseudothecia (160–260 μm in diameter) that are scattered over the leaf spots develop superficially from subcuticular hyphae. Black spines (50–180 μm long) are scattered over the entire pseudothecium. Asci (31–48 × 95–143 μm) are clavate to pyriform. Mature ascospores (11–15 × 29–33 μm) are dark brown, two celled, broadly elliptical, and constricted at the septum, with a coarsely roughened wall enveloped in a thick, gelatinous sheath (Fig. 209). In culture, the fungus produces a dark brown, slow-growing mycelium. No anamorphic stage is known for *P. barriae*.

Disease Cycle and Epidemiology

P. barriae overwinters as subcuticular hyphae in the previous season's leaves and in dead leaves in the duff layer. The reddening of the host cells in early spring may be caused by a transient inhibition of photosynthesis. Pseudothecia develop primarily on dead leaves and are only rarely found on green leaves. The timing of infection is not known, but mature ascospores are present as early as March and as late as October in Washington cranberry beds.

Management

No control measures are currently recommended for Protoventuria early leaf spot.

Fig. 208. Close-up of Protoventuria early leaf spot, caused by *Protoventuria barriae*, on a cranberry leaf. Note the reddening of the host tissue associated with subcuticular hyphae. (Courtesy L. M. Carris; © APS)

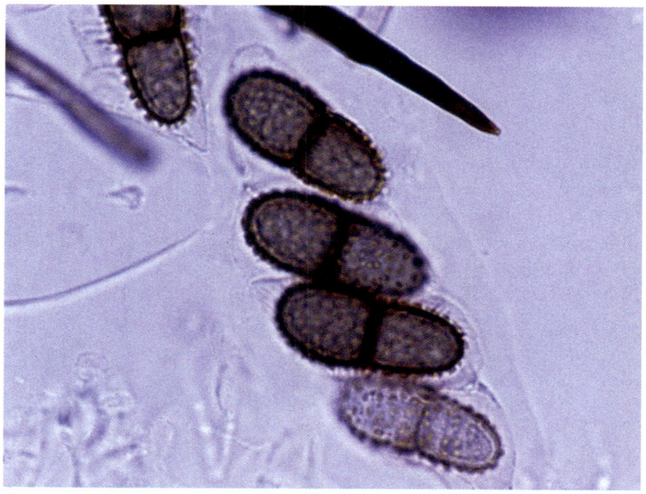

Fig. 209. Ascospores of *Protoventuria barriae*, causal organism of Protoventuria early leaf spot. (Courtesy L. M. Carris; © APS)

Selected Reference

Carris, L. M., and Poole, A. P. 1993. A new species of *Protoventuria* on leaves of *Vaccinium macrocarpon*. Mycologia 85:93-99.

(Prepared by L. M. Carris)

Protoventuria Leaf Spot and Berry Speckle

Protoventuria leaf spot and berry speckle has been reported only from Wisconsin (United States) and British Columbia (Canada), although the fungus is also relatively common in Massachusetts (United States). Neither the leaf spot nor the berry speckle is believed to be particularly harmful. The disease is common only in cranberry beds that have not been treated with fungicides. The speckles on berries are superficial and, except for making berries less attractive for fresh-market sale, do not affect quality even in cold storage.

Symptoms

In early autumn in Wisconsin, small but distinct red to purple lesions appear on the upper surface of current-season leaves. By late spring the following year, the lesions have enlarged but are less distinctly outlined. The large, chlorotic lesions have minute black pseudothecia clustered in the centers. Pseudothecia are most prevalent during July and may have prominent black spines on the upper exposed surface. At this time, affected leaves become more chlorotic, the lesions look like smudgy spots on the upper leaf surface, and many of the affected leaves drop prematurely.

On fruit, speckling appears in midsummer as tiny red lesions on green fruit. Speckles increase in size and number as the season progresses. Lesions on ripe fruit may be pale yellow, dark red, or black and may reach 1–2 mm in diameter; the center is slightly depressed and lighter in color than the margins. More speckles develop around the stem end than around the blossom end of berries.

Causal Organism

The fungus *Protoventuria myrtilli* (syn. *Gibbera myrtilli*) causes this minor leaf and berry disease. The pyriform pseudothecia (50–120 μm in diameter) are somewhat clustered in the center of the chlorotic lesions on the upper surface of leaves. They develop superficially from subcuticular hyphae. The spines that are scattered over the upper surface of the pseudothecia are black, 30–100 μm long, and simple or septate. Asci (9–12 × 40–60 μm) are usually swollen at the lower end. Mature ascospores (4–6 × 12–17.5 μm) are greenish to olivaceous, elliptical, and uniseptate, with cells slightly unequal in size. The ascospore wall is finely roughened.

In culture, the dark mycelium grows very slowly. Consequently, *P. myrtilli* may be overgrown by other fungi and escape detection when isolations are attempted. *P. myrtilli* has no known anamorph.

Disease Cycle and Epidemiology

P. myrtilli overwinters in lesions on cranberry leaves. Pseudothecia mature by midsummer and release ascospores, which are the only known inoculum. The germ tubes from ascospores presumably penetrate the cuticle directly, but no cell collapse is evident in affected tissues. The red color apparently is caused by a reaction by undamaged cells to the presence of the fungus.

In the spring following infection, only a slightly chlorotic lesion remains, and this is often the only evidence of the pathogen. By midsummer, some of the host cells in the center of the lesions have turned brown and collapsed, and by late summer, the entire lesion is brown and many diseased leaves drop.

The fungus is known to sporulate on dead leaves in Massachusetts, but the role of these spores in disease development is not known. The fungus has never been observed to sporulate on infected berries.

Management

Fungicides routinely applied in July and early August to control fruit rot also control infection by ascospores of *P. myrtilli*. Cultivars appear to differ in susceptibility to berry speckle; however, whether they also vary in susceptibility to the leaf spot phase is not known.

Selected References

Barr, M. E. 1968. The Venturiaceae in North America. Can. J. Bot. 46:799-864.

Barr, M. E. 1989. The Venturiaceae in North America: Revisions and additions. Sydowia 41:25-40.

Boone, D. M. 1982. Gibbera leaf spot and berry speckle of cranberry. Univ. Wisc. Ext. Publ. A3193.

Carlson, L. W., and Boone, D. M. 1966. A berry speckle disease of cranberry and its control. Plant Dis. Rep. 50:539-543.

(Prepared by D. M. Boone and P. R. Bristow; Revised by F. L. Caruso)

Pyrenobotrys Leaf Spot

Pyrenobotrys leaf spot appears wherever cranberry is grown but may be more common in the Pacific Northwest. Incidence is usually very low, perhaps because of the fungicides that are used to control other diseases, such as fruit rots. There are some years in which the incidence of the symptoms is higher than normally observed. No appreciable damage occurs even when disease incidence is high.

Symptoms

Conspicuous clusters of black fruiting bodies form on the lower leaf surface during late winter in the Pacific Northwest. These pseudothecia probably do not appear until early summer in growing regions with more severe winters. Irregular, purplish red lesions with diffuse margins may develop on the upper leaf surface, usually over the fruiting bodies (Fig. 210). Leaves with pseudothecia often show no symptoms. By June, some lesions become necrotic and turn brownish red. The discoloration is evident on the lower surface of the leaves in the center of the fruiting body clusters. A few leaves with lesions may turn yellow and fall prematurely. No symptoms are observed on current-season leaves in the autumn as the vines go into dormancy. Speckling of berries does not occur.

Fig. 210. Symptoms and signs of Pyrenobotrys leaf spot, caused by *Pyrenobotrys compacta*, on cranberry in mid-March. Note the purplish red spots on the upper leaf surface (top row) and the pseudothecia of the pathogen on the underside of the leaves (bottom row). A healthy leaf is shown at the right end of each row. (Courtesy P. R. Bristow; © APS)

Causal Organism

Pyrenobotrys leaf spot is caused by the fungus *Pyrenobotrys compacta* (syn. *Gibbera compacta*). Pseudothecia are erumpent through the cuticle on the lower leaf surface. They form in dense, circular clusters (Fig. 210). Few to numerous setae (6 × 30–60 µm) are arranged around the ostiole on the upper half of the pseudothecia. Mature asci (9–12 × 48–66 µm) are swollen at the base, are bitunicate, and have eight spores. Ascospores (3–6 × 14–18 µm) are olivaceous, with a rough surface. They have one septum with two cells of unequal size; the larger cell is broader than the smaller cell. Ascospores are shed in a gelatinous matrix. No anamorph is known for *P. compacta*.

Disease Cycle and Epidemiology

Pseudothecia are most numerous on leaves of the youngest stems. Few arise on leaves of older stems or runners. Ascospores mature in early summer, and infection occurs during the summer or early autumn. The environmental conditions that are required for infection are unknown. The fungus overwinters in infected leaves.

Management

Fungicides applied in July and August to control fruit rots or twig blight also control Pyrenobotrys leaf spot.

Selected References

Bain, H. F. 1926. Cranberry disease investigations on the Pacific Coast. U. S. Dep. Agric. Bull. 1434.

Eriksson, B. 1974. On Ascomycetes on Diapensiales and Ericales in Fennoscandia. 2. Pyrenomycetes. Sven. Bot. Tidskr. 68:192-234.

Shear, C. L., Stevens, N. E., and Bain, H. F. 1931. Fungous diseases of the cultivated cranberry. U. S. Dep. Agric. Tech. Bull. 258.

(Prepared by P. R. Bristow; Revised by F. L. Caruso)

Red Gall

Red gall disease is known mainly in the eastern United States and eastern Maritime Provinces of Canada where cranberries are grown. It is very sporadic in occurrence, but it can cause crop loss when it does appear.

Symptoms

Galls form on the surface of cranberry leaves, young stems, petioles, flowers, and fruit (Fig. 211 left). If the galls are numerous on a given plant part, the infected portion is misshapen and often dwarfed and appears red. The galls are brilliant raspberry or lavender-red and may be as large as 1 mm in height and width. The galls are shaped somewhat like a chalice, and when viewed from above, an outward-flaring rosette of cells resembles a cup with a brilliant yellow resting spore in the base (Fig. 211 right).

Causal Organism

Red gall is caused by the chytrid fungus *Synchytrium vaccinii*. The resting spores are usually solitary but occasionally two are in a cell, almost filling the host cell. Resting spores can be subspherical (86–200 µm in diameter) or ovoid (120–140 ×

Fig. 211. Red gall, caused by *Synchytrium vaccinii*, on a cranberry leaf (left side). Close-up showing the yellow resting spore at the base of a red gall (right side). (Courtesy L. M. Carris; © APS)

150–192 μm); they have a smooth, amber-brown wall (3.8–4.6 μm thick) and brilliant chrome-yellow contents. Germination of the spore has not been observed despite numerous attempts.

Disease Cycle and Epidemiology

S. vaccinii overwinters as a resting spore in the gall tissue. Although not observed as yet, a likely scenario for spread would be as follows: in the spring, the resting spore germinates to form a sorus composed of multiple sporangia; zoospores are formed in the sporangia. Upon release from the sporangia, zoospores are able to swim or be carried in water to new cranberry growth, where infection takes place. Host tissue proliferates in response to the presence of the fungus in a single cell, and a gall is formed. Little is known about the specific conditions needed for zoospore production, zoospore release, or infection of cranberry.

Management

Because the spread of the red gall fungus depends on free water, eliminating low areas where water can stand for prolonged periods aids in the control of the disease. Early withdrawal of the protective winter floodwater reduces the potential for zoospore development and spread by eliminating the constant source of free water over the vines. The use of sprinklers for frost protection also should reduce the spread of *S. vaccinii* by eliminating the need for flooding. Fungicides applied early in the spring also significantly reduce gall numbers.

Selected References

Halstead, B. D. 1889. Some fungus diseases of the cranberry. N. J. Agric. Exp. Stn. Bull. 64.

Karling, J. S. 1958. *Synchytrium vaccinii*. Sydowia Ser. II 12:149-159.

Karling, J. S. 1964. *Synchytrium*. Academic Press, New York.

Rochefort, K., and Zuckerman, B. M. 1967. Control of red-gall of cranberry. Cranberries 31:14, 18.

(Prepared by A. W. Stretch; Revised by L. M. Carris)

Red Leaf Spot

Red leaf spot appears on cultivated American cranberry (*Vaccinium macrocarpon*) and wild small cranberry (*Vaccinium oxycoccos*). The disease has been observed in all cranberry-growing regions, and except for the Pacific Northwest, it generally has caused only minor damage. Severe outbreaks result in premature defoliation and shoot dieback. *Mycosphaerella nigromaculans*, the cause of black spot of cranberry, often colonizes tissues affected by red leaf spot, which leads to further death of vegetative and floral buds.

Symptoms

The most conspicuous symptom of red leaf spot is the presence of circular, bright red spots on the upper surface of leaves produced in the current season (Fig. 212). Less commonly, red spots develop on small, green berries. Spots first become apparent in early summer and are somewhat paler on leaves that are shaded. Diseased leaves are often more round than the typical oblong healthy leaves. Spots are often surrounded by a yellow halo, and the affected leaf tissue is sometimes puckered or blistered. Sometimes spots coalesce so that the entire leaf surface is red. The lower surface of the leaves becomes covered with a white, powdery hymenium of the red leaf spot pathogen. By late summer, spots fade to pink or yellow and may abscise prematurely. In severe cases, the fungus grows through petioles to succulent shoot tips, causing them to become red and distorted (Fig. 213). *M. nigromaculans* often invades shoots and berries affected by red leaf spot, causing the disease black spot. Thus, it is common to see red leaf spot and black spot in the same planting.

Causal Organism

Red leaf spot is caused by the basidiomyceteous fungus *Exobasidium rostrupii* (formerly *E. vaccinii*). Basidia are borne on the hymenium and usually have four sterigmata. Musiform basidiospores (2.0–4.5 × 9–17 μm) are borne on the basidia. Conidia (0.8–1.5 × 5–9 μm) are also shed from the hymenium. On the cranberry plant, basidiospores germinate and either form germ tubes and hyphae or bud off secondary spores. In

Fig. 212. Red leaf spot, caused by *Exobasidium rostrupii*, **A,** in a cranberry field; and **B,** a comparison of healthy leaves (top) and diseased leaves (bottom). (Courtesy P. S. McManus; © APS)

Fig. 213. Cranberry shoot tip with severe red leaf spot, caused by *Exobasidium rostrupii*. (Courtesy P. S. McManus; © APS)

culture, *E. rostrupii* colonies are yeastlike and grow by cell budding.

Disease Cycle and Epidemiology

E. rostrupii overwinters on diseased cranberry shoots, probably as a mycelium. In the spring, either the mycelium or basidiospores infect young, succulent leaves and stems. In tests performed in a growth chamber, symptoms were first visible on the cultivar Ben Lear at 10 days following inoculation and fully expressed in 3 weeks. Following inoculation, guard cells are among the first cells to die, suggesting that stomata serve as an infection court. Infection and disease development are favored by moderate temperatures and prolonged periods of rain, fog, and dew. The disease tends to be more severe in shaded areas of cranberry beds and where the vine canopy is dense. Other aspects of the epidemiology of red leaf spot, such as the dissemination of spores and precise temperatures and moisture requirements, are not known.

Management

In most cranberry-growing regions, the low incidence and severity of red leaf spot does not warrant special control measures. Because lush, succulent vine growth is highly susceptible to the disease, nitrogen fertilizer should be used with care. Because red leaf spot is favored by moisture and low temperatures, adequate water drainage and air circulation are important to minimize disease development. In studies conducted in a greenhouse, cranberry cultivars did not differ significantly in their susceptibility to red leaf spot. In the field, new plantings of the cultivars Ben Lear, HyRed, and Stevens that have been fertilized heavily with nitrogen appear particularly susceptible. In Washington (United States), a Bordeaux mixture has been applied to young shoots to prevent infection. Mancozeb and maneb were effective under greenhouse conditions and may be effective in the field. Fungicides are ineffective if applied after symptoms are widespread in a planting.

Selected References

Nannfeldt, J. A. 1981. *Exobasidium*, a taxonomic reassessment applied to the European species. Symb. Bot. Ups. 23:1-72.

Shear, C. L., Stevens, N. E., and Bain, H. F. 1931. Fungous diseases of the cultivated cranberry. U. S. Dep. Agric. Tech. Bull. 258.

Tontyaporn, S. 1974. Comparison of *Exobasidium vaccinii* from *Vaccinium macrocarpon* with other *Exobasidium* spp. from ericaceous hosts. Ph.D. thesis. University of Wisconsin, Madison.

(Prepared by N. L. Nickerson; Revised by P. S. McManus)

Red Shoot

Red shoot is a minor disease of cultivated American cranberry (*Vaccinium macrocarpon*) and wild small cranberry (*Vaccinium oxycoccos*). Red shoot has been reported in Massachusetts and Wisconsin (United States), the Pacific Northwest, and the Atlantic Provinces of Canada. In Wisconsin, the disease has been observed on the cultivars McFarlin, Pilgrim, Searles, and Stevens. Although not recognized as a distinct disease until 1984, anecdotal reports suggest that red shoot has probably occurred for many decades throughout the cooler cranberry-growing regions. Diseased shoots do not flower but production losses are negligible, since generally less than 1% of shoots are affected. However, the distorted and discolored shoots that result from this disease have sometimes been misidentified as a weed, raising the possibility of the misapplication of herbicides.

Symptoms

Symptoms first appear in the spring on new growth. Affected shoots are stunted and spindly with red leaves that are slightly more round than the typical oblong cranberry leaves. Although leaf shape is altered by red shoot, leaves are not abnormally thickened, as in the case with the related disease rose bloom. Diseased shoots appear singly or in clusters, arising from a single node on vines buried in leaf debris or moss. By midsummer, a white, powdery hymenium of the pathogen is visible on the lower surface of the leaves, and eventually, leaves fade to orange or yellow. Diseased shoots become brittle and break easily from the cranberry vine. However, additional diseased shoots may arise late in the season and in subsequent years from previously infected stems. Symptoms are variable and may be influenced by environmental conditions and cultivar. In Nova Scotia (Canada), diseased shoots often appear etiolated when red shoot develops in shaded areas or where vine growth is dense. In Wisconsin, diseased shoots are often severely stunted and surrounded by healthy shoots (Fig. 214), making red shoot inconspicuous in the field, especially later in the season when leaves have faded from red to yellow.

Causal Organism

Red shoot is caused by the basidiomycetous fungus *Exobasidium perenne*. A fungal hymenium develops on the lower surface of the leaves. Basidia (3.5–5.2 μm wide) have two sterigmata and extend 15–25 μm beyond the leaf surface (Fig. 215A). Basidiospores (1.7–2.6 × 12–22 μm) are hyaline, cylindrical to fusiform, straight to slightly curved, and moderately hooked at the base (Fig. 215B). They are one celled but, upon germination, develop one to three septa (Fig. 215D). Conidia (1.0–1.7 × 9.0–25 μm), which may be present on the hymenium, are hyaline and cylindrical to filiform (Fig. 215C).

Disease Cycle and Epidemiology

E. perenne is systemic in cranberry vines and overwinters in diseased stems. In the spring, the mycelium invades new shoots. By mid- to late summer, spores are shed from hymenia. The infection process and environmental conditions conducive to infection and disease development have not been determined.

Management

Because red shoot is not economically important, control recommendations have not been developed.

Fig. 214. Cluster of shoots with red shoot (left), caused by *Exobasidium perenne*, compared with healthy shoots (right). (Courtesy P. S. McManus; © APS)

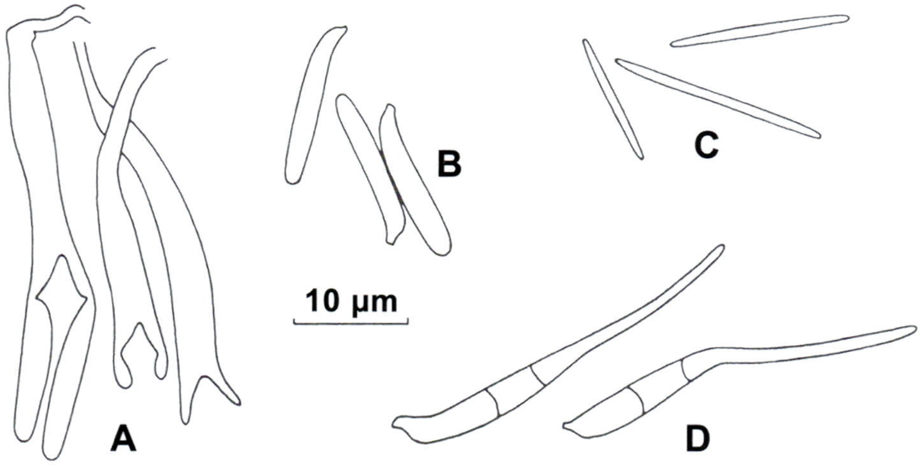

Fig. 215. The basidia (**A**), basidiospores (**B**), conidia (**C**), and germinating basidiospores on water agar (**D**) of *Exobasidium perenne*, the causal organism of red shoot. (Reproduced, by permission, from Nickerson, 1984; © Her Majesty the Queen in Right of Canada, as represented by the Minister of Agriculture and Agri-Food Canada)

Selected Reference

Nickerson, N. L. 1984. A previously unreported disease of cranberries caused by *Exobasidium perenne* sp. nov. Can. J. Plant Pathol. 6:218-220.

(Prepared by N. L. Nickerson; Revised by P. S. McManus)

Ripe Rot

Ripe rot (also called white rot) appears in all major cranberry-growing areas in North America. In Nova Scotia (Canada), it has been reported as the main pathogen of cranberry. The disease is common in stored cranberry fruit and occasionally affects fruit in the field.

Symptoms

The first symptom of ripe rot on fruit is a small, soft area that expands to engulf the entire fruit. Affected fruit appear off-red or reddish orange (Fig. 216). Diseased fruit often become watery soft and squirt fluid when squeezed. These fruit resemble the poppers that form in response to infection by *Godronia cassandrae*. Not all watery-soft fruit are the result of ripe rot; isolation of the causal organism is the only way to obtain positive identification.

Fig. 216. Cranberries with symptoms of ripe rot (left), caused by *Coleophoma empetri*, and healthy cranberries (right). (Courtesy F. L. Caruso; © APS)

Causal Organism

Coleophoma empetri (syn. *Sporonema oxycocci*) causes ripe rot. The fungus forms dark brown pycnidia in diseased leaves and berries and in culture. Pycnidia (50–100 µm in diameter) may be grouped or scattered. They are globose at first but become depressed and disklike when mature (Fig. 217). Pycnidia open by means of wall fragmentation. Conidia (2–4 × 12–18 µm) are hyaline, uniformly cylindrical, and straight (Fig. 218). Conidiophores are simple, oblong, and approximately one-fourth the length of conidia. Hyaline, clavate paraphyses are formed among the conidiophores; they collapse at maturity.

In culture, on cornmeal agar or potato dextrose agar, the fungus starts out whitish but darkens to a grayish brown. The colony is thick with slightly raised, fluffy mycelium. Pycnidia often form a ring near the outer edge of the colony. When berries are cultured, pycnidia readily form at the point where the fruit tissue makes contact with the agar.

Disease Cycle and Epidemiology

The ripe rot pathogen overwinters in berries and infected debris (duff layer) as mycelium or immature pycnidia. Conidia are

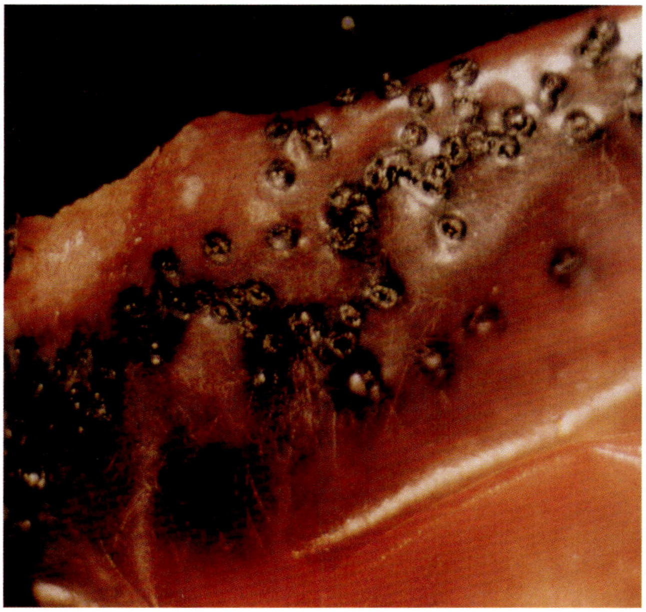

Fig. 217. Pycnidia of *Coleophoma empetri*, the causal organism of ripe rot, on cranberry. (Courtesy J. S. Mika)

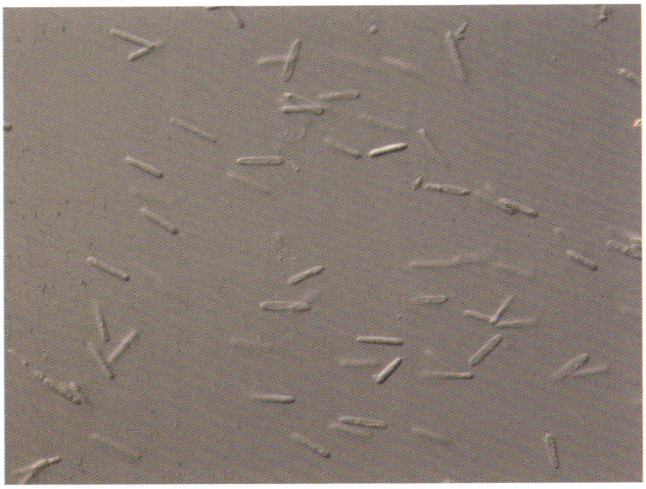

Fig. 218. Conidia of *Coleophoma empetri*, the causal organism of ripe rot. (Courtesy C. Constantelos; © APS)

released and initiate infection during bloom and early berry development. Spores may continue to be present for a significant part of the growing season. As infected fruit mature, hyphae continue to invade the fruit. Rot symptoms do not develop until late in the growing season and mostly in storage. As with *Godronia cassandrae*, *C. empetri* may infect during the harvest flood. Fungal spores in the duff layer float to the top along with the berries, which are susceptible to bruising during harvest, thereby creating an ideal situation for infection. *C. empetri* has also been isolated occasionally from diseased uprights during the growing season. It also has been found on *Rhododendron*, *Leucothoe*, *Stranvaesia*, and *Camellia* spp.

Management

Fungicide applications during bloom and early berry development are an effective control measure. Because conditions during these growth stages are often favorable for spore germination and fungal penetration, fungicide protection must be continuous on susceptible flowers and fruit.

Selected References

Gourley, C. O. 1979. Further observations on cranberry fungi in Nova Scotia. Can. Plant Dis. Surv. 59:15-17.

Limber, D. P. 1955. Studies in the genus *Sporonema*. Mycologia 47:389-401.

McManus, P. S. 2001. Cranberry fruit rot diseases in Wisconsin. Univ. Wis. Ext. Publ. A3745.

Oudemans, P., McManus, P., Levesque, A., Caruso, F., and DeVerna, J. 2005. Cranberry fruit rot: Seeking the sources of inoculum. (Abstr.) Phytopathology 95(Suppl.):S78.

Shear, C. L. 1907. New species of fungi. Bull. Torrey Bot. Club 34:305-317.

Shear, C. L., Stevens, N. E., and Bain, H. F. 1931. Fungous diseases of the cultivated cranberry. U. S. Dep. Agric. Tech. Bull. 258.

(Prepared by C. C. Kusek; Revised by J. S. Mika)

Rose Bloom

Rose bloom was first observed in New England, and the pathogen was described in the early 1900s. At that time, rose bloom was considered to be one of four serious diseases of cranberry, but by 1931, it had declined to only minor significance in the eastern growing area. Minor outbreaks have been noted in Massachusetts (United States) but have been confined to Nantucket island. Currently, rose bloom is a problem only in the Pacific Northwest of the United States. The disease is erratic in occurrence and usually is of only minor economic importance. However, when disease incidence is high, losses can be considerable and control practices should be implemented.

The disease also has been called hypertrophy and Massachusetts false blossom (it was first recorded in Massachusetts). Rose bloom, however, should not be confused with the false blossom disease that is caused by a phytoplasma.

Symptoms

Symptoms of rose bloom first appear after vine growth resumes in the spring following infection during the previous growing season. Infected axillary buds are stimulated to grow into fleshy, abnormal lateral branches (Fig. 219). The hypertrophied tissues on these branches, with swollen pink or light rose-colored leaves, look like miniature roses; hence the common name rose bloom. Healthy axillary buds remain dormant. Usually, buds on 1-year-old wood are affected; occasionally, the terminal bud and buds on older wood are attacked.

Rarely, flowers and berries are affected (Fig. 220). Infected flowers are deformed and enlarge to three or four times normal size. Diseased berries have a deformity at the calyx end, where the swollen flower persists.

Fig. 219. Fleshy, abnormal cranberry branch with rose bloom, caused by *Exobasidium oxycocci*. Branches arise from infected axillary buds on 1-year-old wood. (Courtesy R. S. Byther; © APS)

Fig. 220. Cranberry flowers infected by *Exobasidium oxycocci*, the causal organism of rose bloom. The pathogen has not yet sporulated on the enlarged petals. (Courtesy P. R. Bristow; © APS)

When the fleshy branches first begin to appear, they are pale green; the pink color develops as they enlarge. About 80% of the time, infected uprights have only one abnormal branch; however, as many as eight rose blooms have been observed on a single upright. The branch takes on a white, powdery appearance as the fungus erupts through the epidermis and begins to sporulate (Fig. 221). By the middle of the flowering period, the branches begin to wither; they become dry, brownish black, and hard. Most break off and fall to the ground. Some wilt (flag) until harvest, and a few remain attached throughout the winter and can be seen on the 2-year-old wood the following spring.

Causal Organism

Rose bloom is caused by the fungus *Exobasidium oxycocci*. All species of this genus are plant parasites and stimulate host tissues to form abnormal growths, that is, hypertrophy and possibly hyperplasia. Two other *Exobasidium* spp. cause the red shoot and red leaf spot diseases of cranberry.

Basidia of *E. oxycocci* are elongate-clavate and superficial on affected plant parts; they have two slender sterigmata (four were reported in the original description of the pathogen) (Fig. 222). Basidiospores (2–3 × 13–18 µm) are fusiform, slightly curved, and hyaline and have one to three septa. When mature, they are actively discharged up to 2.3 cm in still air. Basidiospores germinate by budding to produce conidia, which are similar in shape to basidiospores but slightly smaller. Conidia are 1.0–1.5 × 6–10 µm and bacilliform to subfusiform. Germ tubes arise first from the ends and then from the septa of the basidiospores.

In culture, the pathogen grows by continuous budding, with minimal formation of hyphae. Colonies are whitish to cream-colored, raised, and cerebriform, and the margins are undulate to lobate. The sexual stage does not develop in culture.

Disease Cycle and Epidemiology

The disease cycle for rose bloom is very simple. Basidiospores are the primary and only known form of inoculum. In Washington (United States), buds did not become infected when a mixture of conidia and mycelial fragments was used as inoculum during June and early July. There is no repeating secondary cycle, except that, in some years, infected axillary buds on current-year growth break dormancy in July and form abnormal branches with basidiospores in August. The incidence of rose bloom in August is very low and, therefore, inoculum density is also low. Basidiospores in late summer may account for the infection of terminal buds, since these buds are being differentiated at that time. Although it is not known when or for how long buds are susceptible, axillary buds may be resistant by late summer.

Normally, axillary buds are infected by basidiospores and the following spring break dormancy to form the fleshy, abnormal branches that support sporulation. It has been suggested that this fungus–host association is highly evolved because the fungus no longer needs to produce a sporocarp; instead, the fungus-induced abnormal branch serves as a sporocarp. Most of the fungus biomass resides just below and between epidermal cells. Thin, haustorium-like hyphae penetrate mesophyll cells.

In the past, rose bloom was associated with young cranberry beds (less than 10 years old) and areas within beds that were shaded or had excessive vine growth. Recent research has demonstrated that severe outbreaks can occur in beds of any age, even when numbers of upright shoots are near optimum and vine growth is not excessive.

In Washington, most airborne basidiospores were trapped during a 6-week period from early May through mid-June. The dissemination of spores from infected flowers and berries might extend this primary period of sporulation. A second, much smaller peak of basidiospores was recorded in August, but only in years when rose blooms arose from infected axillary buds on the current year's growth. These buds were most likely infected by basidiospores produced in May or June of the same year.

Basidiospores are discharged in a diurnal pattern, with the fewest trapped at midday. It appears that two to four flushes of basidia mature and discharge basidiospores each night. The number of basidiospores trapped per day is positively related to the number of hours of leaf wetness and negatively related to mean daily air temperature, relative humidity, and solar radiation. Basidiospores germinate at temperatures between 10 and 22°C and grow best in culture at 20°C.

The vitality of uprights attacked by *E. oxycocci* is diminished. Yields are reduced by about one-third from a combination of fewer flowers per upright, lower fruit set, and smaller berries on infected uprights. Flowers also develop later on infected uprights. Overall, yield reductions are proportional to the percentage of flowering uprights infected. Infected uprights are less likely than healthy uprights to set a mixed bud for the following year, so the disease has an adverse carryover effect even though the pathogen does not survive systemically in the host.

The chlorophyll content of the pink leaves from abnormal branches is only about 20% of that in healthy leaves; hence, photosynthetic efficiency is low. Transpiration rates for the fleshy branches are high, even though they have no stomata, thus lowering water use efficiency.

Management

Because symptoms do not appear until the spring following infection, the benefits from control measures likewise are not

Fig. 221. The whitish hymenium of *Exobasidium oxycocci*, the causal organism of rose bloom, on the surface of an abnormal, fleshy cranberry branch. (Courtesy R. S. Byther; © APS)

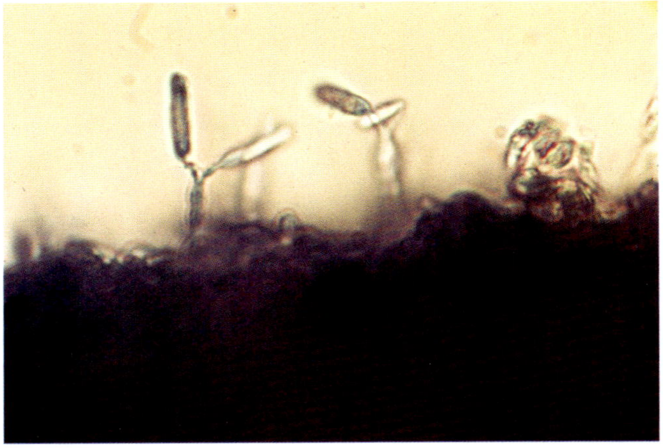

Fig. 222. Basidia and basidiospores of *Exobasidium oxycocci*, the causal organism of rose bloom. Basidia rupture the cuticle as they grow out from between the epidermal cells. (Courtesy P. R. Bristow; © APS)

realized until the following year. A few rose blooms in a bed are not cause for concern, but when they become plentiful, control measures usually are warranted.

To estimate disease incidence in the coming season, dormant uprights can be collected and forced in January or early February. When uprights are collected, as much of the 2-year-old wood as possible should be taken. Within 6 weeks, new growth starts from the terminal bud and abnormal branches develop from infected lateral buds. The rose bloom growths that are produced are pale green rather than pink. To estimate the expected percentage reduction in yield for the coming season for each bed, the percentage of infected uprights in the sample should be divided by three.

Rose bloom can be controlled with a number of fungicides, including Bordeaux mixture, copper hydroxide, mancozeb, and ferbam. Fungicides are applied during May and June when basidiospores are present to protect buds on the new growth and possibly flowers. Two to three applications at 2-week intervals are recommended; the first application should be made in May when the whitish spores are first noticed on the abnormal branches. On farms with mixed cultivars with different early growth patterns, release of basidiospores can occur over a longer time period, thus requiring more extended fungicide applications.

Some cultural practices have also been suggested to reduce the incidence of rose bloom. Soil drainage problems should be corrected to prevent prolonged periods of soil saturation because the disease has been associated with areas where water accumulates on the soil surface. Excessive vegetative growth caused by the overapplication of nitrogen fertilizer should be avoided, and vines should be pruned when necessary. Trees adjacent to problem beds should be removed to avoid shading.

Rose bloom was once thought to be more common on late-maturing cultivars, but cultivar trials in Washington during the 1990s suggest that there is only minor variation in susceptibility. Trials in the 2000s, however, indicated that the cultivar Crimson Queen was prone to rose bloom infection.

Selected References

Bain, H. F. 1926. Cranberry disease investigations on the Pacific Coast. U. S. Dep. Agric. Bull. 1434.

Bristow, P. R., and Windom, G. E. 1989. Effect of rose bloom disease on yield components of cranberry (*Vaccinium macrocarpon* Ait.). Acta Hortic. 241:324-329.

Bristow, P. R., and Windom, G. E. 1991. Epidemiology and control of *Exobasidium oxycocci* (rose bloom) on cranberry. (Abstr.) Phytopathology 81:1182.

Bristow, P. R., Patten, K. D., and Windom, G. E. 2002. Reaction of cranberry genotypes to rose bloom and fruit rots, 2000. Biol. Cult. Tests Control Plant Dis. 17:N02.

Cameron, J. S., and Bristow, P. R. 1992. The impact of rose bloom disease (*Exobasidium oxycocci*) on the photosynthetic biology of cranberry. (Abstr.) HortScience 27:605.

Shear, C. L., Stevens, N. E., and Bain, H. F. 1931. Fungous diseases of the cultivated cranberry. U. S. Dep. Agric. Tech. Bull. 258.

Tontyaporn, S. 1974. Comparison of *Exobasidium vaccinii* from *Vaccinium macrocarpon* with other *Exobasidium* spp. from ericaceous hosts. Ph.D. thesis. University of Wisconsin, Madison.

(Prepared by P. R. Bristow; Revised by K. D. Patten)

Rust

Rust, caused by *Pucciniastrum vaccinii* (syn. *P. myrtilli*), has occasionally been found in Oregon and Washington (United States) and in British Columbia (Canada) on cranberry and the pathogen's alternate host, hemlock. The disease is minor and does not cause significant economic injury to either host. Losses do not warrant control measures. Rust has not been reported in cranberries in eastern North America. Only the uredinial and telial stages have been found on cranberry in western North America, where symptoms are limited to a minor yellow leaf spot. The spermagonial and aecial stages are found on western hemlock on current-year needles, which may be discolored and prematurely shed.

A description of the causal organism and its disease cycle and epidemiology on blueberries can be found under Leaf Rust in the Diseases Caused by Fungi sections for Blueberries (Highbush and Rabbiteye) and Blueberries (Lowbush).

Selected References

Bain, H. F. 1926. Cranberry disease investigations on the Pacific Coast. U. S. Dep. Agric. Bull. 1434.

Hotson, J. W. 1934. Key to the Rusts of the Pacific Northwest. Univ. Wash. Publ. Biol. 3. University of Washington Press, Seattle.

Shaw, C. G. 1973. Host fungus index for the Pacific Northwest. 1. Hosts. Wash. State Agric. Exp. Stn. Bull. 765.

Shear, C. L. 1931. Fungous diseases of the cultivated cranberry. U. S. Dep. Agric. Tech. Bull. 258.

(Prepared by H. S. Pepin; Revised by F. L. Caruso)

Twig Blight

Twig blight has been reported at one time or another from all cranberry production areas, but serious losses have occurred only in Washington and Oregon (United States). The disease has not been observed for many years in Massachusetts, New Jersey, or Wisconsin, even though the early descriptions of the pathogens were based on samples collected in Maine and New York (United States). The pathogens were most likely introduced into the Pacific Northwest on infected cuttings from the east coast of the United States during the early part of this century. Twig blight first became a problem in Washington in the early 1950s and in Oregon during the 1970s. To date, the disease has not become a serious problem in cranberry beds in British Columbia (Canada) because it was effectively controlled when it first appeared in a few beds there during the 1980s.

Symptoms

Although infection occurs during the summer, no symptoms are observed on infected uprights until winter or, more commonly, in early spring, when the foliage on the newest growth (1-year-old wood) turns brown. Infected leaves are dull rather than glossy. The disease initially appears on scattered uprights

Fig. 223. Area of cranberry uprights blighted by twig blight, caused by *Lophodermium oxycocci*, while vines were still dormant. (Courtesy P. R. Bristow; © APS)

or in patches (Fig. 223). In severe outbreaks, large areas or entire beds may be blighted. By late spring, the blighted leaves fade to a bleached tan (Fig. 224) and eventually turn a silvery gray. The fungi kill only 1-year-old wood and do not progress farther down the uprights. Vegetative and fruit uprights and new runners are all susceptible to attack. Immature apothecia (referred to as hysterothecia in early reports) first appear as tiny black specks on the underside of infected leaves beneath the lower epidermis. As apothecia mature, they become elliptical and a median ridge forms. The ridge becomes the median slit, allowing apothecia to open and expose the glistening, grayish tan hymenial layer. Ascospores begin to mature in late April or May. Not all infected leaves on blighted uprights produce apothecia. The fungi do not sporulate on dead stems. Infected leaves wilt and begin to gradually drop by midsummer.

Causal Organisms

Lophodermium oxycocci and *L. hypophyllum* can both cause twig blight, and both are found only on cranberry. No anamorphic stage is known for either species.

Mature apothecia of *L. oxycocci* (250–350 × 500–600 µm) are elliptical, nearly black, and thick walled, with labiate margins that may appear grayish (Fig. 225). Asci (9–10 × 70–90 µm) have eight spores, are clavate, and have a short stipe. Ascospores (4–5 × 50–60 µm) are hyaline, straight or slightly curved, and nonseptate (although oil droplets give the impression of septation). When trapped on the tape of a Burkard volumetric spore trap, ascospores of *L. oxycocci* appear cochleate and the gelatinous sheath is readily apparent. Ascospores germinate with multiple germ tubes. Paraphyses are generally curved to undulate and only occasionally slightly swollen at the apex. They are as long as or slightly longer than mature asci.

Mature apothecia of *L. hypophyllum*, while similar in size to those of *L. oxycocci*, are more oval in shape and lighter in color and have thinner, more membranous walls. The margins of the slit are not labiate. Asci of *L. hypophyllum* have eight spores, are clavate, are 4–5 × 60–80 µm, and have long stipes (usually half as long as the ascus). Ascospores (1.0–1.2 × 25–35 µm) are hyaline and straight or slightly curved. They do not become cochleate on the tape of a spore trap, and their sheath is not obvious. Paraphyses are usually straight, elliptical at the apex, and slightly longer than mature asci.

Mature apothecia of both species open when moistened. Those of *L. hypophyllum* open widely, exposing the entire hymenial layer; those of *L. oxycocci* open less far. Once an apothecium has discharged its spores, the hymenial layer degenerates or disintegrates. In some years, the entire ascocarp breaks away from the leaf, leaving pits with ragged margins in the lower surface of the leaf. Spent apothecia probably break away as a result of repeated openings and closings in response to wetting–drying cycles.

L. hypophyllum may be only a weak parasite of cranberry. While it is common to find only *L. oxycocci* on leaves of blighted uprights, *L. hypophyllum* is rarely, if ever, present in the absence of *L. oxycocci*. Initially, the pathological relationship of *L. oxycocci* was reported as apparently saprophytic or weakly parasitic on leaves. Although *L. oxycocci* is reported to sporulate on both surfaces of the leaf, both species are strictly hypophyllous in the Pacific Northwest. Oregon and Washington are included in the geographic distribution for *L. hypophyllum*, but *L. oxycocci* is limited to Massachusetts, New Jersey, and Wisconsin in older reports.

Disease Cycle and Epidemiology

Ascospores are the only known source of inoculum. Because neither species has a known anamorph, no repeating secondary cycle occurs. Infections remain latent until leaves and stems on the 1-year-old wood are killed after the growing season. Fruiting bodies, which start the cycle again, begin to form the following spring.

Yield reductions are proportional to the percentage of uprights with mixed (fruit) buds blighted. Thus, disease incidence and disease severity are the same. For example, if 15% of the uprights with mixed buds are blighted, then yield is reduced by a similar amount. The fungus also attacks uprights with vegetative buds. The cranberry plant does not compensate for the reduction in the number of flowering uprights by increasing either fruit set or berry size on the remaining healthy uprights. The disease also does not appear to influence floral induction, since the ratio of uprights with mixed and vegetative buds is nearly constant regardless of disease incidence in a given year. Under favorable conditions where the disease is not controlled, more than 80% of the uprights may be affected. Over a period of 8 years at one site in Washington, disease incidence ranged from 20 to 73% in the untreated check plots.

The following information was developed for *L. oxycocci* but probably applies to *L. hypophyllum* as well. The fungus overwinters as mycelium in leaves and stems of infected twigs. The pathogen does not survive or sporulate on dead leaves in the duff layer. Ascospores are liberated from late May until at least early October in Washington and Oregon. The period of peak liberation (June to August) corresponds to the time when the majority of apothecia mature. The spores are forcibly ejected from asci and are disseminated on wind currents. Once ascospores make contact with the host and the gelatinous sheath surrounding the spores has dried, they are not easily removed.

Fig. 224. Close-up of cranberry uprights blighted by twig blight, caused by *Lophodermium oxycocci*, in midsummer. (Courtesy P. R. Bristow; © APS)

Fig. 225. Moistened, fully opened apothecia of *Lophodermium oxycocci*, a causal organism of twig blight, on a cranberry leaf. Note the gray edge of the labium and the exposed hymenium. (Courtesy P. R. Bristow; © APS)

Leaves on new growth are infected during a single limited period (about 30 days long) each summer. The onset of the infection period varies from mid-June to late July but is consistently about 3–4 weeks after the time when 50% of the apothecia contain mature spores. The higher the percentage of infected uprights (disease incidence), the more inoculum is produced during the infection period to infect the new growth.

Ascospore liberation displays a diurnal periodicity. Most ascospores are trapped between 6 a.m. and 1 p.m. (standard time). Moisture for 4–28 h is required before the onset of a spore shower. Either rain or irrigation can satisfy the requirement but not moisture from dew or fog. Water apparently activates a priming mechanism that prepares the ascocarp for spore release. The priming mechanism may be a metabolic rather than a physical process because, for example, rain that falls at 6 a.m., while too late to prepare apothecia for spore release that day, does prepare them for release the following day even if there is no additional rain or irrigation before the spore shower on the following day. Apothecia of both species open within 30 min when moistened with water. Moisture from dew or fog is adequate for opening apothecia. Apothecia may open and close many times a day in response to wetting–drying cycles.

The mechanism that triggers spore release is not known; however, the onset coincides with sunrise and increasing air temperature. A decrease in relative humidity may play a role in the release of spores.

Management

Because symptoms do not appear until the year following infection, the benefits of control practices will also not be seen until the following year. Control is based on protecting new growth with fungicides during the infection period. In the integrated pest management program in Washington and Oregon, scouts monitor the development of ascospores and advise growers when to begin the fungicide program. The first application is made 3–4 weeks after 50% of the apothecia have mature ascospores. Three applications spaced 14 days apart are recommended, especially when the disease is severe. Azoxystrobin, chlorothalonil, febuconazole, maneb, and mancozeb give acceptable control of the disease. Failure to control twig blight with fungicides usually means that the first spray was not applied early enough. Fungicides applied later than mid-August are of little value even though ascospores are still present.

Cultural practices seem to have no impact on the disease, although removing trees to eliminate shade lowers the incidence of twig blight in previously shady areas. When new beds are planted, cuttings should be selected only from beds known to be free of twig blight. New beds with twig blight fill in very slowly until the disease is controlled. Little is known about cultivar reactions to the disease, although the cultivars McFarlin, Stevens, Bergman, Pilgrim, and Crowley are susceptible.

Selected References

Bristow, P. R. 1983. Lophodermium twig blight of cranberry: Relationships between inoculum density, period of host susceptibility, and disease severity. (Abstr.) Phytopathology 73:957.

Bristow, P. R., and Windom, G. E. 1989. Controlling twig blight and fruit rots of cranberry with fungicides, 1987–88. Fungic. Nematic. Tests 44:80.

Bristow, P. R., and Windom, G. E. 1990. Lophodermium twig blight of cranberry: Ascocarp maturity and the onset of the infection period. (Abstr.) Phytopathology 80:886-887.

Moore, B. J., Poole, A. P., and Bristow, P. R. 1980. Control of cranberry twig blight with fungicides, 1978. Fungic. Nematic. Tests 35:44-45.

Shear, C. L., Stevens, N. E., and Bain, H. F. 1931. Fungous diseases of the cultivated cranberry. U. S. Dep. Agric. Tech. Bull. 258.

(Prepared by P. R. Bristow; Revised by K. D. Patten)

Upright Dieback

In the condition known as upright dieback of cranberry, the tips of the upright exhibit dieback. Leaves of uprights first appear desiccated, often with a yellowish cast, then orangish brown, and the uprights and runners die. Disease incidence varies from year to year, but ordinarily upright dieback does not cause appreciable economic loss. Some fruit rot is usually associated with the disease (see Viscid Rot and other fruit rots listed in this compendium).

Symptoms

The typical symptom of upright dieback is tip dieback of the cranberry upright. Symptoms first appear as a desiccation of the upper foliage. Orange coloration or bronzing may develop, and eventually the upright turns brown and dies. The dieback symptoms progress down the stem but are rarely observed to affect runner tissue. Leaves typically are retained during the season of infection, and diseased uprights appear randomly dispersed among healthy uprights in an affected cranberry bed (Fig. 226). Occasionally, particularly severe damage is observed in 1- to 3-year-old cranberry beds, where large patches of uprights can be infected (Fig. 227). Affected uprights may appear on the same runner with healthy uprights, sometimes even distal to them (Fig. 228). In some runners, every upright may be infected (Fig. 229). In Massachusetts (United States), uprights

Fig. 226. Scattered uprights affected by upright dieback, caused by *Phomopsis vaccinii*, in a well-established cranberry bed. (Courtesy F. L. Caruso; © APS)

Fig. 227. Area of vines in a newly planted cranberry bed affected by upright dieback, caused by *Phomopsis vaccinii*. (Courtesy F. L. Caruso; © APS)

die in certain cranberry beds after the withdrawal of the winter flood in March when the plants resume active growth.

The roots of diseased vines are unaffected. If berries are attached to affected uprights in the latter part of the growing season, they wither and desiccate as the upright dies. Crop loss may be significant where 20% of the uprights are infected, although vegetative uprights seem to be affected much more than fruiting uprights.

Causal Organisms

Phomopsis vaccinii (syn. *Diaporthe vaccinii*), and, to a lesser extent, various other yet unidentified *Phomopsis* spp., are the primary causal organisms. Pathogenicity has also been confirmed for *Synchronoblastia crypta*; however, this fungus has rarely been isolated from diseased stems since the late 1980s and is no longer considered a primary causal organism. The sexual stage has not been observed on infected stems or in culture for several decades. *Godronia cassandrae* is regularly cultured from symptomatic uprights and is suspected to also be pathogenic, although this has not been officially confirmed.

P. vaccinii has been found in all the cranberry-growing regions of the United States and several Canadian provinces; *S. crypta* has been reported only in Massachusetts and New Jersey (United States). *P. vaccinii* produces black pycnidia that are 0.3–0.5 mm in diameter on cranberry fruit, but pycnidia are seldom observed on diseased uprights. In culture, pycnidia are readily produced on uprights and fruit and are 1–2 mm in diameter. The pycnidia in culture are partly embedded, thick walled, black, and leathery; they are usually chambered and rupture irregularly to emit whitish to creamy masses of conidia. Conidia (alpha spores) are hyaline, one celled, and elliptical; they typically have two prominent oil globules and are 2–5 × 6–11 µm.

Conidiophores are simple and somewhat spindle shaped, tapering toward the apex; they are 15–25 µm long in young pycnidia and slightly longer in older pycnidia. Scolecospores (beta spores) are occasionally observed; they are usually hooked at one end and are 0.35 × 14–20 µm.

The sexual stage *P. vaccinii* was originally described as synonym *Diaporthe vaccinii*, and both the asexual and sexual forms of this pathogen were originally described in 1931. Other than the original description and a publication in 1940, no reports of the sexual stage exist. Thus, only the asexual stage is currently found on cranberry.

According to the reports and descriptions mentioned above, perithecia of *P. vaccinii* are formed in stromata on fruit and in culture, but on stems, they are separate and without any trace of a stroma. The perithecia produced on stems are found within the bark and the wood, with an eccentric neck protruding through the bark. They are nearly hemispherical and are 0.2–0.4 × 0.3–0.5 mm. The wall of the perithecium is two to several layers thick and carbonous black. On decayed berries in culture, perithecia are produced in stromata, and long perithecial necks protrude in all directions from the folds of the shriveled berry. In media, perithecial stromata are about 1.5–2.0 mm in diameter and are partially embedded. The numerous necks grow to a length of 0.5 mm and are several cells thick; they are heavy walled and black and contain numerous paraphyses. Asci (7–12 × 37–51 µm) are oblong to fusoid and sessile. The apex is thickened and contains a narrow pore. Ascospores (2–3 × 8–12 µm) are irregularly biseriate and ellipsoid, with obtuse ends. They are two celled and slightly constricted at the septum, and each cell is typically biguttulate.

In culture, *P. vaccinii* usually grows out from the ends of the stem sections. The fungal colonies are distinctive and diagnostic; at first white and closely appressed, they are later covered with a somewhat patchy aerial mycelium that may darken to gray.

S. crypta produces pycnidia readily in culture, but they have not been observed on infected uprights in the field. The pycnidia are dark pigmented, are 94–235 µm in diameter, and have distinct ostioles. No paraphyses are present. The cavity is lined with one layer of conidiogenous cells, which are hyaline, pulvinate, and clavate or lobed. Two to 20 or more conidia are formed synchronously on short denticles. The conidia (1.5–2.0 × 6–9 µm) are hyaline, nonseptate, and elliptic to oblong-elliptic or fusiform-elliptic. Conidia are also produced on assimilative hyphae. No teleomorphic stage has been observed on cranberry tissue or in culture.

S. crypta also grows out from the cut ends of infected uprights in culture. Fungal colonies are dark and closely appressed and later appear greasy in nature.

Disease Cycle and Epidemiology

P. vaccinii occasionally can be isolated in the spring from uprights of the previous season. However, the low incidence of the fungus in the spring where dieback was severe late in the preceding summer indicates that this fungus normally does not persist well over the winter in diseased vines, at least in the severe winters of Wisconsin (United States). The fungus

Fig. 228. Cranberry plant with a single upright affected by upright dieback, caused by *Phomopsis vaccinii*. (Courtesy F. L. Caruso; © APS)

Fig. 229. Cranberry plant with a runner with several uprights affected by upright dieback, caused by *Phomopsis vaccinii*. (Courtesy F. L. Caruso; © APS)

can be isolated from current-season stems after July and from flower pedicels and pistils in June and July. *P. vaccinii* infects succulent growth and progresses from leaf tissue into the stem tissue or infects through stem wounds, eventually colonizing vascular tissue. Cranberry plants appear most susceptible to infection in the spring when fresh growth is present, although the uprights and runners are susceptible throughout the season if wounded. At all points in the growing season, wounded plants are more susceptible to infection and often develop more extensive tissue death than nonwounded plants. Generally, infection does not spread into older growth or runners. *P. vaccinii* and various isolates of unidentified *Phomopsis* spp. can be isolated from apparently healthy uprights, which indicates symptomless infections.

Dieback may appear at any time during the growing season, but disease development is apparently favored by warm weather conditions and by periods when vines are under stress from hot or dry conditions and, thus, presumably more susceptible. The fungus grows best under relatively warm weather conditions, so hot summer temperatures generally favor its growth. Vines in the early stages of infection may recover during periods of lower temperatures and favorable moisture conditions. The disease may also appear where vine growth is rank or lush and may be especially severe in young plantings, where large patches of uprights may be affected.

Conditions that favor *P. vaccinii* apparently also favor infection of vines by *S. crypta* and subsequent disease development. Conidia are the sole means by which *S. crypta* infects uprights. This fungus probably overwinters as mycelium in uprights or fruit in the duff layer and produces pycnidia in the spring. Conidia are capable of infecting uprights, blossoms, and fruit.

Other fungi are readily isolated from dying uprights, including *Alternaria* spp., *Aureobasidium pullulans*, *Colletotrichum acutatum*, *Colletotrichum gloeosporioides*, *Gloeosporium minus*, *Cladosporium* spp., *Fusarium* spp., and *Pestalotia vaccinii*.

Management

Cultural practices that encourage vigorous but not rank vine growth should make vines more tolerant or resistant to infection by *P. vaccinii*. Cranberry beds should also be managed to avoid drought stress and plant wounds by protecting plants from freezing injury. Other disease management strategies can include avoiding long periods of leaf wetness, limiting foot traffic, and limiting pruning or other mechanical damage in the spring when vines are particularly susceptible and inoculum levels are likely to be high.

Fungicide applications made at bud break in the spring and later during bloom, when the fungicide also serves to control fruit rot, likely help prevent infection. It seems reasonable that a systemic fungicide would have activity against this disease since *P. vaccinii* grows within the vascular tissues of uprights.

Little is known about the relative resistance or susceptibility of various cultivars to the disease, although most cultivars appear susceptible. In Wisconsin, Searles appears to be more susceptible than other commercially planted cultivars. In Massachusetts (United States), the cultivars Early Black and Howes are both susceptible, and the cultivar Franklin may have some resistance to the disease.

Selected References

Catlin, N. J. 2005. Upright dieback of cranberry, *Vaccinium macrocarpon* Ait.: Causal agents and infection courts. Ph.D. dissertation. University of Massachusetts, Amherst, MA.

Farr, D. F., Castlebury, L. A., and Rossman, A. Y. 2002. Morphological and molecular characterization of *Phomopsis vaccinii* and additional isolates of *Phomopsis* from blueberry and cranberry in the eastern United States. Mycologia 94:494-504.

Friend, R. J., and Boone, D. M. 1968. *Diaporthe vaccinii* associated with dieback of cranberry in Wisconsin. Plant Dis. Rep. 52:341-344.

Shear, C. L., Stevens, N. E., and Bain, H. F. 1931. Fungous diseases of the cultivated cranberry. U. S. Dep. Agric. Tech. Bull. 258.

Uecker, F. A., and Caruso, F. L. 1988. *Synchronoblastia crypta*, a new coelomycetous pathogen of upright stems and fruits of cranberry. Mycologia 80:344-347.

Wilcox, M. S. 1940. *Diaporthe vaccinii*, the ascigerous stage of *Phomopsis*, causing a twig blight of blueberry. Phytopathology 30:441-443.

(Prepared by D. M. Boone and F. L. Caruso;
Revised by N. J. Catlin)

Viscid Rot

Viscid rot appears in all major cranberry-growing areas in North America. It is primarily a storage rot disease but occasionally causes rot in the field. The disease is generally minor, although locally it can cause moderate losses. In New Jersey (United States), for instance, *Phomopsis vaccinii*, the causal organism, was isolated from rotted berries at frequencies up to 53% in clustered locations within the cranberry-growing area.

Symptoms

Fruit with viscid rot are softened and off-colored and may be slightly mottled. In some growing areas and on some cultivars, a viscous substance can be "strung out" from the cut surface of the rotten berry. Isolation of the causal organism is the only way to reliably diagnose this disease.

Causal Organism

Phomopsis vaccinii (syn. *Diaporthe vaccinii*) is the cause of viscid rot. The same fungus causes upright dieback of cranberry. The fungus forms pycnidia in diseased uprights and berries and in culture. Pycnidia vary in shape and are 0.3–2 mm in diameter. Pycnidia are often chambered and rupture with an irregular opening to emit masses of whitish spores.

P. vaccinii produces two types of spores. The alpha conidia (2–5 × 6–11 µm) are hyaline, one celled, and oblong-elliptic, with two oil globules at each end (Fig. 230). Beta conidia, although not as common, may be found in abundance in some isolates. They are filiform and 0.35 × 14–20 µm (Fig. 231).

Perithecia are found very infrequently. They are produced in stromata in fruit and in culture, but in diseased vines, they

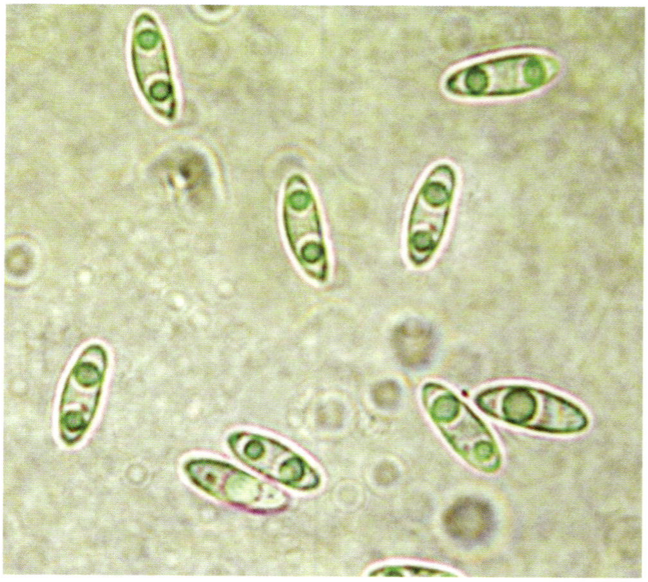

Fig. 230. Alpha conidia of *Phomopsis vaccinii*, the causal organism of viscid rot on cranberry. (Courtesy J. S. Mika)

are separate and without stromata. Perithecia have long, protruding necks and can range from 0.2–0.5 mm in diameter on vines to 1.5–2 mm in culture. Asci are long and contain eight irregularly biseriate, two-celled ascospores. Ascospores (2–3 × 9–12 µm) are ellipsoid, obtuse, and slightly constricted at the septum; each cell is typically biguttulate.

In culture, the mycelium is initially white and appressed on the medium surface. It later turns gray and then black as the embedded stromata form (Fig. 232).

Disease Cycle and Epidemiology

The pathogen overwinters in diseased vines and fruit and in apparently healthy uprights. Ascospore release is believed to begin around bud break and continue up to bloom. Conidia are believed to be released from bloom to harvest. The fungus can be isolated from apparently healthy uprights, flowers, and fruit as early as June, indicating that it is present before symptoms are expressed. In Massachusetts (United States), *P. vaccinii* is often recovered throughout the growing season at higher levels in healthy fruit than in diseased fruit. Fruit are infected during bloom and early berry development. Disease development is favored by high temperatures and when the vines are under stress.

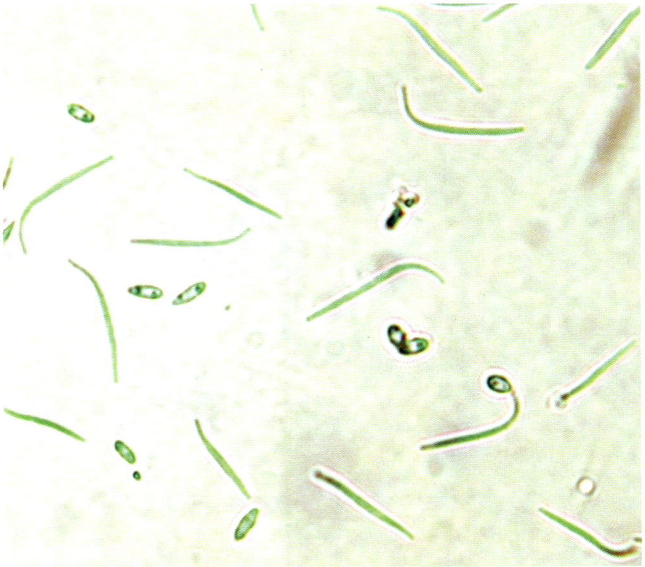

Fig. 231. Beta conidia of *Phomopsis vaccinii*, the causal organism of viscid rot on cranberry. (Courtesy J. S. Mika)

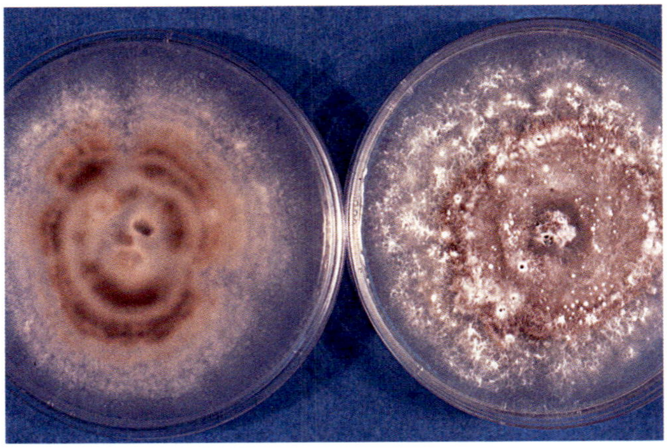

Fig. 232. Cultures of *Phomopsis vaccinii*, the causal organism of viscid rot on cranberry, growing on cornmeal agar viewed from below (left) and from above (right). (Courtesy C. C. Kusek; © APS)

Management

Fungicide applications during bloom and early berry development provide effective control of viscid rot. Because conditions during these growth stages are often favorable for spore germination and fungal penetration, flowers and fruit must be continuously protected with fungicides. Although control through the use of fungicides has been effective, twig dieback caused by *Phomopsis vaccinii* was described in Europe in 2004.

Selected References

Boone, D. M. 1982. Viscid rot and upright dieback of cranberry. Univ. Wis. Ext. Publ. A3195.

Farr, D. F., Castlebury, L. A., and Rossman, A. Y. 2002. Morphological and molecular characterization of *Phomopsis vaccinii* from blueberry and cranberry in the eastern United States. Mycologia 94:494-504.

Gabler, J., Kacergius, A., and Jovaisiene, Z. 2004. Detection of *Phomopsis vaccinii* on blueberry and cranberry in Europe by direct tissue blot immunoassay and plate-trapped antigen ELISA. J. Phytopathol. 152:11-12.

Shear, C. L., Stevens, N. E., and Bain, H. F. 1931. Fungous diseases of the cultivated cranberry. U. S. Dep. Agric. Tech. Bull. 258.

Stiles, C. M., and Oudemans, P. V. 1999. Distribution of cranberry fruit-rotting fungi in New Jersey and evidence for nonspecific host resistance. Phytopathology 89:218-225.

(Prepared by C. C. Kusek; Revised by J. S. Mika)

Yellow Rot

A yellowish rot of stored cranberry fruit was reported in Washington and Oregon (United States) in 1926. In 1929, great losses were attributed to this disease in these two states when cranberries were wet harvested. In cranberry beds in Wisconsin (United States), berries with yellow rot are sometimes found in the harvest "trash." The disease also has been reported in Massachusetts (United States). Usually, however, less than 1% of stored berry losses can be attributed to yellow rot. The pathogen was also reported to cause blossom blight in Washington.

Symptoms

Cranberries with yellow rot usually are a conspicuous, bright yellow-orange (Fig. 233). The yellow coloration is especially bright in berries stored at 15°C or below. In infected berries stored at 24–26°C, disease development is more rapid, and infected berries tend to develop a light brown color. Necrotic tissue is soft and dry. Sporulation on the fruit is infrequent.

Fig. 233. Cranberries with yellow rot symptoms caused by a *Botrytis* sp. (right) and healthy cranberries (left). (Courtesy D. M. Boone)

Causal Organism

The fungus that causes yellow rot is a *Botrytis* sp. of the *Botrytis cinerea* type; although it has been referred to as *B. cinerea*, this species designation should be considered tentative until its teleomorph has been studied.

The *Botrytis* isolate from cranberry grows rapidly on potato dextrose agar at an optimum temperature of 20–24°C. The colony is at first off-white, with loose aerial growth. Later, the mycelial mat becomes gray-brown and somewhat flocculent; it is frequently 0.5 cm or more in height. As sporulation develops, colonies darken because of large, dark conidiophores. Numerous immature white sclerotia form in about 7 days at 20°C and turn black within 10 days. Conidiophores from sclerotia or vegetative hyphae are septate and branched, and each branch is constricted at the juncture. Conidia are ovate or elliptical and vary in length from 9 to 20 μm, depending on the incubation temperature; longer conidia are produced at higher temperatures. Although the teleomorphic stage has not been observed, globose spermatia 2.5–3.0 μm in diameter are produced on branching spermatiophores in cream-colored sporodochia in culture.

Disease Cycle and Epidemiology

Inoculum of a *Botrytis* sp. may come from many sources because the fungus can grow on many different substrates. Cultures isolated from lucerne hay and rotting lettuce were also able to incite yellow rot in cranberry. The fungus has been isolated from senescing tissues of fading corolla lobes, senescing sepals, and dead or dying tissues on shoots, as well as from rotted fruit. No apparent disease was associated with the fungus on blossoms or shoots. However, the fungus has been reported to cause blossom blight in Washington. Occasionally, a *Botrytis* sp. can be isolated from the blossom end of young, healthy fruit, which may indicate that it can become established in senescing, persistent sepals and later invade the fruit from the sepals. However, most berries with yellow rot seem to have been mechanically damaged previous to rotting, so the fungus may invade through wounds rather than through sepals. The pathogens that cause black rot are often also present in the same berry with the yellow rot fungus, which suggests that the causal fungi of both diseases entered through the same wound.

Management

Ordinarily, yellow rot does not cause sufficient economic loss to warrant the use of special methods to control it.

Selected References

Crowley, A. J. 1929. Cranberry growing in Washington. Wash. Agric. Exp. Stn. Bull. 230.

Friend, R. J. 1968. Incidence and pathogenicity of fungi found on cranberry in Wisconsin. Ph.D. thesis. University of Wisconsin, Madison.

(Prepared by D. M. Boone; Revised by F. L. Caruso)

Diseases Caused by Viruses

Cranberries

Ringspot

Ringspot of cranberry is a virus-induced disease similar to red ringspot of blueberry. Like blueberry red ringspot, cranberry ringspot is caused by a caulimovirus and has no known vector. It has been observed in New Jersey, Wisconsin, and Massachusetts (United States) but is not prevalent enough to be economically important. In the few beds where it appears, it does not appear to be increasing in occurrence. The disease has been reported in the cultivars Searles, Howes, and Early Black.

Symptoms

Very striking light-colored, ringlike symptoms develop on affected fruit (Fig. 234). These ring symptoms are especially prominent as fruit begin to ripen and are less apparent when fruit are fully ripe. On berries that are only slightly colored, and even on ripe berries, the area within the rings is often deeper red than the area outside the rings. On the cultivar Early Black, the spots are more like light-colored, circular patches. On the cultivar Searles, the spots are less prominent, but the berries are usually malformed, with some necrosis at the blossom end. In extreme cases, entire berries are necrotic. Faint ringspot symptoms or small red splotches also appear on cranberry leaves as they assume their reddish autumn color (Fig. 235).

Ringspot is a systemic disease, affecting all uprights on a given runner and all berries on a given upright. It does not appear to reduce plant vigor or fruitfulness. The keeping quality of 'Howes' fruit does not appear to be affected, but that of cultivar Searles is markedly reduced. Replicated samples of 'Searles' berries from healthy and diseased vines showed 7 and 33% spoilage, respectively, after storage for 4 months at 8°C.

Causal Agent

The causal agent of ringspot is a virus of the genus *Soymovirus* of the family *Caulimoviridae* of reverse-transcribing, DNA-containing viruses. Virus particles and inclusion bodies have been observed in infected leaf tissue; the inclusion bodies appeared to be similar to those of the *Caulimovirus* type associated with *Blueberry red ringspot virus*. Sequence analysis, using a limited number of samples, confirmed similarity to *Blueberry red ringspot virus*. Attempts to transmit the virus

Fig. 234. Cranberry 'Early Black' fruit with symptoms of ringspot, caused by a soymovirus. (Courtesy J. J. Polashock)

Fig. 235. Cranberry leaves with symptoms of ringspot, caused by a soymovirus. (Courtesy P. R. Bristow; © APS)

to healthy vines have not been successful, possibly because of grafting difficulties. Similarly, attempts to transmit the disease to herbaceous indicator plants by means of tissue grafting or the use of extracts have not given conclusive results.

Disease Cycle and Epidemiology

Ringspot has no known vector. The disease was reported to spread in infected 'Searles' beds in Wisconsin. Otherwise, dissemination appears to be limited to the planting of infected stock. It is unknown whether the disease can spread from cranberry to blueberry and vice versa.

Management

Disease-free planting stock should be used when new cranberry beds are established. Diseased plants should be rogued if possible.

Selected References

Boone, D. M. 1966. Ringspot disease of cranberry. Plant Dis. Rep. 50:543-545.

Kim, K. S., Ramsdell, D. C., Gillett, J. M., and Fulton, J. P. 1981. Virions and ultrastructural changes associated with blueberry red ringspot disease. Phytopathology 71:673-678.

Polashock, J. J., Ehlenfeldt, M. K., and Crouch, J. A. 2009. Molecular detection and discrimination of *Blueberry red ringspot virus* strains causing disease in cultivated blueberry and cranberry. Plant Dis. 93:727-733.

Stretch, A. W. 1987. Ringspot of cranberry. Pages 123-124 in: Virus Diseases of Small Fruits. R. H. Converse, ed. U. S. Dep. Agric. Agric. Handb. 631. U.S. Government Printing Office, Washington, DC.

(Prepared by D. M. Boone; Revised by J. J. Polashock)

Scorch

Blueberry scorch virus (BlScV) was first detected in cranberry in British Columbia (Canada) and subsequently in Oregon and Washington (United States). There were no symptoms associated with the presence of the virus in cranberry; however, cranberry was initially tested as a possible host because of unexpected finding of BlScV in blueberry fields adjacent to cranberry bogs in the Fraser Valley of British Columbia. In a limited survey, BlScV was detected in 12 of 72 cranberry bogs in British Columbia, Oregon, and Washington.

Symptoms

No symptoms have been observed in cranberry plants infected with BlScV, although a limited number of cultivars have been tested.

Causal Agent

The presence of *Blueberry scorch virus* was confirmed in cranberry by enzyme-linked immunosorbent assay and reverse transcription-polymerase chain reaction. See the Scorch section under Diseases Caused by Viruses, Blueberries for other characteristics of BlScV.

Disease Cycle and Epidemiology

Virtually nothing is known about the disease cycle and epidemiology of BlScV in cranberry. The mode of transmission is likely aphids, as has been demonstrated in blueberry. Additionally, as a symptomless host, cranberry plant material could provide an efficient pathway for the movement of BlScV to areas where it has not been detected.

Management

As for other viral diseases of cranberry, plants to be used for propagation should be tested for BlScV and destroyed if found to be infected. Since the virus is not known to cause symptoms in cranberry, no management of the virus in cranberry bogs is recommended at this time. Blueberry fields adjacent to cranberry bogs may be at risk of infection with BlScV from cranberry bogs that have infected plants.

Selected Reference

Wegener, L. A., Punja, Z. K., and Martin, R. R. 2004. First report of *Blueberry scorch virus* in cranberry in Canada and the United States. Plant Dis. 88:427.

(Prepared by R. R. Martin)

Shock

Cranberry shock was first detected in Wisconsin (United States) in 2014. Scarred fruit were collected that tested negative for *Tobacco streak virus* (TSV). This prompted testing for related viruses and those that were negative for TSV were found to be positive for *Blueberry shock virus* (BlShV). Since the first report, shock has also been detected in cultivated cranberry in Massachusetts, New Jersey, and Oregon (United States) and in British Columbia (Canada).

Symptoms

Shock was found to be associated with berry scarring in samples that tested positive BlShV. However, it has not yet been shown whether shock always causes fruit symptoms or whether some BlShV-positive berries might appear healthy as described for TSV in cranberry.

Causal Agent

Shock appears to be caused by *Blueberry shock virus*, as indicated by enzyme-linked immunosorbent assay testing for this virus. However, follow-up data are not yet available for confirmation that BlShV or a closely related virus is responsible for the disease in cranberry. BlShV, like TSV, is in the genus *Ilarvirus* in the family *Bromoviridae*. See the Shock section under Diseases Caused by Viruses, Blueberries for other characteristics of BlShV.

Disease Cycle and Epidemiology

Virtually nothing is known about the disease cycle and epidemiology of BlShV in cranberry. The mode of spread is unknown, as are the timing, development, and persistence of fruit symptoms.

Management

As for other viral diseases of cranberry, plants to be used for propagation and those to be used in breeding should be tested for BlShV and destroyed if found to be infected.

Selected Reference

Wells-Hansen, L. D., Polashock, J. J., Vorsa, N., Lockhart, B. E. L., and McManus, P. S. 2016. Identification of *Tobacco streak virus* in cranberry and the association of TSV with berry scarring. Plant Dis. 100:696-703.

(Prepared by J. J. Polashock)

Tobacco Streak

Tobacco streak virus (TSV) was first detected in cultivated cranberry on vines imported from North America into Scotland in 2001. TSV was detected during routine screening for mechanically transmissible viruses using herbaceous indicator plants. Identification was confirmed by reaction to antisera of TSV isolates from strawberry, red raspberry, and tobacco. Nine cultivars were tested and two (i.e., Pilgrim and AJ) were found to be infected. Three different strains were described based on symptomatology in herbaceous test plants. In 2012, TSV was confirmed for the first time in commercial plantings in Wisconsin (United States). The disease has subsequently been confirmed in Massachusetts and New Jersey (United States).

Symptoms

In the field, TSV is associated with berry scarring and deformation that ranges from minor to severe (Figs. 236 and 237). Symptoms on berries appear shortly after fruit set. Scarring associated with TSV typically is on all or nearly all berries and flowers on a plant. TSV is sometimes associated with a blossom blast symptom that is indistinguishable from blossom blast caused by fungal pathogens such as *Phyllosticta vaccinii*. Attempts to reproduce symptoms through artificial inoculation of cranberry have failed. Some TSV-positive berries appear healthy.

Causal Agent

Tobacco streak virus is the type member of the genus *Ilarvirus* of the family *Bromoviridae*. TSV is known to infect at least 80 different plant species in several different families. The genome of TSV consists of three different single-stranded RNAs. Virus particles detected in sap from cranberry leaves are quasi-isometric and approximately 30 nm in diameter. The virus can be detected by enzyme-linked immunosorbent assay or reverse transcription-polymerase chain reaction using TSV-specific primers.

Disease Cycle and Epidemiology

In plants that had symptomatic fruit the previous season, TSV can be detected in 1-year-old leaves and in new growth in early spring, suggesting that TSV overwinters in plants. The flowers that emerge on these TSV-positive plants appear normal, but their pollen harbors TSV. Likewise, berries appear healthy, but they also test positive for TSV. Thus, TSV in cranberry may be similar to *Ilarvirus* infections of some other hosts, such as *Blueberry shock virus* in blueberry and *Prunus necrotic ringspot virus* in stone fruits, in that a shock phase is followed by plant recovery, despite the presence of the virus in pollen and tissues of recovered hosts. TSV can be mechanically transmitted to herbaceous hosts, but modes of transmission in cranberry are not known. TSV presumably is dispersed in or on infected pollen, but the relative roles of pollinators and other insects in spread and infection is not known. Some other TSV host species become infected when fed on by thrips carrying TSV-infected pollen.

Fig. 236. Relatively subtle symptoms of *Tobacco streak virus* infection of cranberry fruit. (Courtesy P. S. McManus; © APS)

Fig. 237. Relatively severe symptoms of *Tobacco streak virus* infection of cranberry fruit. (Courtesy P. S. McManus; © APS)

Management

Plants for propagation and those used in breeding should be tested for TSV and destroyed if found to be infected. Vine cuttings from cranberry beds where TSV has been identified should not be used to establish new plantings. The long-term impact of infection in established beds has not been determined.

Selected References

Jones, A. T., McGavin, W. J., and Dolan, A. 2001. Detection of *Tobacco streak virus* isolates in North American cranberry (*Vaccinium macrocarpon* Ait.). online. Plant Health Progress. doi: 10.1094/PHP-2001-0717-01-RS

Sdoodee, R., and Teakle, D. S. 1987. Transmission of tobacco streak virus by *Thrips tabaci*: A new method of plant virus transmission. Plant Pathol. 36:377-380.

Tzanetakis, I. E., Mackey, I. C., and Martin, R. R. 2004. Strawberry necrotic shock virus is a distinct virus and not a strain of *Tobacco streak virus*. Arch. Virol. 149:2001-2011.

(Prepared by J. J. Polashock and P. S. McManus)

Diseases Caused by Bacteria

Cranberries

False Blossom

Cranberry false blossom is caused by a leafhopper-vectored phytoplasma (formerly called mycoplasmalike organism or MLO). It occurs naturally in *Vaccinium macrocarpon* and *V. oxycoccos*. The disease threatened the existence of the cranberry industry in the early 1900s when it was spread in diseased vines from Wisconsin (United States) to all cranberry-growing areas in the United States and Canada. The discovery and control of the leafhopper vector, coupled with extensive planting of resistant cultivars, reduced false blossom to an insignificant factor in cranberry production for many years. However, there appears to be a resurgence in this disease in some growing regions. This resurgence has tentatively been attributed to the increased use of more targeted insecticides that has allowed the vector to again multiply in cranberry beds.

Symptoms

Cranberry false blossom is most easily diagnosed at bloom, when the flowers on infected plants assume an upright position because the pedicels are straight rather than arched. The calyx lobes of diseased flowers become enlarged and the petals are short, streaked with red, and tend to be straight or flattened rather than curled as in healthy mature flowers. The stamens and pistils are abnormal, usually resulting in a sterile flower (Fig. 238). Flower buds occasionally exhibit varying degrees of phyllody. In severe cases, normally latent axillary buds are stimulated and give rise to closely spaced branches, resulting in a witches'-broom effect (Fig. 239). The leaves on these branches are closely appressed to the stem. The witches'-broom of uprights may extend above the normal level of uprights, and the foliage may redden prematurely in the autumn. Terminal flower buds during advanced stages of development are enlarged and very susceptible to spring frost injury because they are protected by only one layer of scale leaves rather than the four layers that protect normal buds.

Causal Organism

A phytoplasma has consistently been associated with false blossom-diseased plants; however, the phytoplasma has not yet been grown in pure culture and then reintroduced into cranberry to prove its causal association with diseased plants. The phytoplasma is found in sieve tube elements. It varies in shape from spherical to oval to irregular and in diameter from 80 to 300 nm (Fig. 240). Each body is surrounded by a single unit membrane. Detection by polymerase chain reaction is simple and reliable using universal phytoplasma primers.

Disease Cycle and Epidemiology

Once infected with the false blossom phytoplasma, plants can continue to survive for many years but remain diseased.

The phytoplasma is transmitted solely by the blunt-nosed leafhopper, *Limotettix vaccinii*. In August and September, the leafhopper eggs are deposited lengthwise under the bark of young, tender cranberry stems, where they overwinter. Eggs hatch in late May and early June, and the nymphs go through five instars before becoming winged adults. Adults capable of transmitting the false blossom phytoplasma develop by late June to mid-July. Some evidence suggests that an incubation period is necessary before an adult leafhopper that has fed on diseased plants can transmit the disease. Symptoms may de-

Fig. 238. Cranberry blossoms affected by false blossom, caused by a phytoplasma. (Courtesy J. J. Polashock)

Fig. 239. Witches'-broom symptoms of false blossom in cranberry (left), caused by a phytoplasma, as compared with healthy cranberry (right). (Courtesy J. J. Polashock)

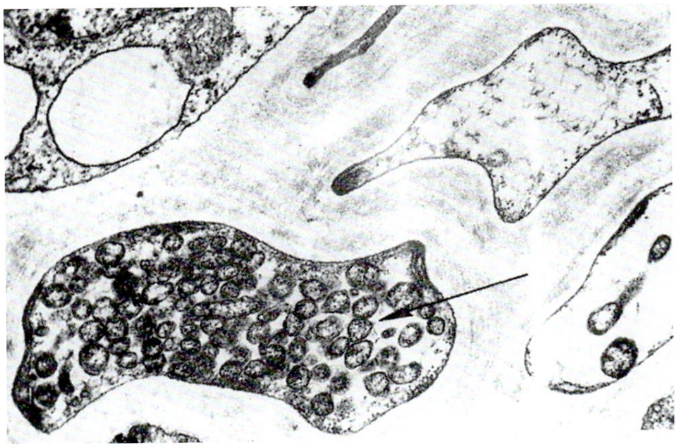

Fig. 240. Electron micrograph of phytoplasmas (arrow) in sieve tubes of a cranberry leaf infected with false blossom. (Courtesy T. A. Chen; © APS)

velop within 1 month after transmission, or they may not become visible for up to 1 year.

Management

Heat therapy and, more recently, certain antibiotics such as tetracycline have been used to control or eliminate phytoplasmas from infected plant tissue. However, these techniques are not practical in cranberry beds and control measures are directed toward preventing the introduction and spread of the disease. The leafhopper vector may be controlled with a suitable, properly timed insecticide.

Late flooding has been shown to cause a higher mortality of plants with false blossom than of healthy plants and thus reduces inoculum sources. Flooding a bed in June, just before the flowers open and after the leafhopper eggs have hatched, has proved effective for leafhopper control, but timing is critical and vine damage is a distinct possibility.

Cultivars reported as field resistant, such as Early Black and McFarlin, were found to be less attractive to the vector than were more susceptible cultivars. A combination of planting newer hybrids with field resistance and effective use of pesticides has dramatically reduced the incidence of this disease in cultivated beds.

Selected References

Chen, T. A. 1971. Mycoplasmalike organisms in sieve tube elements of plants infected with blueberry stunt and cranberry false blossom. Phytopathology 61:233-236.

Dobroscky, I. D. 1931. Studies on cranberry false blossom disease and its insect vector. Contrib. Boyce Thompson Inst. 3:59-83.

Lee, I.-M., Polashock, J., Bottner-Parker, K. D., Bagadia, P. G., Rodriguez-Saona, C., Zhao, Y., and Davis, R. E. 2014. New subgroup 16SrIII-Y phytoplasmas associated with false-blossom diseased cranberry (*Vaccinium macrocarpon*) plants and with known and potential insect vectors in New Jersey. Eur. J. Plant Pathol. 139:393-400.

Stevens, N. E., and Sawyer, W. H., Jr. 1926. The distribution of cranberry false blossom. Phytopathology 16:223-227.

Wilcox, R. B., and Beckwith, C. S. 1933. A factor in the varietal resistance of cranberries to the false blossom disease. J. Agric. Res. 47:583-590.

(Prepared by T. A. Chen; Revised by J. J. Polashock)

Stem Gall

Cranberry stem gall has been reported sporadically in Massachusetts and Wisconsin (United States) and in Quebec (Canada) since the 1980s. Cultivars affected include Ben Lear, Early Black, Howes, McFarlin, Pilgrim, Searles, and Stevens. In Wisconsin, where symptoms were severe and widespread in 1998 and 2002, the disease has been referred to as "stem canker" or "beater damage," under the assumption that the characteristic stem swelling was caused either by pathogenic fungi or by wound callus tissue that developed following injury from harvest machinery. However, the malady has been renamed "stem gall" to better reflect the key symptom. Usually only a small portion of a planting is affected, and yield losses are insignificant. However, following a severe outbreak, vine dieback can create openings in the cranberry canopy where weeds quickly become established. Stem tissues below galls remain healthy and give rise to new lateral shoots, although affected vines require 2–3 years to regain full productivity.

Symptoms

When a cranberry planting is viewed from a distance, the most obvious symptom of stem gall is shoot dieback. The tan to brown discoloration of shoots sometimes forms a pattern in the field, as though related to the movement of harvest equipment (Fig. 241). In other cases, shoot dieback occurs in irregularly shaped patches, especially near the ends of beds where floodwater enters first and exits last. Symptoms at the ends of beds also may be related to areas where harvest machinery is slowed down to turn, resulting in prolonged contact with vines. Although affected shoots begin to decline before cranberry plants flower, shoot dieback becomes most apparent during the flowering and fruit set stages.

Upright shoots and runners affected by stem gall appear swollen and rough with split, peeling bark (Fig. 242). Soft, greenish tan, globular-shaped galls that are approximately 0.5–1.0 mm in diameter emerge from under the bark just prior to bloom (Fig. 243). Galls shrink, become hard, and darken by late July. Galls can be confined to a small portion of the stem or masses of galls can girdle stems and/or extend several centimeters along the length of an upright shoot or runner. Galls are found on stems that are 1 year old and older but apparently do not develop on the current year's growth. By the following spring, affected portions of stems are brittle and break easily from the healthy portions of vines.

At a microscopic scale, cells external to the vascular cambium undergo extensive hypertrophy and hyperplasia (Fig. 244). Thin-walled parenchymatous cells that comprise galls may differentiate from phloem cells. Xylem vessels in galls are narrower and more numerous than the vessels produced in healthy stems.

Causal Organisms

Stem gall is probably caused by bacteria that produce the plant growth hormone indole-3-acetic acid (IAA). The most prominent groups of IAA-producing bacteria isolated from galls are members of the family Enterobacteriaceae and genera *Agrobacterium*, *Pseudomonas*, and *Rhizobium*. The role of IAA-producing bacteria as the cause of stem gall is supported by the following lines of evidence: (i) bacteria isolated from galls cause galls when inoculated onto cranberry plantlets grown in tissue culture; (ii) there is a positive correlation between the ability of bacteria to produce IAA in vitro and to cause galls on cranberry plantlets; (iii) microcavities within galls are densely packed with rod-shaped bacteria; and (iv) the pathological anatomy of galls is consistent with elevated levels of IAA. However, Koch's postulates have not been fulfilled on woody cranberry stems.

Members of the family Enterobacteriaceae are aerobic or facultatively anaerobic and gram negative. Cells are rod shaped (0.3–1.5 × 0.6–6.0 µm) and motile by peritrichous flagella or nonmotile. *Agrobacterium* spp. and *Rhizobium* spp. are aerobic and gram negative. Cells of *Agrobacterium* spp. are rod shaped (0.6–1.0 × 1.5–3.0 µm) and motile by one to six peritrichous flagella. Cells of *Rhizobium* spp. are rod shaped (0.5–0.9 × 1.2–3.0 µm) and motile by two to six peritrichous flagella.

Pseudomonas spp. are aerobic and gram negative. Cells of *Pseudomonas* spp. are rod shaped (0.5–1.0 × 1.5–4.0 µm) and motile by polar flagella. All of these bacterial taxa grow in standard bacteriological media at 20–30°C.

Disease Cycle and Epidemiology

The IAA-producing bacteria believed to cause stem gall inhabit the soil and leaf litter of cranberry beds and can be isolated from cranberry shoots regardless of disease history of a planting. Bacteria are believed to enter plants through wounds (e.g., from harvest machinery or low-temperature injury) or leaf abscission zones. It is not known whether bacteria replicate within the plant; however, microcavities within the galls contain large quantities of bacteria. The production of small, dense xylem vessels in the current year's growth ring suggests that greater-than-normal amounts of IAA are present when the vascular cambium resumes activity in the spring. Where water is not limited, conductance through a stem is less efficient through narrow than through wide vessels. Dieback from stem gall is likely caused by the inability of narrow vessels to sustain tissues distal to the galls and desiccation from loss of bark.

Stem gall bacteria do not elicit a hypersensitive response on tobacco and lack hypersensitive response/pathogenicity (*hrp*) genes, suggesting that they are not classical bacterial pathogens. Rather, they may be saprophytes that invade and multiply within cranberry stems under certain conditions. The difficulty in reproducing symptoms and the sporadic occurrence of stem gall in the field despite the ubiquity of IAA-producing bacteria in cranberry plantings suggest that the environment plays an especially important role in disease development.

Fig. 241. Shoot dieback symptom of cranberry stem gall, caused by indole-3-acetic acid-producing bacteria but appears as though associated with mechanical injury from harvest equipment. (Courtesy P. S. McManus; © APS)

Fig. 242. Rough, swollen cranberry stems with stem gall, caused by indole-3-acetic acid-producing bacteria. (Courtesy P. S. McManus; © APS)

Management

Because of the relative rarity of stem gall, control measures have not been developed. Cultural practices that avoid wounding vines should minimize infection by bacteria. Care should be taken during harvest to prevent mechanical injury. In Wisconsin, the most severe outbreaks have occurred in seasons following mild winters, when vines were not continuously protected by ice. Therefore, vines should be protected during the winter in colder cranberry-growing regions. Affected areas should be irrigated sufficiently to promote the growth of shoots below galled stems. Additional fertilization is not recommended, however, because this might favor weed growth in areas of canopy dieback. Copper compounds are bactericidal, but they have not been tested for the control of stem gall.

Selected References

Best, V. M., Vasanthakumar, A., and McManus, P. S. 2004. Anatomy of cranberry stem gall and localization of bacteria in galls. Phytopathology 94:1172-1177.

Vasanthakumar, A., and McManus, P. S. 2004. Indole-3-acetic acid-producing bacteria are associated with cranberry stem gall. Phytopathology 94:1164-1171.

(Prepared by P. S. McManus)

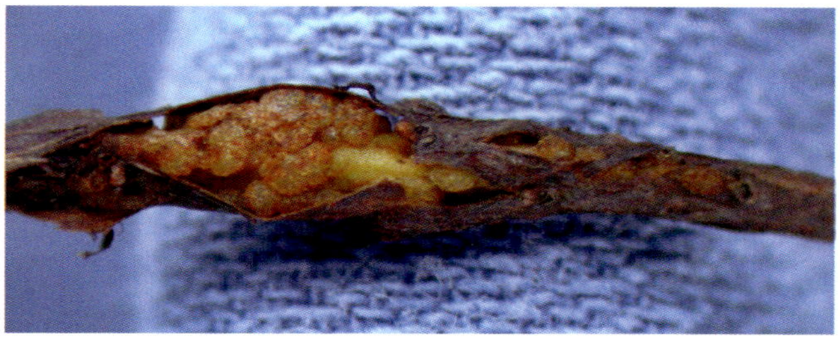

Fig. 243. Galls on a cranberry stem with stem gall, caused by indole-3-acetic acid-producing bacteria. (Courtesy P. S. McManus; © APS)

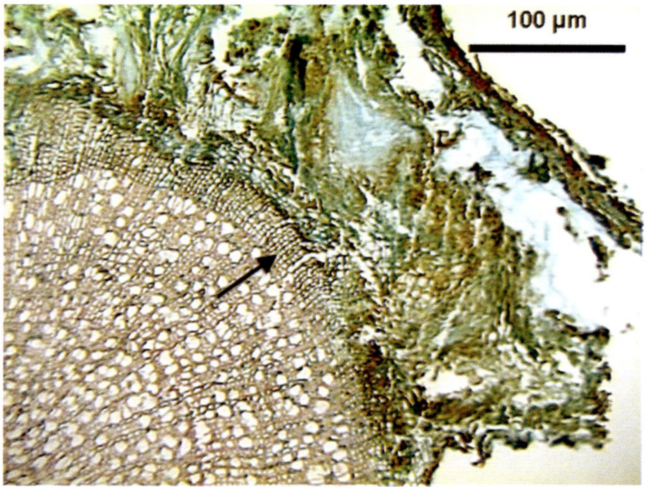

Fig. 244. Cross section of a gall on a cranberry stem with stem gall, caused by indole-3-acetic acid-producing bacteria, showing abnormal cell development, including numerous small vessels in the current year's xylem ring (arrow). (Courtesy P. S. McManus; © APS)

Diseases Caused by Nematodes

Blueberries and Cranberries

Diseases Caused by Nematodes

Many species of stylet-bearing, plant-parasitic nematodes can be found associated with root systems of blueberry and cranberry; however, not all stylet-bearing nematodes are pathogenic. Controlled greenhouse or microplot studies have been used to demonstrate the damage and symptoms caused by only a few nematode species parasitizing blueberry and cranberry. Some nematode species associated with these crops are potentially important as vectors of viruses.

Symptoms

The majority of plant-parasitic nematodes associated with blueberry and cranberry are ectoparasites, which use their stylets to feed on root epidermal and cortical cells but do not enter roots. Most ectoparasites preferentially feed on root tips and undifferentiated elongating cells near root tips, thereby disrupting root elongation, causing stubby roots or witches'-broom symptoms, and reducing overall root growth without necessarily causing noticeable lesions or necrotic areas. Some ectoparasites (e.g., *Hemicycliophora* spp.) can cause terminal root swellings or galls on blueberry and cranberry roots; their feeding kills cortical cells and stimulates pericycle cells to produce apical galls. Root lesion nematodes (*Pratylenchus* spp.) are migratory endoparasites that use their stylets to perforate cell walls and completely enter fine roots, where they move through the cortex while feeding on cortical cell cytoplasm. As these nematodes feed and move through the cortex, they cause localized lesions that can become infection courts for opportunistic fungal pathogens. When roots have become infected by many nematodes, lesions coalesce, causing girdling and dieback of fine roots. The host–parasite relationship of *Pratylenchus* spp. with blueberry is unclear (see below). Root-knot nematodes are sedentary endoparasites. Juveniles enter roots and establish permanent feeding sites. As the female nematodes mature, they become enlarged and root tissue around them swells, resulting in the formation of the characteristic galls or root knots. *Meloidogyne carolinensis* is the only root-knot nematode known to infect blueberry.

The ultimate effect of all plant-pathogenic nematodes is impaired root functioning, and aboveground symptoms of nematode damage are usually nonspecific and manifest only as patchy areas of poor growth that are difficult to distinguish observationally from the effects of poor nutrition or water availability. Plant-parasitic nematodes alone generally do not cause plant death.

Because there are no obvious aboveground symptoms of nematode damage, diagnosing the possible role of nematodes in plant disease depends entirely on analyzing soil and root samples for the presence of plant-pathogenic species. Recommendations for sampling, extraction, and identification protocols are described elsewhere. It is difficult, however, to directly relate nematode population densities to the health of any perennial crop such as blueberry or cranberry. Damage to roots in one year may not be manifest as reduced growth or yield until several years later or after additional stresses, such as drought, are imposed upon the crop. In the specific case of cranberry, runners can extend for meters from mother plants, resulting in spatial separation of nematode populations and where their impacts are manifest. Finally, year-to-year and within-year fluctuations in nematode population densities are considerable and do not always follow simple seasonal cycles. This further obscures relationships between nematode population densities and crop health.

Causal Organisms and Effects on Blueberry

Stubby root nematodes (*Paratrichodorus* spp.) are frequently associated with highbush blueberry in the mid-Atlantic and northeastern United States, in Arkansas, Oregon, and Washington (United States), and in British Columbia (Canada) and with rabbiteye blueberry in Arkansas (Fig. 245). *Paratrichodorus minor* (syn. *Trichodorus christiei*) reduced root growth of young highbush blueberry cuttings by 67% over 6 months under greenhouse conditions. The incidence of *Paratrichodorus renifer* increased 70× over 14 months and reduced yield by approximately 40% on 3-year-old 'Chippewa' blueberry in field microplots in British Columbia. *Paratrichodorus* population densities usually observed under field conditions are seldom as great as those attained in greenhouse pots, however, and experimental data on impacts of *Paratrichodorus* spp. on mature blueberry plants under field conditions are lacking. Most *Paratrichodorus* spp. are vectors of tobraviruses (e.g., *Tobacco rattle virus*, *Pea early browning virus*, and *Pepper ringspot virus*), but none of these viruses appear to be problems in blueberry.

Sheath nematodes (*Hemicycliophora* spp.) have been associated with highbush blueberry in New Jersey and Massachusetts (United States) and very infrequently associated with highbush and rabbiteye cultivars in British Columbia (Canada) and Arkansas (United States), respectively (Fig. 246). Feeding by *Hemicycliophora vaccinium* (identified as *Hemicycliophora gracilis* before 1964) and *Hemicycliophora vidua* (syn. *Hemicycliophora similis*) resulted in the formation of terminal root galls on highbush blueberry. Some nematodes previously identified as *Hemicycliophora similis* were probably *Hemicycliophora zuckermani*.

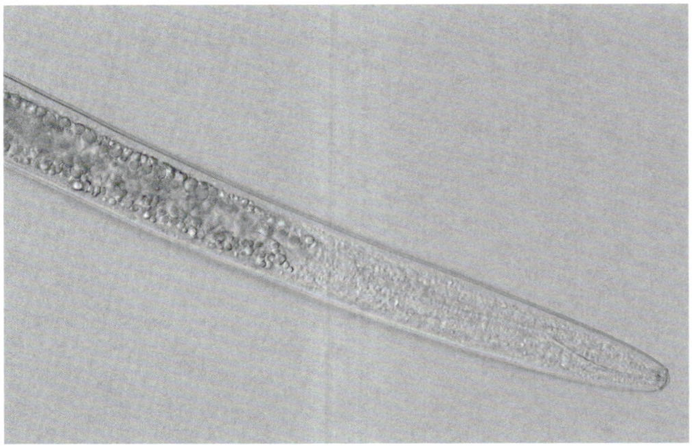

Fig. 245. Anterior region of *Paratrichodorus renifer.* Note the curved ondontostyle characteristic of stubby root nematodes (*Paratrichodorus* spp., *Trichodorus* spp.). (Courtesy T. A. Forge of the Department of Agriculture and Agri-Food, Government of Canada; © Her Majesty the Queen in Right of Canada, as represented by the Minister of Agriculture and Agri-Food Canada)

The dagger nematode, *Xiphinema americanum*, has been found in association with highbush blueberry in Massachusetts, Maryland, Arkansas (highbush and rabbiteye cultivars), Oregon, and Washington (United States) and in British Columbia (Canada). Damage caused directly by this nematode alone has not been demonstrated experimentally, but it is a vector of *Tobacco ringspot virus* and *Tomato ringspot virus*, which are damaging pathogens of blueberry.

One species of root-knot nematode, *Meloidogyne carolinensis*, was found on highbush and lowbush blueberries in 76% of fields surveyed in North Carolina (United States). Although this nematode has been associated with stunted and slow-growing blueberry plants with reduced fruit size and yields on the basis of field observations, experimental data on pathogenicity are lacking.

Root lesion nematodes (*Pratylenchus* spp.) are often found in the root zone of blueberry. However, controlled studies indicate that blueberry is a poor host for *Pratylenchus penetrans*, and populations of *Pratylenchus penetrans* may be sustained in blueberry plantings by feeding on weeds. Impacts of other *Pratylenchus* spp. on blueberry have not been studied.

Merlinius joctus (syn. *Tetylenchus joctus*) has been recovered from blueberry plantings in Massachusetts, New Jersey, and Maryland (United States) and in British Columbia (Canada). In a greenhouse experiment, this nematode increased from 140 to 800 nematodes per 15-cm pot over 6 months but did not cause statistically significant reductions in root growth. *Tylenchorhynchus* spp., which are closely related to *Merlinius* spp., have been found in the root zone of blueberry in Arkansas (United States). Spiral nematodes (*Helicotylenchus* spp., *Rotylenchus* spp.) have been found relatively frequently in the root zone of blueberry in all blueberry-growing regions of North America, but pathogenicities of *Helicotylenchus* and *Tylenchorhynchus* nematodes to blueberry have not been studied experimentally.

Causal Organisms and Effects on Cranberry

Genera of plant-parasitic nematodes found in cranberry beds in North America include *Helicotylenchus*, *Hemicycliophora*, *Merlinius*, *Mesocriconema*, *Paratrichodorus*, and *Tylenchorhynchus*. *Paratrichodorus minor*, *Hemicycliophora vidua*, *Pratylenchus penetrans*, and the ring nematode *Mesocriconema curvatum* (syns. *Criconemoides curvatus* and *Criconemella curvatus*) reduced the mean number of developing runners, runner length, and fresh weight of roots and tops of cranberry plants growing in pots in a greenhouse. *Paratrichodorus minor* and *Hemicycliophora vidua* appear to be the most damaging pathogens. Population densities of these two species increased 150-fold and 50-fold, respectively, over a 5-month period and caused discolored and stunted root systems in a greenhouse experiment; *Hemicycliophora vidua* caused the formation of terminal root galls. *Mesocriconema xenoplax* (syns. *Criconemoides xenoplax*, *Criconemella xenoplax*, and *Macroposthonia xenoplax*) multiplied 16-fold after 150 days and, when coinoculated with *Hemicycliophora vidua*, significantly reduced the numbers of runners and the fresh weights of tops and roots. While *Pratylenchus penetrans* was pathogenic to cranberry in greenhouse pots, it is not found frequently or at high population densities in cranberry bogs. *Helicotylenchus* spp. were recovered frequently from cranberry bogs in New Jersey and Wisconsin (United States) and in British Columbia (Canada). However, pathogenicity of these nematodes has not been assessed under controlled conditions.

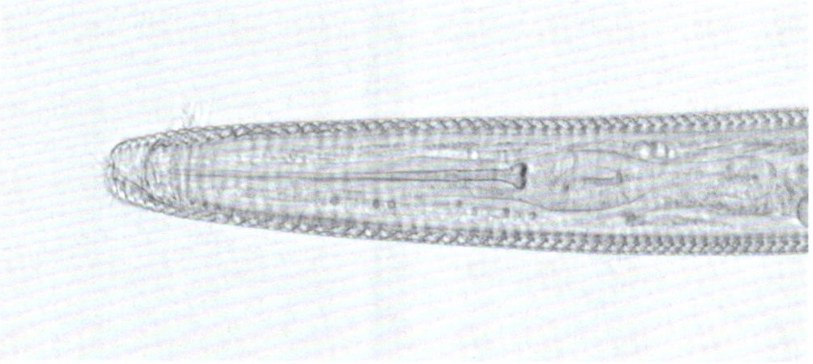

Fig. 246. Anterior region of a sheath nematode (*Hemicycliophora* sp.). Note the coarse annulation, extra cuticle, relatively long robust stylet, and fused metacorpus characteristic of the genus. (Courtesy T. A. Forge of the Department of Agriculture and Agri-Food, Government of Canada; © Her Majesty the Queen in Right of Canada, as represented by the Minister of Agriculture and Agri-Food Canada)

In one field study, spatial variation in population densities of *Hemicycliophora ritteri* (0–3,750 nematodes per 500 cm^3 of soil) and of *Paratrichodorus minor* (0–1,400 nematodes per 500 cm^3 of soil) was not correlated with the numbers or weights of sound or rotten berries or with the extent of damaged vegetation. This research suggests that *Hemicycliophora ritteri* and *Paratrichodorus minor* may not have a great effect on plant health in the field. However, as indicated in the section on general symptoms, demonstration of relationships between nematode population densities and plant health parameters in the field is complicated for perennial crops.

Disease Cycle

Most stylet-bearing nematodes have four juvenile stages and four molts; the first juvenile stage occurs within the egg. Usually, all females, some males (if present), and most second- through fourth-stage juveniles feed on living plant tissues. Life histories of nematodes vary with respect to type and rate of reproduction and length of life cycle. Because *Paratrichodorus* spp. and *Hemicycliophora* spp. are likely the most damaging species affecting blueberry and cranberry, the life cycles of these two species are discussed here.

***Paratrichodorus* spp.** In flooded cranberry beds in the northeastern United States, eggs and juveniles of *Paratrichodorus minor* appear to be the main overwintering stages. Spring populations were made up predominantly of juveniles along with a relatively few females, and only juveniles were present during the summer. Males of *Paratrichodorus minor*, which are rare, were first observed in September, and females appeared in October. In coastal British Columbia (Canada) and Washington (United States), females of *Paratrichodorus* spp. are observed all year. In Massachusetts (United States), overall population densities varied twofold through the growing season, with a peak in the autumn. *Paratrichodorus minor* is capable of rapid population development when environmental conditions are optimal, requiring 143 degree-days (base 10°C) to mature from egg to adult.

***Hemicycliophora* spp.** Life cycles from egg to egg-laying female are usually about 20–30 days. Adult males are rare for most species; for those species with males, the males do not have stylets and do not feed on roots. In a Massachusetts cranberry bog, overall population densities varied twofold through the growing season, with peaks occurring in early summer and again in the autumn.

Management

Blueberry. Only the preplant fumigants methyl isothiocyanate, 1,3-dichloropropene, and *N*-methyldithiocarbamate are currently labeled for nematode control in blueberries.

Populations of *Xiphinema americanum* were significantly reduced (up to 80%) after either a spring or autumn application of fenamiphos at 37.4 L/ha or after a split application of 18.7 L/ha. Fenamiphos at 13.4 kg of a.i. per hectare significantly reduced populations of *Pratylenchus penetrans* to 13% of the level in untreated controls in Oregon (United States). Application of fenamiphos (149 kg/ha) with oxyfluorfen (9.4 L/ha) doubled new growth relative to application of the herbicide alone and significantly reduced *Pratylenchus penetrans* population densities at one of three post-application sample dates. One biological nematicide that was tested did not alleviate the impacts of a mixture of nematodes (sheath, stubby root, and stunt) on the rooting of blueberry cuttings.

Populations of *Paratrichodorus minor*, but not of *Xiphinema americanum*, were significantly lower in mulched compared with nonmulched blueberry plots.

Cranberry. Because cranberries are grown in areas with a high water table adjacent to streams, any chemical treatment increases risks to fish and drinking water, and no nematicides are registered for use on cranberry. Treatment with thionazin (18 kg/ha) significantly increased cranberry yields and reduced nematode levels 89%. Dibromochloropropane and thionazin significantly reduced nematode numbers and increased the number of fruit buds 24 and 34% and the number of uprights 23 and 38%, respectively. Treatment of a cranberry bed in New Jersey (United States) with carbofuran, oxamyl, fenamiphos, or ethoprop did not significantly influence yields or numbers of *Hemicycliophora ritteri* or *Paratrichodorus minor.* Preplant fumigation with 1,3-dichloropropene significantly lowered the numbers of *Tylenchorhynchus* spp. and *Hemicycliophora* spp. and increased the weight of cranberry vines the following year.

Data to date suggests that chemical control of nematodes in mature plantings of blueberry and cranberry is not warranted. Preplant fumigation or sterilization of blueberry and cranberry rooting and nursery beds in order to provide nematode-free nursery stock should be economically feasible. Preplant fumigation may also be economical when blueberry or cranberry beds are replanted or new beds are established on land with a history of nematode problems. No nematode-resistant blueberry or cranberry selections are available.

Selected References

Barker, K. R., and Boone, D. M. 1966. Plant-parasitic nematodes on cranberries in Wisconsin. Plant Dis. Rep. 50:957-959.

Bird, G. W., and Jenkins, W. R. 1964. Occurrence, parasitism, and pathogenicity of nematodes associated with cranberry. Phytopathology 54:677-680.

Brown, D. J. F., Ploeg, A. T., and Robinson, D. J. 1989. A review of reported associations between *Trichodorus* and *Paratrichodorus* species (Nematoda: Trichodoridae) and tobraviruses with a description of laboratory methods for examining virus transmission by trichodorids. Rev. Nématol. 12:235-241.

Clark, J. R., Riggs, R. D., and Moore, J. N. 1987. Plant parasitic nematodes associated with highbush blueberry in Arkansas. HortScience 22:408-409.

Forge, T. A., Pinkerton, J. N., O'Gorman, D. T., Koch, C., and Sweeney, M. 2007. A root-rot and bush-decline syndrome affecting highbush blueberry (*Vaccinium corymbosum*) in the Fraser River valley: Are nematodes involved? (Abstr.) Can. J. Plant Pathol. 29:84.

Forge, T. A., Koch, C., Pinkerton, J. N., and Zasada, I. A. 2009. First report of *Paratrichodorus renifer*, a nematode parasite of highbush blueberry. (Abstr.) Phytopathology 99:S35.

Myers, R. F. 1991. Nematodes associated with dieback disease of cranberries. J. Nematol. 23:629-633.

Zasada, I. A., Pinkerton, J. N., and Forge, T. A. 2010. Plant-parasitic nematodes associated with highbush blueberries (*Vaccinium corymbosum*) in the Pacific Northwest of North America. Int. J. Fruit Sci. 10:123-133.

Zuckerman, B. M. 1962. Parasitism and pathogenesis of the cultivated highbush blueberry by the stubby root nematode. Phytopathology 52:1017-1019.

Zuckerman, B. M. 1964. Studies of two nematode species associated with roots of the cultivated highbush blueberry. Plant Dis. Rep. 48:170-172.

(Prepared by T. A. Forge)

Lingonberry Diseases

Diseases of stems and foliage of lingonberry are not frequent in domesticated stands in the United States, although under circumstances of environmental stress to the host, some fungi can cause alarming defoliation in some years. The paucity of disease problems is likely caused by the small areas in cultivation. In wild stands, disease is generally, with some exceptions, of low incidence and scattered over the landscape. The relatively little disease normally found in native stands may be a function of genetic heterogeneity of the plants and differences in age of members of a population. As the popularity of lingonberry increases and larger areas are planted with more homogeneous populations, disease pressure will likely increase as well. Many of the diseases below are not currently a significant problem in cultivation but may become so.

Red Leaf, Red Shoot, and Dwarfing Redleaf

Exobasidium is a genus of fungal parasites that causes red leaf, red shoot, and dwarfing redleaf of lingonberry. Worldwide, there are at least three *Exobasidium* spp. found on lingonberry. *E. vaccinii*, the most common and widely distributed species, causes red leaf and is also found on blueberry, cranberry, and more than 100 other species of plants, although there appear to be morphological differences between isolates that occur on specific hosts. *E. vaccinii* on lingonberry is found wherever the host grows, except at the highest elevations and in dry wind- or sun-exposed sites. Although it has been found in wild populations, *E. vaccinii* has not been reported from domesticated plantings in the United States. *E. splendidum* causes red shoot but is limited to subarctic areas at the timberline or, in lower latitudes, at high altitudes. It has been reported in wild stands in Norway, Sweden, Greenland, the mountains of central Europe (the former Czechoslovakia, Switzerland, France, and Austria), the Caucasus, and Canada (Baffin Island, Quebec). *E. juelianum* causes dwarfing redleaf. *E. juelianum* is much less widely distributed and is rare within a population of plants. It has been found in Scotland, Wales, Sweden, Finland, Norway, Austria, Germany, Switzerland, and Canada (Quebec). Neither *E. splendidum* nor *E. juelianum* are likely to be seen in cultivation. Infection of lingonberry by *Exobasidium* spp. is generally of low incidence within a population.

Symptoms

Exobasidium spp. cause striking pink or red symptoms in newly infected vegetative or reproductive tissues that later turn white and spongy as the fungus sporulates. Red leaf is the disease caused by *E. vaccinii*, which begins when a young leaf is colonized in a circumscribed spot. The fungus induces abnormal growth of the palisade parenchyma on the upper side of the leaf and of the spongy parenchyma on the lower side of the leaf. Greater hypertrophy on the lower side results in the invagination of the upper surface and the formation of a pouch or pocket gall. These galls may be as great as 1 cm across and are usually red to bright pink with a yellow margin. Symptoms on older leaves are less obvious, resulting only in the formation of a slightly fleshy spot. Infection is usually of only one or a few leaves on a shoot. Infrequently, the growing point may be attacked, which results in a thick, fleshy pink or red shoot (Fig. 247). When infected, floral tissues produce malformed flower buds. On leaves, the lower surface of the pouch gall is at first pinkish and then becomes white as the fungus matures and basidiospores are produced in a hymenial layer. Once sporulation has occurred, the affected leaves wither and die. Infection remains localized and is not systemic, hence no stunting or witches'-brooms appear. Infected tissues are more susceptible to frost damage. *E. vaccinii* has been reported from Britain, Greenland, Denmark, Finland, Norway, Sweden, Russia, Hungary, Poland, Japan, the United States (Maine, New York, Alaska), and Canada (British Columbia, Quebec).

Causal Organisms

E. vaccinii is a basidiomycetous fungus that produces both basidiospores and conidia. The latter have been the result of much confusion, sometimes because affected tissues parasitized by other fungi were then included in descriptions of the pathogen. Microscopic characters of *E. vaccinii* may differ slightly from isolates originating on hosts other than lingonberry; characters may also be influenced by the environment, spore age, and method of observation. The hymenium forms on the underside of leaves with pouch or pocket galls or, as in older leaves, slightly thickened spots. Basidia usually have four sterigmata; basidiospores are 2.0–4.0 × 11–19 μm, slender, musiform (banana shaped), and slightly curved and eventually develop one to three septa. Conidia are 0.8–1.2 × 8–11 μm and bacilliform to subfusiform.

The red shoot pathogen, *E. splendidum*, is not systemic in infected stems—mycelium is limited to the affected tissues. Formation of a hymenium on affected shoots and leaves produces basidia in a powdery white mass. Basidia have two sterigmata; basidiospores are large, averaging 6.0–11.5 × 20–27 μm, and broadly oblong-ellipsoid to subpyriform. *E. splendidum* differs from *E. vaccinii* and *E. juelianum* in that septa never form in the basidiospores and conidia are not known to develop.

The dwarfing redleaf pathogen, *E. juelianum*, produces systemic mycelia and whole plant stunting, witches'-brooms, and dwarfed leaves. Infections are perennial and persistent. The hymenium forms between the veins on the underside of affected leaves. Basidia have four or sometimes five sterigmata. Basidiospores are 2.0–4.0 × 9–14 μm, musiform, and only slightly curved and eventually develop a single septum. Bacilliform conidia are 0.8–1.5 × 5–10 μm.

Disease Cycle

The overwintering stage of *E. vaccinii* is presumed to be basidiospores and conidia, protected by bud scales. Infection happens in the spring when spores (basidiospores or conidia) germinate to form a germ tube and penetrate young leaves. Spores can enter leaves directly through the epidermis or through natural openings. Hyphae grow into the palisade and spongy parenchymas and eventually appear between the mesophyll and lower epidermis. Just prior to hymenial formation, the hyphae emerge through the epidermis either by direct penetration or through stomata. Basidiospores form in a hymenium on the lower leaf surface, this area at first being pinkish and then becoming powdery white. Basidiospores and conidia are formed in profusion. Disease peaks in midsummer, but the disease is thought to be polycyclic since newly diseased shoots can be found in the autumn.

E. splendidum is a monocyclic fungus that probably overwinters as dormant basidiospores on vegetative buds. Infections are initiated in the spring after snowmelt and result in the reddening and thickening (hypertrophy) of young leaves and shoots. Affected tissues become densely covered with the hymenium, which becomes whitish with spores. Conidia are not formed. Only the current season's shoots are affected, and infections are initiated annually. Diseased shoots may be colonized by fungal hyperparasites.

E. juelianum is a systemic pathogen that causes perennial disease, persisting in an infected plant for years. New shoots are diseased from internal infections as they emerge each spring. The initial infection process is similar to that of the other *Exobasidium* spp. discussed above.

Fig. 247. Thickened, reddened shoot and infected leaf caused by *Exobasidium vaccinii*, the causal organism of red leaf, on lingonberry. (Courtesy B. A. Roy; © APS)

Management

Only *E. vaccinii* is likely to appear in cultivated plants. Affected tissues should be removed and destroyed in the spring, before spores develop.

Selected References

Booth, C. 1978. CMI Descriptions of Pathogenic Fungi and Bacteria No. 778. *Exobasidium vaccinii*. CAB International, Kew, Surrey, England.

Nannfeldt, J. A. 1981. *Exobasidium*, a taxonomic reassessment applied to European species. Symb. Bot. Ups. 23:1-72.

Pehkonen, T., Ranta, H., Tolvanen, A., and Laine, K. 2002. The frequency of the fungal pathogen *Exobasidium splendidum* in two natural stands of the host *Vaccinium vitis-idaea* in the subarctic timberline area. Arct. Antarct. Alp. Res. 34:428-433.

(Prepared by M. L. Putnam)

Gibbera Canker

Gibbera canker is found throughout the range of lingonberry, including the United States (Alaska, Washington, New Hampshire), England, Scotland, northern Ireland, Italy, Austria, Bulgaria, Germany, Poland, Denmark, Sweden, Norway, Finland, and Russia. It is relatively common but has not been reported from domesticated stands. In wild populations, *Gibbera vaccinii* was found on 15–28% of the shoots monitored in a study plot in Russia (Republic of Karelia).

Symptoms

Infection of shoots results in the formation of a canker, which causes dieback of tissues above the infection. This is a disease of mature stems, since shoots 3 years old or older are more often affected than are younger stems.

Causal Organism

Gibbera vaccinii (Ascomycota, Pleosporales) also infects cranberry and *Vaccinium caespitosum* (dwarf bilberry), in which it causes a minor leaf spot. The synonym is *Brachysporium vaccinii*. *G. vaccinii* ascocarps are subcuticular or deeper, erumpent from a blackish brown crustlike stroma that is 180–250 µm in diameter. Ascocarps are grouped within the stroma and covered with short, stiff, black-brown setae that are 30–55 µm. The wall of the ascocarp is 22–27.5 µm wide and consists of several layers of brown, polygonal cells. Asci are bitunicate, 10–13 × 65–92 µm, and oblong to cylindrical with numerous pseudoparaphyses. Ascospores are elliptical, green to olivaceous brown, and two celled and have a central, or slightly higher, constricted septum. The spores are 7.5–10 × 13–17.5 µm and have pointed or rounded ends.

Disease Cycle

Little is known of the life cycle of the pathogen. Presumably, ascospores initiate infection in the spring. Older stems are often infected near the ground and the fungus eventually girdles the shoot. Tissues distal to the canker die back, resulting in reduced fruit yield. Sporulation occurs on affected leaves and shoots.

Management

Gibbera canker has not yet been observed in domesticated stands. Dead stems should be pruned below the canker and infected material should be removed from the site.

Selected References

Barr, M. E. 1968. The Venturiaceae in North America. Can. J. Bot. 46:799-864.

Paal, T. 1991. Pests and disease of the cowberry (*Vaccinium vitis-idaea*). Aquilo Ser. Bot. 31:35-39.

Paal, T. 2006. Lingonberry (*Vaccinium vitis-idaea* L.) research in Estonia: An overview. Acta Hortic. 715:203-217.

(Prepared by M. L. Putnam)

Mummy Berry

Mummy berry is found in the United Kingdom, Norway, Sweden, and Russia (Leningrad Region) and is likely widespread where there are wild populations. Mummy berry has not been reported from plants in cultivation.

Symptoms

Mummy berry affects current-season shoots and fruits; the latter turn brown or black, shrivel, and fall prematurely. On shoots, the fungus colonizes one side of the stem more extensively than the other, causing the shoot tip to bend in a shepherd's crook. The petiole of leaves on affected branches may initially show a brown discoloration before the entire shoot shrivels and dies.

Causal Organism

Mummy berry is caused by a member of the order Helotiales (Ascomycota), *Monilinia urnula*. The synonym is *Monilia urnula*. Sporodochia are yellowish or white en masse and are formed on the current-year shoots or on the underside of leaves. Conidiophores are 10–12 × 140–160 µm, monilioid, sometimes straight, subhyaline, and branched near the base. Conidia are formed in chains, easily disrupted, aseptate, thin walled, lemon shaped, with one (sometimes two) papillate apical caps and one basal cap, and average 18–25 × 20–35 µm, although the length may be as great as 42 µm and the width as little as 16 µm. Conidia are joined by spindle-shaped, two-celled disjunctors 2–2.8 × 5–8 (sometimes 9) µm. Apothecia, which arise from mummified fruit on the ground, are cupulate to discoid initially, at times becoming recurved. The hymenial layer consists of linear multicellular paraphyses and asci containing eight aseptate, oblong ascospores that are of two sizes: either narrow ellipsoid or subcylindric, both with rounded ends. The four larger spores are 1.5 µm longer than the others and are 5.6–9 × 14–17 µm.

Disease Cycle

Flowers are infected in the spring from wind-blown conidia that germinate to penetrate the floral stigmata. Hyphae ramify throughout developing fruit, which wither and drop to the ground. In the spring of the following year, stalked apothecia form in the fallen fruit mummies. Ascospores ejected from the apothecia are carried by wind to stems and leaves of developing shoots, which are penetrated directly by the pathogen. The fungus is necrotrophic and kills cells in advance of colonization. Cambial tissues of infected stems shrivel and separate from the wood. As colonization continues, a stromalike cushion is formed from which hyphae emerge through the cuticle. Conidia are formed the same season in a white, powdery coating of the leaves and shoots and may initiate one secondary infection within the same year in which the ascospores are produced. Conidia are aromatic and are likely attractive to insects, which may aid in conidial dispersal.

Management

Mummy berry has not been found in domesticated plants and is found only in wild populations.

Selected References

Batra, L. R. 1983. *Monilinia vaccinii-corymbosi* (Sclerotiniaceae): Its biology on blueberry and comparison with related species. Mycologia 75:131-152.

Mel'nik, V. A., and Carris, L. M. 2001. *Monilia urnula* in Leningrad Region (Russia). Mikol. Fitopathol. 35:29-33.

Woronin, M. 1889. Über die Sclerotienkrankheit der Vaccineenbeeren. Mem. Acad. Imp. Sci. St. Petersbourg, 7 Ser. Vol 36:1-49 (Reviewed in: J. Mycol. 1889. 5:39-41).

(Prepared by M. L. Putnam)

Witches'-Broom Rust

Witches'-broom rust is probably the most easily recognized disease of lingonberry because of the bloated appearance of affected stems. This rust is uncommon in cultivated plants and is a sporadic problem in wild populations. Witches'-broom rust also is found on other *Vaccinium* spp., including blueberry, and additional information on the fungus, its disease cycle, and epidemiology can be found in Witches'-Broom under Diseases Caused by Fungi, the Blueberries (Lowbush) section. On lingonberry, witches'-broom rust is found in North America in Canada (Saskatchewan) and the United States (Alaska, the north central sections, and northeastern sections) and in Europe in Bulgaria, Finland, Germany, Poland, Norway, Russia, Sweden, and the United Kingdom. The alternate hosts are true firs, on which it causes needle rust. On firs, the rust is found in the United States, Canada, Bulgaria, Denmark, England, Germany, Switzerland, Finland, Japan, and Russia.

Symptoms

Young diseased shoots are swollen to several times their natural diameter because of the hypertrophy of infected cortical cells. Multiple spongy shoots arise from infected tissue, resulting in a witches'-broom. Affected shoots vary in appearance depending on the stage of infection, being initially a pinkish brown, then shiny chestnut brown, and finally progressing to a dull brown without gloss (Fig. 248). Elongation of the internodes pushes leaves farther apart than is normal. The lower leaves are usually drastically reduced in size,

Fig. 248. Witches'-broom rust, caused by *Pucciniastrum goeppertianum*, on huckleberry (*Vaccinium parvifolium*), which is similar in appearance to witches'-broom rust on lingonberry. (Courtesy B. A. Roy; © APS)

whereas the younger upper leaves are shortened and rounded at the tips. Affected shoots produce little, if any, fruit and die prematurely.

Causal Organism

Witches'-broom rust is caused by the basidiomycete *Pucciniastrum goeppertianum*, a heteroecious rust fungus. This species is sometimes placed in the genus *Calyptospora* as *Calyptospora columnaris*. The alternate hosts are *Abies* spp., including *A. amabilis*, *A. balsamea*, *A. concolor*, *A. lasiocarpa*, and *A. magnifica*, on which needle rust occurs. See Witches'-Broom under Diseases Caused by Fungi, the Blueberries (Lowbush) section for additional information.

Disease Cycle

Young lingonberry shoots are infected in the spring. The fungus colonizes the cortex, causing enlargement of the parenchymal cells during the second year following infection. Teliospores form in the epidermal cells and germinate within the epidermis to form basidia above the epidermis. The basidiospores infect *Abies* spp. Depending on the latitude and elevation, aecia appear in late spring through late summer on needles that were infected the previous year or, in some cases, on current-year needles. No uredia are produced. The fungus is perennial in lingonberry.

Management

This is an infrequent problem in cultivation. Affected plants should be removed, but reinfection is likely if the alternate host is nearby.

Selected References

Arthur, J. C. 1934. Manual of the Rusts in the United States and Canada. Purdue Research Foundation, Lafayette, IN.

Ziller, W. G. 1974. The Tree Rusts of Western Canada. Can. For. Serv. Publ. No. 1329.

(Prepared by M. L. Putnam)

Leaf Rust

Leaf rust is a minor problem on both lingonberry and its alternate hosts, which are species of hemlock (*Tsuga* spp.). This disease has not been a problem on cultivated plants and is only occasionally found in wild populations. Leaf rust on lingonberry is found throughout its range in Canada, the United States (Alaska, Maine, New Hampshire), the United Kingdom, Europe, the Asian parts of Russia, Japan, Korea, and China. On hemlock, the fungus is found in Alaska, Idaho, and Washington (United States), in British Columbia (Canada), and in Japan.

Causal Organism

Leaf rust is caused by the basidiomycete *Pucciniastrum vaccinii*; some authors have placed this fungus in the genus *Naohidemyces* as *Naohidemyces vaccinii*. *Pucciniastrum vaccinii* also causes rust in lowbush and highbush blueberry and cranberry; further information can be found in Leaf Rust under Diseases Caused by Fungi, the Blueberries (Highbush and Rabbiteye) and Blueberries (Lowbush) sections, and in Rust under Diseases Caused by Fungi, the Cranberries section.

Symptoms

On lingonberry, leaf yellowing coincides with the formation of uredia, and affected leaves eventually abscise. Uredinia appear in pustules on the underside of the leaves. Telia are formed on other *Vaccinium* spp. and are rarely found on lingonberry. Pycnia and aecia are formed on the alternate hosts, which include *Tsuga diversifolia*, *Tsuga heterophylla*, and *Tsuga mertensiana*. Affected needles are chlorotic and prematurely cast. Aeciospores are produced on the underside of the needles in shallow pustules in two rows on either side of the midrib.

Management

Leaf rust is not currently a problem on cultivated plants.

Selected References

Allen, E., Morrison, D., and Wallis, G. 1996. Common Tree Diseases of British Columbia. Canadian Forest Service, Victoria, British Columbia.

Arthur, J. C. 1934. Manual of the Rusts in the United States and Canada. Purdue Research Foundation, Lafayette, IN.

(Prepared by M. L. Putnam)

Foliar *Phytophthora* Diseases

Phytophthora spp. are members of the family Peronosporaceae (Oomycota, Peronosporales). In lingonberry, *Phytophthora* spp. have been found in cultivated stands and in nurseries, generally when grown in association with ornamentals.

"*Phytophthora citricola*" complex

In Oregon (United States), "*Phytophthora citricola*" has been recovered several times from cultivated 'Red Pearl' lingonberry plants showing twig dieback and leaf necrosis. Since these findings, "*P. citricola*" has been found to be a species complex. Leaf lesions begin as light brown spots that turn into mahogany-colored to purple-black spreading lesions of irregular shape with no associated chlorosis or other discoloration (Fig. 249). Infection spreads from leaves into petioles, twigs, and shoots, resulting in cankers and the reddening of leaves distal to the canker. A report of "*P. citricola*" isolated from lingonberry in a Polish nursery was likely *P. plurivora*, based on primer sequences used to identify the organism. An isolate originally identified as *P. inflata*, isolated in Scotland from a single lingonberry plant grown in association with rhododendrons in an ornamentals nursery, was later identified as *P. plurivora*.

P. plurivora is morphologically very similar to *P. citricola* sensu stricto and other members of the "*P. citricola*" complex but may be distinguished by having a lesser length-to-breadth ratio of the sporangia (1.43 ± 0.19) and a chrysanthemum-type colony morphology on potato dextrose agar. Other morphological features overlap with related species, and sequencing of various gene regions may be the most reliable method of identifying isolates. In general, the group members are homothallic, with primarily paragynous antheridia located close to the oogonial stalk, and have spherical, smooth, terminally borne oogonia that turn yellowish with age. Oospores are globose, nearly plerotic, and hyaline or yellow. Sporangia are formed abundantly in water and vary in shape from ovoid to limoniform, obpyriform, ellipsoid, or obclavate, with broad flat papillae (semipapillate) and sometimes more than one apex. Sporangia are persistent on simply or sympodially branched sporangiophores.

Phytophthora ramorum

P. ramorum has been found on lingonberry plants in nurseries in Poland, and it is suspected that the infection originated in rhododendrons grown at the same nursery. Susceptibility of lingonberry to *P. ramorum*, as determined by lesion size produced in the inoculation of detached leaves, varies depending on the cultivar tested.

Phytophthora kernoviae

P. kernoviae has a broad host range and is considered more aggressive on some plants than *P. ramorum*. While *P. kernoviae* has not been found in natural or cultivated stands of lingonberry, the host has been found to be highly susceptible to *P. kernoviae* in detached leaf inoculations.

Importance

Given the international movement of plant materials and the growing popularity of lingonberry, additional *Phytophthora* spp. are likely to be found affecting this plant. Infection of lingonberry produced in ornamentals nurseries should be of concern, since they could potentially be a source of *Phytophthora* inoculum when planted in landscapes or production fields. *P. ramorum* and *P. inflata* (part of the "*P. citricola*" complex) have been found on nursery-grown rhododendrons in Finland, a country that has abundant wild lingonberry populations. *P. ramorum* and "*P. citricola*" have been recovered from the soil and organic material found on the exterior of pots containing ornamental plants from a nursery in Washington (United States), and *P. ramorum* has been recovered from vehicle tires and the shoes of people that have passed through infested areas in California (United States). *P. ramorum* produces abundant chlamydospores that, under experimental conditions, have been found to remain viable for over 1 year. These findings highlight the importance of strict sanitation and good nursery scouting and testing programs to prevent the spread of *Phytophthora* spp. from nurseries.

Fig. 249. Foliar blight symptoms, caused by *Phytophthora* spp. in the "*Phytophthora citricola*" complex, on lingonberry. (Courtesy J. Parke; © APS)

Selected References

Dart, N. L., and Chastagner, G. A. 2007. High recovery rate of *Phytophthora* from containerized nursery stock pots at a retail nursery highlights potential for spreading exotic oomycetes. Online. Plant Health Progress doi: 10.1094/PHP-2007-0816-01-BR

Davidson, J. M., and Shaw, C. G. 2003. Pathways of movement for *Phytophthora ramorum*, the causal agent of sudden oak death. Sudden Oak Death Online Symposium. doi: 10.1094/SOD-2003-TS

Jung, T., and Burgess, T. I. 2009. Re-evaluation of *Phytophthora citricola* isolates from multiple woody hosts in Europe and North America reveals a new species, *Phytophthora plurivora* sp. nov. Persoonia 22:95-110.

Lilja, A., Rytkönen, A., Kokkola, M., Parikka, P., and Hantula, J. 2007. First report of *Phytophthora ramorum* and *P. inflata* in ornamental rhododendrons in Finland. Plant Dis. 91:1055.

Wiejacha, K., Trzewik, A., and Orlikowska, T. 2005. Genetic differences between isolates of *Phytophthora citricola* obtained from ornamental nurseries in Poland. Phytopathol. Pol. 35:157-162.

(Prepared by M. L. Putnam)

Miscellaneous Foliar Fungi

Numerous fungi have been reported in association with foliage, primarily in noncultivated stands in Scandinavia and other subarctic regions of Europe and Asian Russia. Nearly 50 leaf-spotting fungi are listed in the USDA's Systematic Mycology and Microbiology's online database, to which the following could be added: *Godronia foliicola* (Ascomycota, Helotiales), which causes a stem disease in Saskatchewan (Canada) and Switzerland; and *Seimatosporium vaccinii* (Ascomycota, Xylariales), reported from Sweden. *Coccomyces leptideus* (Ascomycota, Rhytismatales) has been found in Europe and North America on dead leaves and is of uncertain pathogenicity.

Fungi Causing Leaf Spots

In the United States, foliar diseases of cultivated lingonberry are generally minor in incidence, although they can be locally damaging in some years. In Oregon, *Allantophomopsis lycopodina* (Ascomycota, Helotiales) was isolated from twig tips and leaves of the cultivar Korralle. Affected plants looked somewhat sunburned overall, although symptoms were not limited to south- or west-facing foliage. Infected leaves had brown, spreading lesions originating at the base or leaf margins (Fig. 250). Shoot tips were blackened and dying, and attached fruit were black and shriveled. *Allantophomopsis lycopodina* is a fungus that also causes a storage rot (black rot) of cranberry fruit. In lingonberry, the fungus causes disease on plants that are stressed from water shortage and likely other environmental factors, such as excessive heat or when planted on shallow soils. Lingonberry plants cultivated in Oregon are not managed for disease. Incidence of all foliar diseases is low, and keeping plants in good health seems to be sufficient to prevent most problems.

Allantophomopsis cytisporea (Ascomycota, Helotiales) has been isolated from leaves of wild lingonberry in Sweden and was found causing defoliation in the cultivars Ida, Sanna, Koralle, and Sussi grown in Slovakia. This fungus also causes a postharvest disease of cranberry fruit (see Black Rot under Diseases Caused by Fungi, Cranberries section).

Mycosphaerella stemmatea (Ascomycota, Capnodiales) is a common leaf-spotting fungus found in Italy, Germany, Poland, the former Czechoslovakia, Lithuania, Denmark, Norway, Sweden, Finland, and Latvia. Although widespread, it usually does not have a significant impact on the plant since spotted leaves often remain attached. In some years, incidence may affect as many as 80% of the plants in a stand but, in general, disease does not develop to this extent. Young lesions have brown centers with reddish purple margins; the spots later turn ashen gray. Spots average 1 cm in diameter or less, although several adjacent spots may merge. In Sweden, pycnidia appear in late summer in the center of the sunken leaf spots, followed by ascocarps, which mature in May and June.

Powdery mildew. Powdery mildew caused by *Podosphaera myrtillina* (Ascomycota, Erysiphales) is widespread on *Vaccinium* spp., but is found on lingonberry in Hungary, Germany, Norway, and Sweden. The mycelium grows on both sides of the leaves, appearing thin and poorly developed. Ellipsoidal conidia (oidium) with fibrosin bodies form sparsely on the grayish mycelium in chains, averaging 14–16 × 27–33 µm, with a range of 11–18 × 20–40 µm. Chasmothecia are scattered over the leaves, singly or in clusters, are subglobose, and are fairly small, averaging 80–95 µm in diameter, with a range of 66–110 µm. There are sometimes as few as three and as many as twelve appendages per chasmothecium (average is five to ten), generally arising from the upper half or middle of the ascoma. Appendages are dichotomously branched three to four times (sometimes as many as six) and are knoblike. There is one ascus, containing eight ascospores per ascoma.

Various fungal pathogens, including *Coleophoma empetri* and *Phomopsis columnaris* have been associated with leaf spots and twig blights on lingonberry, but the symptoms do not allow these diseases to be distinguished without a laboratory diagnosis (Fig. 251). *Cladosporium*, *Botrytis*, *Rhizoctonia*, *Pestalotiopsis*, and *Volutella* spp. have also been recovered from necrotic leaves of plants with stem dieback in Oregon but are considered secondary.

Fungi Associated with Stem Dieback

Phacidium vaccinii (Ascomycota, Heliotiales) has been reported on lingonberry from Belgium, England, France, Germany, Switzerland, and Sweden, although it is found over a wider geographical area on other ericaceous plants such as *Rhododendron* spp., cranberry, and evergreen huckleberry (*Vaccinium ovatum*). The synonym is considered to be *Ceuthospora anomala*. Pycnidia are stromatic, immersed, and rounded, then erumpent, the epidermis splitting and peeling back into irregular teeth. Conidiomata are black or pale brown and up to 1,000 µm in diameter, with as many as 30 locules, each of which may be variable in shape, 55–70 µm in diameter, up to 215 µm deep, and each with its own ostiole up to 16 µm in diameter. Conidia are elongate-subcylindrical, straight or slightly curved, blunt at both ends or sometimes tapering below, aseptate, hyaline, 1.5–2 (sometimes 3) × 8–15.5 µm, with an irregular to funnel-shaped, mucoid, apical appendage. The ascigerous stage produces ascocarps that are scattered in the epidermis, 0.5 mm in diameter, about 0.3 mm high, and circular, and they open by splitting up radially. The asci are 8 × 35–42 µm, are club shaped on short stalks, have eight ascospores that average 3–3.5 × 10–13 µm, and are ellipsoidal to fusiform, nonseptate, hyaline, and uniseriate.

Phacidium vaccinii causes dieback of stems and entire plants in high-density wild populations in boreal regions. In Sweden, the conidial state is more common and is present in the spring and early summer, with the teleomorph appearing in late summer and autumn.

Phomopsis columnaris (Ascomycota, Diaporthales) was isolated in Oregon from leaf lesions and dying stems of cultivated 'Korralle' lingonberry. Affected plants had a progressive shoot dieback that eventually killed whole plants, affecting nearly 50% of a 2-ha field. The causality of the fungus to the dieback has not been proven, although the fungus was clearly associated with dying shoots. *Phomopsis columnaris* differs from other members of the genus in having multicellular vertically aligned conidiophores. Conidia are hyaline, ellipsoidal to oval, with ends slightly rounded, and 3.6–4.9 × 6.4–9.4 µm. Pycnidia are subcuticular, scattered to confluent, uniloculate, black, broadly spherical to flattened, 200–330 × 440–840 µm, and open by a single ostiole at the end of a beak often less than 60 µm long, but up to 260 µm. Unlike most other *Phomopsis* spp., *Phomopsis columnaris* has only a single, large oil globule per alpha conidium. Beta conidia were absent. This fungus was found only on a single cultivar grown domestically and has not been reported from wild populations.

Snow mold. *Eupropolella vaccinii* (Ascomycota, Heliotiales) causes snow mold of lingonberry and bearberry (*Arctostaphylos uva-ursi*), although there is some evidence of fungal host specialization. The fungus has been reported on lingonberry from Switzerland, Norway, Sweden, and Finland and is likely more widespread in the northern range of lingonberry where there is prolonged snow cover. This disease has not been reported from domesticated fields but is found in wild populations, especially those in logged areas that had previously been shaded. *Eupropolella vaccinii* infects leaves and stems during the winter under the snow, eventually killing plants, and is capable of causing up to 70% stand mortality. After snowmelt, affected foliage is covered with a dense gray-green mycelium that frequently disappears after the first rain. Infected leaves are usually found cast on the ground or clinging to dead plants in mid- to late spring.

Fig. 250. Lingonberry leaves inoculated with *Allantophomopsis lycopodina* (top row), a causal organism of leaf spots, compared with healthy leaves (bottom row). (Courtesy M. L. Putnam; © APS)

Fig. 251. Leaf spots caused by *Coleophoma empetri* on lingonberry. (Courtesy M. L. Putnam; © APS)

Selected References

Eriksson, B. 1969. On Ascomycetes on Diapensiales and Ericales in Fennoscandia. 1. Discomycetes. Symb. Bot. Ups. 19:1-80.

Eriksson, B. 1974. On Ascomycetes on Diapensiales and Ericales in Fennoscandia. 2. Pyrenomycetes. Svensk Bot. Tidskr. 68:192-234.

Farr, D. G., Castlebury, L. A., Rossman, A. Y., and Putnam, M. L. 2002. A new species of *Phomopsis* causing twig dieback of *Vaccinium vitis-idaea* (lingonberry). Mycol. Res. 106:745-752.

Grove, W. B. 1935. British Stem and Leaf Fungi (Coelomycetes). Volume I. Cambridge University Press, London.

Gustavsson, B. A. 2001. Genetic variation in horticulturally important traits of fifteen wild lingonberry *Vaccinium vitis-idaea* L. populations. Euphytica 120:173-182.

Heluta, B. P., and Minter, D. W. 1998. *Podosphaera myrtillina*. IMI Descriptions of Fungi and Bacteria No. 138. CAB International, Kew, Surrey, England.

Kurkela, T. 1974. The association of *Eupropolella vaccinii* (Rehm) v. Höhn. and winter mortality of some ericaceous plants. Karstenia 14:97-101.

Minter, D. W. 1998. *Coccomyces leptideus*. IMI Descriptions of Fungi and Bacteria No. 1291. CAB International, Kew, Surrey, England.

Nag Raj, T. R. 1993. Coelomycetous Anamorphs with Appendage-Bearing Conidia. Mycologue Publications, Waterloo, Ontario, Canada.

Putnam, M. L. 2005. *Allantophomopsis lycopodina*—A new aerial pathogen of lingonberry (*Vaccinium vitis-idaea*). Plant Pathol. 54:248.

Radaitiene, D., and Bandzaitiene, Z. 2000. Fungal diseases of lingonberry (*Vaccinium vitis-idaea* L.) in Lithuania. Pages 42-43 in: Abstracts of Plenary Session Lectures. www.agr.unideb.hu/novved/ttnvf/proceedings/Abstracts2000.pdf

Rodríguez, M., and Minter, D. W. 1998. *Meliola niessleana*. IMI Descriptions of Fungi and Bacteria No. 1356. CAB International, Kew, Surrey, England.

Simala, D. 2007. Cultivation of selected lingonberry varieties in mountain regions in Slovakia. Ekológia trávneho porastu VII—medzinárodná vedecká konferencia, 2007, Banska Bystrica, Slovakia.

(Prepared by M. L. Putnam)

Little Leaf

Little leaf was first observed in 1972 in south Sweden in a field planted previously with selected wild plants. It has been seen in wild lingonberry in Sweden, but it seems to be more widespread in cultivated plants in south and central Sweden. Similar symptoms have also been observed in wild plants of blueberry (*Vaccinium myrtillus*) in the archipelago outside Stockholm. The causal organism has been graft transmitted from infected to healthy lingonberry plants. It is not known how widespread the disease occurs, but it was observed in the former Czechoslovakia in the 1950s, Finland in the 1980s, and the United States in the 1990s.

Symptomatic plants have the usual symptoms associated with diseases caused by phytoplasma-like organisms.

Symptoms

Infected plants have reduced leaf size; stunting; proliferation, absence, or malformation of flowers; absence of berries; and premature leaf reddening in late summer. However, plants still retain good vegetative development. Shoots from infected runners develop the same symptoms, and these runners can become bushy and winding tendrils. Similar symptoms are found in *Vaccinium myrtillus*.

Causal Organism

Transmission electron microscopy of ultrathin sections of diseased lingonberry tissues revealed the presence of a phytoplasma in the sieve tubes of the symptomatic leaves. Cells are pleomorphic with a diameter of 170–800 nm. Phytoplasmas were numerous in diseased plants but were not detected in symptomless plants.

The causal organism is easily graft transmitted. Grafted plants develop symptoms during the following growing season.

Disease Cycle and Epidemiology

The vector for the dispersal of the little leaf phytoplasma is unknown, but small cicadae have been suggested.

Management

There is no recommended management for little leaf, except that diseased plants and their runners should be removed from the field as soon as possible and destroyed.

Selected References

Blattny, C. 1955. Einige Ergebnisse der pflanzlichen Virusforschung in der Tschechoslowakischen Republik. Kongressber. Pflanzenschutzkongr. 47-65.

Bremer, K. 1981. Witches' broom disease of *Arctostaphylos* and *Vaccinium* species in Finland. Ann. Agric. Fenn. 20:188-191.

Tomenius, K., and Ahman, G. 1983. Mycoplasma-like organisms in the phloem of little leaf diseased plants of *Vaccinium vitis-idaea* and *V. myrtillus* in Sweden. Swed. J. Agric. Res. 15:205-209.

(Prepared by F. L. Caruso)

Part II. Noninfectious Disorders

Nutritional and Other Abiotic Disorders

Blueberries

Cold Injury

Blueberries can be injured by low temperatures at various times of the year. Low temperatures during the dormant season commonly damage stems and buds (winter injury) (Fig. 252). Some injury to flower buds happens regularly in colder regions. Injury may be observed when buds are sliced open with a sharp knife or razor blade. Dead flower primordia are brown to black, whereas healthy primordia are white or light colored (Fig. 253). Usually all primordia in a bud are killed, but some buds may contain dead and healthy primordia. These buds may produce only one or two flowers in the spring, compared with the normal eight to twelve. The apical flower buds on a fruiting lateral are usually injured before the basal buds. As growth begins, the twig tissue adjacent to the dead flower buds also dies. Low winter temperatures may also damage cambial tissue at the base of larger stems, particularly those weakened by previous injury. These stems may initiate normal growth but die during the following season.

When freeze injury happens during the bloom period, the corollas of flowers turn brown and persist on the bush longer than those of healthy, pollinated flowers (Fig. 254). Not all

Fig. 253. Partially damaged flower bud of 'Draper' blueberry caused by cold injury during the acclimation phase of dormancy. The photo was taken in January, about 5 weeks after a temperature of –17°C, in Corvallis, Oregon (United States). (Courtesy B. C. Strik; © APS)

Fig. 252. Winter injury to a highbush blueberry shoot. (Courtesy E. J. Hanson; © APS)

Fig. 254. Blueberry flowers with freeze injury. Note the browning of the corolla and the wilting stem. (Courtesy M. Longstroth; © APS)

Fig. 255. Young blueberry leaves with freeze injury. (Courtesy M. Longstroth; © APS)

flowers with browning on the corollas caused by freeze injury are killed; some may develop into fruit if the damage is partial. Spring freezes may also injure young leaves (Fig. 255). In some cases, growth ceases and injured leaves and apical tissues turn dark brown or black. Very late spring freeze events may injure green berries. After a freeze, severely affected fruit often show a faint blue or purple discoloration and may shrivel (Fig. 256A). Severely injured fruit often have internal browning and necrosis (Fig. 256B). Frost rings can form where the freeze injury is marginal, such as around the calyx cup (Fig. 256C). These rings can result in sunken areas of the fruit, rusting, or even cracking of the berry around the caylx cup.

(Prepared by E. J. Hanson and M. Longstroth)

Fig. 256. A, Faint blue-purple discoloration and initial fruit shrivel resulting from freeze injury to green highbush blueberry fruit. **B,** Internal necrosis resulting from freeze injury to green highbush blueberry fruit. **C,** Frost ring. Purple discoloration indicating freeze injury around the calyx of green highbush blueberry fruit. (Courtesy M. Longstroth; © APS)

Nutritional Disorders

Symptoms are often useful in diagnosing specific nutritional problems. However, symptoms of only a few nutrient deficiencies have been reported appearing naturally on blueberry under field conditions. Most nutrient deficiencies and toxicities have been described only as they appear on young plants when the nutrient supply is modified under experimental cultural conditions. It is important to recognize that symptoms that develop in the field may be different from those induced under controlled conditions. Most nutrient deficiencies cause similar symptoms on highbush, lowbush, and rabbiteye blueberries. Early experiments suggest that, in a field soil of sand covered with peat, the order of deficiency would be nitrogen, potassium, phosphorus, magnesium, boron, calcium, sulfur, iron, and manganese.

Inaccurate diagnoses are quite common because different nutritional stresses may result in similar symptoms. Tissue analysis is an important diagnostic tool that is also valuable in confirming diagnoses based on visual appearance. For blueberry, leaf tissue collected during the growing season is most commonly used for analysis. Many universities and commercial laboratories offer tissue analysis services.

pH

The efficiency of nutrient uptake of the highbush, lowbush, and rabbiteye blueberries is largely dependent on soil pH. Nutrient availability varies with pH, and deficiency problems are often not a function of nutrient level in the soil but improper pH. The blueberry requires an acid soil. Optimum growth is on soils with a pH range of 4.3 to 4.8, although a soil pH range of 4.0 to 5.2 is often satisfactory. Rabbiteye blueberry can thrive with a slightly wider pH range. The most obvious symptom of improper pH is iron chlorosis (see Iron below), which happens when the pH climbs too high and iron is no longer in a form available to the blueberry plant. Soils that are high in manganese or aluminum should be maintained at the high end of the pH range to prevent toxicity of these nutrients.

Nitrogen

Symptoms of nitrogen deficiency have been observed in the field, have been induced experimentally, and are similar on highbush, lowbush, and rabbiteye blueberries. Symptoms are likely to develop on plants with leaf tissues containing less than 1.5% nitrogen. Nitrogen deficiency is characterized by a general stunting and reduction in shoot growth. Few new stems are initiated, and bushes seldom produce more than one flush of growth per season. Nitrogen-deficient plants produce short shoots, usually with fewer flower buds, so yield is reduced.

Leaves are smaller than normal and pale green (chlorotic), in contrast to the lush, dark green growth of adequately fertilized plants. The chlorosis is uniform across the leaf, with no mottling or pattern. The older or lower leaves usually turn pale before the younger leaves at the top of shoots. Leaves on severely deficient highbush blueberry plants may develop small, reddish, necrotic spots. Lowbush blueberry leaves develop a red tint near the margins, which spreads toward the midrib as the deficiency becomes more severe. Nitrogen-deficient rabbiteye blueberry plants develop very pale green leaves, usually with red margins. Stems also may develop a pink to red coloration. Leaves of nitrogen-deficient plants often turn red and abscise prematurely.

Excessive nitrogen results in abundant vigorous shoots and large, dark green leaves. Bushes may produce numerous growth flushes, and the last flushes often occur too late in the season to allow time for the plants to properly harden off before winter. The tips of these shoots and occasionally whole stems are winter killed. Bushes that receive too much nitrogen often produce small berries that are low in acidity and soluble solids. Berries may ripen late or over an extended period of time. Excessive nitrogen reduces yields.

Phosphorus

Symptoms of phosphorus deficiency have been observed on young, potted plants when the supply of phosphorus was removed but have not been reported on field plants. Symptoms are associated with plants containing less than 0.1% phosphorus in leaves. Affected plants are stunted and produce short shoots with small, dull leaves. Leaves develop a dark green to purple color, which may be most pronounced on the margins and tips of leaves exposed to direct sun (Fig. 257). Basal leaves are the first to exhibit this color. In highbush blueberry, leaves may lie unusually flat against the stem. Lowbush blueberry leaves first appear dark green and are smaller than normal. Older leaves then develop a purple color along the margins, which eventually spreads to form large interveinal patches. Twigs on phosphorus-deficient plants are narrow and may also be a red-purple color.

Symptoms of excessive phosphorus have been induced on young plants in sand culture. An interveinal or general chlorosis of the basal leaves may appear, followed by abscission of all but the terminal leaves. Very high levels of phosphorus have been shown to inhibit iron uptake and may induce iron deficiency.

Fig. 257. Phosphorus-deficient highbush blueberry plant. Note the dark green to purple coloration. (Courtesy H. J. Amling; © APS)

Potassium

Applications of potassium have increased the yields of highbush and rabbiteye blueberries under field conditions. Visible deficiency symptoms are rare but have been reported in New Jersey (United States) on sandy soils low in organic matter. Deficiency symptoms or yield responses may be expected when leaf potassium levels fall below 0.35% (highbush) or 0.4% (rabbiteye).

Highbush blueberry leaves may develop several symptoms, ranging from scorching along the margin to cupping and curling (Fig. 258). Basal leaves may develop necrotic spots near the leaf margin. These spots grow together and extend to the edge of the leaves. Younger leaves near the shoot tips may develop interveinal chlorosis similar to that caused by iron deficiency. Potassium deficiency may also cause a dieback of shoot tips.

The younger leaves on potassium-deficient rabbiteye bushes develop an interveinal, mottled chlorosis. Small necrotic spots then develop over the surface of the leaves, followed by marginal scorching, curling, and eventually leaf abscission. Older leaves may develop a distinct "Christmas tree" appearance from the contrasting colors of the interveinal areas and the tissue adjacent to the main veins.

Deficiency symptoms on lowbush blueberry plants resemble those on rabbiteye blueberry plants (Fig. 259). Young leaves develop interveinal chlorosis similar to that caused by iron deficiency. Leaf margins then turn red, and a necrotic scorching follows.

Calcium

Symptoms of calcium deficiency have been observed only when the calcium supply to plants has been experimentally removed. On highbush blueberry, young leaves at the shoot

Fig. 258. Potassium-deficient highbush blueberry leaves with marginal scorching. (Courtesy H. J. Amling; © APS)

Fig. 259. A potassium-deficient lowbush blueberry leaf with marginal scorching and chlorosis. (Courtesy W. J. Kender; © APS)

terminal develop an interveinal chlorosis and are often stunted and cupped, with a scorching near the tip (Fig. 260). Shoots do not elongate normally, so a rosette of leaves is produced. Older leaves may first develop a speckling near the margins, which eventually develops into a scorching. Symptoms are associated with leaf calcium levels below 0.3% and appear most often when pH levels fall below 3.5.

Rabbiteye blueberry plants first develop a marginal chlorosis of both young and older leaves. This is followed by interveinal chlorosis, marginal necrotic spots, and eventually tip burn, cupping, and abscission. Symptoms on rabbiteye blueberry plants become apparent when leaves contain less than 0.2% calcium. Symptoms are similar on lowbush blueberry but develop first on midshoot leaves. The marginal necrotic spots coalesce into a reddish brown scorching. Leaves curl and eventually abscise as the deficiency progresses.

Magnesium

Symptoms of magnesium deficiency have been induced experimentally and are observed in commercial highbush blueberry plantings when leaf levels of magnesium fall below 0.2%. Symptoms include a distinctive pattern of discoloration between the main veins or along the margins of leaves; affected regions may turn yellow to bright red, while tissue adjacent to the main veins remains green (Fig. 261). This often results in a "Christmas tree"-shaped green region in the middle of the leaves. This red color may be brightest under intense light; dull brown necrotic areas may predominate under low light conditions. Leaf margins may also become necrotic on severely affected bushes. Symptoms normally develop later in the growing season, with the leaves at the base of young, vigorously growing shoots first showing symptoms. Severely affected leaves may abscise, with the older leaves dropping first. Young leaves at the tip of shoots are seldom affected.

Symptoms are similar on highbush, rabbiteye, and lowbush blueberries. Symptoms become apparent on rabbiteye plants when leaves contain less than 0.15% magnesium.

Sulfur

Sulfur deficiency has not been reported on field-grown blueberry. Symptoms induced on young, experimentally deprived plants are similar to those caused by nitrogen deficiency. Highbush blueberry leaves develop a uniform chlorosis (no pattern or mottling). Symptoms of sulfur deficiency tend to be confined to younger leaves, whereas nitrogen deficiency causes symptoms first on older leaves. Sulfur-deficient rabbiteye leaves first develop a mottling and then become chlorotic and bleached in appearance (Fig. 262).

Boron

Symptoms of boron deficiency have been induced on young plants experimentally and were reported on highbush blueberry plants in the field. Affected plants often first exhibit a dieback of shoot tips similar in appearance to winter injury, and their leaves contain less than 20 ppm of boron. Normal and severely affected branches may be on the same bush. Leaves close to aborted shoot tips develop a mottled chlorosis and cupped shape. Fruit set may be reduced on boron-deficient plants. Flower and vegetative buds on severely affected plants may not open in the spring or may develop normally for some time before dieback of the shoot tip. Boron deficiency may also cause sunken, necrotic lesions on highbush blueberry fruit (Fig. 263). Cultivars may differ in their boron requirements and in the severity of symptoms. Deficiencies are most likely to develop following dry seasons, since drought inhibits boron uptake.

The first symptoms of boron deficiency induced on lowbush blueberry plants are small necrotic spots on young, expanding leaves. Leaves develop a distorted shape and may curl downward. The shoot terminals die (Fig. 264), and lateral shoots subsequently emerge. Both leaf size and shoot growth are reduced.

Symptoms of boron toxicity may be induced by excessive applications. On highbush blueberry plants, a mottled, interveinal

Fig. 260. Calcium deficiency symptoms induced on a young highbush blueberry plant. (Courtesy H. J. Amling; © APS)

Fig. 261. Magnesium-deficient highbush blueberry. (Courtesy C. J. Rajzer; © APS)

Fig. 262. Sulfur deficiency in rabbiteye blueberry. (Courtesy J. M. Spiers)

chlorosis appears first on the basal leaves but then spreads acropetally (Fig. 265). Scorching develops along leaf margins and then spreads inward until the entire leaf becomes necrotic.

Iron

Symptoms of iron deficiency are most common in highbush and rabbiteye blueberries where the soil pH is too high. However, symptoms also can appear when the pH is within the optimum range. Deficiency symptoms are not closely associated with leaf iron concentrations. Inadequate water drainage and excessive soil phosphorus levels may accentuate the problem. Symptoms include a distinct pattern of chlorosis between the leaf veins (Fig. 266). Main veins and many minor veins remain green. Chlorotic tissue may range from light yellow to a bronzy gold. Symptoms normally appear first on young leaves near the shoot tips. Shoot growth and leaf size are reduced considerably. The symptoms may be more pronounced on one branch or section of the bush. Bushes may set fruit that are small and do not fully ripen.

Copper

Copper deficiency has not been reported on field-grown blueberry, but plants grown under experimentally deficient conditions develop symptoms resembling the interveinal chlorosis caused by manganese or iron shortages. As the deficiency progresses, shoot dieback may occur.

Copper toxicity causes chlorosis on terminal leaves of highbush blueberry. The midrib of these leaves remains green, and affected leaves are easily detached.

Manganese

Manganese deficiency causes symptoms similar to those caused by inadequate iron. Leaves at the shoot terminals develop a distinctive interveinal chlorosis. Affected leaves may also develop necrotic regions near the margin in addition to isolated spots throughout the leaf. Lowbush blueberry may develop minute necrotic flecks between the veins on young leaves. The flecks enlarge into spots surrounded by chlorotic tissue. Since the supply of both manganese and iron is reduced when the soil pH is too high, plants may exhibit symptoms of both deficiencies, which further complicates diagnosis.

Fig. 263. A necrotic lesion on an 'Elliott' highbush blueberry fruit caused by boron deficiency. (Courtesy A. Pergher; © APS)

Fig. 264. Dieback of lowbush blueberry shoot tips caused by boron deficiency. (Courtesy E. J. Hanson; © APS)

Fig. 265. Boron toxicity symptoms in highbush blueberry. (Courtesy H. J. Amling; © APS)

Fig. 266. Highbush blueberry leaves with chlorosis caused by iron deficiency (bottom row) and healthy leaves (top row). (Courtesy E. J. Hanson; © APS)

Fig. 267. Manganese toxicity in highbush blueberry. (Courtesy H. J. Amling; © APS)

Toxic levels of manganese may result when some soils are acidified for blueberry culture. Excessive manganese causes an interveinal chlorosis (Fig. 267) that develops first on the terminal leaves and then progresses basipetally. Small necrotic spots may also develop on the leaves.

Zinc

Zinc deficiency has not been reported under field conditions. Symptoms induced experimentally include a chlorosis on young leaves resembling that caused by iron and manganese shortages, except that zinc-induced chlorosis appears uniformly across the leaf, without the interveinal pattern associated with iron and manganese deficiencies. Symptoms may develop early in the season.

Selected References

Amling, H. J. 1958. Influence of nutrient-element supply on leaf composition and growth of highbush blueberry (*Vaccinium corymbosum* L.) with special reference to importance of sampling date on leaf and fruit composition of field grown blueberries. Ph.D. thesis. Michigan State University, East Lansing.

Coville, F. V. 1910. Experiments in Blueberry Culture. U. S. Dep. Agric. Bull. 193.

Eck, P. 1988. Blueberry Science. Rutgers University Press, New Brunswick, NJ.

Kender, W. J., and Anastasia, F. 1964. Nutrient deficiency symptoms of the lowbush blueberry. Proc. Am. Soc. Hortic. Sci. 85:275-280.

Kramer, A., and Schrader, A. L. 1942. Effects of nutrients, media, and growth substances on the growth of the Cabot variety of *Vaccinium corymbosum*. J. Agric. Res. 65:313-328.

Mikkelsen, D. S., and Doehlert, C. A. 1950. Magnesium deficiency in blueberries. Proc. Am. Soc. Hortic. Sci. 55:289-292.

Minton, N. A., Hagler, T. B., and Brightwell, W. T. 1951. Nutrient-element deficiency symptoms of the rabbiteye blueberry. Proc. Am. Soc. Hortic. Sci. 58:115-119.

Spiers, J. M. 1983. Elemental leaf content and deficiency symptoms in rabbiteye blueberries: 1. Nitrogen. J. Plant Nutr. 6:1059-1071.

Spiers, J. M. 1983. Elemental leaf content and deficiency symptoms in rabbiteye blueberries: 2. Calcium and magnesium. J. Plant Nutr. 6:1073-1086.

Spiers, J. M. 1984. Elemental leaf content and deficiency symptoms in rabbiteye blueberries: 3. Phosphorous and potassium. J. Plant Nutr. 7:1567-1581.

Woodbridge, C. G., and Drew, R. H. 1960. A boron deficiency dieback in highbush blueberries. Plant Dis. Rep. 44:855-857.

(Prepared by E. J. Hanson)

Pesticide Toxicity Injury

Most herbicides used in blueberry production are applied to the soil. These preemergence herbicides kill some established weeds and inhibit the establishment of others. All preemergent herbicides can injure blueberry if rates are sufficiently high. The likelihood of injury is generally minimal if directions on the herbicide label are followed. In general, injury is more likely if high rates are used, if the soil is sandy and/or low in organic matter, or if the plants are young or weak.

Some postemergence herbicides are also used on blueberry. These are applied directly on emerged weeds and either kill the contacted tissue directly or are absorbed and translocated throughout the weed. Glyphosate and paraquat are two common postemergent herbicides. To prevent injury on blueberry, they should be applied in a manner that prevents contact with susceptible tissues.

Glyphosate

Glyphosate is a postemergence herbicide that may be absorbed by blueberry leaves and green stems. Glyphosate is not mobile in soils but moves readily in plants. As a result, symptoms may develop soon after plants are exposed, or different symptoms may appear the following season. Treated plants may exhibit chlorosis of the terminal leaves within a few weeks, often followed by a dieback of shoot tips for 3 or 4 weeks after exposure. Different symptoms develop the next spring, particularly if plants were exposed in late summer or autumn. Budbreak is often delayed, and leaves are small and narrow, particularly at the base of the plant (Fig. 268). Additional shoot tips and stems may die. Lateral buds below aborted tips may develop, resulting in a witches'-broom appearance. Some branches may be severely affected, while others are normal.

Diuron

Diuron can injure blueberry if it is applied at high rates on coarse-textured soils. The first symptom is a chlorosis along the leaf veins (Fig. 269). Severely affected leaves develop general chlorosis accompanied by necrotic regions.

Hexazinone

Hexazinone can injure blueberry if it is soil-applied at excessive rates or if leaves or young tissues are directly treated. Applications to blueberry foliage cause leaves to abscise. Symptoms

Fig. 268. Highbush blueberry plant with small, narrow, straplike leaves caused by excessive glyphosate applied the previous season. (Courtesy E. Hanson; © APS)

caused by excessive soil applications may develop slowly. Leaf tips and margins turn dark brown and become necrotic (Fig. 270), and the rest of the leaf becomes chlorotic. Leaves often abscise as the necrotic areas spread. On lowbush blueberry, the basal leaves are affected first, and symptoms progress toward the shoot tips. On severely affected bushes, later growth flushes may also defoliate, and bushes may eventually die.

The severity of symptoms varies across the field. The chemical may be carried in surface runoff so that bushes in low areas may be most affected. Damage may also be greater in regions where the soil is sandier or lower in organic content.

Norflurazon

Norflurazon injury causes leaf veins to become chlorotic and the veins often develop a pink tint (Fig. 271). The bleached appearance of the veins is in striking contrast to the normal green of the surrounding tissue. Symptoms are most often observed first on the leaves of young, basal shoots or branches. Symptoms may develop during the year of application but progress further the following year.

Paraquat

Paraquat is a postemergent herbicide that causes necrotic spotting on the leaves and on the green bark of young canes. Symptoms appear within a day or two of exposure (Fig. 272).

Fig. 269. Highbush blueberry plant with chlorosis along the veins and necrotic patches caused by excessive diuron levels. (Courtesy E. Hanson; © APS)

Fig. 270. Highbush blueberry plant showing marginal necrosis resulting from hexazinone injury. (Courtesy E. Hanson; © APS)

Simazine

Excessive simazine causes an interveinal chlorosis similar to symptoms of iron deficiency (Fig. 273). Severely affected leaves may develop a general chlorosis of the tissue near the tip of the leaf, followed by necrosis of the leaf tip and margin.

Fig. 271. Highbush blueberry plant with chlorosis and pink tinting along the veins caused by excessive use of norflurazon. (Courtesy E. Hanson; © APS)

Fig. 272. Necrotic spotting on the leaves and green canes of highbush blueberry caused by excessive paraquat levels. (Courtesy E. Hanson; © APS)

Fig. 273. Highbush blueberry plant with interveinal yellowing and general chlorosis of the leaf tips caused by excessive simazine applications. (Courtesy M. P. Pritts; © APS)

Fig. 274. Highbush blueberry plant with yellowing of the leaf veins caused by excessive terbacil application. (Courtesy M. P. Pritts; © APS)

Fig. 275. Highbush blueberry plant with abnormal twisting of the shoots and red coloration on the leaves resulting from excessive exposure to 2,4-D ester. (Courtesy M. Longstroth; © APS)

Terbacil

Terbacil toxicity causes an interveinal chlorosis similar to that caused by excessive simazine (Fig. 274). The chlorosis may be accompanied by blotchy necrotic regions on the leaf tip and margins. Whole leaves may become necrotic on severely affected bushes.

2,4-D

2,4-D can cause distorted, twisted growth on young shoots and red coloration on young leaves. Ester formulations of 2,4-D are volatile and can cause injury even if the spray is applied to the ground beneath the blueberry bushes (Fig. 275).

(Prepared by E. J. Hanson)

Salt Injury

Blueberries are sensitive to high salt levels in the soil resulting from excessive fertilizer applications or from irrigation water high in salt. Symptoms appear first as an interveinal chlorosis, followed by necrosis along the leaf margins, resembling the scorching caused by acute water shortage. Bushes next to roads treated with de-icing salt may exhibit severe dieback of shoots and death of flower buds during the winter (Fig. 276). Symptoms result from salt spray deposited on branches by wind and are nearly identical to cold injury.

(Prepared by E. J. Hanson)

Water Stress

Blueberry is sensitive to both inadequate and excessive soil moisture. Young, succulent shoots and leaves wilt readily under dry conditions. If drought persists, leaf margins and tips may become necrotic and scorched (Fig. 277). This scorching is similar in appearance to salt injury.

Temporary flooding or waterlogging of soil during the growing season may cause shoot tips to wilt. Inadequate drainage in the field causes a general reduction in vigor and productivity and is often accompanied by a slight chlorosis of the leaves.

(Prepared by E. J. Hanson)

Fig. 276. Injury to blueberry leaf and flower buds resulting from the drift of de-icing salt spray from an adjacent highway. (Courtesy E. Hanson; © APS)

Fig. 277. Necrosis of blueberry leaves resulting from an acute water deficit. (Courtesy E. Hanson; © APS)

Cranberries

Frost Injury

Frost injury is a serious hazard in cranberry production. Injury to the buds and vegetative structures may happen in the spring and berries may be damaged in the autumn. Sensitive tissues are protected through the use of sprinkler irrigation utilizing systems that deliver at least 0.25 cm of water per hour. Prior to the release of dormancy in early spring, temperatures as low as –17°C are tolerated. As dormancy is released, but prior to the start of new growth, the terminal bud that contains the inflorescence withstands temperatures of –3 to –4°C depending on the cultivar. Flowers and unopened flower buds of all cultivars are damaged at temperatures below –1°C.

Spring frost injury can be observed within the flower buds, appearing as a light brown center in cut buds within 24 h of injury. The centers of such buds become black within 2 weeks of injury. In some cases, buds are only partially damaged. Such buds, when cut, exhibit brown spots (damaged flower initials; Fig. 278) or browning only in the center. In such cases, the bud develops into an upright with fewer-than-normal flowers or with a full complement of flowers with no apical vegetation (umbrella bloom), respectively. If the entire bud or the apical meristem is damaged by frost, the plants may recover through the production of side shoots; however, such shoots bear no crop in that year. Young vegetative growth exposed to frost turns brown, and the tissue is necrotic (Fig. 279). Diagnosis of spring frost injury well after the event depends on the observation of umbrella blooms (no apical vegetation above the fruit) and side shoot production, since injured apical tissues become necrotic and may fall away.

Immature berries are sensitive to temperatures below –1°C. However, as the berry loses its green color, frost tolerance increases. At the white berry stage, cranberry fruit withstands –2°C. At full mature red color, tolerance of –5°C is common with as low as –7°C tolerated by some cultivars at full maturity. Berries damaged by autumn frost events become opaque and soft. Such berries do not burst readily when squeezed, however.

(Prepared by I. E. Demoranville;
Revised by C. J. DeMoranville)

Fig. 278. Damaged cranberry flower bud initials caused by frost injury. (Courtesy C. J. Demoranville; © APS)

Fig. 279. A cranberry upright with necrosis of the vegetative shoots caused by frost injury. (Courtesy D. Schiffhauer; © APS)

Hail Injury

Hail injury is most serious during the flowering period in June and July, when blossoms and flower buds may be severed or battered so severely that fruit set is impaired. In August, developing, immature fruit may be bruised, punctured, or detached from the plant if petioles are broken (Fig. 280). Hard, brown scar tissue forms on punctured fruit, making it unsuitable for the fresh market, but such fruit can be salvaged for processing. Leaves may also be damaged by hail, but the plants

Fig. 280. Immature cranberry fruit with bruised areas caused by hail. (Courtesy F. L. Caruso; © APS)

Fig. 281. Injury in a wavy pattern in a cranberry bed, likely caused by lightning. (Courtesy F. L. Caruso; © APS)

usually recover with no problems. In very rare instances, hail injury has an adverse effect on the formation of terminal buds for the following season.

(Prepared by I. E. Demoranville; Revised by C. J. DeMoranville)

Lightning Injury

Occasionally, an area of dead, injured plants appears in a cranberry bed for no obvious reason. Both aboveground and belowground parts of the plants are dead, and no pathogens or insects are evident. This type of problem can be caused by a lightning strike. Damage may be expressed as a solid area of dead plants, or plants are killed in wavy patterns that radiate from a central axis (Fig. 281). The central point of the injury is usually a metal sprinkler head.

(Prepared by F. L. Caruso)

Nutritional Disorders

All plants require specific mineral elements in specific quantities to complete growth and development. Cranberry is unique in that the amounts of these required elements are much smaller than those for most crop plants. The reason for this is that cranberry has adapted through evolution for growth on acid, sandy soils. These soils have low nutrient content and the plants in the family Ericaceae, such as cranberry and blueberry, that evolved on them have correspondingly modest nutrient needs. Further, cranberry is a perennial plant with the capacity to store and reuse nutrients in old leaves, woody stems, and roots. A unique and important feature of cranberry is that it maintains its leaves over the winter. These leaves also serve as a nutrient source when the plants resume growth in the spring.

Nutritional disorders appear in plants when the concentrations of required nutrients are too high (luxury consumption/nutrient toxicity), too low (nutrient deficiency), or out of balance. Research has defined the critical concentrations for nutrients in cranberry shoot tissue. The critical concentration is defined as the concentration of a specific nutrient associated with 90% of optimal plant growth and crop productivity. If the concentration drops below this level, growth and production may be severely impaired since that nutrient becomes the growth-limiting factor.

Tissue testing, in cranberry the analysis of shoot tip samples, can determine whether nutrients are within the optimum range and enables the detection and correction of borderline deficiencies in the absence of visible symptoms. Visual symptoms are associated with nutrient deficiency in cranberry. However, for the most part, the symptoms for the deficiency of specific elements are not well defined and cannot be used as the basis for diagnosis in the absence of a tissue test. Further, by the time these symptoms are visible, the deficiency is often quite severe and production significantly reduced. Generally, healthy vine appearance and optimal yield is indicative of adequate nutritional status.

Instructions for the collection and handling of cranberry plant tissue samples and tissue test standards can be found in the publication *Cranberry Tissue Testing for Producing Beds in North America*, listed below in the Selected References.

Nitrogen

Nitrogen is the plant nutrient that is most often limiting in soils. While the nitrogen requirement of cranberry is low compared with that of many crops, the cultivation of cranberry in sandy soils makes it necessary to add nitrogen fertilizer to meet the plant's nutritional requirements. Rates of 15–65 kg/ha/year are applied in split applications and adjusted depending on plant vigor and variety. Excessive nitrogen application has been associated with poor cropping, rank vegetation including runner production, and an increase in the incidence of fruit rot disease.

Nitrogen deficiency in cranberry is readily recognizable; it results in reduced growth and abnormal color development (Fig. 282). The reduced growth appears as shortened uprights with small leaves. The abnormal color is seen in light green leaves and red pigmentation on leaf margins and on new stem tissue. Symptoms are first visible in the oldest tissues.

The critical nitrogen concentration for cranberry is near 0.80% on a dry matter basis. If tissue testing is used as a means of evaluating the nitrogen status of a crop, a concentration of 0.90% or below suggests a need for additional nitrogen fertilization. The optimum concentration range of 0.90 to 1.10% is narrow, and higher concentrations are not unusual, particularly for hybrid cultivars such as Stevens which often show 1.30% in productive beds. Although excessive vine growth and poor productivity can occur with excessive nitrogen fertilization, nitrogen toxicity per se has not been reported.

A significant feature of cranberry that separates it from most other crops is its specific need for the ammonium (NH_4^+) form of nitrogen and its limited ability to use the nitrate (NO_3^-) form. Use of an ammonium form of nitrogen fertilizer meets this requirement.

Phosphorus

The acidic soils common to cranberry culture tend to have high concentrations of iron, aluminum, and magnesium. When

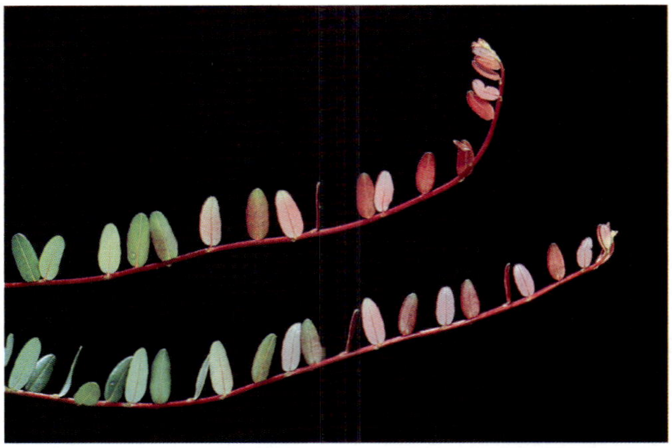

Fig. 282. Nitrogen deficiency in cranberry. Note the reduced growth and abnormal color in the uprights. (Courtesy L. A. Peterson; © APS)

phosphorus fertilizer (as orthophosphates) is added to these soils, it forms insoluble compounds with these naturally occurring metal cations and becomes unavailable to cranberry plants. Therefore, cranberry growers make frequent applications of phosphorus fertilizer to keep some phosphate ions in the soil solution to support plant uptake. Previous research in Massachusetts (United States) has shown a yield increase with the application of phosphorus fertilizers, but there was no observable increase in yield with phosphorous additions beyond 20 kg/ha/year.

The critical tissue concentration for phosphorus in cranberry is 0.08%, which is about one-half that of many other crops. Sampling tissues at various times during the growing season shows that the phosphorus concentration decreases with time, indicating a dilution factor, with growth rate exceeding uptake rate of this nutrient. Thus, the time of sampling is important; however, it is equally important that the phosphorus concentration not go below the critical concentration at any time during the season.

Although response to phosphorus fertilization has been reported, surveys involving the sampling and analysis of a large number of cranberry beds for plant phosphorus composition showed that more than 90% of the samples were in or above the optimum range (0.1–0.2%). This indicates that, in most cases, applications of phosphorus were adequate to support plant growth.

Phosphorus deficiency reduces the growth rate of cranberry and causes dark green leaves and a slight reddening of the stem and leaves of the shoot (Fig. 283). These symptoms have been observed only under greenhouse conditions when phosphorus availability was experimentally controlled.

Potassium

Potassium is most frequently applied to cranberry in complete N-P-K fertilizers. Potassium in these fertilizers is generally in either the sulfate or the chloride form. Potassium sulfate is the preferred product for cranberry production, since research has confirmed that cranberries are sensitive to chloride.

The critical concentration for potassium in cranberry is 0.3%, which (as with phosphorus) is much lower than that for other crops. Data from tissue samples from cranberry beds in Wisconsin and Massachusetts (United States) show potassium concentrations well above the 0.3% level. It is, therefore, not surprising that potassium deficiency symptoms have not been observed in field situations, since the survey data illustrate that cranberry soils are adequately supplied with available potassium by common fertilization practices. Normal values for potassium in cranberry tissue are 0.40–0.75% and may be greater because of luxury consumption if application rates are excessive.

Potassium deficiency induced under controlled greenhouse conditions first appears as a reddening of the leaf margins at the tip of the shoot. Growth is retarded and, eventually, leaves become necrotic and the plant dies (Fig. 284).

Calcium

Calcium is a nonmobile nutrient element. It moves up into a plant but, once incorporated into a tissue within a plant structure such as a leaf, it cannot be remobilized. Nitrogen, phosphorus, and potassium, in contrast, are mobile elements and move freely up and down the plant. Because of this lack of mobility, calcium deficiency appears as a marginal reddening of the apical leaves, with relatively rapid death of the growing tip (Fig. 285). Calcium deficiency symptoms can be induced on cranberry plants in controlled conditions but have not been observed in the field. Even in the acid soils that are common for cranberry production, calcium is typically present in sufficient quantities to ensure that it is not a limiting factor. The normal concentration for calcium in cranberry shoots is 0.3–0.8%.

Magnesium

No magnesium-associated nutritional disorders have been reported for cranberry under field conditions. Controlled studies have shown that the critical concentration for magnesium in cranberry tissue, although not defined specifically, is 0.10% or less. Analyses of plant samples from a large number of beds in Massachusetts and Wisconsin (United States) showed that all magnesium concentrations were in excess of the critical level,

Fig. 284. Potassium deficiency in cranberry. Note the reddening of the leaf margins. (Courtesy L. A. Peterson; © APS)

Fig. 283. Phosphorus deficiency in cranberry. Note the reddening of the stems and leaves. (Courtesy L. A. Peterson; © APS)

Fig. 285. Calcium deficiency in cranberry. Note the marginal reddening of the apical leaves. (Courtesy L. A. Peterson; © APS)

indicating a sufficient supply of this nutrient element despite its infrequent use as a cranberry fertilizer. The normal concentration for magnesium in cranberry shoots is 0.15–0.25%.

Symptoms of induced magnesium deficiency in cranberry (greenhouse conditions) appear as a dark mottling and distortion of older leaves on the shoot, followed by the development of marginal necrotic spots on these leaves. In time, the affected leaves drop off. Injury progresses from the base of the shoot to the tip.

Sulfur

Sulfur deficiency is not a common problem in field situations, perhaps because sulfur originates from a number of sources. Sulfur (as sulfate) is an ingredient of many fertilizer products, including potassium sulfate (0-0-50), ammonium sulfate (21-0-0), ordinary superphosphate (0-20-0), and magnesium sulfate. Sulfur dioxide in the air is a sulfur source for plants; leaves either absorb it directly or it is dissolved in atmospheric moisture and deposited on plants and soils. Mineralization of soil organic matter is also a source of sulfur.

The critical concentration for sulfur in cranberry is 0.08%. Deficiency symptoms that appear when sulfur is at or below the critical concentration are very similar to those for nitrogen deficiency (Fig. 286). Plant analysis data from surveys of cranberry beds indicate that sulfur concentrations are significantly above the critical level. This result is consistent with the absence of sulfur-related nutritional disorders in cranberry beds. The normal concentration for sulfur in cranberry shoots is 0.08–0.25%.

Micronutrient Deficiencies

The availability of micronutrients in soil is very closely related to the inherent content and to soil pH. A decrease in soil pH (increase in acidity) increases the availability of micronutrients such as copper, zinc, manganese, boron, and iron. Molybdenum is less available in acid soils than in soils that are near neutral pH. Considering the acidic soils typically used for cranberry production, disorders associated with micronutrient deficiencies should be minimal. This conclusion is supported by data from plant analysis surveys of cranberry-growing areas, which generally indicate that the supply of micronutrients is adequate for optimum cranberry production.

Although micronutrient disorders in cranberry have been researched, a definitive set of symptoms for each micronutrient has not been published; therefore, visual observations may not be reliable for diagnostic purposes.

The symptoms of micronutrient deficiencies that were observed by L. A. Peterson when cranberries were grown in hydroponic culture without the micronutrient of concern in the nutrient solution are described below. In all cases, extended periods of time elapsed after the plants were deprived of a micronutrient before any symptoms appeared.

Fig. 286. Sulfur-deficient cranberry. (Courtesy L. A. Peterson; © APS)

Zinc. Deficiency symptoms included mild leaf chlorosis, reddening of the stem tips and margins of the youngest leaves, and development of terminal rosettes as a result of reduced internode growth (Fig. 287). Zinc concentration in affected tissue was 12 ppm or less.

Boron. Deficiency symptoms appeared at the growing point, and this structure eventually died. Terminal leaves were small and slightly chlorotic, with some reddening of the margins. Boron concentration in affected tissue was 10 ppm or less.

Copper. Deficiency first appeared as a slight chlorosis and reddening of the youngest leaves. As the deficiency progressed, entire runners died rapidly, becoming light tan and necrotic (Fig. 288). Copper deficiency has also been associated with "monkey-faced" berries. Copper concentration in affected tissue was 3 ppm or less.

Manganese. About 20 weeks passed before any symptoms were observed on rooted cranberry cuttings in hydroponic nutrient solution minus manganese. The relatively high manganese concentration in the cuttings provided a manganese source for this extended period and reflects the high manganese availability in low-pH cranberry soils. Symptoms appeared as distorted chlorotic and necrotic leaves beginning about 2 cm from the growing tip (Fig. 289). Manganese concentrations in tissue of 10 ppm or less were associated with this disorder.

Iron. Deficiency appeared as a chlorosis of new growth, decreasing in intensity from the tip to the base of the shoot. Symptoms were associated with tissue iron concentrations of 25 ppm or less.

Molybdenum and chlorine. No disorders associated with a deficiency of these micronutrients have been reported.

Fig. 287. Zinc deficiency in cranberry. Note the chlorosis and the development of terminal rosettes. (Courtesy L. A. Peterson; © APS)

Fig. 288. Copper deficiency in cranberry. Note the dead runners. (Courtesy L. A. Peterson; © APS)

Luxury Consumption and Micronutrient Toxicity

Concentrations of manganese in excess of 500 ppm are commonly seen in cranberry shoot tissues. However, manganese toxicity has not been reported. Generally, highly elevated manganese levels in cranberry tissue have been associated with poor drainage and soil anoxia. Bed drainage should be evaluated in such situations.

Many cranberry beds have high levels of soil iron. In such beds, tissue iron levels may be elevated. There has been no report of iron toxicity in cranberry.

Field and greenhouse research has shown that cranberries are sensitive to chloride. When exposed to irrigation water containing more than 100 mg of chloride per milliliter, cranberries produced more runners and fewer flowers as compared with control plants. In such plants, tissue chloride concentrations were greater than 0.05%.

Nutritional Imbalances

Yellow vine or yellow vine syndrome (YV) is a leaf disorder that manifests as visual symptoms consisting of marginal and interveinal chlorosis (Fig. 290). Usually, the symptoms arise first in the older leaves and then proceed up the stem into the new growth. The most common time for the symptoms to become severe is around fruit set, when demand for resources in the plant is high and dependence on sprinkler irrigation is greatest.

The YV symptoms are most likely caused by nutritional imbalances in the cranberry plants. However, empirical evidence supports the hypothesis that the nutrient imbalance is secondary to root problems caused by abiotic stress. Most often, water stress is the primary causal condition (too much or too little) but herbicide stress may be a contributing factor at some sites. Water stress alone or in combination with herbicide stress early in the season results in poor root development. Nutritional imbalances result secondarily.

Cranberry beds with patches of YV have been found to have soil water content in the YV areas that is either much higher or much lower than that in the surrounding green areas. Additionally, in greenhouse experiments, plants subjected to very shallow or very deep water table conditions developed YV. The consistent finding in the field has been that the rooting depth in YV areas is shallower than that in unaffected areas. As demand for water and nutrients peak during the hottest and driest months of July and August, YV may appear in patchy distributions that mirror an uneven distribution of sprinkler irrigation.

In most cases, YV appears in areas that were too wet early in the season, leading to poor rooting depth. Rooting depth can be improved by keeping the bed well drained early in the season. When the water table is closer than about 15 cm below the surface, root development and root function is impaired.

Fig. 289. Manganese deficiency in cranberry. Note the distorted necrotic and chlorotic leaves. (Courtesy L. A. Peterson; © APS)

Fig. 290. Cranberry leaves affected by yellow vine syndrome, most likely caused by nutritional imbalances. (Courtesy C. J. Demoranville; © APS)

Selected References

Davenport, J., DeMoranville, C., Hart, J., Patten, K., Peterson, L., Planer, T., Poole, A., Roper, T., and Smith, J. 1995. Cranberry tissue testing for producing beds in North America. Oreg. State Univ. Ext. Serv. Publ. EM 8610.

Davenport, J., DeMoranville, C., Hart, J., and Roper, T. 2000. Nitrogen for Bearing Cranberries in North America. Oreg. State Univ. Ext. Serv. Publ. EM 8741.

DeMoranville, C. J., and Davenport, J. R. 1997. Phosphorus forms, rates, and timing in Massachusetts cranberry production. Acta Hortic. 446:381-388.

Eck, P. 1990. The American Cranberry. Rutgers University Press, New Brunswick, NJ.

Greidanus, T., Peterson, L. A., Schrader, L. E., and Dana, M. N. 1972. Essentiality of ammonium for cranberry nutrition. J. Am. Soc. Hortic. Sci. 97:272-277.

Roper, T., Davenport, J., DeMoranville, C., Marchand, S., Poole, A., Hart, J., and Patten, K. 2004. Phosphorus for bearing cranberries in North America. Univ. Mass. Ext., Univ. Wisc. Ext., Wash. State Univ., and Oreg. State Univ.

(Prepared by C. J. DeMoranville)

Oxygen Deficiency Injury

Historic research in Massachusetts (United States) by H. Bergman determined that a lack of dissolved oxygen in the winter flood water was the cause of injury to cranberry plants, resulting in leaf drop and reduced yield potential. Bergman stated that oxygen deficiency injury may happen when oxygen levels in the winter flood water drop below 4 mg/L (full oxygenation = 10 mg/L or more). Bergman further stated that the lack of light penetration when snow or sand covered the ice led to poor photosynthesis, and it was the lack of photosynthesis that led to poor oxygenation in the water. The recommended remedy was to remove water from under the iced-over flood if light penetration was poor.

Following the removal of the winter flood, the following symptoms may be observed: loss of leaves, death of stems, death of terminal buds, poor flower development, decreased fruit set, and undersized fruit. These symptoms have been attributed to oxygen deficiency during the winter flood (Fig. 291).

Research in both Massachusetts and Wisconsin (United States) has led to a reexamination of Bergman's theories and recommendations. This work showed that cranberry requires very little light for photosynthesis during the winter and the light that penetrates snow or sand may be sufficient for this purpose. Further, even with 23 cm of snow on the ice, oxygen in the water beneath remained at 8 mg/L or greater. In Wisconsin, covering ice with black cloth, sand, or snow did not lead to leaf drop or crop reduction in the plants below the treatments. In

Fig. 291. Cranberry bed with an area of leaf drop caused by oxygen deficiency injury. The plants in the background and on the left have retained their leaves. (Courtesy F. L. Caruso; © APS)

Massachusetts, plants held flooded in darkness and low oxygen levels did not show reduced carbohydrate content or leaf drop.

As a wetland plant, cranberry can survive periods of poor oxygenation during flooded conditions. In particular, the plants can tolerate low oxygen levels in saturated soil. However, survival under these conditions depletes carbohydrate reserves. Plants with poor carbohydrate reserves caused by large crops, lack of sunshine the previous autumn, or other stresses may have less ability to tolerate low-oxygen stress and may show injury in the spring. In those cases, failure to prevent oxygen deficiency can result in leaf drop, inability of blossoms to set fruit, and crop reduction. However, any risk associated with using a winter flood is far outweighed by the benefit of protection from winter injury.

Certain management actions ensure that leaf drop potential is minimized. Water should be removed from beneath ice as soon as practical and care should be taken to avoid leaving vines trapped in shallow water beneath the ice. If water is being held beneath ice to expedite sanding, oxygen levels in the underlying flood should be monitored and water removed at an action level of 3–5 mg/L.

Selected References

Franklin, H. J., Bergman, H. F., and Stevens, N. E. 1943. Weather in cranberry culture. Mass. Agric. Exp. Stn. Bull. 402.

Vanden Heivel, J., Roper, T., and Altweis, J. 2006. Good cranberry management requires winter flooding. Fruit Grow. News 45(12):9-11.

(Prepared by C. J. DeMoranville)

Pesticide Toxicity Injury

Herbicides

Dichlobenil. Dichlobenil is a strong inhibitor of root growth and shoot growth in many types of germinating weed seedlings. It is effective against many species of grasses, rushes, and sedges, as well as broadleaf weeds. This herbicide is very important because it controls a wide range of weed species, including nutsedge and dodder.

Dichlobenil can injure cranberry vines by inhibiting root formation in newly planted cuttings and by reducing root growth in mature plants. Stems become brittle and break easily, and plant growth is stunted. The most commonly observed symptoms of dichlobenil injury are the reddening of old leaves, unusually small developing fruit, and interveinal chlorosis (Fig. 292). Injury in the field is often visible as a swath of red vines in a straight line. This pattern is caused by uneven aerial applications of the herbicide that result in an overlap of the material. Small quantities of fertilizer and frequent irrigation are helpful in the recovery of injured plants.

Norflurazon. Norflurazon is a preemergence herbicide that restricts chlorophyll production in several of the grasses, rushes, and sedges that grow in cranberry beds. Injury symptoms in cranberry plants are very distinctive. First, the stems and leaves of the new growth become bright pink, then the leaves turn white between the veins, and then the leaf margins turn brown (Fig. 293). The new growth is stunted and the leaves are close together, forming a witches'-broom. Injury is more common where there are water puddles in low areas for several days. The risk of injury is higher if the herbicide is used in consecutive growing seasons. Some varieties are more susceptible to injury than others.

Napropamide. Napropamide inhibits the root growth of grasses, rushes, and sedges in cranberry beds. It also inhibits the sprouting of nutsedge tubers. No indications of injury to

Fig. 292. Cranberry uprights with injury from the herbicide dichlobenil. (Courtesy H. A. Sandler; © APS)

Fig. 293. Cranberry upright with pink coloration and the interveinal whitening of leaves induced by injury from the herbicide norflurazon. (Courtesy F. L. Caruso; © APS)

cranberry vines have been observed from applications of twice the approved usage rate. Rates higher than 9 kg/ha inhibit the rooting of cuttings in newly planted beds.

Simazine. Simazine acts as an inhibitor of photosynthesis. It controls broadleaf weed seedlings and the sprouting of underground stems of some grasses in cranberry beds. Cranberry can be injured if excessive rates are used or at the recommended rates if the vines are in a weakened condition. The earliest indication of injury is a browning of the margin at the tips of new leaves. The necrosis moves down the leaves and eventually also affects the old leaves. New leaves that are partly brown from the tip down while the other leaves are green are a good indicator of simazine injury. Irrigation is helpful in the recovery of affected plants.

Glyphosate. Glyphosate is a nonselective herbicide used as a postemergence solution in mid- to late summer; it is wiped on tall weed species that extend above cranberry vines. The herbicide is nonvolatile, so vines are injured only if the solution drips onto or otherwise comes in contact with the vines. Injury begins as a yellowing followed by necrosis of the leaves. This condition is generally very localized, and weeds that are present show similar symptoms.

2,4-D. 2,4-D is used as a hand-applied wipe application over the vines for tall woody and herbaceous weeds during the summer. The danger of vine injury arises mainly from the volatility of the herbicide. 2,4-D should not be used under hot, humid conditions. Affected leaves on cranberry vines turn brown very quickly, usually within 48 h. The stems may grow abnormally or twist as they recover from the injury.

Copper sulfate. Copper sulfate is commonly used as an algaecide in flood water at rates that are not harmful to cranberry vines. An extreme overdose kills cranberry vines, but the usual symptoms of injury are a bright red color on one or more of the older leaves on an affected upright.

Clopyralid. Clopyralid is a postemergence herbicide that can be applied as a wipe or spray. It controls annual and perennial broadleaf plants in the pea and aster families. Its mode of action is not completely understood but is similar to other auxin-type herbicides. Injury includes epinastic bending and twisting of the stems and petioles, stem swelling, and leaf cupping (Fig. 294). Injured cranberry vines are often described as gray because the leaves are cupped, exposing the lighter underside of the leaf. Cranberry vines can be very slow to recover from clopyralid injury.

Insecticide

Chlorpyrifos. Insecticides normally do not cause any injury when they are applied through the chemigation system. However, when they are applied aerially in low-volume injection systems, they may be deposited in high concentrations on cranberry vines. Leaves may develop necrosis after 48 h as discrete lesions (Fig. 295), which may coalesce. Developing berries exhibit necrotic areas that develop some russeting as the fruit continues to mature.

Fig. 294. Cranberry uprights with leaf cupping induced by injury from the herbicide clopyralid. (Courtesy H. A. Sandler; © APS)

Fungicide

Chlorothalonil. Like insecticides, chlorothalonil can injure fruit when it is applied in highly concentrated aerial sprays. Symptoms include reddish patterns on young fruit and necrosis of developing berries, similar to that described for chlorpyrifos injury (Fig. 296).

Fig. 295. Cranberry leaves with necrotic areas caused by injury from the insecticide chlorpyrifos. (Courtesy F. L. Caruso; © APS)

Fig. 296. 'Stevens' cranberries with russeting caused by injury from the fungicide chlorothalonil. (Courtesy P. S. McManus; © APS)

Selected Reference

Caruso, F. L. 2009. Evaluation of Bravo for control of field and storage rot of cranberries and injury to the fruit, 2008. Plant Dis. Manage. Rep. 3:SMF010.

(Prepared by H. A. Sandler and F. L. Caruso)

Physiological (Sterile) Breakdown

Postharvest storage losses of cranberry fruit of 30–40% have been experienced in some years. Losses vary from year to year, from bed to bed, and even from one section of a bed to another. Losses during refrigerated storage result from a combination of physiological breakdown and fungal decay. Physiological breakdown is also called sterile breakdown or sterile rot because no microorganisms can be isolated from the softened berries. What some call low-temperature breakdown, which happens when berries are stored at –1.1 to 1°C, is included in physiological breakdown.

Stage of Fruit Maturity

Cranberries are physiologically mature when the daily respiration rate for fruit reaches a preharvest or preclimacteric minimum level. For the cultivar McFarlin in Washington (United States), this happens at about 11 weeks after full bloom. Respiration of fruit gradually but steadily declines from midsummer onward. Fruit does not fully color and reach commercial maturity until 2–3 weeks after physiological maturity. Storage losses are less when cranberries are harvested at physiological maturity than when they are picked at color and °BRIX levels more suitable for commercial usage.

The blossom period varies by region. In eastern growing regions, it tends to be concentrated within a few weeks, but in the Pacific Northwest, it frequently spreads over a period of 3–6 weeks. Fruit from flowers with early anthesis mature well before those with late anthesis. Consequently, the early-ripening berries (20–30%) are beyond physiological maturity by the time of commercial harvest. These overripe berries tend to be softer and hence more sensitive to bruising and have a higher incidence of physiological breakdown than the late-ripening fruit. Less fruit is lost when berries are firm and are picked before they are overripe. The respiration rate of stored fruit increases with time in storage. The time of harvest for maximum storage potential is thus out of phase with the most desirable color and °BRIX development.

Fruit for the fresh market are either dry harvested or wet harvested with equipment designed to cause minimal damage to fruit. Traditional water reel wet-harvested fruit are not suitable for the fresh market. Studies on keeping quality, however, have been conducted with water-harvested fruit. The keeping quality of water-harvested fruit is inferior to that of dry-harvested fruit even when the berries are dried before storage. The rotating water reel of the harvester causes considerable bruising of the fruit. The longer berries remain in the flood water, the more fungal decay and physiological breakdown happens in storage. Water immersion time has a greater impact on physiological breakdown than on decay. Regardless of the harvest method, fresh berries should be dried before storage.

Bruising

Bruising plays a very large role in physiological breakdown. Bruised fruit breaks down more rapidly in cold storage than does undamaged fruit. Bruising accentuates low-temperature breakdown; more bruised fruit soften during storage for 14 days at 0°C than at either 8 or 20°C. Even bruises too small to be visible to the unaided eye are sufficient to speed up the breakdown process. Injuries can happen at any time during harvesting, separating, sorting, and packing. Berries with unnoticed injuries are important because they are mistakenly recognized as marketable fruit at the time they go into storage or are packaged for retail markets. Bruising also increases the rate at which fruit dries out and shrivels.

Mechanical harvesters injure berries as they are being picked. For example, with one type of picker–pruner, major injury can happen when the first elevator paddle touches the berries. The paddle strikes the berries from the top and pinches them before carrying them up the elevator. The front elevator of this picker was moved forward and down on later models to minimize the pinching and high-velocity impact between the first elevator paddle and the berries. Berries sustain little additional injury as they move up the elevator but are bruised again when they drop into the sack. The farther berries drop, the greater the injury. The injury rate for berries on the bottom of the sack is nearly double that of those on the top. The effects of impact and compression bruising are cumulative. Prestorage injury levels of mechanically harvested fruit were related to breakdown after 5 weeks of storage but not after 8 weeks.

Modern cleaners and separators can also damage berries. The impact injuries associated with the cleaning process may account for most of the increase in storage losses. For Furford-picked fruit (Furford Picker Co., Grayland, WA), storage losses are less when fruit are stored in incidental leaves and stems than when they are cleaned to remove these items before refrigerated storage. This suggests that the cleaning process further injures the fruit.

At temperatures above 7°C, bruise-induced softening is attributed to autocatalytic production of polygalacturonase. As mentioned above, berries held at 0°C also soften, but only trace amounts of the enzyme were detected in berries that softened at that temperature. This result suggests that more than one mechanism may cause physiological breakdown or that there is an interaction between chilling injury and bruise-induced softening.

Bruising increases fungal decay as well as sterile breakdown. Often more than a single fungus can be isolated from a decayed berry. Similarly, fungal decay and physiological breakdown may occur in the same berry. Most fungi are recovered from the blossom end rather than the stem end of the berry. This and other evidence strongly suggests that rot fungi enter via the flower parts during or shortly after bloom. Latent or quiescent infections become established in the developing fruit. Fungicides applied during the growing season, especially during and shortly after bloom, reduce rot caused by fungi but have no impact on physiological breakdown.

Temperature

Temperature has a major influence on losses in storage. Cranberries are subject to chilling injury; low-temperature breakdown happens when fruit are stored at temperatures between –1.1 and 1.0°C for longer than 2 weeks. The berries become water-soaked and rubbery, lack normal berry luster, are somewhat lighter in color, and feel somewhat sticky. Red pigment, which is normally restricted to the skin, leaches into the flesh (pulp), turning the normally white flesh pink. The cultivar Early Black appears to be more sensitive to low-temperature injury than the cultivars Searles or Howes. Bruised fruit softens faster during storage at 0°C than at either 8 or 20°C. Some of these same symptoms are noted for physiological breakdown at storage temperatures above freezing.

At temperatures above the chilling injury threshold, temperature is directly related to fruit respiration and inversely related to keeping quality. Overall losses are minimized when fruit is stored at 2–4°C (Fig. 297). Losses from fungal decay are slightly higher in this temperature range than at 0°C, but physiological breakdown is minimized.

Low temperatures do not ameliorate the harmful effects of poor handling. The higher the quality of fruit going into storage, the lower the losses during storage. Losses caused by fungal rots in storage increase with rising temperature, especially at 10°C or above (Fig. 297). Most fungi grow very slowly below 10°C (one notable exception is *Godronia cassandrae*, which causes end rot). Fruit stores best when the field heat has been removed within 2 days after harvest.

In unrefrigerated common storage where the temperature of fruit fluctuates between 4 and 20°C, the fungal rot portion of total losses is slightly greater than the physiological breakdown component. For fruit held in refrigerated storage at temperatures of 0.5–3.7°C, both fungal decay and physiological breakdown are reduced if the fruit is allowed to warm to 21°C for 1 day out of each 4 weeks of storage; however, intermittent warming of stored fruit may be difficult to do without causing condensation.

Atmospheric Control

The benefit to cranberry of modified storage atmosphere has not been consistent. Earlier studies indicated that neither elevating the concentration of carbon dioxide nor reducing the oxygen level had much impact on either fungal decay or physiological breakdown. Storing cranberries in nitrogen increased sterile breakdown while it decreased losses caused by fungal rots. More recent work suggests that elevated carbon dioxide concentrations decreased bruising, physiological breakdown, and decay of berries, with the optimal storage atmosphere being 30% carbon dioxide plus 21% oxygen.

Air should move freely but slowly through stored fruit. Rapidly moving air dries out the fruit and may lead to shriveling. Ethylene gas does not accumulate as long as there is some air movement. Oxygen concentrations below 8% have a deleterious effect on cranberries but do not occur in practice, even when fruit is stored in 10-barrel containers.

Relative humidity of 70% has been the standard practice in cranberry storage, but an increase to 90% has recently been recommended. Increasing the relative humidity above 90% did not provide any further reduction in cold storage losses.

Cultivar Sensitivity

Cranberry cultivars differ in their vulnerability to physiological breakdown. The cultivars Franklin, Pilgrim, and Stevens exhibit less physiological breakdown and fungal rot in storage than do the cultivars Early Black and Wilcox. The cultivar Ben Lear is only slightly sensitive to physiological breakdown but is susceptible to fungal rots. There is no data on the susceptibility of newer cultivars of cranberry (e.g., HyRed, Crimson Queen, Mullica Queen, DeMoranville, Scarlett Knight, or Gryleski) to physiological breakdown.

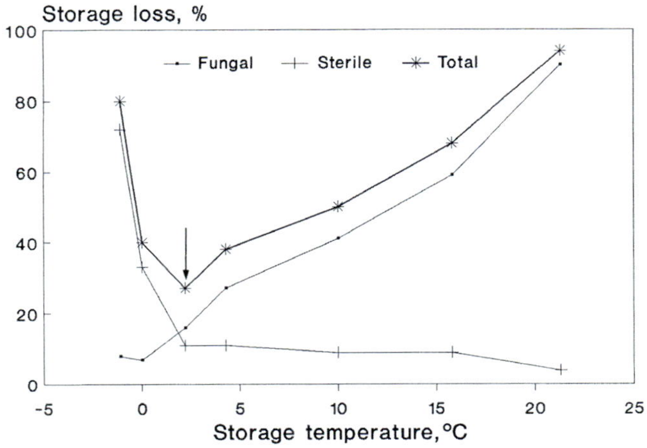

Fig. 297. Effect of storage temperature on the spoilage of fresh cranberries caused by fungi, by physiological (sterile) breakdown, and by a combination of both. The arrow indicates the temperature associated with minimum total loss (about 2°C). (Courtesy P. R. Bristow; © APS; Data from Wright et al., 1936)

Recommendations for Minimizing Storage Losses

The following practices help ensure good keeping quality and minimal physiological breakdown of fresh cranberries. Excessive use of nitrogen fertilizer, which generally leads to soft fruit, should be avoided. Handling should be minimized to reduce bruising. Fruit should be stored as dry as possible. Fruit should not be screened. A temperature of 2–4°C should be maintained. Relative humidity should be kept at about 90%. Adequate air circulation around fruit and storage containers should be ensured. Fruit should be packed and repacked in a cool room.

Selected References

Anderson, R. E., Hardenburg, R. E., and Vaught, H. C. 1963. Controlled-atmosphere storage studies with cranberries. Proc. Am. Soc. Hortic. Sci. 83:416-422.

Bristow, P. R., and Windom, G. E. 1985. The impact of machine-harvesting and fungicides on rot and physiological breakdown occurring in cold-stored cranberries. (Abstr.) Phytopathology 75:1285.

Ceponis, M. J., and Stretch, A. W. 1983. Berry color, water-immersion time, rot and physiological breakdown of cold-stored cranberry fruits. HortScience 18:484-485.

Davis, D. C., and Shawa, A. Y. 1983. Reducing injury during mechanical harvesting of cranberries. J. Am. Soc. Hortic. Sci. 108:444-447.

Doughty, C. C., Patterson, M. E., and Shawa, A. Y. 1968. Storage longevity of 'McFarlin' cranberry as influenced by certain growth retardants and stage of maturity. Proc. Am. Soc. Hortic. Sci. 91:192-204.

Forney, C. F, 2004. Postharvest handling and storage of fresh cranberries. HortTechnology 13:267-272.

Graham, S. O., Patterson, M. E., and Allen, B. 1967. Bruising as a predisposing factor in the decay of stored cranberries. Phytopathology 57:497-501.

Gunes, G., Liu, R. H., and Watkins, C. B. 2002. Controlled-atmosphere effects on postharvest quality and antioxidant activity of cranberry fruits. J. Agric. Food Chem. 50:5932-5938.

Hruschka, H. W. 1970. Physiological breakdown in cranberries—Inhibition by intermittent warming during cold storage. Plant Dis. Rep. 54:219-222.

Patterson, M. E., Doughty, C. C., Graham, S. O., and Allen, B. 1967. Effect of bruising on post-harvest softening, color and changes in detection of polygalacturonase enzyme in cranberries. Proc. Am. Soc. Hortic. Sci. 90:498-505.

Stretch, A. W., and Ceponis, M. J. 1986. Fungal and physiological breakdown in six cranberry cultivars following water harvesting and cold storage. HortScience 21:265-267.

Wright, R. C., Demaree, J. B., and Wilcox, M. S. 1936. Some effects of different storage temperatures on keeping of cranberries. Proc. Am. Soc. Hortic. Sci. 34:397-401.

(Prepared by P. R. Bristow; Revised by K. D. Patten)

Salt Injury

Cranberry vines appear to be fairly tolerant of occasional exposure to salt. Greenhouse studies confirm that if salt exposure is rare, flushing of the soil via natural rainfall and production flooding cycles prevent adverse effects. However, salt injury has been observed under specific conditions.

Exposure to Salt-Laden Winds

Damage to cranberry vines has been reported following exposure to salt-laden winds during hurricanes. Symptoms are bronzing of the foliage.

Incursion of Highly Saline Water

Tidal surges. Occasionally, a bed is briefly flooded by hurricane tides. Vines may be injured or killed in low areas where this salt water is slow to drain (Fig. 298). High salt concentrations have been found in the soil to a depth of 0.9–1.2 m where leaching of the soil did not occur. Salt injury may be corrected or minimized by flushing with clean irrigation water or by applying a mixture of one part calcium sulfate to three parts potassium magnesium sulfate (500 lb of the mixture per acre).

The severity of injury from salt water depends on several variables. Vines that are disturbed by harvesting before submergence are more seriously injured than vines from which fruit had not yet been harvested. Areas with poor drainage or where soils have a low organic matter content and areas of immature vines are more prone to salt injury.

Road runoff. The most common cause of acute salt intrusion is runoff from highways treated with salt for melting ice and snow. In places where road runoff continually drains into marginal ditches, vines may die along the edges of the ditch or, in extreme cases, at a distance of 5 m or more out into the bed. Another potential adverse impact of road salt is runoff into the cranberry irrigation source (see Chronic Exposure to Moderate Salt Loads in Irrigation Water).

Chronic Exposure to Moderate Salt Loads in Irrigation Water

In greenhouse studies with sodium chloride, potassium chloride, and sodium sulfate, adverse effects to cranberry were seen after long-term exposure to 100 ppm of chloride (approximately 165 ppm of sodium chloride) in irrigation water. Adverse effects included elevated sodium and chloride levels in the plants and soil, visible symptoms on the leaves (Fig. 299), and vegetative growth stimulation. Yield was depressed in some chloride treatments but the effect was inconsistent. When chloride exposure increased to 250 ppm, symptoms increased and flowering was suppressed, indicating that yield suppression would be likely.

Fig. 298. Cranberry vines with necrosis caused by salt injury in a low spot of a bed that was flooded by salt water during a hurricane. (Courtesy F. L. Caruso; © APS)

Fig. 299. Individual cranberry upright with symptoms of stress caused by high levels of sodium chloride. (Courtesy C. J. Demoranville; © APS)

Table 1. Proposed Levels of Sodium and Chloride for Cranberry Plants and Soil

	Proposed Standard Levels of Sodium	Proposed Acceptable Levels of Chloride
Cranberry soil	Up to 15 ppm	7 ppm
Cranberry shoot tissue	~0.06% dry weight	~0.05% dry weight
Cranberry root tissue	~0.03% dry weight	~0.05% dry weight

In soil sorption/desorption studies, treatment with salts at concentrations of chloride greater than 250 ppm resulted in significant, only partially reversible, changes to cranberry soil chemistry. At 250 ppm, recovery occurred after desorption.

Based on these research results, there is good indication that a cranberry water supply containing chloride at 250 ppm or greater would be cause for concern and would indicate the potential need for remedial action. Symptoms, soil chemistry changes, and possible growth stimulation at 100–125 ppm of chloride indicate that a cranberry water supply containing 100 ppm of chloride or greater for extended periods, e.g., two consecutive months during the summer when the need for irrigation is greatest, might also be cause for concern.

Based on an examination of control samples in the greenhouse and laboratory and of field samples from uncontaminated sites, the levels for sodium and chloride proposed in Table 1 are expected to be acceptable for cranberry plants and soil. Levels consistently greater than these would be cause for concern. Damage to cranberry plants, consisting of visible symptoms on the leaves, has been previously reported at a site with 15 ppm of chloride in the soil.

Selected Reference

Chandler, F. B., and DeMoranville, I. E. 1959. The harmful effect of salt on cranberry bogs. Cranberries 24(8):6-9.

(Prepared by C. J. DeMoranville)

Sunscald and Heat Injury

Sunscald in cranberry, sometimes called scalding, refers to fruit injury that results from solar radiation exposure. Incidence of sunscald is greatest on large berries at the top of the canopy when ambient temperatures are above 32°C. In the cooler growing conditions in the Pacific Northwest, cranberries may be vulnerable to sunscald at temperatures as low as 26°C when the relative humidity is low and solar radiation levels are high. Although this disorder is most commonly associated with direct exposure to sunlight, some heat injury to fruit can also happen with lightly overcast skies under conditions of high radiation reflectivity. Sunscald is common when vine coverage is thin, especially in areas where soil surface temperatures are high because of sunlight exposure. Scald has been observed in all of the principal areas where cranberries are grown and may cause crop losses of 25% or more. Berries can also be scalded when puddling occurs prior to a hot day and berries either sit directly in the superheated water or are situated directly above rapidly evaporating water during hot days.

Symptoms

Sunscald injury is evident as circular or oval, soft, light-colored lesions on the shoulder or side of the fruit directly exposed to solar radiation (Fig. 300). Secondary pathogens rarely invade these lesions, except on smaller berries injured when they are most susceptible in late summer. However, fungi that have infected the fruit and remain as latent pathogens can rapidly grow through the injured tissue, resulting in fruit rot. Sunscald on fruit happens most frequently in mid- to late August, during the final stages of berry swelling and early anthocyanin color development.

Heat injury to cranberry plants can happen throughout the period of active upright growth and flowering (Fig. 301). Blossom blast, the failure of blossoms to set fruit, may develop at temperatures above 30°C accompanied by low humidity and high winds. Under similar conditions, desiccation and heat damage to new shoots and leaves during active growth before bloom can result in upright shoot dieback and can also suppress growth, reduce flowering, and lower productivity.

Management

Sunscald and heat injury can be substantially reduced by the use of overhead sprinkler irrigation systems for evaporative cooling. Judicious irrigation during periods of high temperatures can greatly reduce the damage to plants or fruit while minimizing problems of soil moisture saturation. When humidity is low, evaporative cooling can reduce plant canopy temperatures below ambient air temperatures by 3–5°C.

Fig. 300. 'Searles' cranberries with sunscald injury. (Courtesy S. N. Jeffers; © APS)

Fig. 301. A mature 'Stevens' cranberry planting with heat stress injury. (Courtesy S. N. Jeffers; © APS)

Selected References

Croft, P. J. 1992. Meteorological investigations and prediction of scald. Cranberries 56:6-7, 18.

Croft, P. J. 1995. Field conditions associated with cranberry scald. HortScience 30:627.

Cross, C. E. 1983. Weather as it relates to cranberry production and quality. Pages 52-58 in: Modern Cranberry Production. Mass. Coop. Ext. Serv. Bull. SP-126.

Eck, P. 1990. The American Cranberry. Rutgers University Press, New Brunswick, NJ.

Shawa, A., Shanks, C. H., Bristow, P. R., Shearer, M. N., and Poole, A. P. 1984. Cranberry Production in the Pacific Northwest. Pac. Northwest Ext. Publ. 247.

(Prepared by E. J. Stang; Revised by F. L. Caruso)

Winter Injury

Cranberry vines may be injured or killed by severe winter weather. The injury is classified as a physiological drought or desiccation in which moisture lost from the vines because of wind and evaporation cannot be replaced because of freezing in the root zone. The common term used to describe this injury is "winterkill." Such injury can happen within 3 days if the root zone is frozen to a depth of 10 cm, the air temperature is below freezing, and there are winds of 10 mph or greater. It is thought that 2–3 days with temperatures below –7°C are enough to freeze the soil. Injury may happen anytime from early December until mid-March.

The symptoms of winterkill injury are leaf discoloration (bronze-orange hue) and eventual leaf drop (Fig. 302). In severe cases, the terminal bud is destroyed and the plant may die back to ground level. Plants recover with regrowth from uninjured tissues but no fruit are produced during that year and crop suppression may continue into the following year depending on the severity of the injury.

Injury is prevented by protecting vines with a winter flood that should be in place when conditions favorable for winter injury exist. Beds that have not been harvested and new plantings (first year) are less susceptible to winter injury but should still be protected in severe conditions.

(Prepared by C. J. DeMoranville)

Fig. 302. Cranberry vines with necrosis of the leaves and stems caused by winter injury. Note the rusty color on the vines at the upper left and center. The surrounding vines were not affected. (Courtesy University of Massachusetts Cranberry Experiment Station)

Part III. Blueberry Certification

The nursery industry is primarily concerned with two issues: trueness-to-type and pathogens. Currently available genotyping technologies allow for the rapid identification of plant material to cultivar. Trueness-to-type is more a logistics endeavor within the nursery system than a matter of finding expertise for cultivar identification based on phenotypical characteristics, as was the case in the past. For this reason, trueness-to-type will not be discussed here; the focus instead will be on pathogen control within the system.

It is well documented that diseased plants do not establish well in the field, produce less and lower-quality fruit, and have a shorter lifespan, affecting the bottom line of the producers. All pathogens of blueberry mentioned in this compendium may appear on nursery stock. Some, such as *Botrytis* and *Pseudomonas* spp., are so widespread that they can become established in fields from natural sources. Others, such as phytoplasmas and viruses, often become established in a field by means of infected planting stock. Clean stock and certification programs are designed primarily to eliminate this second group of pathogens.

When infected by systemic viruses, bacteria, viroids, or phytoplasmas, plants are never cured even though infections may be symptomless. Over time, vectors may feed on infected material and transmit those pathogens to other plants.

The growing environment or the pathogen strain–cultivar combination may lead to latent infections, making identification of infected plants based on visual symptoms virtually impossible. It is therefore important to develop guidelines that will safeguard nursery stock from systemic viruses, bacteria, viroids, or phytoplasmas. Those guidelines come in the form of publicly supervised certification programs, implemented to ensure the health status of plants and ensure that buyers obtain propagation material of the highest possible quality.

Certification programs for the production of systemic pathogen-tested blueberry stock require the following two major elements: reliable sources of clean material and propagation of clean stock.

Reliable Sources of Propagation Material

Before the establishment of any certification program, it is imperative to develop the guidelines that ensure the quality of the propagation material. On the pathogen side, this starts with the development of a list of target pathogens or agents that should be eliminated from advanced selections or other sources of material before they enter into a clean plant system. In general terms, this includes phytoplasmas, systemic bacteria, and all viruses and viroids other than those that do not cause disease individually or in combination with other viruses (cryptic or amalgaviruses). A list of all targeted pathogens of blueberry in the draft United States National Certification Program (available at http://ncpnberries.org) as it stood in 2015 is found in Table 2.

One of the major components of those guidelines is the introduction of the "Generation" or "G" designation for plants as they move through the propagation pipelines. The "G" designation eliminates confusion between parties involved in the

TABLE 2. Targeted Pathogens of the Drafted United States National Certification Program

	Regional Occurrence and Risk in the United States[a]				
Organism	**Upper Midwest**	**South Central**	**Northeast**	**Southeast**	**California**
Viruses					
Blueberry latent spherical virus (BLSV)	5	5	5	5	5
Blueberry leaf mottle virus (BLMV)	1	4	3	4	5
Blueberry mosaic associated virus (BlMaV)	2	2	2	4	1
Blueberry necrotic ring blotch virus (BNRBV)	5	5	5	1	5
Blueberry red ringspot virus (BRRSV)	2	1	1	1	5
Blueberry scorch virus (BlScV)	2	4	1	4	4
Blueberry shock virus (BlShV)	3	4	1	4	2
Blueberry shoestring virus (BSSV)	1	4	2	3	4
Blueberry virus A (BVA)	1	5	5	5	5
Cherry leaf roll virus (CLRV)	5	5	5	5	5
Peach rosette mosaic virus (PRMV)	3	4	3	4	5
Strawberry latent ringspot virus (SLRSV)	5	5	5	5	5
Tobacco ringspot virus (TRSV)	1	1	1	1	5
Tomato ringspot virus (ToRSV)	1	3	1	1	5
Phytoplasma					
Blueberry stunt phytoplasma	2	1	1	1	5
Bacteria					
Rhizobium radiobacter	1	1	1	1	5
Xylella fastidiosa	4	4	4	1	5

[a] Risk codes: 1 = moves in the environment, 2 = present at low levels, 3 = reported to occur, 4 = not known to occur, and 5 = no data available. These ratings refer to the presence of the pathogen in blueberry, not in other crops.

certification process and streamlines the process. Figure 303 illustrates how the "G" designation translates to the several terms used by the nursery industry in various crops and countries to depict the same propagation material.

Genotypes identified as new cultivars are designated as provisional G1. They are moved from the field to a clean plant center where they are rigorously tested for all targeted pathogens (Fig. 304). This may include large-scale sequencing followed by bioinformatic analysis for the detection and/or discovery of new viruses and viroids, polymerase chain reaction (PCR) and its variants, and/or immunological techniques such as enzyme-linked immunosorbent assay or dot-blot. Table 3 summarizes the laboratory methods routinely used to detect and identify those agents. Readers may consult the discussions of these pathogens in this compendium for details regarding their detection, method of spread, and economic importance. If plants are found infected by one or more of the targeted pathogens, they go through thermotherapy, cryotherapy, and/or chemotherapy. Details of therapy procedures and media used for blueberry tissue culture can be found in the Selected References below. Those procedures may have to be repeated multiple times until targeted pathogens are eliminated from the material. Once plants are verified free of all target pathogens and are DNA fingerprinted to verify their trueness-to-type, they are designated as Generation 1 or G1.

G1 plants are housed in public or private facilities that have the infrastructure (e.g., screenhouses, screened greenhouses) to keep the plants free of the targeted or any other systemic pathogens. G1 material is the starting point for any future propagation of plants to be use in certification programs.

Propagation of Clean Stock from G1 plants

Systemic pathogens of blueberry naturally infect plants in the field by means of specific nematode and insect vectors and by pollen spread, depending on the pathogen in question. To shield plants from natural infections, propagation material should be maintained using stringent guidelines to restrict virus movement. Fields should be several hundred meters away from blueberry production fields or wild *Vaccinium* spp.—although this may not always be practical. Field soils with a history of plant-parasitic nematodes should be avoided or fumigated before planting to reduce the population of these pathogens. Vector insects should be controlled in nursery fields (for more details, visit http://ncpnberries.org).

Propagation may be done by traditional means with cuttings or by tissue culture. Once a propagule is taken from a G1 plant (excluding propagation of tissue culture material), it is designated as G2 and may be maintained in tissue culture or a screenhouse/screened greenhouse that has been inspected and approved for this purpose. G2 plants serve as stock material maintained by the nursery under conditions outlined in the certification standard. After every propagation cycle, the generation designation changes: G3 when propagated from G2 plants or G4 when propagated from G3 plants. Material that is designated as G4 is only available for sale and cannot serve as propagation material.

One of the major changes proposed in the draft certification guidelines is the incorporation of testing to minimize the possibility of nursery-driven disease epidemics. The proposed scheme requires testing of G2 material every 3 years for all targeted pathogens. For G3 and G4, material is to be tested for the

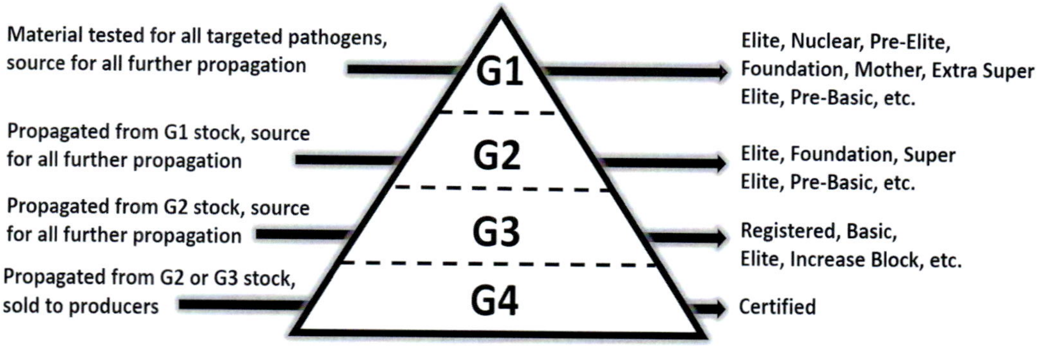

Fig. 303. Diagram illustrating the G terminology for the stages of blueberry plant propagation from clean plant centers (G1) through the propagation steps in nurseries enrolled in certification programs. (Adapted from Gergerich et al., 2015; © APS)

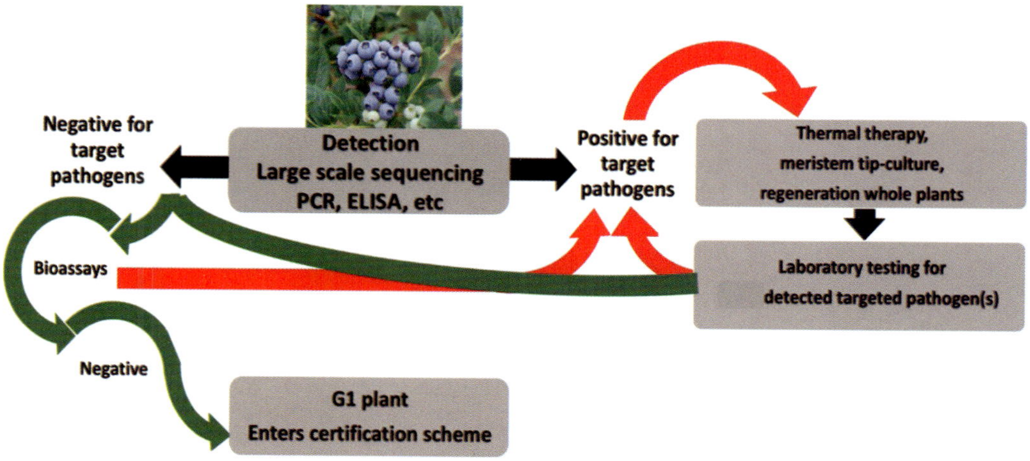

Fig. 304. The journey of provisional G1 blueberry material, including laboratory testing, bioassays, therapy, and meristem tip culture, before it enters a certification scheme in the United States. (Adapted from Gergerich et al., 2015; © APS)

TABLE 3. Available Laboratory Tests for All Targeted Pathogens of the Drafted United States National Certification Program

	Taxonomy/Genus	Laboratory Detection[a]
Viruses		
Blueberry shock virus (BlShV)	*Ilarvirus*	ELISA, RT-PCR[b]
Blueberry latent spherical virus (BLSV)	*Nepovirus*	RT-PCR
Blueberry leaf mottle virus (BLMV)	*Nepovirus*	ELISA, RT-PCR
Blueberry mosaic associated virus (BlMaV)	*Ophiovirus*	RT-PCR
Blueberry necrotic ring blotch virus (BNRBV)	*Cilevirus*	RT-PCR
Blueberry red ringspot virus (BRRSV)	*Soymovirus*	ELISA, PCR
Blueberry scorch virus (BlScV)	*Carlavirus*	ELISA, RT-PCR
Blueberry shoestring virus (BSSV)	*Sobemovirus*	ELISA, RT-PCR
Blueberry virus A (BVA)	*Closterovirus*	RT-PCR
Cherry leaf roll virus (CLRV)	*Nepovirus*	ELISA, RT-PCR
Peach rosette mosaic virus (PRMV)	*Nepovirus*	ELISA, RT-PCR
Strawberry latent ringspot virus (SLRSV)	Unassigned genus in the family *Secoviridae*	ELISA, RT-PCR
Tobacco ringspot virus (TRSV)	*Nepovirus*	ELISA, RT-PCR
Tobacco streak virus (TSV)	*Ilarvirus*	ELISA, RT-PCR
Tomato ringspot virus (ToRSV)	*Nepovirus*	ELISA, RT-PCR
Phytoplasma		
Blueberry stunt phytoplasma	16SrI (Aster yellows group)	PCR
Bacteria		
Rhizobium radiobacter		ELISA, PCR
Xylella fastidiosa		ELISA, PCR

[a] ELISA = enzyme-linked immunosorbent assay; and RT-PCR = reverse transcription-polymerase chain reaction.
[b] Also quantitative PCR.

most prevalent (canary) viruses in the area where a nursery is located. Sampling is to be performed to allow for identification of a 1% infection rate at 95% confidence for G3 plants and of a 5% infection rate at 95% confidence for G4 plants.

More certification programs may be established in the future to accommodate the soaring demand for the best quality propagation material that will result in longer-lived blueberry fields and increased profitability.

Selected References

Chandler, M. J., and Draper, A. D. 1986. Effect of zeatin and 2iP on shoot proliferation of three highbush blueberry clones in vitro. HortScience 21:1065-1066.

Converse, R. H., ed. 1987. Virus Diseases of Small Fruits. U. S. Dep. Agric. Agric. Handb. 631. U. S. Government Printing Office, Washington, DC.

Converse, R. H., and George, R. A. 1987. Elimination of mycoplasmalike organisms in Cabot highbush blueberry with high-carbon dioxide thermotherapy. Plant Dis. 71:36-38.

Eck, P. 1988. Blueberry Science. Rutgers University Press, New Brunswick, NJ.

Gergerich, R. C., Welliver, R. A., Gettys, S., Osterbauer, N. K., Kamenidou, S., Martin, R. R., Golino, D. A., Eastwell, K., Fuchs, M., Vidalakis, G., and Tzanetakis, I. E. 2015. Safeguarding fruit crops in the age of agricultural globalization. Plant Dis. 99:176-187.

Ho, T., and Tzanetakis, I. E. 2014. Developing a virus detection and discovery pipeline using next generation sequencing. Virology 471-473:54-60.

Martin, R. R., and Tzanetakis I. E. 2015. Control of virus diseases of berry crops. Adv. Virus Res. 91:44-81.

Martin, R. R., Polashock, J. J., and Tzanetakis, I. E. 2012. New and emerging viruses of blueberry and cranberry. Viruses 4:2831-2852.

Schaper, U., and Converse, R. H. 1985. Detection of mycoplasmalike organisms in infected blueberry cultivars by the DAPI technique. Plant Dis. 69:193-196.

(Prepared by I. E. Tzanetakis and R. R. Martin)

Part IV. Pests

Blueberries (Highbush and Rabbiteye)

Aphids

There are at least six species of aphids (Hemiptera: Aphididae) that attack blueberry. Some species simply devitalize the plant, while others are disease vectors. The four principal species include the blueberry aphid, *Illinoia pepperi* (present in Michigan, United States); *Illinoia azaleae* (present in New Jersey, United States) (Fig. 305); the (western) blueberry aphid, *Ericaphis fimbriata* (syn. *Fimbriaphis fimbriata*); and the green peach aphid, *Myzus persicae*. The significance of aphid pests in a blueberry field depends on the species, the population pressure, and the presence of vectored diseases in or near the field.

Life History and Description

Overwintering eggs can be found on the twigs and protected areas of recent-growth wood. Eggs hatch in the spring. Immatures feed on tender new growth, usually on the underside of leaves at the top or bottom of blueberry bushes. The first and subsequent generations are all apterous (wingless) females, with stem mothers giving live birth to more females. As populations build, some alate or winged forms develop and spread out to more bushes to start new colonies. There may be four or more generations per growing season. Most alates are present in the autumn when males are produced, and overwintering eggs result from mating. Variations in the life cycle can occur. The green peach aphid has many alternate hosts and can spend much of the season on potatoes, peppers, or weed hosts before mating and laying eggs on woody tissue. Aphids are soft-bodied insects and adults are, on average, about 2 mm long and light to dark green. They have piercing-sucking mouthparts and two siphunculi (cornicles) that protrude to the rear from the sixth abdominal segment. A caudal segment protrudes from the rear of the abdomen and can be used to help identify species, as can the length and morphology of the cornicles and antennae. Nymphs resemble adults but are smaller and wingless.

Fig. 305. Aphids (*Illinoia azaleae*) on blueberry. (Courtesy D. Polk; © APS)

Damage

All aphid species suck sap from tender growth and new shoots, especially developing terminal foliage. Under heavy populations, a sooty mold can develop on the honeydew secreted by the aphids. This is usually of only minor importance, since growers seldom allow populations to build up to densities that result in sooty mold growth. Of more importance is the fact that many aphids function as disease vectors. In blueberry, at least three species can transmit *Blueberry scorch virus* (BlScV) and its several strains. *Ericaphis fimbriata* is the principal vector in the Northwest (United States), while *Illinoia azaleae* is the principal vector in New Jersey (United States) and probably also a vector of *Blueberry shoestring virus* (BSSV). *Myzus persicae* can also transmit BlScV.

Management

Since disease transmission is usually the main concern in commercial production areas, only very low aphid populations should be tolerated, especially if BlScV or BSSV are known problems. Aphids may be present while bushes are in bloom, but populations do not start to build up until after bloom. Monitoring should begin as soon as bees are removed after bloom and continue through at least the first picking. Sampling should be biased in new terminal growth and data recorded as the percent of terminals infested with aphid colonies. Where disease transmission is an issue, a colony should be defined as a minimum of one to two aphids, either nymphs or adults. Treatment is justified if greater than 10% of terminals are infested with live aphids. While apterous forms cannot fly from bush to bush and spread disease, it is still important to manage them. Harvesting, especially with a machine, can spread even nonwinged aphids down a row, magnifying the effects if the aphids previously fed on a BlScV-infected plant. Roguing of diseased and damaged bushes should be done at the end of the season. In isolated fields, where BlScV is not an issue, higher populations can be tolerated, and lady beetles, lacewings, syrphid flies, or other biological controls may be allowed to suppress the population. However, high populations of some predators may pupate on the berries and result as insect contaminants at harvest.

Selected References

Antonelli, A., Elsner, E., and Shank, C. 1992. Arthropod management. Pages 55-75 in: Highbush Blueberry Production Guide. M. P. Pritts and J. F. Hancock, eds. NRAES-55. Northeast Regional Agricultural Engineering Service, Ithaca, NY.

Demchak, K., coordinator. 2006. Mid-Atlantic Berry Guide for Commercial Growers AGRS-97. Pennsylvania State University.
Martin, R. R., Bristow, P. R., and Wegener, L. A. 2006. Scorch and shock: Emerging virus diseases of highbush blueberry and other *Vaccinium* species. Acta Hortic. 715:464-467.

(Prepared by D. Polk and C. Rodriguez-Saona)

Blueberry Blossom Weevil (Cranberry Weevil)

The blueberry blossom weevil (also known as the cranberry weevil), *Anthonomus musculus* (Coleoptera: Curculionidae), is native to North America and ranges from Ontario (Canada) and New England south to Florida (United States) and west to the Rocky Mountains. The blueberry blossom weevil has a wide host range of cultivated and native plants. This insect is a major pest of highbush blueberry in New Jersey and a major pest of cranberry in Massachusetts (United States). In other cranberry-growing regions, except for the west, cranberry weevil is present but rarely is present at treatable levels. In the wild, this beetle is commonly observed on black huckleberry and the flowers of chokeberry.

Life History and Description

The life history of this weevil is not fully understood. A single generation was reported in 1926 and in 1948. However, in 1992, there was said to be evidence of at least two complete generations in New Jersey and Massachusetts and two strategies for overwintering within the population. The first strategy may be adult cold hardiness, and a second strategy may be to spend the winter in pupal diapause (quiescent state). Although there are several unanswered questions about its life history, it is clear that a portion of the blueberry blossom weevil population overwinters in the adult stage under debris and fallen leaves in wooded areas. In New Jersey, adults leave their winter sites and congregate on blueberries when blueberry buds begin to swell in late February or early March. In Massachusetts, adults emerge from overwintering sites in late April and move from wild and cultivated blueberry plants to cranberry plants in late May through early June. Weevils hide in the trash on cold, windy days, coming out only on sunny, warm days. Adults (Fig. 306) are about 1.5 mm long, brown with a few white markings on the wings, and seldom fly; but when they do, they can make a single flight up to 75 feet. After mating, females insert a single egg (0.5 mm long) inside a bud at the base of the anthers among the stamens. Each female may lay 50 or more eggs in her lifetime, which can last 13 months. Egg laying may continue for 3 weeks. The eggs hatch in 3–9 days. Larvae are white, C-shaped, legless, and about 3 mm long. Infested blossoms soon fall to the ground. The larvae develop and pupate in the blossom buds. They mature in 10–14 days. When the beetle emerges (in late May in New Jersey blueberry and in late June in Massachusetts cranberry), it eats its way out of the bud near the calyx lobes. The entire cycle (egg to adult) takes about 2 months.

Fig. 306. Blueberry blossom weevil adult (*Anthonomus musculus*) on blueberry. (Courtesy D. Polk; © APS)

Damage

Injury results from both adult and larval feeding. Adults puncture both the leaf and flower buds with their chewing mouthparts. Heavily damaged flower buds fail to open and injured leaf buds produce an abnormal cluster of dwarfed foliage. Females completely sever the pedicel with their mouthparts after laying eggs in large, fully developed buds. But on smaller buds, females may not sever the bud or may only partially cut the pedicel, creating a point of weakness. Larvae feed within the buds (Fig. 307), consuming the flower parts (the pistil and the stamens) and causing the infested blossoms to turn from pink to brownish orange, remain unopened, and fall from the plant. A study that simulated weevil damage showed that cranberry may compensate for low to moderate levels of flower bud destruction by the development of other flowers on the peduncle. Later in the growing season, before they go into hibernation, the feeding of adult weevils causes numerous small, irregular holes in the leaves. Adults may also feed on small berries.

Management

Monitoring should start at bud swell and continue through bloom. In cranberry, adult weevils are monitored with a sweep net. The action threshold has been an average of 4.5 weevils per set of 25 sweeps. In blueberry, adults are collected by shaking the foliage onto a 0.9-m^2 white beating tray. Adults drop down onto the cloth. Another method is to count the number of blueberry flower clusters with feeding punctures. Treatment is justified if at least five adults are found per bush or if at least 20% of the terminals show feeding signs. Plants should be sampled for weevils on calm, sunny days and monitoring should be intensified near wooded areas. New studies show that lures baited

Fig. 307. Cranberry weevil larva (*Anthonomus musculus*) developing inside a cranberry flower bud. (Courtesy A. L. Averill; © APS)

with male pheromone with the addition of host plant volatiles may be utilized in a trapping system for adults.

If populations are small and localized, spraying only border areas along the woods should be considered. Eliminating overwintering sites in blueberry fields with good pruning and the elimination of weeds also helps control this pest.

Selected References

Doehlert, C. A., and Tomlinson, W. E., Jr. 1947. Blossom weevil on cultivated blueberries. N. J. Agric. Exp. Stn. Bull. 504:1-8.

Franklin, H. J. 1948. Cranberry insects in Massachusetts. Mass. Agric. Exp. Stn. Bull. 445:56-60.

Lacroix, D. S. 1926. The life history and control of the cranberry weevil, *Anthonomus musculus* Say (Coleoptera: Curculionidae). J. Econ. Entomol. 19:819-829.

Long, B. B., and Averill, A. L. 2003. Compensatory response of cranberry to simulated damage by cranberry weevil (*Anthonomus musculus* Say) (Coleoptera: Curculionidae). J. Econ. Entomol. 96:407-412.

Mechaber, W. L. 1992. Ecology of *Anthonomus musculus*: Hostplant finding and oviposition by cranberry weevil. Ph.D. dissertation. Tufts University, Medford, MA.

Mechaber, W. L., and Chew, F. S. 1991. Rewriting the natural history of cranberry weevil. Cranberries Mag. 55:5-8.

Szendrei, Z., Malo, E., Stelinski, L., and Rodriguez-Saona, C. 2009. Response of cranberry weevil (Coleoptera: Curculionidae) to host plant volatiles. Environ. Entomol. 38:861-869.

Szendrei, Z., Averill, A., Alborn, H. T., and Rodriguez-Saona, C. 2011. Identification and field evaluation of attractants for the cranberry weevil, *Anthonomus musculus* Say. J. Chem. Ecol. 37:387-397.

(Prepared by C. Rodriguez-Saona and D. Polk)

Blueberry Bud Mite

The blueberry bud mite, *Acalitus vaccinii* (Acari: Eriophyidae), is an important pest infesting both cultivated and wild plants in the genus *Vaccinium*. Many species of lowbush and highbush blueberry plantings, mostly in mild winter climates of eastern North America, are hosts to the pest.

Life History and Description

The blueberry bud mite belongs to a group of microscopic mites known as eriophyid mites. The white, cigar-shaped mites are extremely small (about 200 μm) with two pairs of legs bunched toward the wider head end. The two body regions are the mouthparts (gnathosoma) and the rest of the body (idiosoma), which resembles the abdomen of an insect. With their styletlike chelicerae, the mites feed on the epidermal region of the plant tissue, transferring a substance or toxin that causes the deformation of plant growth. Like other eriophyids, blueberry bud mites live together in large clusters and reproduce rapidly within folds of plant tissue. At 19°C, 15 days are required for mites to develop from oviposition to the adult stage. The female is known to lay up to 200 eggs. In Florida and other southeastern states (United States), the population has been observed to peak as early as February and to experience a decline during the hot summer months of June through September. All four life stages (egg, protonymph, deutonymph, and adult) are present year-round in the preferred large and succulent scales of flower buds and blossoms.

Damage

The blueberry bud mites remain almost continuously in the protective confines of the bud. Here, they feed on the developing leaf and floral parts, causing the tissue to become roughened and warty in appearance. Persistent feeding by large numbers of mites results in the reddening and swelling of the base of the bud scales. Buds appear rosetted and may desiccate and fail to open (Fig. 308). Affected berries appear blistered and malformed (Fig. 309). Summer generations cause retarded leaf growth, reducing the vegetative growth that impacts the following year's crop. An average of 66% of bud infestation was recorded for highbush blueberry in Georgia (United States). The symptoms are often confused with winter injury and are the initial indication of a problem since the mites are not visible without magnification.

Management

Infestations of the blueberry bud mite are challenging to manage. It is difficult to detect infected plants, the number of registered chemical control options is limited, and it is difficult to reach the mites within the buds. Although some blueberry cultivars are less likely to be infested, there are no fully resistant cultivars identified. Recommendations for the control of blueberry bud mite include pruning older canes and removing infected branches. Abamectin and light horticultural oils are effective. The fungal acarine parasite, *Hirsutella thompsonii*, has also been associated with population decline of the blueberry bud mite. Therefore, management programs should consider

Fig. 308. Injury by blueberry bud mites (*Acalitus vaccinii*) results in rosetted and desiccated blueberry buds and swelling of the base of bud scales. (Courtesy O. Liburd; © APS)

Fig. 309. Blistering caused by blueberry bud mites (*Acalitus vaccinii*) on blueberry. (Courtesy O. Liburd; © APS)

H. thompsonii when selecting fungicides for foliar disease control in blueberry.

Selected References

Isaacs, R., Morrone, V., and Gajek, D. 2004. Potential acaricides for management of blueberry bud mite in Michigan blueberries. Hortic. Technol. 14(2):18-21.

Sorensen, K. A. 1994. Blueberry bud mite and its control. N. C. State Univ. Insect Note #B-4.

Weibelzahl, E., and Liburd, O. E.. 2009. Epizootic of *Acalitus vaccinii* (Acari: Eriophyidea) caused by *Hirsutella thompsonii* on southern highbush blueberry in north-central Florida. Fla. Entomol. 92:601-607.

Weibelzahl, E., and Liburd, O. E. 2010. Blueberry bud mite, *Acalitus vaccinii* (Keifer) on southern highbush blueberry in Florida. EDIS-ENY858/IN844. UF/IFAS Extension, University of Florida, Gainesville, FL.

(Prepared by E. Weibelzahl and O. E. Liburd)

Blueberry Gall Midge (Cranberry Tipworm)

Larvae of *Dasineura oxycoccana* (Diptera: Cecidomyiidae) feed on and kill meristematic tissue of developing buds in blueberry and cranberry, where the pest is commonly named blueberry gall midge and cranberry tipworm, respectively. Recent research carried out in British Columbia (Canada) strongly suggests that populations of *D. oxycoccana* on cranberry and highbush blueberry (*V. corymbosum*) are reproductively isolated and may represent two cryptic species. Populations originating from the two hosts produce and respond to different sex pheromones, do not mate, and show genetic divergence.

The blueberry gall midge is an important insect pest of rabbiteye blueberry in the southeastern United States. Larvae injure both floral and vegetative buds, and a high infestation can limit the plant's ability to produce sufficient fruit during subsequent seasons. Larval feeding can reduce floral bud development and fruit loss by 80% if left unmanaged. A survey of northern highbush blueberry in Michigan (United States) showed the gall midge to be widespread, but damage was only found in vegetative buds. The midge has been reported in lowbush blueberry in Nova Scotia (Canada) and highbush blueberry in the Northwest, but its pest status is not established.

Cranberry tipworm appears in all cranberry-growing regions and has the greatest impact on yield in northern areas, such as Canada and Maine and northern Wisconsin (United States), where the interval for upright recovery is shortened. The effects of tipworm injury on cranberry production in Massachusetts (United States) from one year to the next are not clear, since significant reduction in tipworm injury at the plot level (using insecticide) did not increase flower and fruit production during the following growing season.

Life History and Description

Dasineura oxycoccana is a tiny, short-lived fly, approximately 2 mm long (Fig. 310). It has long, slender legs, delicate hairy wings, and globular antennae. Females are slightly larger than males, with long ovipositors. Climatic conditions and latitude influence the number of generations within a particular region. In the south, there are several generations per year in blueberry. In cranberry, adults emerge in May just as the leaves of the bud begin to elongate. There are at least three to five generations, with populations persisting until late summer. Eggs are laid on the inner leaves of the developing bud. The eggs are transparent when they are first laid, becoming darker just prior to hatching. Under ideal conditions, eggs hatch within 2–3 days. The larva has sicklelike mouth hooks used to scrape the inner surface of succulent leaves. It develops through three instars that are characterized by changes in color: transparent (first instar), white (second instar), and orange (third instar) (Fig. 311). The pupal cocoon is constructed of closely spun white silk, often within the cupped leaf. Larvae of the final generation drop to the ground to pupate and overwinter in the upper layers (1–3 cm) of the soil.

Fig. 310. Adult cranberry tipworm (*Dasineura oxycoccana*) on cranberry. (Courtesy J.-F. Landry of the Department of Agriculture and Agri-Food, Government of Canada; © Her Majesty the Queen in Right of Canada, as represented by the Minister of Agriculture and Agri-Food Canada)

Fig. 311. Third instars of cranberry tipworm (*Dasineura oxycoccana*) on cranberry. (Courtesy D. L. Mahr; © APS)

Damage

Injury happens when larvae feed on developing buds. In blueberry, symptoms of infestation include curled, cupped leaves (younger leaves). The larvae feed on the meristematic cells within the flower and vegetative buds, resulting in necrotic tissues. Tip burn on older leaves is often misdiagnosed as frost damage. Rabbiteye floral and leaf buds are very susceptible to infestations. Heavy injury can limit the flowering potential for subsequent seasons. In cranberry, larval feeding induces cupping of the developing bud (Fig. 312). Later, the tip turns brown and breaks off. Tipworm-injured cranberry uprights may resume vegetative growth by the activation of lateral meristems and form a side shoot. Removing developing fruit from tipworm-injured uprights increased the production of side shoots, suggesting that partitioning limited available resources between the potential sinks (developing fruits and lateral axillary buds) plays a role in determining the regrowth response. Studies using the girdling of tipworm-injured uprights showed that the tolerance (as measured by fruit biomass) of cranberry uprights to tipworm feeding depends on the physiological integration among the different sources of photosynthate and not just the upright's current-season growth. In the growing regions of Maine and Massachusetts, as the season progresses, those tipworm-injured uprights that lack side shoots at the onset of dormancy do not produce flowers during the next growing season.

Management

Plastic emergence traps (25 cm in diameter) with transparent tops provide an effective monitoring tool for adults. These traps can detect midges early in the season so that management action can be taken before populations reach damaging levels. The population can also be monitored by collecting known numbers of buds into a plastic bag for emergence observation. The bag must be kept at room temperature (22°C) with a 14-h photoperiod. Over a period of 14 days, larvae exit the buds into the plastic bag if the buds are infested. Unbaited yellow sticky traps are ineffective for monitoring adult populations in blueberry. In cranberry, monitoring every few days is initiated 1 week after first shoot elongation (the onset of egg laying) by taking 50–100 uprights per bed and examining each microscopically to determine the stages represented in the sample. This allows timing of an insecticide application to target young larvae, which are the most susceptible.

Fig. 312. Cupped leaves on cranberry caused by the larval feeding of cranberry tipworm (*Dasineura oxycoccana*), also known as blueberry gall midge. (Courtesy D. L. Mahr; © APS)

Management of blueberry gall midge with insecticides is extremely difficult because the larvae are hidden inside the buds. Two applications of a spinosyn at 14-day intervals at stage 2 bud development (prior to bud scales separation) may reduce larval infestations in blueberry. There are a few undescribed platygastrids in the genera *Synopeas*, *Platygaster*, and *Inostemma* that appear to play a role in reducing blueberry gall midge infestations in blueberry plantings. In cranberry, predatory larvae of *Toxomerus marginatus* are also often seen in midseason. Heavily fertilized cranberry beds with lush vegetative growth may promote tipworm population numbers. On the other hand, regular sanding (particularly if a deeper layer) of the cranberry bed suppresses the emergence of adults. This tactic is most effective when done area wide, since fly movement among beds allows reinfestation from unsanded to sanded beds.

Selected References

Cook, M. A., Ozeroff, S. N., Fitzpatrick, S. M., and Roitberg, B. D. 2011. Host associated differentiation in reproductive behavior of midges (Cecidomyiidae) on cranberry and blueberry. Entomol. Exp. Appl. 141:8-14.

Dernisky, A. K., Evans, R. C., Liburd, O. E., and Mackenzie, K. 2005. Characterization of early floral damage by cranberry tipworm (*Dasineura oxycoccana* Johnson) as a precursor to reduced fruit set in rabbiteye blueberry (*Vaccinium ashei* Reade). Int. J. Pest. Man. 51:143-148.

Roubos, C. R., and Liburd, O. E. 2010. Pupation and emergence of blueberry gall midge, *Dasineura oxycoccana* (Diptera: Cecidomyiidae), under varying temperature conditions. Fla. Entomol. 93:283-290.

Roubos, C. R., and Liburd, O. E. 2010. Evaluation of emergence traps for monitoring blueberry gall midge (Diptera: Cecidomyiidae) adults and within field distribution of midge infestation. J. Econ. Entomol. 103:1258-1267.

Sampson, B. J., Rinehart, T. A., Liburd, O. E., Stringer, S. J., and Spiers, J. M. 2006. Ecology of endoparasitoids (Hymenoptera) attacking cranberry and blueberry tipworms, *Dasineura oxycoccana* Johnson and *Prodiplosis vaccinii* (Diptera: Cecidomyiidae). Ann. Entomol. Soc. Am. 99:113-120.

Sarzynski, E. M., and Liburd, O. E. 2003. Techniques for monitoring blueberry gall midge (Diptera: Cecidomyiidae) in rabbiteye and southern highbush blueberries. J. Econ. Entomol. 96:1821-1827.

(Prepared by O. E. Liburd and A. L. Averill)

Blueberry Maggot Fly

Blueberry maggot fly, *Rhagoletis mendax* (Diptera: Tephritidae), is one of the most serious insect pests in commercial blueberry owing to a zero tolerance of infested fruit in many production areas. Entire loads of blueberries may be rejected at the point of sale. The distribution of this species in the eastern United States is as far south as Florida, Georgia, and Alabama; as far west as New York, Pennsylvania, and Tennessee; and as far north as Maine, New Hampshire, Vermont, and Michigan. In Canada, it is known from New Brunswick, Nova Scotia, Ontario, Prince Edward Island, and parts of southern Quebec. Some authorities suggest that blueberry maggot fly is found wherever blueberry is grown east of the Rocky Mountains. This species has been reported infesting lowbush blueberry (*Vaccinium angustifolium*), sour top blueberry (*V. myrtilloides*), hillside blueberry (*V. pallidum*), highbush blueberry (*V. corymbosum*), rabbiteye blueberry (*V. ashei*), deerberry

(*V. stamineum*), lingonberry (*V. vitis-idaea*), black huckleberry (*Gaylussacia baccata*), dangleberry (*G. frondosa*), and dwarf huckleberry (*G. dumosa*).

Life History and Description

The adult is 3–5 mm long (females, approximately 4.75 mm) (Fig. 313). The female abdomen is pointed and black with four white crossbands. The thorax is black with a small, backward-pointing, white projection (scutellum). The male fly is somewhat smaller than the female and has a rounded abdomen with only three white crossbands. The wings (wingspan approximately 8 mm) of both sexes are clear and patterned with black bands, characteristic of many members of the Tephritidae. Separation of adult blueberry maggot flies from apple maggot flies (*R. pomonella*) depends upon measurements of wing band ratios, ovipositor length, and genitalia. However, wing band patterns can be used to separate the blueberry maggot fly from the walnut husk fly, cherry fruit fly, and black cherry fruit fly. The larvae are pale white and translucent (body is conical with the anterior end tapered and the posterior end possessing paired spiracles). The maggots develop in the host fruit (Fig. 314), require three instars, and attain a final length of 5–6 mm prior to pupation.

The life history and phenology of the blueberry maggot fly varies according to geographic local, indicating that this insect is phenotypically variable. Most populations in North America have an annual life cycle, but in Maine and possibly other locations, about 85% of the population has an annual life cycle, while approximately 10 and 5% of the population spends two and three or even four winters as a pupa, respectively. In addition, in New Jersey, populations have been found that exhibit a delayed late summer emergence. In general, the life cycle is as follows. In the spring or early summer of the year, just prior to or at the time of the onset of fruit ripening, flies emerge from the puparium in the soil (from a depth of 2–15 cm). Initial emergence and the rate of emergence for the population is dependent upon soil temperature and has been modeled using degree-days (median emergence = 934.3 degree-days, threshold = 4.7°C in New Jersey; and median emergence = 971 degree-days, threshold = 6.1°C in Maine). In many locales, emergence of the entire population happens over a 3- to 5-week period, depending upon temperature and possible soil moisture level.

Upon emergence, flies mate and reproductively maturing females search for and imbibe food (an approximately 7- to 10-day preoviposition period). During and after reproductive maturation, males and females move to locations with ripe fruit. This is especially significant in lowbush blueberry, where fields are usually managed on a 2-year cropping cycle. Flies migrate from fields that are in the vegetative phase to fields that are in the cropping or fruit production phase. Females lay eggs in predominantly ripe fruit; normally, a single egg is placed in a single noninfested fruit just under the fruit surface. Females can lay 25–100 eggs during their reproductive life (15–45 days, depending upon temperature and other weather conditions). Eggs incubate for 3–10 days before larvae hatch. Larvae feed on the fruit pulp and complete development, always in the same fruit within which the egg was laid. Development under field conditions takes 17–30 days, depending upon temperature. When larval development is complete, third instars drop to the soil, burrow, and pupate. Pupation commences in early July in the southern limit of its distribution and may continue through to late September in the northern limit of its distribution. While the fly is fairly cryptic, predation is known to occur by jumping spiders (Salticidae). Larvae are parasitized by wasps (family Braconidae: *Opius melleus*, *O. ferrugineus*, and *Diachasma alloeum*), with *Opius* spp. parasitism rates of 0–40% in Maine and New Jersey. The parasitoids apparently use fruit volatiles for long-range search. Larvae and pupae are attacked by ants (Formicidae), larval members of the family Elateridae, and/or ground beetles (Carabidae). Predation rates of sentinel prey placed in fields are 0–15%.

Fig. 313. Blueberry maggot fly adult (*Rhagoletis mendax*) (right) and apple maggot fly (*Rhagoletis pomonella*) (left). Both flies are identical is size distribution and this image shows the size range (3–5 mm; average = 4.75 mm) they can exhibit. (Courtesy F. A. Drummond; © APS)

Damage

Blueberry maggot fly is a direct fruit pest, but only feeding by the larvae results in fruit damage (there has not been any report of damage solely caused by fruit puncture during oviposition). Damage is generally not noticeable until larvae are

Fig. 314. Blueberry broken open to show a blueberry maggot fly larva (*Rhagoletis mendax*). (Courtesy F. A. Drummond; © APS)

Fig. 315. Blueberries damaged by blueberry maggot fly larvae (*Rhagoletis mendax*). (Courtesy F. A. Drummond; © APS)

in the third instar and much of the pulp has been consumed. At this point, the berries loose turgidity and collapse and shrivel (Fig. 315). Flies are sensitive to desiccation, so high-damage years tend to be those that have moderate levels of rainfall. In addition, fields that are irrigated or have high weed densities that offer shade tend to be associated with higher levels of infestation. Flies are attracted to, or reside longer in, heavily weed-infested sections of fields, areas that have high fruit density, and low spots in fields; therefore, these areas tend to be associated with higher levels of fruit infestation. In lowbush blueberry, because of the 2-year production cycle, the majority of flies invade fruit-bearing fields each year from field edges that support wild host plants or from nearby production fields that were in the fruit production cycle the previous year. In this system, the flies enter the field and then appear to search randomly for suitable fruit for oviposition. On average, flies move 10 m per day in a random direction. This results in the population density of flies being high along field edges and decreasing toward the field interior. The resulting damage pattern caused by this behavior is an inverse relationship of infestation level and distance into the field interior.

Management

The current state of pest management for blueberry maggot fly relies almost exclusively upon insecticide applications. Factors such as time to harvest, type of harvest (hand picked or mechanical), market (fresh/frozen/export), and restricted-entry interval/preharvest interval determine the insecticide selection. Treatments are applied either as a prophylactic strategy or as an integrated pest management strategy in which treatments are based upon economic thresholds. In both cases, monitoring population incidence or relative abundance forms the basis of decision-making. Monitoring blueberry maggot flies is based upon the use of baited visual sticky traps, either yellow flat-panel sticky traps (Fig. 316) or red or green spherical sticky traps. Traps are usually deployed with an ammonium acetate bait to enhance attractiveness. Some extension specialists recommend yellow panel traps early in the season and spherical traps later in the season, when female flies are sexually mature. Traps are used either to detect the presence of flies in the field, upon which insecticide applications are initiated, or to estimate economic thresholds, below which treatment is not recommended. Thresholds range from two flies per trap per week in highbush blueberry to 10 flies per trap in lowbush blueberry. Trap placement recommendations are along the field edge (less than 8 m) in lowbush blueberry and split between the field edge (less than 20 m) and interior of the block or field in highbush blueberry. Trap density ranges between one trap per 4 ha in lowbush blueberry and one trap per 2 ha in highbush blueberry. It is also recommended that traps be placed in weedy sections of fields and in areas adjacent to abandoned fields or natural areas that support wild host plants.

Fig. 316. Blueberry maggot fly (*Rhagoletis mendax*) (center) captured on a yellow sticky trap used to monitor population incidence. (Courtesy F. A. Drummond; © APS)

The most common application strategy in highbush blueberry is to treat the entire field or block if trap captures suggest control. Several strategies are employed in lowbush blueberry. In Canada, treatment of emerging flies in adjacent vegetative phase fields is practiced. In Maine, field perimeter applications are the norm. Deployment of traps in a grid and selective spot treatment adjacent to traps that exceed threshold levels is practiced in small fields. The use of insecticide-treated spheres appears to be promising in highbush blueberry but has not been consistently effective in lowbush blueberry. Trapping flies with sticky traps has not been shown to be effective in lowbush blueberry, even when traps are deployed every 3 m along field edges.

Biological control is not currently an option for blueberry maggot fly. Cultural controls have a range of efficacies. Variety selection in highbush and rabbiteye blueberries offers potential to reduce damage through the mechanism of "escape" by planting early-maturing varieties. Whether this strategy offers a long-term solution in light of the variability in fly emergence time remains to be seen. The use of a harvesting machine to harvest all fruit left after hand harvesting is suggested as a means of eliminating late-season oviposition sites. It is also suggested that weed control may result in a less favorable habitat for fly survival in blueberry fields. An effective cultural control for blueberry maggot fly in isolated lowbush blueberry fields is to manage entire fields on the same production cycle. Many growers split fields so that half of the field is bearing fruit every year. However, this results in a habitat where flies can move back and forth from emergence sites (vegetative sections) to fruit-bearing oviposition sites every year. Another cultural control that may not be feasible for most growers is to produce blueberries in geographic areas in North America not yet inhabited by the blueberry maggot fly.

The challenge for the future, in regards to management of blueberry maggot fly, is to maintain a high level of control in a zero-tolerance market with reduced-risk insecticides that are often characterized by rapid environmental degradation.

Selected References

Collins, J. A., and Drummond, F. A. 2004. Field-edge based management tactics for blueberry maggot in lowbush blueberry. Small Fruits Rev. 3(3/4):283-293.

Gaul, S. O., Estabrooks, E. N., and McRae, K. B. 2002. Integrated pest management of *Rhagoletis mendax* (Diptera: Tephritidae) in lowbush blueberry using vegetative management. J. Econ. Entomol. 95(5):958-965.

Liburd, O. E., Alm, S. R., and Casagrande, R. A. 1998. Susceptibility of highbush blueberry cultivars to larval infestation by *Rhagoletis mendax* (Diptera: Tephritidae). Environ. Entomol. 27:817-821.

Stelinski, L. L., and Liburd, O. E. 2001. Evaluation of various deployment strategies of imidacloprid-treated spheres in highbush blueberries for control of *Rhagoletis mendax* (Diptera: Tephritidae). J. Econ. Entomol. 94:905-910.

Teixeira, L. A. F., and Polavarapu, S. 2002. Postdiapause development and prediction of emergence of female blueberry maggot (Diptera: Tephritidae). Environ. Entomol. 30(5):925-931.

(Prepared by F. A. Drummond and J. A. Collins)

Blueberry Tip Borer

Blueberry tip borer, *Hendecaneura shawiana* (Lepidoptera: Tortricidae), is a tortricid moth that has one generation per year. It is distributed across eastern North America and is a sporadic pest of blueberry that can reduce fruit potential.

Life History and Description

Adult moths are brown with orange marks near the tip and a silver patch along the hind margin, with a 9- to 15-mm wingspan. Moths are active after bloom and females lay a small translucent egg on succulent shoots (usually the current-year's growth). The young larvae are elongate and pink-brown, growing to 12 mm long. They bore into the shoot near the egg after hatching and feed in the center of the stem toward the base of the plant (Fig. 317). Development continues into the autumn. Mature larvae overwinter, pupate the next spring, and emerge toward the end of bloom.

Fig. 317. Blueberry tip borer larva (*Hendecaneura shawiana*) in a blueberry stem. (Courtesy R. Isaacs; © APS)

Fig. 318. Symptoms on blueberry of blueberry tip borer (*Hendecaneura shawiana*) infestation. (Courtesy R. Isaacs; © APS)

Damage

A small, darkened pinhole can be seen where the larva entered the stem. The most obvious symptom is that the feeding causes the stem to wilt above the entry point (top 20–25 cm), which happens while fruit are sizing. The leaves dry up from the tips and shoot tips may bend over, resembling primary mummy berry infection (Fig. 318). Shoots infested by this insect lose multiple fruit or leaf buds above the point of entry, thus reducing the potential for next year's growth and yield. Secondary growth may also impede ripening and affects pruning decisions. Levels of infestation can vary from a few shoots with symptoms to fields with extensive populations, and symptoms are seen more commonly in some varieties.

Management

Pruning damaged tips when they are observed can help minimize populations. Cut well below the shoot tip to ensure larvae are collected. While burning or burying of these canes is recommended in some states, researchers in North Carolina (United States) found that larvae did not survive in drying shoots left on the ground. The standard spray program used for fruitworms or other post-bloom insect pests normally keeps blueberry tip borer under control.

Selected Reference

Milholland, R. D., and Meyer, J. R. 1984. Diseases and arthropod pests of blueberries. N. C. Agric. Res. Serv. Bull. 468.

(Prepared by R. Isaacs)

Cherry Fruitworm and Cranberry Fruitworm

Cherry fruitworm, *Grapholita packardi* (Lepidoptera: Tortricidae), and cranberry fruitworm, *Acrobasis vaccinii* (Lepidoptera: Pyralidae), are direct pests of *Vaccinium* spp. in eastern North America. Both species attack blueberry and, because they co-occur temporally, a key management challenge is distinguishing between the two insects. *Acrobasis vaccinii* also attacks cranberry; because cranberry fruitworm is clearly the most problematic insect pest in cranberry, it is also covered separately in that section. Both fruitworms are native to North America and feed on a wide range of host plants. For cherry fruitworm, hosts include apple, plum, cherry, rose, blueberry, hawthorn, and peach. For cranberry fruitworm, hosts include cranberry, blueberry, and huckleberry. Without decisive management, these pests can cause economic injury to commercial berry crops.

Life History and Description

Cherry fruitworm and cranberry fruitworm undergo a single generation per year. In blueberry, adult emergence for each begins at the start of bloom, peaks shortly after petal fall, and gradually declines over a 3- to 4-week period. The cranberry fruitworm moth is approximately 11 mm long and forewings are grayish brown with two identifying whitish triangles, while adult cherry fruitworm are grayish black and 5–6 mm long (Fig. 319). After mating female egg laying coincides with early fruit set, peaking in the weeks following 100% petal fall. Both species oviposit primarily within the calyx cup area of the fruit, although the cranberry fruitworm tends to lay eggs more strictly along the calyx rim. The eggs of the cranberry fruitworm are irregular shaped and initially appear as raised, pale green scales. As they develop, cranberry fruitworm eggs go through a progressive change of color from creamy white-green to yellow, with an oval orange streak appearing as the embryo forms. Later, within 24 h of hatching, the developing neonate's head capsule can be seen as a darkened spot within the egg. After hatch, the egg casing is clear white. The eggs of cherry fruitworm are flat, with a round to oval outline, and appear opaque. Upon maturity, the dark head capsule of the developing neonate is clearly visible within the egg.

Under normal spring weather conditions, eggs hatch in 3–4 days. Cranberry fruitworm larvae generally migrate out of the calyx cup and enter the berry adjacent to the stem bundle or, to a much lesser extent, on the side or shoulder of the fruit. Cherry fruitworm tend to enter fruit in the calyx cup area where they hatched. Cranberry fruitworm larvae are pale yellowish green and reach a length of 15 mm when mature (Fig. 320). Cherry fruitworm larvae (Fig. 321) appear more pink and reach a length of about 8 mm; they can be distinguished from cranberry fruitworm by the presence of an anal comb. Mature larvae leave the berries and move to overwintering sites. Cranberry fruitworm larvae form a hibernaculum within 2 mm of the soil surface, using soil and plant debris. Cherry fruitworm larvae form a hibernaculum aboveground, in a sheltered area on the bush or in surrounding plants. Larvae remain within the hibernaculum until spring and then pupate as diapause requirements are met.

Damage

Both fruitworm species are considered direct pests because they enter and feed within fruit. One major difference between them is the number of fruit that are infested within a single life cycle. A cranberry fruitworm may infest up to six berries, depending on size, while a cherry fruitworm only enters one or two berries to complete development. Larval feeding on green fruit typically causes premature coloring of the berries and eventual fruit abortion (Fig. 322). This characteristic makes infested fruit readily identifiable. On blueberry clusters, while feeding and moving between berries, cranberry fruitworm larvae produce a characteristic rusty red mass of webbed frass, filling the space in and around the fruit cluster. Cherry fruitworm larvae may silk berries together when moving between fruit, but the frass remains inside the fruit and the webbing is not as conspicuous. Single berry injury from cranberry fruitworm and cherry fruitworm can impact yield if in high enough numbers, but by harvest time in blueberry, the larvae have typically left and damaged fruit has fallen. Cluster damage from the cranberry fruitworm in blueberry, however, often remains on the plant and can be a serious contaminant problem at fruit pack-out.

Management

In blueberry, monitoring adult emergence of both species with pheromone-baited traps provides a relative estimate of population level and distribution within the fields and gives an early indication of when field scouting should begin. A model has

Fig. 319. Adult cherry fruitworm (*Grapholita packardi*) on a sticky trap. (Courtesy R. Isaacs; © APS)

Fig. 321. A cherry fruitworm larva (*Grapholita packardi*) in a blueberry. (Courtesy R. Isaacs; © APS)

Fig. 320. A cranberry fruitworm larva (*Acrobasis vaccinii*) on blueberry. (Courtesy R. Isaacs; © APS)

Fig. 322. Cherry fruitworm (*Grapholita packardi*) infestation on highbush blueberry. (Courtesy S. Van Timmeren; © APS)

been developed for cranberry fruitworm in blueberry to predict the onset of egg laying, based on tracking adult emergence with pheromone-baited traps. Scouting for fruitworm eggs is the most reliable indicator of the timing and location of egg hatch. Scouting for eggs should begin after adult emergence and when there are some early fruit available for adult oviposition. Initial scouting should begin along the perimeter rows of berry fields that either have a history of fruitworm pressure or are adjacent to woods with native host plants. Fruitworm eggs tend to be laid in a clustered distribution across fields, and adults show preference for the largest berries in a cluster.

Clean cultivation can reduce the overwintering populations of cranberry fruitworm within a field, but under high pressure, this is not likely to be a stand-alone tactic. A number of biocontrol agents, including parasitic wasps and flies in the families Trichogrammatidae, Ichneumonidae, and Braconidae parasitize egg and larval stages of cranberry fruitworm and cherry fruitworm. The parasitoid *Trichogramma minutum* is known to infest up to 20% of cranberry fruitworm eggs in unmanaged blueberry plantings. Selecting insecticides that have lower toxicity to these parasitoids (e.g., insect growth regulators, spinosyns, and *Bacillus thuringiensis*) conserve natural enemies.

When using insecticides to control fruitworms, both timing and coverage of the fruit are critical. Some insecticides (e.g., insect growth regulators) are primarily active on eggs and hatching larvae and should be applied during the early egg-laying period to ensure contact with eggs. Since early egg hatch often happens before petal fall is complete, only insecticides that are safe to pollinators should be used during this period. After 100% petal fall, a greater selection of insecticides is available for control. Contact nerve poisons, such as organophosphates, synthetic pyrethroids, and carbamates, are active on all fruitworm life stages. Insecticides from the neonicotinoid, spinosyn, oxidiazine, and diamide classes are primarily active on eggs and larvae and their performance is enhanced with ingestion.

Selected References

Isaacs, R., Wise, J. C., and Garcia-Salazar, C. 2009. Developing integrated fruitworm control strategies for blueberry in preparation for pesticide restrictions. Acta Hortic. 810:259-264.

Mallampalli, N., and Isaacs, R. 2002. Distribution of egg and larval populations of cranberry fruitworm (Lepidoptera: Pyralidae) and cherry fruitworm (Lepidoptera: Tortricidae) in highbush blueberries. Environ. Entomol. 31:852-858.

Sarzynski, E. M., and Liburd, O. E. 2004. Effect of trap height and within-planting location on captures of cranberry fruitworm (Lepidoptera: Pyralidae) in highbush blueberries. Agric. For. Entomol. 6:199-204.

Wise, J. C., Jenkins, P., VanderPoppen, R., and Isaacs, R. 2010. Activity of broad spectrum and reduced-risk insecticides on various life stages of cranberry fruitworm (Lepidoptera: Pyralidae) in highbush blueberry. J. Econ. Entomol. 103:1720-1728.

(Prepared by J. C. Wise)

Flower Thrips

There are several species of flower thrips (Thysanoptera: Thripidae) that attack blueberry throughout the United States, including the eastern flower thrips, *Frankliniella tritici*, found east of the Rocky Mountains, and the western flower thrips, *F. occidentalis*, found throughout the United States and Canada. In Florida (United States), a species of flower thrips, *F. bispinosa*, locally called Florida flower thrips, causes significant damage to southern highbush and rabbiteye blueberries. In the northeast, the blueberry thrips *F. vaccinii* and *Catinathrips kainos* are minor leaf pests in lowbush and highbush blueberries (see Blueberry Thrips under the Blueberries (Lowbush) section below).

Life History and Description

Thrips have punch-and-suck mouthparts and fringed wings. Members of the genus *Frankliniella* are approximately 1–1.3 mm long and yellow to brown, often with gray blotching across the abdomen (Fig. 323). Males are generally smaller and paler than females. Eggs are cylindrical and laid within blueberry tissues, making them difficult to see. Thrips progress through two actively feeding larval instars and two inactive instars (often called prepupa and pupa). The prepupae and pupae each resemble the adult morph, except that the wings are represented by developing wing buds. Most flower thrips can disperse over long distances and are known to have an extensive host range. However, the range of hosts on which they can survive and reproduce is usually much smaller than the variety of hosts on which they can feed but not reproduce.

Damage

Both leaf and flower thrips damage blueberry plants when they feed. In the northeastern United States, *F. vaccinii* and *C. kainos* feeding causes the distortion and discoloration of leaves in small, isolated patches in some blueberry plantings. Flower thrips, including *F. bispinosa*, feed on all parts of the blueberry flower, including the petals, ovary, pistil, and calyx. This can lead to scarred or distorted fruit and even fruit abortion. Female thrips also cause damage when they lay their eggs in the leaf and flower tissues. This damage is not as severe as feeding damage, but it can also lead to fruit scarring. Thrips can also cause significant damage to the fruit, particularly in areas where fruit are touching or closely clustered. This can result in silvery blemishes resembling a halo.

Management

Thrips are difficult to manage because they are very mobile. However, an economic threshold has been developed for flower thrips for specific rabbiteye blueberry varieties, suggesting that a population can be suppressed once it exceeds the threshold. There are several methods available for monitoring flower thrips populations, including the use of sticky traps (white or blue), tapping infested flowers over a white surface, and dipping infested flowers into alcohol. Grower preferences and crop specifics may dictate which monitoring technique works best in a particular area. Spinetoram is the insecticide of choice in areas where flowering thrips are problematic. When spinetoram is sprayed late in the evening or early in the morning and allowed to dry, it tends to have a less detrimental effect on pollinators. Several predatory insects may prove to be successful in suppressing thrips populations, including the minute pirate bug

Fig. 323. Adult of *Frankliniella bispinosa*, a flower thrips. (Courtesy O. E. Liburd; © APS)

(*Orius insidiosus*) and *Amblyseius swirskii*. The fungal pathogen *Beauveria bassiana* has shown potential for the control of *F. occidentalis* and may be a potential agent for control of other thrips species in the future. Cultural methods for the control of thrips species include eliminating alternate hosts near the main crop, particularly if the blooming periods overlap.

Selected References

Arévalo, H. A., and Liburd, O. E. 2007. Horizontal and vertical distribution of flower thrips in southern highbush and rabbiteye blueberry fields, with notes on a new sampling method for thrips inside the flowers. J. Econ. Entomol. 100:1622-1632.

Arévalo, H. A., Fraulo, A. B., and Liburd, O. E.. 2006. Key to the most common species of thrips found in early-season blueberry fields in Florida and southern Georgia. EDIS. Univ. Fla. Inst. Food Agric. Sci. Ext. ENY-836 (IN679).

Arévalo, H. A., Fraulo, A. B., and Liburd, O. E. 2009. Management of flower thrips in blueberries in Florida. Fla. Entomol. 92:14-17.

Moritz, G., Morris, D., and Mound, L. 2001. Thrips ID: Pest thrips of the world. CSIRO Publishing, Collingwood, Victoria, Australia.

Rhodes, E. M., and Liburd, O. E. 2011. Flower thrips (Thysanoptera: Thripidae) dispersal from alternate hosts into southern highbush blueberry (Ericales: Ericaceae) plantings. Fla. Entomol. 94:311-320.

Rhodes, E. M., Liburd, O. E., and Grunwald, S. 2011. Examining the spatial distribution of flower thrips in southern highbush blueberries utilizing geostatistical methods. Environ. Entomol. 40:893-903.

Rodriguez-Saona, C. R., Polavarapu, S., Barry, J. D., Polk, D., Jornsten, R., Oudemans, P. V., and Liburd., O. E. 2010. Color preference, seasonality, spatial distribution and species composition of thrips (Thysanoptera: Thripidae) in northern highbush blueberries. Crop Prot. 29:1331-1340.

(Prepared by O. E. Liburd)

Japanese Beetle

The Japanese beetle, *Popillia japonica* (Coleoptera: Scarabaeidae), was first reported in North America in New Jersey (United States) in 1916. Since then, the readily available food for adults, the abundance of turf for larvae, and the low level of natural enemies have enabled this insect to spread across eastern North America. It is currently established in all states east of the Mississippi River, except Florida (United States), and also in Ontario (Canada), with sporadic outbreaks in more western states. Quarantine measures are in place to prevent the artificial spread of this species in airplanes or on nursery stock.

Life History and Description

Adult Japanese beetles are about 13 mm long (Fig. 324). The head and thorax are metallic green, and the rest of the upper surface is copper colored. Five tufts of white hairs on both sides of the lower surface and a sixth pair on the tip of the abdomen are distinguishing features of this beetle. Eggs are 1–2 mm in diameter, spherical, and white. Females lay the eggs 5–10 cm below the soil surface, generally in areas with grass and moist soil. The larvae are 2–3 mm when newly hatched, and fully grown larvae are about 30 mm long (Fig. 325). Larvae are C-shaped and white, with a dark head capsule, and three pairs of legs. A V-shaped arrangement of two rows of seven hairs on the ventral side of the anal segment distinguishes the Japanese beetle grub from other white grubs (see figures within Phyllophaga under the Cranberries section below), and a hand lens is needed to see this clearly. The larvae feed on roots, preferring grasses, small seedling plants, and other organic matter in the soil. When larvae are numerous, their feeding can cause turf to die in patches, and in severe cases, the grass can be easily pulled back like a carpet. Secondary damage from moles and raccoons who feed on the grubs is a good indicator of grub infestation. There is usually one generation per year, although a full life cycle may take 2 years in the coldest regions of Japanese beetle distribution. Japanese beetle larvae overwinter as third instars at a depth of 10–15 cm, moving down in the soil to avoid freezing. In the spring, as the soil warms up, the larvae move closer to the surface and begin feeding on roots again. Larvae then pupate, and about 2 weeks later, adults emerge to start the life cycle again. In the southern United States, adult emergence may start as early as mid- to late May, while in northern regions of its distribution, Japanese beetles emerge in early July. Females mate immediately after emergence from the soil, but they usually move to other sites to feed before laying the first eggs. Females lay 40–60 eggs over a 2- to 3-week period. Eggs are laid in batches of five to ten, with batches distributed near feeding sites. Preferred egg-laying sites are moist soil with grassy cover. Eggs hatch 1–2 weeks after egg deposition, and the small larvae feed on nearby roots and other organic matter.

Damage

Adult Japanese beetles feed on the foliage and fruit of about 300 plant species, including fruits, field and garden crops, ornamentals, and many wild plants. Adult Japanese beetles

Fig. 324. Adult Japanese beetle (*Popillia japonica*) feeding on blueberry. (Courtesy R. Isaacs; © APS)

Fig. 325. Japanese beetle larva (*Popillia japonica*). (Courtesy R. Isaacs; © APS)

skeletonize blueberry foliage, consuming the soft tissue between the veins and leaving a lacelike skeleton. Males and females fly into blueberry fields and congregate on the top of plants exposed to sunlight, so the upper canopy is usually defoliated more severely and bushes at the edge of fields are more infested than those in the center. Feeding starts on foliage while berries are green, but this is not of great economic importance. Far more damaging is that, as the fruit become ripe, beetles aggregate on fruit clusters, and their feeding can reduce yield in severe infestations and open fruit to pathogens. Machine-harvested fields are at risk of having a contaminated harvest because beetles are easily shaken off the bush and collected with the fruit.

Management

The presence of adult Japanese beetles on bushes in midsummer is often the first sign that this pest has become established in a farm. In small plantings, beetles can be physically removed from the plants early in the morning, when they are less active. Monitoring traps baited with a pheromone/plant odor lure are highly effective at attracting beetles, but should be used with caution. These can provide an early warning of beetle emergence but can also attract beetles to a field without removing all of them from the population, so the traps are not recommended for control.

Cultural controls can help make crop fields and nearby areas less suitable for Japanese beetle. All attractive host plants (e.g., wild grape, Virginia creeper, brambles, and sassafras) should be removed from the blueberry field and nearby areas to reduce the attraction of beetles. Reestablishment of these weeds can be prevented by hoeing, using herbicide, or mulching. Adult Japanese beetles prefer to lay eggs in grassy areas, so cultivated fields have a much lower abundance of Japanese beetle than fields with grass or fields with pasture nearby. Removing grass in row middles during the period of adult activity is particularly important. If fields are not suitable for bare ground throughout the season, a combined winter rye cover crop sown after harvest and tilled into the soil before beetle activity the next spring can provide some soil structure while minimizing beetle egg laying. Nongrass cover crops that create an unattractive ground cover are another alternative to bare ground, providing some structure for machinery to pass over and reduces the risk of erosion. Recent studies comparing the levels of feeding on fruit and foliage of different highbush blueberry cultivars suggested differences in susceptibility, and these results have potential use in breeding programs.

Sampling soil for Japanese beetle larvae in and around infested fields can help determine where beetles are coming from. Larvae are relatively easy to sample because of their size and light color. Samples should be taken in grassy areas around and within fields once the soil warms in the spring, when fully grown larvae are feeding near the soil surface. To sample for larvae, defined areas of sod (e.g., 0.3 m^2) in suspected sources of beetles should be cut and the soil and roots should be shaken through for grubs. By keeping a record of the position and number of grubs per square foot from a number of sites, samples can be compared over time and between fields.

Biological control cannot be relied upon for the complete control of Japanese beetle but can help suppress populations by varying amounts, depending on location. Tiphiid wasps attack larvae in the ground and a tachinid fly parasitizes adults of Japanese beetle. Ants and predatory ground beetles feed on eggs and young larvae, while mammals dig under turf to feed on grubs. Some bird species can also eat beetles off bushes. Grubs may become infected with nematodes, viruses, or bacteria. Symptoms of these infections include mushy grubs or grubs turning yellow or red. Infection rates vary from about 1% of larvae in newly infested areas to 20–30% in areas where they are established. Milky spore disease and various species of nematode are commercially available biocontrol options. These are generally considered less effective in areas with cold winters and they are expensive.

Controlling adults by applying insecticides to bushes may be necessary to prevent the contamination of harvested fruit or to protect fruit from damage. Insecticides with some repellent action, such as pyrethrins or pyrethroids, can help ensure that beetles are not killed and stay on the fruit to be harvested. Control is greatest if good coverage of the whole plant canopy is achieved. Beetles are strong flyers and can fly into field edges immediately before harvest. Insecticides with short preharvest intervals may then be required to get beetles off the bushes before fruit is harvested.

Soil-applied insecticides are an option for the control of grubs, thus reducing beetle abundance on bushes during the next year. This approach can be particularly useful in fields with grass between bushes or around the field, where clean cultivation (discussed above) is not feasible. These preventative insecticides must be applied before beetle emergence so the residue is present in the soil to target young grubs soon after hatching, and they should be irrigated into the turf immediately after application. These treatments are highly effective where they are applied but beetles from outside treated areas can fly into the fields, so monitoring and supplemental controls may be needed.

Selected References

Potter, D. A., and Held, D. W. 2002. Biology and management of Japanese beetle. Annu. Rev. Entomol. 47:175-205.

Szendrei, Z., and Isaacs, R. 2006. Survey of Japanese beetle management practices in Michigan blueberry. HortTechnology 16:82-88.

Szendrei, Z., Mallampalli, N., and Isaacs, R. 2005. Effect of tillage on abundance of Japanese beetle, *Popillia japonica* Newman (Col., Scarabaeidae), larvae and adults in highbush blueberry fields. J. Appl. Entomol. 129:258-264.

Van Timmeren, S., and Isaacs, R. 2009. Susceptibility of highbush blueberry cultivars to cranberry fruitworm and Japanese beetle. Int. J. Fruit Sci. 9:23-34.

(Prepared by R. Isaacs)

Leafrollers

Leafrollers (Lepidoptera: Tortricidae) are the larvae of omnivorous lepidopterans that use a large number of forest and fruit trees and shrubs as hosts. The name leafroller is derived from their habit of rolling leaves for shelter. On blueberry, at least four species have been identified: oblique-banded leafroller (OBLR), *Choristoneura rosaceana*; red-banded leafroller (RBLR), *Argyrotaenia velutinana*; three-lined leafroller, *Pandemis limitata*; and gray-banded leafroller, *Argyrotaenia mariana*.

Life History and Description

Of these four species, only OBLR is a major pest of blueberry. It has a broad host range and also infests apple, pear, cherry, plum, peach, rose, raspberry, gooseberry, currant, strawberry, and many other wild hosts. Early in the growing season, these brown or green worms feed on floral buds, blossoms, and leaves (Fig. 326). When full grown (4.5–5 cm long), the larvae seal up the leaf shelter and form a glossy brown case, and 1–2 weeks later, the adults emerge from the shelter. Adults are about 5 cm long and vary in color from brown to yellow (Fig. 327). In most fruit-growing regions of North America, OBLR undergoes two generations per year. Larvae of the summer generation feed on leaves, green berries, and ripe berries.

RBLR causes little or no damage in blueberries. However, there are occasional outbreaks mistakenly attributed to OBLR. In addition to blueberry, RBLR has a broad range of hosts, including cherry, peach, plum, grape, and other small fruits, as well as ornamental crops. RBLR moths have rusty red or red-brown bands extending from about the middle of the front edge of the wing to a point about two-thirds of the distance from the wing base on the rear edge. This band forms a V-shape when the moth is at rest (Fig. 328). Larvae are a fairly uniform green with a green head capsule (Fig. 329), reaching a length of about 17 mm. If larvae have fed on fruit, they develop a yellow tint. In the Great Lakes region, RBLR completes two generations: one from mid-April to May and the second from mid-June to July. RBLR overwinters as a pupa in the ground cover. Adults emerge in the spring around the green tip stage of the plant. Females lay eggs primarily on the blueberry canes, mainly at the pink bud growth stage. Larvae of this generation are present from early May to mid-June.

The three-lined leafroller is bivoltine and has been found attacking blueberry in Michigan (United States) (Figs. 330 and 331). Of minimal importance, its presence in blueberry was not known until 2004, when moths were found in pheromone traps used to monitor OBLR.

Fig. 326. Oblique-banded leafroller larva (*Choristoneura rosaceana*). (Courtesy A. Agnello)

Fig. 327. Oblique-banded leafroller moth (*Choristoneura rosaceana*). (Courtesy L. Gut; © APS)

Fig. 328. Red-banded leafroller moth (*Argyrotaenia velutinana*). (Courtesy L. Gut; © APS)

Fig. 329. Red-banded leafroller larva (*Argyrotaenia velutinana*). (Courtesy C. Garcia-Salazar; © APS)

Fig. 330. Three-lined leafroller moth (*Pandemis limitata*). (Courtesy C. Garcia-Salazar; © APS)

Fig. 331. Three-lined leafroller larva (*Pandemis limitata*). (Courtesy C. Garcia-Salazar; © APS)

Damage and Management

Leafrollers are often a problem since they contaminate machine-harvested fruit. Most processors and consumers have a zero-tolerance policy for this pest. Pheromone-based monitoring systems have been developed to improve pest control timing. In general, control of OBLR is at pink bud, full bloom, 100% petal fall, and preharvest cover sprays. Control before bloom is based on field scouting. Used at egg hatch, *Bacillus thuringiensis*-based products effectively manage first-instar larvae. However, their efficacy declines as larval size increases.

RBLR management is similar to that for OBLR. Further, applications at petal fall against fruitworms take care of most leafroller problems. Regarding biological control, leafroller eggs are parasitized by *Trichogramma minutum*, and a number of parasitoids attack larvae and pupae. Larvae are also subject to infection by a granulosis virus.

Selected References

Elsner, E. A., and Whalon, M. E. 1985. Common Blueberry Insect Pests and Their Control. Mich. State Univ. Ext. Bull. E-1863.

Howitt, A. J. 1993. Common Tree Fruit Pests. Mich. State Univ. Ext. Serv. NCR 63.

Hull, L. A., Pfeiffer, D. G., and Biddinger, D. J. 1995. Mid-Atlantic Orchard Monitoring Guide NRAES-75. Northeast Regional Agricultural Engineering Service, Ithaca, NY.

(Prepared by C. Garcia-Salazar)

Oriental Beetle

The oriental beetle, *Anomala orientalis* (Coleoptera: Scarabaeidae), was first recorded on the United States mainland in a New Haven, Connecticut, nursery sometime before 1920 and is now known to be found throughout the mid-Atlantic states as far south as North Carolina and as far west as Ohio (and also in Hawaii). Beetles have also been found in Maine, West Virginia, Tennessee, and South Carolina, but it is not clear whether established populations exist in these states. Long-distance dispersal of this species has been facilitated by the shipment of infested nursery stock. The larvae can cause considerable damage to the roots of many plant species, but there is little or no feeding in the adult stage. This insect has become one of the most serious pests in New Jersey blueberry and the number of confirmed infestations of cranberry beds in Massachusetts has increased over time.

Life History and Description

The adult beetle is about 9–10 mm long and is usually straw colored with black markings (Fig. 332). The pattern and extent of the markings may vary widely among individuals, with some beetles (especially females) being almost entirely straw colored. Adults may also have metallic green or red markings on the head and pronotum. There are three larval instars, with the third instar reaching a length of 20–25 mm. The larvae can be distinguished from other scarab larvae by the rastral patterns on the underside of the abdomen; in the oriental beetle, the raster has two more-or-less parallel rows of setae, with 10–16 setae in each row (see figures within Phyllophaga under the Cranberries section below).

The oriental beetle typically has one generation per year, with a small portion of the population taking 2 years. The larvae overwinter in the soil and migrate toward the soil surface when temperatures reach about 10°C in the spring. The larvae then feed for 1–2 months before pupating. Adult emergence begins in June to July and may continue for about 2 months. Adult behavior is cryptic and dispersal is typically limited. In the absence of a synthetic sex pheromone, adults may be difficult to locate.

Peak daily flight and mating activity occur between 6:00 and 10:00 p.m. Virgin females begin calling (i.e., releasing sex pheromone) upon emerging from the soil. Mated females immediately burrow into the soil to lay eggs and do not surface until most, if not all, of their eggs have been laid. Females lay an average of 25 eggs, which are deposited singly in the soil and hatch after 2–3 weeks.

Damage

Adult females may occasionally feed on flowers of various ornamental plants, including roses, hollyhocks, phlox, dahlias, and Japanese irises, but they are not considered pests in this life stage. However, the larvae can cause considerable damage by feeding on the roots of many plant species, including turf, ornamentals, cranberry, blueberry, and strawberry. Infestations of the oriental beetle are typically patchy, which is likely in part because females generally call, mate, and oviposit within a very small radius from their point of emergence from the soil. In cranberry (Fig. 333), patches of damaged vine may radiate from year to year as subsequent generations of larvae move lat-

Fig. 332. Adult oriental beetle (*Anomala orientalis*). (Courtesy A. L. Averill; © APS)

Fig. 333. Areas of oriental beetle (*Anomala orientalis*) infestation in cranberry have turned reddish orange as a result of root feeding by the larvae. (Courtesy A. L. Averill; © APS)

erally to adjacent, undamaged vines. Some cranberry bogs may not be infested by oriental beetle grubs despite the presence of large numbers of grubs and adults in field margins; this might be at least partly attributable to the limited dispersal tendencies of the beetle. In blueberry, feeding injury on the root system leads to lowered vigor and yield and, if destruction is severe enough, to death of the bush.

Management

The neonicotinoid imidacloprid is effective against first-instar larvae, but its relative ineffectiveness against later instar larvae means that it must be applied prophylactically. Pheromone-baited traps may be used for monitoring, but care must be taken to determine whether oriental beetle grubs are present in the crop before treatment is considered. In cranberry, a summer flood (with loss of yield in that year) would likely be effective in suppressing the oriental beetle.

Sex pheromone-mediated mating disruption shows promise in both blueberry and cranberry. Work in blueberry has demonstrated successful disruption with extremely low application rates, which could make the cost of mating disruption comparable with using insecticides.

The oriental beetle has several natural enemies, including the introduced parasitoid wasps *Tiphia vernalis*, *Tiphia popilliavora*, and *Scolia manilae*. The latter species dramatically reduced beetle populations in Hawaii, but the wasp cannot overwinter in the Northeast. Commercially available entomopathogenic nematodes are not effective. A recently discovered species (*Steinernema scarabaei*) can exhibit very high mortality against third-instar larvae in the lab and field, but difficulties with mass rearing of this nematode currently limit its use.

Selected References

Facundo, H. T., Villani, M. G., Linn, C. E., Jr., and Roelofs, W. L. 1999. Temporal and spatial distribution of the oriental beetle (Coleoptera: Scarabaeidae) in a golf course environment. Environ. Entomol. 28:14-21.

Friend, R. B. 1929. The Asiatic beetle in Connecticut. Conn. Agric. Exp. Stn. Bull. 304:585-664.

Koppenhöfer, A. M., and Fuzy, E. M. 2009. Long-term effects and persistence of *Steinernema scarabaei* applied for suppression of *Anomala orientalis* (Coleoptera: Scarabaeidae). Biol. Control 48:63-72.

Polavarapu, S., Koppenhöfer, A. M., Barry, J. D., Holdcraft, R. J., and Fuzy, E. M. 2007. Entomopathogenic nematodes and neonicotinoids for remedial control of oriental beetle, *Anomala orientalis* (Coleoptera: Scarabaeidae), in highbush blueberry. Crop Prot. 26:1266-1271.

Rodriguez-Saona, C. R., Polk, D. F., and Barry, J. D. 2009. Optimization of pheromone deployment for effective mating disruption of oriental beetle (Coleoptera: Scarabaeidae) in commercial blueberries. J. Econ. Entomol. 102:659-669.

Wenninger, E. J., and Averill, A. L. 2006. Mating disruption of oriental beetle (Coleoptera: Scarabaeidae) in cranberry using retrievable, point-source dispensers of sex pheromone. Environ. Entomol. 35:458-464.

(Prepared by E. J. Wenninger)

Plum Curculio

A native insect to North America, the plum curculio, *Conotrachelus nenuphar* (Coleoptera: Curculionidae), is a well-known pest on pome and stone fruits but can also be a pest in highbush blueberry. Found primarily throughout eastern and midwestern growing regions, where populations are large, there can be significant fruit damage.

Life History and Description

Plum curculio adults overwinter in leaf litter in woods and hedgerows and in other protected places. Nearby orchards may also provide overwintering sites. They become active in the spring, usually just prior to bloom. Activity and mating is temperature dependent. They can be seen when average temperatures reach 10–15.5°C for at least 3–4 days or when above 24°C for at least 2 days. Adults first feed on developing buds, flower parts, and setting fruit. Adults are about 5-mm-long, grayish brown weevils with long snouts about one-fourth to one-third the length of the body (Fig. 334). There are mottled gray markings on the back and two prominent humps on the backs of the wing covers. Eggs are laid 1–2 weeks after emergence. The female eats a small hole in the fruit, deposits an egg in the hole, and makes a crescent-shaped cut below the egg. This creates a C-shaped egg scar. Eggs hatch in 2–12 days, with larvae feeding in the fruit for 1–3 weeks. Larvae are curved, yellowish white to cream colored, and legless, with a brown head capsule and a light brown shield behind the head. Fully grown larvae are about 6–7 mm long. Full-grown larvae exit the fruit and burrow 1–2 in. into the soil, where they construct cells in which to pupate about 2 weeks later. About 2–3 weeks after pupation (4–5 weeks after entering the soil), new adults emerge and feed on late-season hosts, usually not blueberries, before finding overwintering sites. The insect is univoltine in the Northeast but can be bivoltine in Virginia (United States) and further south and multivoltine in the far south.

Fig. 334. Plum curculio adult (*Conotrachelus nenuphar*) and a crescent-shaped oviposition scar on a blueberry. (Courtesy D. Polk; © APS)

Damage

Since the insect usually overwinters outside cultivated fields, most adult activity in the spring is located near field borders, especially near wooded areas. Early feeding signs on the flowers, shucks, and developing fruit, as well as egg scars, can be present in commercial plantings, especially on outside rows near overwintering sites. Because host phenology can be closely synchronized to feeding and oviposition, early- and midseason cultivars (i.e., Bluetta, Weymouth, sometimes Duke) are often more heavily damaged than late-season cultivars. Therefore, early-season cultivars planted near wooded borders are most likely to receive damage. Infested early cultivars are often harvested with internal larvae. Larval injury eventually causes premature ripening and fruit drop. While a second brood may develop in southern growing areas, they are not present during fruit development, so they never appear in commercial blueberry plantings. Untreated or wild hosts (e.g., crabapple, wild plum, blueberry, and other wild fruits) also sustain the above injuries.

Management

While some trapping systems have been tried in other fruit crops, no satisfactory attractants or traps are currently available for plum curculio in blueberry. Monitoring should be concentrated to field borders in early-season cultivars for the presence of adults and egg scars. Since blueberry can set fruit for several weeks, the first fruit set can also be the first fruit damaged while bees are still active. Since no effective insecticides can be used during this time, management can become a challenge. Effective insecticides should be applied immediately after bee removal, and if needed, one or two more times. In northern areas, adult activity generally tapers off after mid- to late June. Since early cultivars may contain larvae, special care should be taken on the packing line. This is usually unnecessary for mid- to late-season cultivars, since any infested fruit will have dropped off and larvae exited into the soil.

Selected References

Hull, L. A., Pfeiffer, D. G., and Biddinger, D. J. 1995. Apple—Direct pests. In: Mid-Atlantic Orchard Monitoring Guide NRAES-75. Northeast Regional Agricultural Engineering Service, Ithaca, NY.

Leskey, T. C., Pinero, J. C., and Prokopy, R. J. 2008. Odor-baited trap trees: A novel management tool for plum curculio (Coleoptera: Curculionidae). J. Econ. Entomol. 101:1302-1309.

Mampe, C. D., and Neunzig, H. H. 1967. The biology, parasitism, and population sampling of the plum curculio on blueberry in North Carolina. J. Econ. Entomol. 60:807-812.

Polavarapu, S., Kyryczenko-Roth, V., and Berry, J. 2004. Phenology and infestation patterns of plum curculio (Coleoptera: Curculionidae) on four highbush blueberry cultivars. J. Econ. Entomol. 97(6):1899-1905.

Reissig, W. H., Nyrop, J. P., and Straub, R. 1998. Oviposition model for timing insecticide sprays against plum curculio (Coleoptera: Curculionidae) in New York State. Environ. Entomol. 27:1053-1061.

(Prepared by D. Polk)

Blueberry Scales

Many species of scale can be found on highbush and rabbiteye blueberries, but three are most often reported: Putnam scale, *Diaspidiotus ancylus*, which is a species of armored scale (Hemiptera: Diaspididae); and two soft scale species, lecanium scale, *Parthenolecanium corni*, and terrapin scale, *Mesolecanium nigrofasciatum* (Hemiptera: Coccidae). Lesser snow scale, *Pinnaspis strachani*, an armored scale, is the primary scale pest in blueberries grown in the southern United States. Other species that infest blueberry include oystershell scale, *Lepidosaphes ulmi* (cosmopolitan); azalea bark scale, *Acanthococcus azaleae*; cottony maple scale, *Pulvinaria innumerabilis*; Japanese wax scale, *Ceroplastes japonicus*; oak eriococcus, *Eriococcus quercus*; parlatoria-like scale, *Pseudoparlatoria parlatorioides*; and red bay scale, *Acutaspis perseae.* Scale infestation is most common in older plantings where excessive old wood is retained during pruning, during very hot and dry years, and in fields where pesticides have killed natural enemy populations.

Life History and Description

Scales are small insects with a high reproductive capacity that allows a rapid buildup of populations in infested fields. Female scale insects are round and yellow but secrete a waxy covering of varying size over themselves that may be light gray to dark gray or brown, depending on the species, and may be smooth or rough, depending on the species. Scales typically spend the winter as fully developed females on stems and canes. During the spring and early summer, they produce highly mobile yellow crawlers that disperse and colonize the surrounding foliage and fruit. Upon finding a suitable feeding site, the long needlelike mouthparts are inserted into plant tissue to feed and the scale begins to form the waxy scale covering. Some species are variable in their appearance, depending on where they settle, e.g., Putnam scale is almost white when settled on fruit or leaves but is dark brown or gray when on stems. Submitting samples to a diagnostic clinic is advised for accurate identification of the scale species.

Damage

High infestations of scale can reduce plant vigor and, in severe cases, plants may be killed. Scale infestation on fruit at harvest can make berries unmarketable as fresh fruit (Fig. 335), and even a single scale per fruit can distort berry growth. Some species release large amounts of honeydew, thus providing a substrate for sooty molds that can contaminate fruit and reduce photosynthesis.

Management

Removing and destroying old canes is a highly effective preventative measure to avoid scale infestation. Regularly pruned fields are much less likely to experience scale infestations. Early detection is important to avoid high populations, and a hand lens can help detect scale infestations on wood. Loose bark should be examined for overwintering females in March at multiple sites through each field. To determine when crawlers emerge for timing control sprays, double-sided sticky tape should be wrapped around canes during the late dormant pe-

Fig. 335. Putnam scale (*Diaspidiotus ancylus*) on blueberries. (Courtesy R. Isaacs; © APS)

riod in regions where active pest populations are suspected. By examining and changing tape every week and examining under a microscope, the level of activity over time can be determined. Later in the season, scales stand out more clearly on the underside of waxy green leaves and on coloring berries, where they may be surrounded by a circular discoloration.

Scale insects are subject to predation and parasitism by a wide range of other insects, including ladybeetles and lacewings. Parasitic wasps are effective biocontrol agents and their presence can be monitored with a hand lens, owing to the fact that the wasp leaves a small circular hole when it exits the scale. Growers should aim to conserve natural enemies where possible through selection of more selective insecticides. Use of oil early in the season to kill the overwintering females or use of a selective growth regulator later in the season (discussed below), timed to control the crawlers, can help maintain scale biocontrol agents.

Dormant sprays of oil or lime sulfur are timed to control the overwintering female scale before reproduction. Thorough coverage is essential to penetrate the crevices in the bark, so 190–380 L of water per 0.4 ha are needed and applications should be made between dormancy and 6-mm leaf opening. Growers should use 7.6–11.4 L per 0.4 ha of a 70-sec superior-type oil labeled and sold specifically as an insecticide. Powdered or liquid lime sulfur applied in the dormant stage for disease control also helps control scale. Both oil and lime sulfur may be phytotoxic if used after this growth stage, and they should also not be applied together or within 24 h of cold weather. Applications on still days when temperatures are above 10°C avoids plant injury. A more recently developed option is the insect growth regulator pyriproxifen, which is a highly effective scale control. This can be added to oil in the dormant timing or applied when crawlers first emerge.

Selected References

Miller, D. R., and Davidson, J. A. 2005. Armored scale insects of trees and shrubs. Cornell University Press, Ithaca, NY.

Polavarapu, S., Davidson, J. A., and Miller, D. R. 2000. Life history of the Putnam scale, *Diaspidiotus ancylus* Putnam, (Hemiptera: Coccoidea: Diaspididae) on blueberries (*Vaccinium corymbosum*, Ericaceae) in New Jersey, with a world list of scale insects on blueberries. Proc. Entomol. Soc. Wash. 102:549-560.

Rosen, D. 1990. Armored scale insects: Their biology, natural enemies and control. Volume B. Elsevier, Amsterdam.

(Prepared by R. Isaacs)

Sharp-Nosed Leafhopper

The sharp-nosed leafhopper, *Scaphytopius magdalensis* (Hemiptera: Cicadellidae), is one of several species of leafhoppers that feed on blueberry. This species is, however, of considerable importance because it is the primary vector of a phytoplasma that causes blueberry stunt disease. Sharp-nosed leafhoppers are common throughout the eastern United States, where they feed and reproduce on blueberry, huckleberry, cranberry, and other plants in the family Ericaceae. They are abundant in abandoned blueberry fields and in wooded habitats surrounding blueberry fields where wild blueberry and huckleberry are common in the ground cover.

Life History and Description

In the southern part of its range, the sharp-nosed leafhopper completes three generations a year, while in the northern states, it completes only two generations a year. The eggs overwinter inside fallen leaves. In southern states, eggs hatch around the middle of March (about the time when leaf buds begin to open), with peak adult populations appearing near the end of May, the middle of July, and the beginning of October. In northern states, eggs hatch in mid- to late spring, with the first-generation adults appearing by late May to early June. Peak adult populations from the second generation appear mid- to late August to early September. Nymphs are ivory in color and, as they grow, they develop a white or cream hourglass-shaped mark on the dorsal surface. Nymphs reach adult stage after completing five instars. Adults are 5 mm long, brownish gray, and wedge shaped (Fig. 336). Both nymphs (Fig. 337) and adults have a distinct pointed extension of the head shaped like the horn of an anvil. Adults are capable of long flights and tend to disperse out of the woods into cultivated plantings during the spring generation and back into the woods during the autumn generation. Only motile adults transmit the disease.

Damage

Sharp-nosed leafhoppers cause little direct injury to blueberry plants. They are most important when they feed on stunt-infected plants, acquire the stunt pathogen, and transmit the disease to healthy plants in subsequent feedings. Rabbiteye blueberry is less preferred for feeding and oviposition than is highbush blueberry.

Management

Unbaited blueberry maggot fly yellow sticky traps can be used to monitor leafhoppers and help time insecticide sprays when necessary. These traps should be hung vertically directly on the bushes 0.5 m above the ground. Both nymphs and adults can be collected with a sweep net; however, sweep net sampling

Fig. 336. Sharp-nosed leafhopper adults (*Scaphytopius magdalensis*). Note the mottling on the wings and the pronounced point of the head. (Courtesy D. Polk; © APS)

Fig. 337. Sharp-nosed leafhopper nymph (*Scaphytopius magdalensis*). (Courtesy D. Polk; © APS)

tends to give highly variable estimates of population density. Sharp-nosed leafhoppers are quite susceptible to insecticides registered for use in blueberry. The first generation is often controlled with sprays targeted for plum curculio, aphids, and cranberry fruitworm. Treatment decisions for the following generations should be based on individual population levels and the history of stunt disease. Because adults migrate from woods, monitoring should be intensified in, and sprays should be directed to, the perimeter of fields to control migrants carrying the disease. It is also important to remove all plants that show symptoms of stunt disease. Removal of bushes should be done after insecticide treatment to avoid the movement of leafhoppers from infested to healthy plants, thereby facilitating spread of the disease.

Selected References

Etzel, R. W., and Meyer, J. R. 1986. Resistance in blueberries to feeding and oviposition by the sharpnosed leafhopper, *Scaphytopius magdalensis* Provancher (Homoptera: Cicadellidae). J. Econ. Entomol. 79:1513-1515.

Hutchinson, M. T. 1955. An ecological study of the leafhopper vectors of blueberry-stunt. J. Econ. Entomol. 48:1-8.

Meyer, J. R. 1984. Life history of the sharpnosed leafhopper [*Scaphytopius magdalensis* (Provancher)] and four related species in southeastern North Carolina. J. Ga. Entomol. Soc. 19:72-87.

Tomlinson, W. E., Jr., Marucci, P. E., and Doehlert, C. A. 1950. Leafhopper transmission of blueberry stunt disease. J. Econ. Entomol. 43:658-662.

(Prepared by C. Rodriguez-Saona and D. Polk)

Winter Moth and Bruce Spanworm

The winter moth, *Operophtera brumata* (Lepidoptera: Geometridae), has become a pest of blueberry and cranberry in eastern North America and is a sporadic pest of these *Vaccinium* crops in western North America. Native to Europe, winter moth was introduced to North America, where it has been a major threat to hardwood forests in Nova Scotia (Canada), shade trees in British Columbia (Canada), and filbert orchards in Oregon (United States). The congeneric species, Bruce spanworm, *O. bruceata*, superficially resembles winter moth and has a similar host range but is native to North America. Adults and eggs of both species are adapted for winter survival.

Life History and Description

Winter moth adults (Fig. 338) emerge from pupation sites in October and November. The wingless females begin emerging about 1 week after the first males. Bruce spanworm adults (Fig. 339) emerge slightly later in the autumn than do winter moth adults. Females of both species climb trees or bushes and stop about 1–2 m above the ground to release pheromone, which attracts males for mating. Pheromone communication among winter moths operates at temperatures as low as 4°C. The threshold for male winter moth flight is about 5°C. After mating, winter moth females lay 70–300 eggs in cracks on bark and under lichens and mosses. Winter moth eggs can survive temperatures of –31 to –37°C without being frozen and killed. Larvae within the eggs begin to develop at about 4°C. It has been shown that Bruce spanworm can persist in areas with much colder winters than can winter moth.

In southwestern British Columbia and in Massachusetts (United States), larvae hatch from eggs in mid-March through April. Larvae are tiny (1 mm or 0.04 in. long) at emergence and tend to disperse, which they do by spinning down from the foliage on a silken thread and then drifting along on air currents. Larvae are especially likely to disperse if they emerge before food is available, that is, before buds have begun to swell. Once a larva settles on a bud, it tunnels in and feeds on developing leaves or petals. If second instars run out of food, they disperse by drifting with the wind. Later instars tie leaves together with silken threads and feed on tender leaf tissue. (In cranberry, larvae make tents similar to those made by black-headed fireworm larvae.) All five instars are pale to dark green, with three pale white longitudinal lines along their sides (Fig. 340). Larvae move by "inching": arching their bodies to move the posterior end forward and then propping with two pairs of hind (pro-) legs to lift and advance the head and three pairs of front (true) legs. When final-instar larvae (about 13 mm or 0.5 in. long) stop feeding, they ease down on a silken thread to the ground and then tunnel 1–2 cm into the soil and spin a cocoon in which they pupate. The pupal stage typically lasts from May until October or November. There is one generation of winter moths (and Bruce spanworms) per year.

Damage

By feeding as described above, larvae weaken or kill developing apical shoots of cranberry and blueberry. Serious economic damage results when large numbers of blossoms are entirely consumed, which can prevent large patches of cranberry or blueberry plants from producing crop. In British Co-

Fig. 338. Winter moth adult (*Operophtera brumata*). (Courtesy J. T. Troubridge of the Department of Agriculture and Agri-Food, Government of Canada; © Her Majesty the Queen in Right of Canada, as represented by the Minister of Agriculture and Agri-Food Canada)

Fig. 339. Bruce spanworm adult (*Operophtera bruceata*). (Courtesy J. T. Troubridge of the Department of Agriculture and Agri-Food, Government of Canada; © Her Majesty the Queen in Right of Canada, as represented by the Minister of Agriculture and Agri-Food Canada)

lumbia and in Massachusetts, larvae drift from infested trees onto adjacent cranberry beds.

Management

Monitoring for winter moth adults can be done from October until January by using wing-type sticky traps or universal moth traps baited with the pheromone (*Z*,*Z*,*Z*)-1,3,6,9-nonadecatetraene. Traps placed in or near cranberry or blueberry plantings attract male winter moths and male Bruce spanworms. Adults of the two species can be distinguished by wing characters or by mitochondrial DNA.

In the spring, larvae of winter moth and Bruce spanworm can be detected during visual examination of blueberry buds and during visual or sweep net monitoring of cranberry. An appropriately labelled insecticide, used according to directions on the label, can be applied to areas where larvae have been detected.

Larvae of winter moths and Bruce spanworms are preyed upon by spiders, lacewing larvae, and ladybird beetles. In British Columbia, there are at least three native species of parasitic wasps in the family Ichneumonidae (*Agrypon provancheri*, *Phobocampe* spp., and *Trichistus crassus*) and one native parasitic fly in the family Tachinidae (*Cyzenis pullula*) that attack larvae of both species. In Oregon, six species of native parasitoids are known to attack winter moth larvae. Diseases kill a small proportion of larvae. Ground beetles and small rodents are important predators of winter moth and Bruce spanworm pupae.

Winter moths have been extensively studied in Britain, where winter moth populations peak every 9–10 years. Population cycles are mostly caused by three factors: the degree of synchrony between winter moth egg hatch and host plant bud burst; predation on pupae in the soil; and the impact of parasitoids and disease. If egg hatch happens before bud burst, first instars starve or disperse by drifting to other host plants. If hatch happens after bud burst, larvae tend not to disperse. Plants that leaf out early (e.g., apple) have higher densities of winter moth than those that leaf out later (e.g., oak). In the soil, pupae are heavily preyed upon by generalist predators, especially beetles. Predation is heaviest, and most effective, where there are many winter moth pupae. Parasitism by the European parasitoids *Cyzenis albicans* (Tachinidae) and *Agrypon flaveolatum* (Ichneumonidae) causes less mortality than the first two factors (typically less than 10%) but becomes important during the decline phase of an outbreak.

Studies in Nova Scotia and British Columbia suggest that population cycles in Canada are similar to those in Britain and are complicated by pesticide applications in agricultural situations. The introduced parasitoid *Cyzenis albicans* was associated with decline of winter moth populations in Canada. This parasitoid is presently being released in New England to control winter moth populations there.

Fig. 340. Winter moth larva (*Operophtera brumata*). (Courtesy J. T. Troubridge of the Department of Agriculture and Agri-Food, Government of Canada; © Her Majesty the Queen in Right of Canada, as represented by the Minister of Agriculture and Agri-Food Canada)

Selected References

Cuming, F. G. 1961. The distribution, life history, and economic importance of the winter moth, *Operophtera brumata* (L.) (Lepidoptera, Geometridae) in Nova Scotia. Can. Entomol. 93:135-142.

Elkinton, J. S., Boettner, M. S., Gwiazdowski, R., Hunkins, R. R., Callahan, J., Scheufele, S. B., Donahue, C. P., Porter, A. H., Khrimian, A., Whited, B. M., and Campbell, N. K. 2010. Survey for winter moth (Lepidoptera: Geometridae) in northeastern North America with pheromone-baited traps and hybridization with the native Bruce spanworm (Lepidoptera: Geometridae). Ann. Entomol. Soc. Am. 103:135-145.

Elkinton, J. S., Lance, D., Boettner, G., Khrimian, A., and Leva, N. 2011. Evaluation of pheromone-baited traps for winter moth and Bruce spanworm (Lepidoptera: Geometridae). J. Econ. Entomol. 104:494-500.

Fitzpatrick, S. M., Troubridge, J. T., and Peterson, B. 1991. Distribution of the European winter moth, *Operophtera brumata* (L.), and Bruce spanworm, *O. bruceata* (Hulst), in the lower Fraser Valley, British Columbia. J. Entomol. Soc. B. C. 88:39-45.

Roland, J., and Embree, D. G. 1995. Biological control of the winter moth. Annu. Rev. Entomol. 40:475-492.

Troubridge, J. T., and Fitzpatrick, S. M. 1993. A revision of the North American *Operophtera* (Lepidoptera: Geometridae). Can. Entomol. 125:379-397.

(Prepared by Sheila M. Fitzpatrick)

Blueberries (Lowbush)

Blueberry Sawfly

Blueberry sawflies (Hymenoptera: Tenthredinidae) are common in lowbush blueberry fields in Maine (United States) and eastern Canada but do not generally cause significant injury. Sawfly as a pest appears to be less common on highbush and rabbiteye blueberries. Because of this, the discussion below mostly refers to their biology and management in lowbush blueberry. Two species of sawfly larvae feeding on blueberry leaves were collected in a field in Maine. Nineteen species were identified from lowbush blueberry fields in New Brunswick (Canada); however, only three (i.e., *Neopareophora litura*, *Pristiphora idiota*, and a *Pristiphora* sp.) were found to feed on lowbush blueberry, the remainder were associated with other plant species growing in or near blueberry fields. *Neopareophora litura* is the most common blueberry sawfly in maritime Canada.

Life History and Description

Adults are 6 mm long, black, and generally wasplike in appearance, but they do not have the narrow waist usually associated with a wasp. They have membranous wings that they hold flat over their body. The slow-moving, caterpillar-like larvae (Fig. 341) are usually the color of the blueberry foliage (grass green) (*Neopareophora litura*), which makes them difficult to detect directly on the foliage. Larvae of *Pristiphora idiota* are similar in appearance but with a conspicuous white dorsal stripe. Fully grown larvae are 9–15 mm long. Pupae, which are found in the litter layer, are enclosed in a cocoon camouflaged with dead leaves or other debris that adheres to the sides.

The most common species, *Neopareophora litura* and *Pristiphora idiota* have only one generation a year. The larvae of all species overwinter within cocoons. The overwintered larvae change to prepupae shortly before the adults emerge. The emergence period varies with the species and the season. The females of all the species are parthenogenic and begin oviposition immediately after emergence. In Maine, blueberry sawfly adults lay eggs in early to late May inside developing, still folded, leaf whorls of lowbush blueberry. The eggs hatch and larvae emerge within 1–2 weeks. The length of the feeding period ranges from 4 to 6 weeks, depending on species. After reaching maturity, the larvae move to the ground and spin cocoons in the loose duff and litter, where they spend the winter. Several species of ichneumon parasitoids attack blueberry sawfly species and are thought to be responsible in maintaining low population densities during most years.

Damage

The young larvae, which are rarely seen, are white or flesh colored with black heads and black legs and feed on leaf tissue inside the developing whorl, killing the new leaves and filling the interior of the whorl with their excrement, a flakey brown material. Older, green larvae appear on the blueberry foliage in late May or early June, when leaves are well developed. The larvae continue to feed on the foliage until mid- to late June. When feeding, they usually coil their body over the edge of the leaf and chew around the edge. Injured leaves have a scalloped appearance.

Management

Although not usually an economic concern to growers, in large numbers, larvae may cause damage by defoliating the plants. Infestations are usually to isolated areas within a field. Sweep net sampling is used to monitor for the presence of late instars. Early-instar sawflies, which feed inside the developing lead buds, are difficult to detect and not collected in sweep net samples. In Maine, a threshold of 50 larvae collected in ten 180° sweeps with a 12-in.-diameter sweep net is considered the action threshold. The blueberry sawfly is susceptible to many of the insecticides used in blueberry pest management, but control can be challenging since sawfly larvae are often not detected by scouting until bloom or just prior to bloom. Early scouting in late April and early May for adult sawflies does not provide growers with an indication of which fields have potential for high densities of larvae and thus defoliation. There is a tendency for outbreaks to happen more often in fields pruned by mowing. Pruning by burning (in the autumn or spring) is a cultural control and reduces the densities of overwintering larvae and prepupae, provided the leaf litter is ignited and thoroughly burned.

Selected References

Neilson, W. T. A. 1955. Sawflies collected from low-bush blueberry fields in New Brunswick (Hymenoptera: Tenthredinidae). Can. Entomol. 87:356-357.

Neilson, W. T. A. 1958. Notes on life-histories of sawflies (Hymenoptera: Tenthredinidae) common in low-bush blueberry fields in New Brunswick. Can. Entomol. 90:229-234.

Phipps, C. R. 1930. Blueberry and huckleberry insects. Maine Agric. Exp. Stn. Bull. 356:107-232.

Yarborough, D. E., and Drummond, F. 2001. Integrated Crop Management Field Scouting Guide for Lowbush Blueberries. Fact Sheet 204. Univ. Maine Ext. No. 2275.

(Prepared by F. A. Drummond and J. A. Collins)

Fig. 341. Blueberry sawfly larvae (*Neopareophora litura*) on velvet leaf lowbush blueberry (*Vaccinium myrtilloides*). (Courtesy F. A. Drummond; © APS)

Blueberry Spanworm

The blueberry spanworm, *Speranza argillacearia* (syn. *Itame argillacearia*) (Lepidoptera: Geometridae), is one of a group of closely related species, all in the genus *Speranza* (*Itame*), that feed on lowbush blueberry (*Vaccinium angustifolium*), half-highs (*V. angustifolium* × *V. corymbosum*), and sour top blueberry (*V. myrtilloides*). In Quebec (Canada), it is one of the earliest season foliage feeders on cranberry but is not reported as problematic in other regions. In lowbush blueberry, although many years may pass between infestations, the frequency of blueberry spanworm outbreaks has increased with the change of pruning practices from burning to mowing. Closely related species of spanworms may also cause infestations; these species are presumed to have life cycles, behaviors, and damage potential similar to that of the blueberry spanworm.

Life History and Description

The blueberry spanworm spends the winter in the egg stage in leaf litter in blueberry fields. Egg hatch happens over a 3- to 4-week period, depending upon temperature. First to third instars are 3–20 mm long and are dark gray to black with a series of white bands encircling the body. Fully grown last-instar (fourth-instar) larvae are about 20–25 mm long and are yellowish orange with longitudinal rows of black spots, which look like continuous black stripes (Fig. 342). Blueberry span-

Fig. 342. Blueberry spanworm larva (*Speranza argillacearia*) on blueberry. (Courtesy F. A. Drummond; © APS)

worm larvae have three pairs of thoracic legs and two pairs of fleshy prolegs at the posterior of the abdomen. Larvae begin feeding in mid-April (in Maine [United States]) on developing leaf and flower buds. Later in the season, larvae feed on leaves and tender stems. Blueberry spanworm larvae are crepuscular and often appear on plants in large numbers in the evening and prior to dawn. During the day, larvae drop down on silk threads and hide in the leaf litter at the base of plants. Mature larvae go through a wandering prepupal stage in which they leave the plant, walk along the ground for a short distance, burrow into the leaf litter, and pupate. Pupae are 10 mm long and dark brown or black. Pupation usually happens between full bloom and fruit set. Adults emerge from the middle of June through July, depending upon the temperature. They are delicate, light gray-brown moths with a wingspan of 23–29 mm. The adults are generally most active on calm warm evenings, but they readily fly when flushed from the foliage. Because the moths are weak fliers, they are especially common in wind-protected weedy areas or wind breaks. There is only one generation per year. Periods of regional outbreaks have been as long as 37–38 years in New Hampshire (United States) and 8–10 years in Maine.

Damage

Infestations of blueberry spanworm may be confined to isolated areas or damage may be widespread. Large numbers may completely defoliate areas in both crop and pruned fields. Early in the season, the larvae damage the berry crop by eating flower buds and blossoms. Later, larvae chew out notches on developing leaves. The first sign of a severe infestation in a pruned field is an area devoid of, or with slower-developing, sprouts. Signs of feeding on developing shoots at ground level or below the soil surface can be detected. With severe outbreaks, extensively damaged areas can appear burned, with only twigs and stems remaining.

Management

Monitoring of blueberry spanworm is performed with a 30-cm-diameter sweep net. The threshold in a crop-year field is 10 larvae per ten 180° sweeps and five or more per ten sweeps in a prune-year field. Larvae feed from late April through late June. Monitoring in crop-year fields should begin in mid-April to early May (bud break). In prune-year fields, monitoring should commence when blueberry sprouts begin emerging. Pruning fields by burning the leaf litter reduces overwintering egg populations. Blueberry spanworm larvae are some of the most susceptible insect pests to insecticides registered for blueberry. Organophosphate, carbamate, pyrethroid, spinosyns, and neonicitinoid insecticides all are effective synthetic insecticides. In addition, various growth regulators and *Bacillus thuringiensis* toxins are also effective as control measures that may be used once larval populations exceed threshold levels. These latter classes of insecticides are recommended if spanworm populations are detected late in a crop field during bloom.

Natural enemies are suspected of regulating populations and the frequency of outbreaks. An unidentified polyhedrosis virus disease is prevalent in populations prior to population crashes. Ichneumon parasitoids of the genus *Ophion* are also common. Generalist predators, such as ants, ground beetles, and harvestmen, have also been found to consume blueberry spanworm.

Selected References

Collins, J. A., and Forsythe, H. Y., Jr. 1995. Blueberry Spanworm. Maine Wild Blueberries Fact Sheet No. 197. Univ. Maine Ext. No. 2371.

Crozier, L. 1995. The Blueberry Spanworm. Lowbush Blueberry Fact Sheet. Nova Scotia Department of Agriculture and Marketing, Truro, NS.

Drummond, F. A., and Groden, E. 2000. Evaluation of entomopathogens for biological control of insect pests of lowbush (wild) blueberry. Maine Agric. For. Exp. Stn. Tech. Bull. 172.

Forsythe, H. Y., Jr., and Flanders, K. L. 1982. The blueberry spanworm in 1981. Maine Life Sci. Agric. Exp. Stn. Misc. Rep. No. 262.

Maund, C., and Savoie, B. 1996. Blueberry Spanworm. Fact Sheet C.2.2.0. New Brunswick Department of Agriculture and Aquaculture, Fredericton, NB, Canada.

(Prepared by D. E. Yarborough, F. A. Drummond, and J. A. Collins)

Blueberry Thrips

Blueberry thrips (Thysanoptera: Thripidae), also known as lowbush blueberry thrips, is composed of an assemblage of several species that affects the leaves on blueberry plantings. Two of the most important species are *Frankliniella vaccinii* and *Catinathrips kainos*, both considered native to North America and specialists of blueberry. These leaf-dwelling thrips are more common on lowbush blueberry than on tall species, such as rabbiteye or southern highbush blueberries.

Life History and Description

See Flower Thrips under the Blueberries (Highbush and Rabbiteye) section for other general information on this group. Thrips associated with lowbush blueberry tend to be lemon yellow. Using their single mandible as a hammer, thrips break the plant cells and then feed on the exposed cell contents. *F. vaccinii* was observed for the first time in Maine (United States) in 1926, and it has been reported affecting wild *Vaccinium* spp. throughout the east coast from Maine to Georgia (United States). Active adults of this species are more common during the summer, between mid-July and late August. Nymphs and pupae are collected from the middle of June to the beginning of August on leaf tissues in Maine. This species overwinters in the soil as an adult and reemerges in the spring (mid-May to mid-June in Maine) to lay eggs. *Catinathrips kainos* was differentiated from *Catinathrips vaccinophilus* in 1967 by O'Neill. *C. kainos* was reported in Maine and Massachusetts (United States) and in New Brunswick (Canada), and *C. vaccinophilus* was reported to be present in New York and Pennsylvania (United States). Unlike *F. vaccinii*, active adults of *C. kainos* are observed during the late spring to early summer (end of May to beginning of July). However, immature stages of both species are observed on blueberry leaves between mid-June and early August. The phenology of *C. kainos* suggests that it overwinters as a late immature stage and the adults emerge in late spring. This explains the high density of adults early in the season. Less common species within the blueberry thrips complex are *F. caudiseta* and *C. similis*, among others.

Damage

Their small size and cryptic behavior makes them difficult to spot before symptoms begin to appear in the field. Injury is more commonly seen in vegetative fields than in fruit-bearing fields. In part, this may be because of the ease in damage detection in vegetative fields that have lower canopy density early in the season. However, it is also hypothesized that success in feeding, and thus resulting damage, is associated with a high proportion of rapidly growing and expanding leaves in vegetative fields. Affected leaves do not unfold properly and remain curled around the stem, affecting photosynthesis and the development of flower buds. Thrips develop inside the curled leaves, which protect the insect from natural enemies and contact insecticides. Thrips injury can be recognized as a patch of underdeveloped plants that have reddish curled leaves (Fig. 343) and a lack of fruit during the crop cycle. Populations usually do not reach significant levels, but if left unmanaged, they can cause economic

Fig. 343. Reddish and curled leaves caused by blueberry thrips (*Frankliniella vaccinii* or *Catinathrips kainos*) injury to lowbush blueberry. (Courtesy F. A. Drummond; © APS)

damage and even total crop loss in heavily infested sections of a field. During late summer and early autumn, blueberry thrips leave the leaf curls and burrow into the ground to overwinter.

Management

Monitoring is the key to efficient management. At the end of the summer, damage symptoms in fruit-bearing fields should be mapped so that control tactics can be initiated the following season when the field is in the vegetative stage. In mapped infestations, blue sticky cards should be placed on stakes in mid-May, 8–10 cm from the ground surface, and be checked every 3 days. At the first sign of lemon yellow thrips or the appearance of the first curled leaves, the area should be treated and applications repeated at 1-week intervals if thrips continue to emerge. If blue traps are not available, chemical treatment should be applied when emerging blueberry sprouts are 0.6–1.2 cm long, and a repeat application should be made when sprouts are 2.0–2.5 cm long. These chemical treatments should be applied only in the area that has been marked as infested during the previous season. Neonicotinoid soil applications, made presprout, are effective during the vegetative year. A cultural control tactic is a delayed burn. Thrips are allowed to emerge and colonize the blueberry plants during the vegetative year and then are destroyed by burning the field ("hard" burn). The burn should be performed no later than June 1–15 (warm or cool spring, respectively), or the following year's flower bud production will be significantly reduced.

Selected References

Collins, J. A., and Drummond, F. A. 2010. Blueberry thrips control, 2009. Arthropod Manage. Tests 35:C8.

Collins, J. A., Forsythe, H. Y., and Yarborough, D. E. 1995. Blueberry thrips *Frankliniella vaccinii* Morgan and *Catinathrips kainos* O'Neill. Maine Wild Blueberries Fact Sheet 202. Univ. Maine Ext. No. 2373.

Drummond, F., Annis, S., Smagula, J. M., and Yarborough, D. E. 2009. Organic Production of Wild Blueberries. I. Insects and Disease. Acta Hortic. 810:275-286.

Langille, R. H., and Forsythe, H. Y. 1972. Biology of the blueberry thrips *Frankliniella vaccinii* and *Catinathrips kainos* in Maine. Can. Entomol. 104:1781-1786.

Lathrop, F. H. 1942. The blueberry thrips. J. Econ. Entomol. 35:198-201.

Yarborough, D. E., and Drummond, F. 2007. Insect control guide for wild blueberries. Maine Wild Blueberries Fact Sheet 209. Univ. Maine. Ext. No. 2001.

Yarborough, D. E., Peronto, M., and Collins, J. 1993. Integrated Crop Management Field Scouting Guide for Lowbush Blueberries. Fact Sheet 204. Univ. Maine Ext. No. 2275.

(Prepared by H. A. Arevalo, F. A. Drummond, and J. A. Collins)

Red-Striped Fireworm

The red-striped fireworm, *Aroga trialbamaculella* (Lepidoptera: Gelechiidae), is considered a minor pest of lowbush blueberry, cranberry, locust, oaks, and various other *Vaccinium* crops in the northeastern United States and eastern Canada. The larva feeds on leaves (specifically, the leaf surface, outer epidermis, and into the mesophyll) and, in some cases, on flower buds.

Life History and Description

The young larvae have light green bodies with dark-colored heads. As they develop, older larvae are characterized by red lines running longitudinally along their bodies, becoming darker with time (Fig. 344). Fully developed larvae are approximately 1 cm long and may have a maroon coloration. Moths are approximately 1.25 cm long and have a dark brown or black body, a white face, and white spots on the forewings and legs. Red-striped fireworm is univoltine and overwinters as a larva in leaf litter and debris in blueberry fields. In early spring, they pupate, and after 1–2 weeks, adults emerge in early April to late May. They are agile fliers and difficult to detect in the field. The moths lay eggs individually on blueberry plants. They are active from mid-May to early August, with peak populations in July. Eggs are laid on the leaves. Soon after egg hatch, the larvae feed on surfaces of leaves and web together several leaves, where they reside through late summer. As the larvae grow, more leaves are added to the initial web. These webbed shelters facilitate food gathering and provide protection from predators, parasitoids, and insecticides. In the autumn, the last instar moves to the leaf litter to overwinter, where it is susceptible to spider predation.

Damage

In heavily infested fields, close to 50% of the lowbush blueberry stems may be infested and associated with webbed leaves. Historically, the larvae have been implicated in reducing the number of flower buds in lowbush blueberry, although, this effect has not been found in Maine (United States), where several years of investigation could not demonstrate a reduction in yield caused by red-striped fireworm infestation. Concern about red-striped fireworm is mostly because of contamination of the harvested crop with larvae. Red-striped fireworm are also found in harvesting machinery and in packing and processing lines. Machinery that moves debris, such as harvesters, is a potential source for new infestations. Contamination of the processed crop is of particular concern in export markets.

Management

No thresholds or treatments have been developed for red-striped fireworm management, nor have any major natural en-

Fig. 344. Webbed blueberry leaf shelter opened to expose a red-striped fireworm larva (*Aroga trialbamaculella*). (Courtesy F. A. Drummond; © APS)

emies or other population-regulating factors been reported. However, it has been demonstrated that the most effective timing for applications of insecticides is between egg hatch and the commencement of leaf webbing. The most effective cultural method is burning of fields during pruning. A "hard" burn is necessary to expose the overwintering larvae to lethal temperatures. It is recommended (in Maine) that leaf litter be well ignited to provide high levels of mortality. In general, fields that are most likely to be infested are mowed during the pruning stage. Applications of insecticide for blueberry maggot fly often control the red-striped fireworm, except when perimeter treatments are used and large sections of the field interior is not treated. Some of the new insecticides, with translaminar leaf penetration, offer potential as good control options for this pest where contamination of the harvested fruit is a serious concern.

Selected References

Collins, J. A., and Drummond, F. A. 2007. Red-striped fireworm control, 2006. Arthropod Manage. Tests 32:C4.

Collins, J. A., and Forsythe, H. Y., Jr. 1994. Red-striped fireworm (*Aroga trialbamaculella* Cham.). Maine Wild Blueberries Fact Sheet 205. Univ. Maine Ext. No. 2284.

Drummond, F. A., and Groden, E. 2000. Evaluation of entomopathogens for biological control of insect pests of lowbush (wild) blueberry. Maine Agric. For. Exp. Stn. Tech. Bull. 172.

Drummond, F., Annis, S., Smagula, J. M., and Yarborough, D. E. 2009. Organic Production of Wild Blueberries. I. Insects and Disease. Acta Hortic. 810:275-286.

Wood, G. W. 1972. Effects of feeding by the red-striped fireworm on lowbush blueberry production. J. Plant Sci. 52(3):397-398.

Yarborough, D. E., Peronto, M., and Collins, J. 1993. Integrated Crop Management Field Scouting Guide for Lowbush Blueberries. Fact Sheet 204. Univ. Maine Ext. No. 2275.

(Prepared by H. A. Arevalo, F. A. Drummond, and J. A. Collins)

Fig. 345. Strawberry rootworm beetle (*Paria fragariae*) on blueberry. (Courtesy J. A. Collins; © APS)

Fig. 346. Feeding injury by adult strawberry rootworm (*Paria fragariae*) on blueberry. (Courtesy J. A. Collins; © APS)

Strawberry Rootworm

Strawberry rootworm, *Paria fragariae* (Coleoptera: Chrysomelidae), is more commonly a pest of strawberry but can also damage blueberry plants in Maine (United States).

Life History and Description

Adult beetles are about 3 mm long and oval. The color may vary from brown with four darker blotches on the back to solid black (Fig. 345). The larvae are white and slightly longer than the adult and are found in the soil. Adults overwinter in protected sites, such as ground litter, mulch, and soil crevices, and become active in early May. Peak beetle numbers appear between late May and early June. Eggs are laid during this period and developing larvae burrow into the ground, where they feed on the roots of strawberries and other related plants (*Rubus* spp.). They pupate in the soil. The next generation of adults emerges from mid-July through August and feeds on foliage for the remainder of the season until overwintering.

Damage

Adult strawberry rootworms eat holes in the leaves of lowbush blueberry (Fig. 346). When beetles are abundant, the leaves are riddled with holes, giving the plants a ragged appearance.

Management

Although strawberry rootworms appear in many fields, this insect is not usually an economic concern to growers. Infestations are usually confined to isolated areas within a field, especially areas with broadleaf weeds. Sweep net sampling is used to monitor for the presence of adult beetles. Adults are primarily crepuscular; therefore, monitoring activities are most effective when done at dusk. In Maine, a threshold of 50 adults collected in ten 180° sweeps with a 30-cm-diameter net is considered damaging. Strawberry rootworm adults are susceptible to most insecticides used in lowbush blueberry; however, control can be difficult if the adult stage overlaps with bloom. Control tactics generally target the overwintering generation of adults, but the summer adults can be targeted also. A cultural control is the removal of all wild strawberry and bramble larval host plants, although this may not be possible in some locales.

Selected References

Collins, J. A., and Drummond, F. A. 2004. Strawberry rootworm control, 2003. Arthropod Manage. Tests 29:C6.

Collins, J. A., Forsythe, H. Y., Jr., and Yarborough, D. E. 1996. Strawberry rootworm. Maine Wild Blueberry Fact Sheet 199. Univ. Maine Ext. No. 2370.

Drummond, F. A., Yarborough, D., Smagula, J., and Annis, S. 2008. Organic Wild Blueberry Production. Wild Blueberry Fact Sheet No. 304, Ext. Bull. No. 2029.

Yarborough, D. E., and Drummond, F. 2001. Integrated Crop Management Field Scouting Guide for Lowbush Blueberries. Fact Sheet 204. Univ. Maine Ext. No. 2275.

(Prepared by F. A. Drummond and J. A. Collins)

Cranberries

Black Vine Weevil

The black vine weevil, *Otiorhynchus sulcatus* (Coleoptera: Curculionidae), has been present in North America for almost 200 years. It is a cosmopolitan pest with a very broad host range. Black vine weevil is established around the world in nurseries, greenhouses, gardens, and many horticultural crops in Mediterranean, temperate, and northern climates. It is known to feed on broad-leaved plants in the families Ericaceae, Pinaceae, Primulaceae, Rosaceae, Saxifragaceae, Taxaceae, Vitaceae, and Onagraceae and on perennial grasses. Black vine weevil has long been a pest of cranberry on the west coast of North America and has more recently become a pest in Massachusetts (United States) but is not a pest of Wisconsin (United States) cranberry. Two other weevils in the genus *Otiorhynchus* are occasional pests of cranberry. Strawberry root weevil, *Otiorhynchus ovatus*, and clay-colored weevil, *Otiorhynchus singularis*, are found in west coast cranberry; strawberry root weevil has recently been reported in Massachusetts cranberry.

Life History and Description

The adult black vine weevil is sometimes called a snout beetle because the anterior portion of the head is projected forward into a snout shape. Adults are 8–11 mm long, flightless, and black with tiny tufts of orange hairs on the fused elytra covering the abdomen (Fig. 347). All adults are female and lay fertile eggs without previously mating.

Adults emerge in June and early July from pupae (Fig. 347) in the soil. At first, adults are white like the pupa, and they remain in the soil while the cuticle hardens and changes color to brown and then black. After the cuticle has hardened, adults hide at the base of plants or in the trash layer during the day. At night, adults move up to feed on the foliage of host plants, chewing semicircular notches 1–5 mm wide along leaf edges. On cranberry, a feeding notch can look more like a tear that encompasses the entire tip of the cranberry leaf. Adult feeding damage does not usually kill or weaken the plant but can be used as an indicator that weevils are present. After 4–7 weeks of feeding, adults lay several hundred fertile eggs in the soil at night. The type of plant foliage consumed influences the length of the preoviposition period and the number of eggs laid. Weevils with the shortest preoviposition period, such as those feeding on blackberry, lay the most eggs. Egg laying continues as long as moderate conditions last, often into the autumn. The time to egg hatch is temperature and humidity dependent, ranging from 40 days at 10°C to 10 days at 25°C.

Fig. 347. Larva, adult, and pupa (left to right) of black vine weevil (*Otiorhynchus sulcatus*). (Courtesy S. M. Fitzpatrick of the Department of Agriculture and Agri-Food, Government of Canada; © Her Majesty the Queen in Right of Canada, as represented by the Minister of Agriculture and Agri-Food Canada)

Black vine weevil larvae (Fig. 347) are white with a pink or yellow hue, depending on their food plant. They are legless, unlike cranberry girdler and scarab larvae, have a hard brown head capsule, and are often curled in the shape of the letter C. The first three or four instars feed on the fine roots of cranberry. Later instars (fifth, sixth, and sometimes seventh; up to 9 mm long) form chambers around themselves in the soil near larger roots, and from these chambers, they feed on bark and woody conductive tissue. Larvae feed from the time of hatch in the summer until the final instar in the autumn or until feeding is interrupted by low temperatures. Larvae pupate, or finish feeding and then pupate, in the spring as temperatures increase. Pupae are soft and white, with structures that will become the antennae, wings, and legs of the adult apparent along the sides of the pupal body.

Black vine weevil has one generation per year, but some adults overwinter and are present in two consecutive years. In areas with mild winters, overwintering adults begin laying eggs earlier than adults of the current year.

The life histories of the strawberry root and clay-colored weevils are similar to black vine weevil. Adult strawberry root weevils are all flightless females, are shiny black with reddish brown antennae and legs, and are about half the size of black vine weevil (4–5 mm long). The clay-colored weevil, which is intermediate in size between the strawberry root weevil and the black vine weevil, has a mottled, clay-colored exterior that may look black when wet. Clay-colored weevil adults emerge very early in the spring, while the other two species are still larvae or pupae. Egg laying by parthenogenic clay-colored weevils precedes the other two species, and pupation happens in the summer.

Damage

Weevils are most prevalent in dry-pick cranberry beds or in beds that are not flooded during the winter, because long periods of flooding drown the larvae. Beds surrounded by preferred weevil host plants, such as blackberry or salal, can be infested by adults throughout the summer and autumn, particularly if there are no water-filled ditches between the headlands and the cranberry bed. Adult feeding (called notching) does not harm cranberry or other host plants but serves as an indication that weevils are present.

Feeding by weevil larvae during late summer, autumn, and spring can seriously damage or kill cranberry vines. In late summer and autumn, newly hatched and young larvae feed on the fine roots and begin chewing bark from woody runners in the trash layer and in the soil. Most of the damage is done in the spring, when older larvae feed on bark and surface tissue of woody runners. Weevil feeding is similar to feeding by cranberry girdler larvae, except that weevils tend to chew contiguous areas of bark, whereas girdler larvae tend to make deep gouges into the conductive tissue. Weevil larvae do not leave obvious frass pellets.

The foliage of vines damaged by weevil larvae appears reddish brown in early spring. Patches of reddish brown vines along the edge of a cranberry bed are typical of weevil damage, because adults walking into the bed lay eggs along the edges before moving toward the middle of the bed.

Management

Patches of black vine weevil damage are usually first detected in early spring. Weevil larvae can be found by rolling back the damaged vines and searching through the top 5 cm of soil.

Entomopathogenic nematodes in the genera *Steinernema* or *Heterorhabditis* are commercially available for the control of black vine weevil. Heterorhabditids cruise in search of insect hosts, whereas steinernematids wait and then ambush the host. Heterorhabditids are generally more effective than steinernematids against black vine weevil. Low temperatures in early spring and late autumn limit the effectiveness of many species of nematodes, so late spring is the optimum time to apply nematodes for weevil control. Application in the summer against newly hatched larvae has limited success. Nematode application in the autumn, before soil temperatures fall below 12°C, may provide some control.

Adult weevils can be detected by sweep net sampling the cranberry bed after dusk during late June and July. Sweep net sampling can be done earlier in the spring if the presence of overwintering adults is suspected. An appropriately labeled insecticide applied on a warm evening kills many of the adults. Insecticides used during June or July is most effective if applied before adults have laid most of their eggs, that is, during the preoviposition feeding period.

Holding the flood for 2–4 weeks after harvest drowns some weevil larvae, but others survive and continue feeding in the spring. In habitats where the flood is not held during the winter and where alternate host plants grow beside cranberry beds, black vine weevil can be expected to be an annual pest.

Selected References

Averill, A. L., and Sylvia, M. M. 1998. Curculionid soil weevils. Pages 78-81 in: Cranberry Insects of the Northeast. A Guide to Identification, Biology and Management. University of Massachusetts Cranberry Experiment Station, Amherst, MA.

Cowles, R. S. 1995. Black vine weevil biology and management. J. Am. Rhododendron Soc. 49:94-97.

Cram, W. T., and Pearson, W. D. 1965. Fecundity of the black vine weevil, *Brachyrhinus sulcatus* (F.), fed on foliage of blueberry, cranberry and weeds from peat bogs. Proc. Entomol. Soc. B. C. 62:25-27.

Georgis, R., Koppenhöfer, A. M., Lacey, L. A., Belair, G., Duncan, L. W., Grewal, P. S., Samish, M., Tan, L., Torr, P., and Van Tol, R. 2006. Successes and failures in the use of parasitic nematodes for pest control. Biol. Control 38:103-123.

Maier, C. T. 1981. Reproductive success of the black vine weevil, *Otiorhynchus sulcatus* (F.), fed different foliar diets and evaluation of techniques for predicting fecundity. Environ. Entomol. 10:928-932.

Shanks, C. H., Jr. 1991. Root weevils. Page 64 in: Compendium of Raspberry and Blackberry Diseases and Insects. M. A. Ellis, R. H. Converse, R. N. Williams, and B. Williamson, eds. American Phytopathological Society, St. Paul, MN.

(Prepared by S. M. Fitzpatrick)

Black-Headed Fireworm

The black-headed fireworm, *Rhopobota naevana*, is a key pest in most cranberry-growing regions of North America. Larvae of this tortricid moth damage cranberry by feeding on buds, foliage, and fruit. Fireworm populations increase slowly but exponentially if left untreated, resulting in infestations that take growers by surprise. Eggs adhere to leaf undersides and are readily transported with pruned or mowed vines to new plantings as far away as Latvia. In North America, black-headed fireworm is reported only from cranberry, except for one record of larvae feeding on evergreen huckleberry, *Vaccinium ovatum*, in Washington (United States). European records of *R. naevana* moths collected in Britain or the Azores refer to the holly tortrix, which has similar genitalia but is larger than the North American *R. naevana*. The taxonomic relationship between the North American *R. naevana* and the European *R. naevana* has not been determined.

Life History and Description

The black-headed fireworm overwinters as an elliptical, flattened, light yellow egg (approximately 0.7 × 0.9 mm) in diapause. Egg hatch begins in the spring, usually early April in British Columbia (Canada) and mid- to late April in Wisconsin (United States), as cranberry plants come out of dormancy. Hatching of this first generation of eggs is complete by late April, mid-May, or early June, depending on the region and the accumulated heat units in a given year. Neonate first-generation larvae (approximately 1 mm long) first mine into old leaves and feed and then crawl up to the growing tip of uprights, where they tunnel into unopened buds or feed within expanding buds. The synchrony between plant phenology and hatch time greatly influences the survival of these first instars: survival can be five times greater on expanding buds than on old leaves and buds just breaking dormancy. As larvae grow, they make a tentlike shelter by webbing new leaves together with silk and then they feed on apical meristem tissue within the tent. A feeding larva (Fig. 348) moves among plant tips, making five or six tents before it finishes feeding and pupates. During the 3–5 weeks of feeding and development from neonate to pupa, a larva molts through five instars and reaches a length of about 8 mm. All instars have shiny black or dark brown heads. Bodies may be yellow, light or dark green, or brownish, depending on whether the larva has stopped feeding to molt or is actively feeding. Larvae have three pairs of front (true) legs and five pairs of hind (pro-) legs. Prepupae drop to the ground and pupate within a cocoon of silk and soil; occasionally, pupation happens within a tent on the plant.

The first generation (flight) of adults begins in late May or early June in warm or coastal areas and a few weeks later in cooler regions. The adults are small moths (approximately

Fig. 348. Black-headed fireworm larva (*Rhopobota naevana*) on cranberry. (Courtesy J. T. Troubridge of the Department of Agriculture and Agri-Food, Government of Canada; © Her Majesty the Queen in Right of Canada, as represented by the Minister of Agriculture and Agri-Food Canada)

6 mm long) and are dark grayish brown with darker transverse bands across the folded wings (Fig. 349). The span of open forewings is 11–13 mm. Male and female moths are similar in appearance. When at rest with wings folded, moths take the shape of a narrow isosceles triangle. When disturbed, moths make short flights among the plant tips. Mate-finding flights are most obvious from late afternoon until dusk.

Summer, second-generation eggs (approximately 0.6 × 0.8 mm; Fig. 350) are laid on leaf undersides and hatch within 2–15 days after oviposition in British Columbia and Wisconsin. Second-generation larvae feed on succulent leaves and meristem tissue at the tips of uprights and runners. Larvae develop quickly on growing foliage during long, warm summer days. Second-generation larvae can begin feeding in tents at plant tips and then burrow into green cranberry fruit and complete development there. Following pupation, adults of the second generation fly, mate, and lay eggs from late June through September. Most eggs laid by second-generation moths enter diapause, but 5–20% hatch, giving rise to third-generation larvae. These larvae feed on foliage and fruit unless halted by harvest floods or inclement weather. In warm, coastal areas, third-generation larvae may pupate and emerge as adults in late autumn or early winter.

Fig. 349. Black-headed fireworm moth (*Rhopobota naevana*). (Courtesy S. M. Fitzpatrick of the Department of Agriculture and Agri-Food, Government of Canada; © Her Majesty the Queen in Right of Canada, as represented by the Minister of Agriculture and Agri-Food Canada)

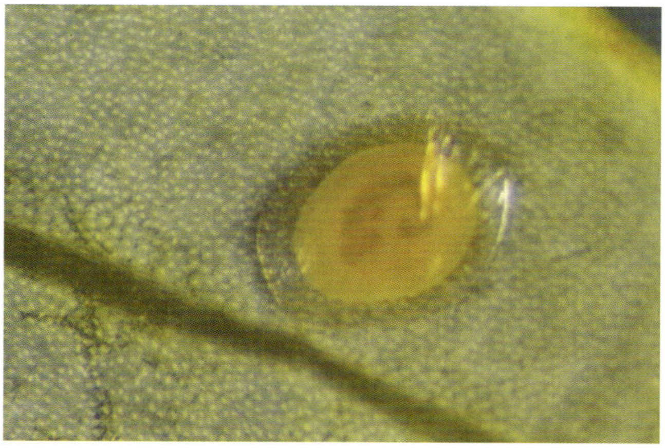

Fig. 350. Black-headed fireworm egg (*Rhopobota naevana*) on the underside of a cranberry leaf. (Courtesy S. M. Fitzpatrick of the Department of Agriculture and Agri-Food, Government of Canada; © Her Majesty the Queen in Right of Canada, as represented by the Minister of Agriculture and Agri-Food Canada)

Damage

All damage is done by feeding larvae. First-generation larvae destroy apical buds that would otherwise produce blossoms. Cranberry plants often recover by activating lateral buds, but the current year's blossom is lost. The most serious economical damage is done by second- and third-generation larvae, which consume plant tissue rapidly during warm weather. One larva can web together several uprights and feed on new leaves and then web a berry against the uprights and destroy the berry by feeding inside and filling it with frass. Damage by second- and third-generation larvae ruins the current year's fruit and prevents uprights from setting a fruit bud for the following year. An infestation of second-generation larvae can go unnoticed until leaves of the damaged uprights desiccate and turn brown. Patches of such uprights can be extensive (0.2 ha), giving a scorched or burned appearance to large sections of a cranberry bed. This hallmark gives the fireworm its name.

Management

The risk of black-headed fireworm infestations can be reduced by ensuring that vines or plants used to establish new plantings originate from beds with few or no fireworms. In young beds, the amount of fertilizer applied should not be excessive. Larvae that develop on foliage fertilized with large amounts of nitrogen become adults that can be more fecund than usual.

To control fireworm populations in new or established beds, monitoring for first-generation larvae should begin just before vines break dormancy. Examine buds on warm edges of the beds should be examined, looking for tiny entry holes and frass. Once first instars have been detected, monitoring should be continued by visually searching for tents or by taking sweep net samples through the vines. An appropriately labeled insecticide, used according to directions on the label, can be applied to areas where larvae have been detected. Pheromone lures and traps can be used to monitor adults of the first flight. Monitoring for second-generation larvae should begin 3–4 weeks after the first moth is caught in a pheromone trap. Visual searches or sweep net samples will detect second-generation larvae, and a labelled insecticide that is not toxic to pollinators can be applied if necessary. Pheromone traps will detect the second flight of moths, which usually overlaps the presence of second- and third-generation larvae in the beds.

Cultural and biological options for the control of black-headed fireworm vary from region to region. In Wisconsin and Massachusetts, beds are sometimes reflooded in June to drown neonates. In British Columbia, the indigenous egg parasitoid *Trichogramma sibericum* is most active where insecticide use is minimal. In all regions, pheromone-based mating disruption, using synthetic pheromone released from baggies, aerosol dispensers, or microscopic capsules, can be used to prevent or delay fireworm mating.

Selected References

Apenite, I., and Cinitis, R. 2006. Major pests in cranberry plantations in Latvia. J. Fruit Ornamental Plant Res. 14:133-136.

Cockfield, S. D., Fitzpatrick, S. M., Giles, K. V., and Mahr, D. L. 1994. Hatch of blackheaded fireworm (Lepidoptera: Tortricidae) eggs and prediction with temperature-driven models. Environ. Entomol. 23:101-107.

Fadamiro, H. Y., Cossé, A. A., Dittl, T., and Baker, T. C. 1998. Suppression of mating by blackheaded fireworm (Lepidoptera: Tortricidae) in Wisconsin cranberry marshes by using MSTRS™ devices. J. Agric. Entomol. 15:377-386.

Fitzpatrick, S. M. 2006. Delayed mating reduces fecundity of black-headed fireworm, *Rhopobota naevana*, on cranberry. Entomol. Exp. Appl. 120:245-250.

Henderson, D. E., Li, S. Y., and Prasad, R. 2002. Chapter 49. *Rhopobota naevana* (Hübner), blackheaded fireworm (Lepidoptera: Tortricidae). Pages 242-246 in: Biological Control Programmes in Canada, 1981-2000. P. G. Mason and J. T. Huber, eds. CABI Publishing, Oxfordshire, United Kingdom.
Shanks, C. H., Jr., Dapsis, L. J., and McDonough, L. M. 1990. Chapter 8. Monitoring blackheaded fireworm (Lepidoptera: Tortricidae) populations in cranberry bogs with sex pheromone traps. Pages 131-139 in: Monitoring and Integrated Management of Arthropod Pests of Small Fruit Crops. N. J. Bostanian, L. T. Wilson, and T. J. Dennehy, eds. Intercept, Andover, England.

(Prepared by S. M. Fitzpatrick)

Blunt-Nosed Leafhopper

The blunt-nosed leafhopper, *Limotettix vaccinii* (syn. *Scleroracus vaccinii*) (Hemiptera: Cicadellidae), is the principal vector of a phytoplasma that causes cranberry false blossom. Cranberry false blossom threatened the entire cranberry industry in the early 1900s. The disease was most problematic in New Jersey (United States), where the cranberry industry was almost eliminated. This insect ranges from Nova Scotia and Quebec (Canada) south to New Jersey and Maryland (United States) and west into Minnesota and Iowa (United States). It is not found on cranberry bogs of the Pacific Coast, and this probably accounts for the failure of cranberry false blossom to spread when introduced there. In Massachusetts (United States), it is seldom seen on commercial bogs but is a dominant species on wild and abandoned cranberry stands. The disease was almost extinct in New Jersey for most of the second half of the last century. However, changes in pest management practices (low insecticide usage) in the last 2–3 years have caused an increase in populations of this insect in New Jersey cranberry. Blunt-nosed leafhopper becomes established in areas where insecticides are eliminated, followed by the appearance of false blossom disease. The blunt-nosed leafhopper has several acceptable food plants, all in the family Ericaceae. It is abundant on leatherleaf, dwarf huckleberry, and fetterbush, which seem to be favorite food plants. This leafhopper does not move around much, and colonization of bogs happens slowly.

Life History and Description

This insect completes one generation a year and overwinters as an egg. The eggs are cylindrical with rounded ends, watery whitish in color, and about 1 mm long. They are generally thrust lengthwise under the bark of the more tender parts of the cranberry stems. On Cape Cod, eggs begin to hatch from late May to the first half of June, and most of them are hatched by the end of June. Holding the winter flood until late May delays the hatch by about 7–10 days. The nymphs go through five instars in about 1 month. At first, they are whitish and about 1.5 mm long. Their color varies as they grow, some being yellow, some blackish, and some greenish gray. The adults begin to appear early in July; the males emerge a week or more before the females. They are most abundant in late July and early August. The numbers of this species start to diminish by the first week in August. The adults have a characteristic blunt head and vary from light yellowish gray to dark brown (Fig. 351). The females are somewhat larger than males, averaging about 4 mm long.

Damage

Nymphs and adults get their food by sucking the plant juices of the cranberry with their piercing-sucking mouthparts. This direct injury is, however, not ordinarily noticeable on the bog. They are most important as vectors of cranberry false blossom. Blunt-nosed leafhopper prefers to feed on certain cultivars, such as Howes, Centennial, and most native Jerseys rather than on the cultivars Early Black or McFarlin.

Fig. 351. Adult blunt-nosed leafhopper (*Limotettix vaccinii*) on cranberry. (Courtesy D. Polk; © APS)

Management

The most important point in blunt-nosed leafhopper management is locating the infestation. Monitoring for this insect is accomplished best with an insect sweep net. Flooding is an effective method to control leafhoppers. Flooding a bog for 24 h just prior to bloom and just after egg hatch kills the nymphs. This approach must be carried out under proper conditions because there is high risk to the vines and crop. Still, there might be late hatching eggs that may survive the flooding.

Selected References

Beckwith, C. S., and Hutton, S. B. 1929. Cranberry false blossom and the blunt-nosed leafhopper. N. J. Agric. Exp. Stn. Bull. 491:1-16.
Beckwith, C. S., and Hutton, S. B. 1929. Life history notes on some leafhoppers that occur on New Jersey cranberry bogs. J. N. Y. Entomol. Soc. 37:425-427.
Dobroscky, I. D. 1931. Studies on cranberry false blossom disease and its insect vector. Contrib. Boyce Thompson Inst. 3:59-83.
Franklin, H. J. 1950. Cranberry insects in Massachusetts. Mass. Agric. Exp. Stn. Bull. 445:8-10.
Wilcox, R. B., and Beckwith, C. S. 1935. The false-blossom disease of cranberries. N. J. Agric. Exp. Stn. Bull. 348:1-4.

(Prepared by C. Rodriguez-Saona and D. Polk)

Coptodisca Leafminer

Coptodisca leafminer, *Coptodisca negligens* (Lepidoptera: Heliozelidae), sometimes referred to as cranberry leafminer, has a limited distribution among cranberry-growing areas, being found only in Wisconsin, Massachusetts, and Connecticut (United States). Coptodisca leafminer was first recorded as a pest of cranberry in northern Wisconsin during the 1930s. Similar small but sporadic outbreaks were also reported in southern and central Wisconsin in the 1960s, the 1980s, and more recently, the 2006–2007 growing seasons. In 1988, it was reported that Coptodisca leafminer was an increasing pest in both Massachusetts and Connecticut. Although Coptodisca leafminer is generally not considered an important economic pest, outbreaks that do happen are likely caused by either a lack of natural enemies, resistance to insecticides, or the decrease in use of broad-spectrum insecticides.

Life History and Description

Coptodisca leafminer is univoltine. Eggs are laid in July where they overwinter and spend most of their lives. Eggs are

Fig. 352. Coptodisca leafminer (*Coptodisca negligens*) injury to cranberry. (Courtesy L. D. Kummer; © APS)

yellowish in appearance and laid singly within the leaf surfaces by piercing the underside of the leaf. Oviposition sites can be seen as elevated, red scars on the bottom side of the cranberry leaves. During the following spring, larvae feed inside the leaf surfaces by "mining" the plant tissue in a circular-like pattern.

Larvae are legless and light brown and approximately 3 mm long when full grown. Only one larva survives in most infested leaves, even when multiple eggs are laid within a single leaf. When mining larvae meet, one larva typically kills the other. Larvae undergo five instars but feed only during the first four, probably starting in May. In June, the fourth instar completes feeding. The fifth instar is actually a prepupa and does not actively feed on cranberry tissue. It constructs an elliptical case by sewing the upper and lower leaf surfaces together with silk, except for an emergence slit. The oblong, elliptical cases where the larvae pupate are approximately 3.5 mm long. These pupal cases fall out of the leaf surface, leaving an elliptical hole (Fig. 352). Infested leaves eventually turn brown around the hole where the feeding occurred and then drop to the ground. Numerous leaves with single oblong holes in them and/or silken pupal cases may be picked up in the spring using a sweep net.

Pupation takes place in mid-June to early July. Pupae are dark mahogany brown and approximately 2.5 mm long. Moth flight is generally very short lived. Moths begin to emerge in late June, with peak emergence during full bloom, primarily the first 2 weeks of July.

Damage and Management

Injured leaves turn brown around the hole (Fig. 352) and drop off the vine. One study showed that second-year leaves abscise more often and earlier when damaged by Coptodisca leafminer larvae. Severe infestations can cause significant leaf drop similar to winter injury (e.g., deprivation of dissolved oxygen in flood waters). Although work has not been conducted to determine whether Coptodisca leafminer reduces yield in cranberry, significant leaf loss can impact photosynthesis and the total amount of stored carbohydrates within the plant.

Because of the sporadic nature of Coptodisca leafminer, attempts at management have been scant. It is probable that new chemistries with systemic and lepidopteran activity will impact populations. Most likely a secondary pest of cranberry, Coptodisca leafminer are normally thought to be kept in check by their existing natural enemies.

Selected References

Averill, A. L., and Sylvia, M. M. 1998. Page 66 in: Cranberry Insects of the Northeast—A Guide to Identification, Biology and Management. University of Massachusetts Cranberry Experiment Station, Amherst, MA.

Boush, G. M., and Anderson, R. J. 1967. *Coptodisca* sp. (Lepidoptera: Heliozelidae) attacking cranberries in Wisconsin. J. Econ. Entomol. 60(5):1472.

Goldsworthy, V. 1934. Insect control in Wisconsin. Wisconsin Cranberry Growers Association Report.

Maier, C. T. 1988. Life cycle of *Coptodisca negligens* (Lepidoptera: Heliozelidae) on cranberry. J. Econ. Entomol. 81:497-500.

Maier, C. T. 1989. Accelerated abscission of cranberry leaves damaged by the leafminer, *Coptodisca negligens* (Lepidoptera: Heliozelidae). Environ. Entomol. 18:773-777.

Rogers, L. M. 1931. Cranberry culture in Wisconsin in 1931. Wisc. Cranberry Growers Assoc. Rep. 45:19-20.

Roper, T. R., and Klueh, J. S. 1996. Movement patterns of carbon from source to sink in cranberry. J. Am. Soc. Hortic. Sci. 121:846-847.

(Prepared by T. G. Dittl)

Cranberry Blossomworm

The cranberry blossomworm, *Epiglaea apiata* (Lepidoptera: Noctuidae), is found throughout the northeastern United States from Maine to Illinois. This insect is one of the three most important pests of cranberry in New Jersey (United States); however, it is considered a minor problem in Massachusetts (United States) and in Quebec and British Columbia (Canada) and is not known to infest cranberry in other growing states. Other hosts include blueberry and leatherleaf.

Life History and Description

The cranberry blossomworm has one generation a year. Eggs are laid singly, mostly in October, in fallen leaves and pieces of dead vines littering the bog floor. Eggs are pale yellow to brown and 0.7 mm in diameter. These eggs overwinter and begin to hatch after the middle of May on bogs drained in April or hatch in late May and early June on bogs where winter water is held until after mid-May. Eggs hatch over a 4-week period. Young larvae are green, but older larvae become reddish brown with a whitish stripe along each side of the body. Mature larvae become pale brown and nearly 38 mm long (Fig. 353). The larvae mature in July on bogs where winter water is removed in early spring. Larvae become nocturnal by late May. An unusual feature of the life cycle is the presence of a long prepupal stage, which lasts 30–60 days. Pupation takes place in the ground or deep trash in late July and early August. The pupae are brown and about 16 mm long. The moths start to emerge in September and are active until late November. On bogs drained late of winter water, pupation may take place in August or early September and moth emergence begins in later September and continues through early October. The moths vary greatly in color and have a wingspan of about 38 mm (Fig. 354). Females lay 100–200 eggs.

Fig. 353. Late-instar cranberry blossomworm larva (*Epiglaea apiata*) on cranberry. (Courtesy A. L. Averill; © APS)

Fig. 354. Cranberry blossomworm adult (*Epiglaea apiata*) on cranberry. (Courtesy A. L. Averill; © APS)

Damage

Young larvae nibble the leaves, beginning at the margin and working toward the midrib, or bore into the buds, spoiling them for fruit production. Fruit production is further decreased as older larvae nip off buds and blossoms and dropping them to the ground, often cutting off more blossoms than they can consume. Unlike spotted cutworm and armyworm, they rarely cut off leaves. Each mature larva can fully destroy 100 blossoms over a 3-week period. The impact of cranberry blossomworm on fruit production relates to the time of flood removal. Early flood removal in April results in larvae developing in mid-May, when buds and blossoms are abundant. If the flood is held later, larvae are not present until late May to early June, and the impact on crop production is minimized.

Management

Young instars are readily picked up in daytime sweep net sampling. However, later in the season, monitoring for this pest requires nighttime (9 p.m. to 12 a.m.) sweep net sampling at the end of May for those bogs where winter water is removed in early spring. At this time, most larvae are third or fourth instars. The female sex pheromone of this insect was identified as a blend of (*Z*)-9-hexadecenyl acetate, (*Z*)-9-tetradecenyl acetate, and tetradecyl acetate at a 65:2:33 ratio, respectively. Current use of the sex pheromone to monitor and control this pest in commercial farms is limited. Further studies need to correlate male trap captures and larval infestation. Late water floods can control cranberry blossomworm. Flooding late in May, when most eggs have hatched, for 12 h destroys this pest if the winter water was drained from April until the first week of May. If bogs were drained earlier, hatch would also happen earlier so the flood could be moved up to May 15–20, reducing danger to developing flower buds. Historically, flooding for 10 h about June 10 was also done.

Selected References

Beckwith, C. S. 1929. The blossom worm, a cranberry pest. J. N. Y. Entomol. Soc. 37:409-416.

Franklin, H. J. 1948. Cranberry insects in Massachusetts. Mass. Agric. Exp. Stn. Bull. 445:27-28.

Landry, J.-F., Roy, M., and Turcotte, C. 2000. Insectes ravageurs de la canneberge au Quebec. Guide d'identification. Agriculture and Agri-Food Canada, Ottawa, ON, Canada.

Tomlinson, W. E., Jr. 1948. Cranberry fireworms and blossom worm. N. J. Agric. Exp. Stn. Bull.

Zhang, A., and Polavarapu, S. 2003. Sex pheromone of the cranberry blossom worm. J. Chem. Ecol. 29:2153-2164.

(Prepared by C. Rodriguez-Saona and D. Polk)

Cranberry Fruitworm

Cranberry fruitworm, *Acrobasis vaccinii* (Lepidoptera: Pyralidae), is an internal fruit feeder on cranberry (Fig. 355). It is found throughout North America and Canada where cranberry is found and is commonly observed in wild cranberry stands. It is one of the most serious and challenging insect pests in all cranberry regions, with the exception of the west coast and New Jersey (United States). Cranberry fruitworm enjoys a high pest status owing to several factors. It is very difficult to monitor egg infestation, and moth flight may last many weeks. Larval infestation becomes apparent only after the larva has entered green berries and they begin to prematurely redden. Once inside the fruit, insecticide treatments are ineffective and the larva is protected from natural enemies. Pressure in commercial settings is typically high and there is a very low threshold for injury by handlers. Thus, most growers must treat the beds to meet the acceptable injury level set by handlers. Until recently, management has been limited to broad-spectrum organophosphate and carbaryl sprays that negatively impact natural enemies.

Life History and Description

Additional details on the life history and description of this species are found in Cherry Fruitworm and Cranberry Fruitworm under the Blueberries (Highbush and Rabbiteye) section. In early to mid-June in Massachusetts (United States) cranberry beds, moths begin to emerge on bogs (Fig. 356). Peak flight typically is in the first 2 weeks of July and a second small peak may follow in several weeks; scattered flight may continue through August or even September. Moths tend to emerge from hibernacula in late afternoon and hide among the vines during the day. The egg-laying pattern is clumped, with greater frequencies of eggs than expected near weedy areas and the bog margin. Eggs are typically placed inside the rim of the calyx cup. Neonate larvae enter the fruit in less than 2 h, typically leaving the calyx, moving to the top of the berry, and aligning their body along the pedicel as they eat into the top of the fruit. The entry is barely visible to the unaided eye. On

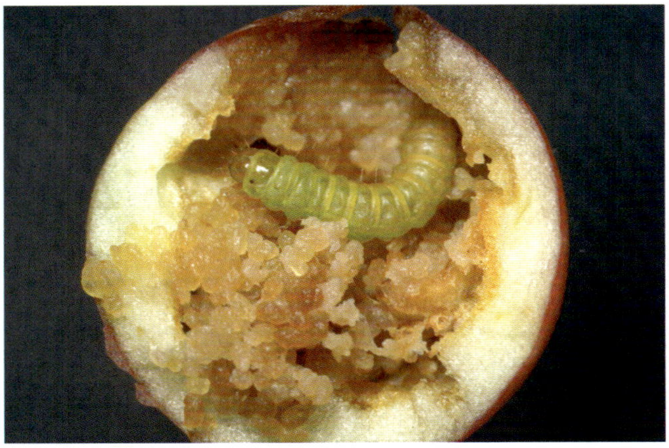

Fig. 355. Larva of cranberry fruitworm (*Acrobasis vaccinii*) in a cranberry, showing frass accumulation. (Courtesy A. L. Averill; © APS)

Fig. 356. Adult cranberry fruitworm (*Acrobasis vaccinii*) on cranberry. (Courtesy A. L. Averill; © APS)

Fig. 357. Cranberries (center) that were infested by cranberry fruitworm (*Acrobasis vaccinii*). After the larva departs, fruit shrivel but remain on the upright. (Courtesy A. L. Averill; © APS)

rare occasions, larvae remain in the calyx and enter here. Early instars excavate the seeds first and do not move until much of the fruit flesh is consumed. Larvae space out uniformly among berries as a result of aggressive interactions; when meetings occur, the larger larva typically ends up occupying the fruit. All first- and second-instar larvae construct a silken closure at the surface of the berry where they entered the fruit; most intermediate instars also do so, but the final (fifth) instar very rarely does. The larva sometimes moves from one berry directly into another at the point of berry contact. As the larva fills the berry with frass, it turns red prematurely, providing a clear sign of infestation. The previously infested berries often cling to the vines as husks until the following year.

Females are polyandrous and, along with males, move throughout the bog habitat; they are active in the forest understory and high in trees surrounding the cranberry bed in late evening. Studies show that, in Massachusetts, counts of *A. vaccinii* in the woods surrounding a bed are often considerably higher than those on the bog.

Damage

Peak populations appear at fruit set, and yield is decreased by the several berries required by each larva to complete development. Early-injured berries gradually shrivel like raisins (Fig. 357) and are seldom problematic at harvest. Berries injured by later-season infestation may be picked up at early harvest dates and the presence of larvae may lead to the devaluation or rejection of loads by the handler.

Management

Male flight may be monitored with wing traps baited with a blend of (*E*,*Z*)-8,10-pentadecadien-1-ol and (*E*)-9-pentadecen-1-ol acetate. Cranberry fruitworm males tend to fly higher above the bog than other moth species and studies showed that, at peak flight in July, traps placed 0.5–1.0 m above vine level resulted in higher counts than those for traps placed at vine level. However, trap data provide little useful information regarding spray timing in cranberry; there is no consistent relationship between moth activity and oviposition in cranberry. The earliest emerging females appear well before berries have set; significant egg laying does not happen until pinheads have begun to enlarge (approximately 3 mm). To monitor for eggs, labor-intensive visual inspections of berry samples have been employed in integrated pest management programs. A methodology to detect a less-than-1% egg infestation with a high degree of precision was developed in Massachusetts and involves collecting 50 randomly picked berries per acre with action thresholds based on sample size. The lack of adequate and efficient monitoring techniques, coupled with the fact that cranberry fruitworm pressure is routinely much higher than the tolerance for infested fruit, has lead to prophylactic spray recommendations based on crop phenology. Two sprays, approximately 10 days apart, are recommended when a given number of days has elapsed after 50% out-of-bloom (when half of the flowers have set). Conventional broad-spectrum insecticides are effective, but low-gallonage applications of insect growth regulators and the newer spinosyn-based products are effective alternatives. Selective insecticides aid in the conservation of natural enemies, particularly *Trichogramma* egg parasitoids that are abundant in mid- to late season. In Massachusetts, the braconid *Phanerotoma franklini* is found in large numbers on untreated bogs. It attacks the fruitworm's egg stage but then the wasp larva develops in the host larva and emerges the following spring.

Every third year, a late water flood in cranberry (a 1-month reflooding of the beds in mid-April) can eliminate cranberry fruitworm populations. Lengthy autumn floods are also effective at suppressing populations. Preliminary work with seeded hibernacula of cranberry fruitworm on beds flooded for 3–4 weeks in September to October showed substantially lower moth emergence, but more work is required to determine whether there is a concomitant stress on the flooded vines. In any flooding scenario, beds may be repopulated by moths from neighboring untreated beds, so it is prudent to consider flooding on an area-wide basis or to monitor berries for eggs on a regular basis.

Selected References

Averill, A. L., and Sylvia, M. M. 2008. Insect management. Pages 107-124 in: Cranberry Production: A guide for Massachusetts. H. A. Sandler and C. J. DeMoranville, eds. Univ. Mass. Publ. CP-08.

Averill, A. L., Sylvia, M. M., Kusek, C. S., and DeMoranville, C. J. 1995. Flooding in cranberry to minimize insecticide and fungicide inputs. Am. J. Altern. Agric. 12:50-54.

DeMoranville, C. J., Sandler, H. A., Shumaker, D. E., Averill, A. L., Caruso, F. L., Sylvia, M. M., and Pober, D. M. 2005. Fall flooding for management of cranberry fruitworm (*Acrobasis vaccinii*) and dewberry (*Rubus hispidus*) in Massachusetts cranberry production. Crop Prot. 24:999-1006.

McDonough, L. M., Averill, A. L., Davis, H. G., Smithhisler, C. L., Murray, D. A., Chapman, P. S., Voerman, S., Dapsis, L. J., and Averill, M. M. 1994. Sex pheromone of cranberry fruitworm *Acrobasis vaccinii* Riley (Lepidoptera: Pyralidae). J. Chem. Ecol. 20(12):3269-3279.

Sharma, N. R. 2005. Adult activity and host plant utilization in cranberry fruitworm *Acrobasis vaccinii* Riley (Lepidoptera: Pyralidae). Electronic Doctoral Dissertation. Univ. Mass. Amherst Paper AA13163706.

(Prepared by A. L. Averill)

Cranberry Girdler

The cranberry girdler, *Chrysoteuchia topiaria* (Lepidoptera: Pyralidae), is indigenous to North America and belongs to the group of grass-infesting crambid moths commonly called sod webworms. Cranberry girdler appears in a variety of habitats, from high-mountain parks to intensively cultivated monocultures in central and coastal regions. This insect has many hosts in addition to cranberry and can be an economic pest in turf, seed grass fields, and conifer nurseries. Cranberry girdler gets its name from the damage done by larvae, which chew on and girdle the woody subterranean runners that transport water and nutrients to the leaves.

Life History and Description

Cranberry girdler has one generation per year. In the autumn, fully grown larvae stop feeding and become prepupae that enter diapause and overwinter in cocoons in the soil. The cocoon is made of substrate materials such as soil, sand, and detritus held together and lined with silk. Pupation takes place in May and early June but, in the Northeast, can be delayed until July if cranberry beds are flooded until late May. Temperature thresholds and degree-day summations for development from prepupa to adult probably vary among western coastal, midwestern, and northeastern cranberry-growing regions. Adults emerge from cocoons as early as mid-May and are active until August, depending on the region. The moth is about 1.3 cm long, including the prominent labial palps that extend like a snout in front of the eyes. The gray and straw-colored forewings have three dark dots at the outer edge and are fringed with silver (Fig. 358). Moths often rest at the tips of the uprights, with the snout pointed toward the ground and the wings folded close to the body. Females are larger and whiter than males and, if disturbed, often drop to the soil, where they remain motionless or scurry away on their side.

Females release pheromone between 0700 and 1100 hours as they perch at the tips of the uprights. Mating occurs within 24 h of emergence. Females move down toward the soil after mating and drop several hundred nonadhesive eggs onto the trash layer. Neonate larvae (1.5 mm long) hatch 4–18 days after oviposition, depending on temperature. Neonates move down to feed on succulent tissue in dark places. They may tunnel into the crown of grassy weeds, feed on fungi and detritus in the trash layer, or feed on fine roots just below the soil surface. Larvae develop through at least five instars. The type of food changes and the amount consumed increases with successive instars. Larvae progress from feeding on small amounts of fine roots to chewing considerable gouges in the woody bark and inner conductive tissue of runners in the soil. All instars have three pairs of front (true) legs and five pairs of hind (pro-) legs, brown head capsules, and dingy white bodies bearing sparse hairs (Fig. 359). Fully grown larvae are approximately 1.3 cm long. They stop feeding and make cocoons in September, October, or early November, depending on the region.

Damage

Young larvae feeding on fine roots weaken the cranberry vines. More obvious vine stress and death is caused by older larvae feeding on the bark and inner conductive tissue of woody runners in the trash layer and in the soil.

The first sign of damage by girdler larvae usually appears in the summer as a patch or patches of vines with orange leaves interspersed among vines with healthy green leaves. Patches are usually circular and vary in diameter from 0.5 m to many meters, depending on the number of larvae feeding. If the damaged vines are traced back into the soil, feeding gouges in the woody runners are apparent and there are few or no fine roots. Feeding gouges often originate on the lower side of the runner or where two runners cross. A larva might be found near the feeding site; if not, the brown or orange frass pellets excreted by feeding larvae are an indication that the culprit is cranberry girdler, not a root weevil or a rodent, which also feed on woody tissues. Weevils do not leave frass pellets. Rodents usually clip runners on an angle to make runways or burrows in the beds.

Young cranberry beds may show stress earlier than older beds, because a small amount of feeding can sever the thin woody runners produced by young cranberry plants. Plant stress may appear earlier in weed-free beds than in weedy beds, because young larvae in weedy beds may feed on weed crowns and roots for several instars before moving to cranberry roots. In new plantings started from rooted cuttings, feeding gouges may be seen at and several millimeters above the crowns. In old, established beds, feeding gouges may be on the underside of a woody runner meters away from the affected upright.

An infestation of cranberry girdler larvae can occur anywhere in a cranberry bed. A damaged area usually appears circular in the first year. In subsequent years, the circle of stressed and dead vines enlarges and satellite patches of damage appear around it. This pattern of spread results from late-instar larvae crawling outward to find new feeding sites and from egg-laden females flying only short distances to drop eggs. Severe infestations can cost a grower several years' worth of crop.

Management

Cranberry girdler moths can be expected to fly into cranberry beds from grassy dikes and neighboring fields, particularly after grass is mowed. Larvae can crawl in from grassy areas if there is no ditch between the grass and the cranberry bed. In newly created beds, moths can emerge from pupae remaining in the soil from the previous crop or habitat. In June and July, it is helpful to walk through the beds and count the

Fig. 358. Male cranberry girdler adult (*Chrysoteuchia topiaria*) on cranberry. (Courtesy S. M. Fitzpatrick of the Department of Agriculture and Agri-Food, Government of Canada; © Her Majesty the Queen in Right of Canada, as represented by the Minister of Agriculture and Agri-Food Canada)

Fig. 359. Fully grown larva of cranberry girdler (*Chrysoteuchia topiaria*). (Courtesy S. M. Fitzpatrick of the Department of Agriculture and Agri-Food, Government of Canada; © Her Majesty the Queen in Right of Canada, as represented by the Minister of Agriculture and Agri-Food Canada)

number of girdler moths that fly up. This helps identify patches where most egg laying will take place.

Prevention is the optimal strategy for controlling cranberry girdler. Prevention means reducing the habitat and food supply of young larvae and ensuring that cranberry runners and roots are vigorous enough to withstand a small amount of feeding damage. Both of these objectives can be attained by applying a layer of sand (1.3–5.0 cm) to cranberry beds during the dormant season every 3 years. Neonates hatching from eggs dropped on the sand layer will not find succulent plant tissue to eat, and it will be difficult for larvae to tunnel through to the root zone below. Sand stimulates the production of fine roots from many points along the runners, so cranberry plants will be anchored and fed from a network of roots. The sand should contain few weed seeds, and grassy weeds particularly should be removed from the beds. The use of herbicides that retard cranberry root growth should be minimized.

Entomopathogenic nematodes can be applied in midsummer to control developing larvae. It is recommended to apply nematodes 2–4 weeks after the number of cranberry girdler males caught in pheromone traps has reached a peak. This is a general guideline that may be improved in the future. The cranberry bed should be thoroughly moist before, during, and after application. It is helpful to work with a pest management expert when applying nematodes.

Flooding the cranberry bed for 48–72 h drowns many girdler larvae if the flood is applied before girdler larvae stop feeding and make cocoons. The flood water should be deep enough to cover all vines and weeds that floating larvae can cling to. Flooding is an effective technique for nonbearing beds but increases the risk of fruit rot in bearing beds.

Patches of vines killed by cranberry girdler should be removed and replanted. Cranberry girdler moths should be expected to emerge in these areas and a grower should be prepared to apply sand and/or entomopathogenic nematodes for the control of larvae.

Selected References

Averill, A. L., and Sylvia, M. M. 1998. Cranberry girdler. Pages 84-86 in: Cranberry Insects of the Northeast. A Guide to Identification, Biology and Management. University of Massachusetts Cranberry Experiment Station, Amherst, MA.

Fitzpatrick, S. M. 2007. Survival of submerged larvae of cranberry girdler, *Chrysoteuchia topiaria*, in the laboratory. Crop Prot. 26(12):1810-1816.

Kamm, J. A., Fisher, G. C., McDonough, L. M., and Shanks, C. H., Jr. 1990. Chapter 9. Cranberry girdler as a pest of grass seed, cranberry and nursery tree production in the Pacific Northwest. Pages 141-150 in: Monitoring and Integrated Management of Arthropod Pests of Small Fruit Crops. N. J. Bostanian, L. T. Wilson, and T. J. Dennehy, eds. Intercept, Andover, England.

Roberts, S. L., and Mahr, D. L. 1986. Development of cranberry girdler, *Chrysoteuchia topiaria* (Lepidoptera: Pyralidae) in relation to temperature. Great Lakes Entomol. 19:85-90.

Tashiro, H. 1987. Lepidopteran pests: Family Pyralidae. Pages 70-90 in: Turfgrass Insects of the United States and Canada. Cornell University Press, Ithaca, NY.

(Prepared by S. M. Fitzpatrick)

Cranberry Root Grub

The cranberry root grub, *Lichnanthe vulpina* (Coleoptera: Glaphyridae), is a native species in the family of bumble bee scarab beetles. This is a riparian species, and larvae are adapted for survival in soil habitats periodically inundated by rain or flooding. The larva has a flattened ventral surface with projecting lobes (Fig. 360) and together with dense hydrofuge hairs, these morphological traits lead to the formation of a plastron (physical gill) bubble when submerged. The plastron provides an exchange surface for oxygen renewal and carbon dioxide loss; in addition, it serves to minimize water entry through permeable areas of the cuticle. *Lichnanthe vulpina* has a long history as a cranberry root-feeding pest, but only in Massachusetts (United States), where it was reported nearly 100 years ago. A study showed that of cranberry beds known to be infested with root-feeding insects, most beds had a mixed species infestation, and more than half had populations of cranberry root grub.

Life History and Description

The cranberry root grub beetles are 13–16 mm long. The male's body is covered with long fox-red hair and females are covered with shorter, more sparse lemon yellow hairs; the elytra are brown (Fig. 361). Adults are active aboveground from mid-June to mid-July during bloom, emerging from the soil on early, humid mornings. Males fly freely over the bog and adjacent areas and are difficult to distinguish from bumblebees. Females tend to come to the surface and rest on the top of the cranberry vines or weeds. They produce a sex pheromone and, after mating, quickly return to the soil, where they lay eggs singly approximately 7–8 cm deep. In the spring, larvae feed on roots, typically within 8–10 cm of the surface, and as the bog becomes drier over the season, they go deeper and may be found approximately 10–25 cm deep in July and August. The larval stage consists of three instars. The final instar is about 2.5 cm long and forms a pupal cell in the soil. Prepupae form before mid-May and pupation takes place in early June.

Fig. 360. Cranberry root grub larvae (*Lichnanthe vulpina*): second and third instars. (Courtesy A. L. Averill; © APS)

Fig. 361. Cranberry root grub adults (*Lichnanthe vulpina*), female (left) and male (right) on cranberry. (Courtesy A. L. Averill; © APS)

Damage

Cranberry root grub larvae feed on the roots and, in a Massachusetts survey, sampling of infested beds showed an average of six to seven larvae per 30-cm^2 plot. Larvae of all sizes (6–25 mm) are typically found together in the soil. Infestations are patchy within a bed, which is likely (in part) because of females remaining within a small radius from their point of emergence from the soil. Infested beds may be weakened and not show overt evidence of damage until a period of drought stress. As the vines die, an area first shows an orange coloring; the infestation radiates outward. Adults do not feed on cranberry but have been documented feeding on pollen of meadowsweet (*Spiraea latifolia*) in Massachusetts.

Management

A summer flood (mid-May to July) that destroys the crop for the year is effective. The sex pheromone released by the female has been identified and could be used for monitoring. A test of trap vane colors indicated that traps with green or black vanes maximized target male catch while minimizing nontarget the catch of important pollinators. Cranberry root grub larvae have been found feeding in the grass near the infested bog, suggesting that beds that have been renovated to remove an infestation may have a constant threat of reinvasion.

Selected References

Carlson, D. C. 1980. Taxonomic revision of *Lichnanthe* (Burmeister) (Coleoptera: Scarabaeidae). Coleopterist's Bull. 34:177-208.

Dunn, J. B., and Averill, A. L. 1996. Survey of soil scarab insects in Massachusetts. Cranberries 60:9-12, 22-23.

Robbins, P. S., Zhang, A., Averill, A. L., Linn, C. E., Jr., Roelofs, W. L., Sylvia, M. M., and Villani, M. G. 2006. Sex pheromone of the cranberry root grub *Lichnanthe vulpina*. J. Chem. Ecol. 32:1573-1561.

Villani, M. G., Allee, L. L., Diaz, A., and Robbins, P. S. 1999. Adaptive strategies of edaphic arthropods. Annu. Rev. Entomol. 44:233-256.

(Prepared by A. L. Averill)

Hoplia

Hoplia equina (Coleoptera: Scarabaeidae) is a small native species of scarab beetle that was fairly recently discovered as a root-feeding problem in Massachusetts (United States) cranberry during the 1990s. The species was only correctly identified approximately a decade later. The adult has not been observed feeding on the vines, but other *Hoplia* spp. are known to feed as adults. It has not been reported in other growing areas.

Fig. 362. *Hoplia equina* adult on cranberry. (Courtesy A. L. Averill; © APS)

Life History and Description

Adults emerge and are day active at bloom, from mid-June to mid-July. The adult (Fig. 362) is dull brown and approximately 9 mm long. The larva undergoes three instars and is a C-shaped white grub that can be distinguished from the other white grubs found in cranberry soils by the Y-shaped anal slit and V-shaped pattern of stout hairs on the underside of the abdomen (see figures within Phyllophaga below, particularly Fig. 365E). Observations suggest that larvae may take 2 years to complete development, but further work is required to determine the life cycle of this species.

Damage

Larval feeding on fine roots and the bark of larger roots leads to vine discoloration in the early stages and subsequent vine death. Vines injured by larvae that are not yet dead tend to show dormant coloration earlier in the autumn than do surrounding healthy vines and to retain dormant coloration later in the spring. Injured vines may be easily pulled up like a carpet, although this is also found with most other root-feeding insects.

Management

The neonicotinoids imidacloprid and clothianidin are effective against first-instar larvae, but its relative ineffectiveness against later-instar larvae means that it must be applied prophylactically. Pheromone-baited traps may be used for monitoring. In cranberry, a summer flood (with loss of yield in that year) would likely be effective in suppressing larval populations.

Selected References

Weber, D. C., Robbins, P. S., and Averill, A. L. 2005. *Hoplia equina* (Coleoptera: Scarabaeidae) and non-target capture using 2-tetradeconone-baited traps. Environ. Entomol. 34:158-163.

Zhang, A., Robbins, P. S., Averill, A. L., Weber, D. C., Linn, C. E., Jr., Roelofs, W. L., and Villani, M. G. 2003. Identification of the female-produced sex pheromone of the scarab beetle, *Hoplia equina*. J. Chem. Ecol. 29:1635-1642.

(Prepared by A. L. Averill)

Phyllophaga

Phyllophaga anxia (Coleoptera: Scarabaeidae), a serious root-feeding pest of cranberry, belongs to a group commonly called the May and June beetles, because the majority of adult flight activity in this genus is during those months. *P. anxia* is the most widely distributed of the approximately 215 *Phyllophaga* spp. in the United States and Canada.

Life History and Description

The adults of *P. anxia* are large (20–22 mm), with shiny black to chestnut-brown elytra (Fig. 363). In the northeastern and north-central United States and Canada, adult flights begin in early to mid-May and may last until the end of June. Adults

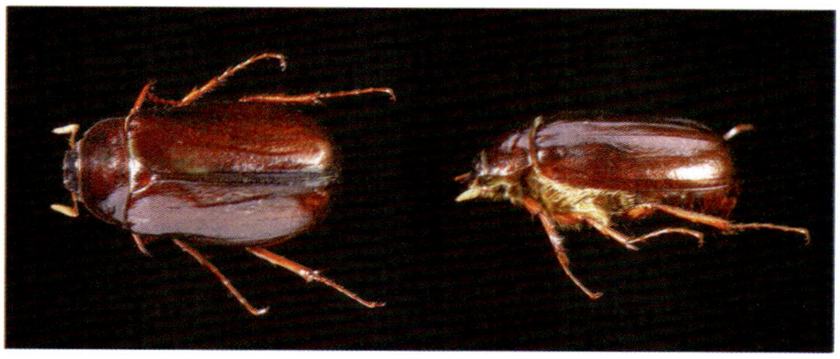

Fig. 363. *Phyllophaga anxia* adults. (Courtesy J. Ogrodnick; © APS)

emerge from the soil at dusk and feed on the foliage of a variety of plants, including oak, poplar, rose, aspen, basswood, willow, chokecherry, alder, walnut, birch, and cherry. Mating takes place at night. Mated females deposit an average of about 55 eggs during their lifetime but may lay up to 100. After egg hatch, first instars feed on decaying vegetation, fungal hyphae, and root hairs. After 6–8 weeks, they molt into second instars, which begin root feeding. At the end of year 1, the second instars burrow deeper into the soil to overwinter. In the spring of year 2, they feed for a short time before the final molt. This third instar is the most damaging, owing to its large size. Upon attaining sufficient size (which may occur in either year 2 or 3), the third instar creates a cavity in the soil and pupation takes place in August, followed by eclosion of the adult beetle within the soil cavity. Adults remain in the cavity and emerge to fly in early to mid-May of the following year. Thus, the life cycle takes 3–4 years, depending on local conditions and latitude.

Damage

P. anxia is a serious pest of cranberry in both Massachusetts and Wisconsin (United States). The larvae (commonly called white grubs) feed on the roots of the cranberry plants, both reducing yield and thinning the vine density, thus predisposing thinned areas to weed invasion.

Management

Good management programs require correct pest identification. In Wisconsin, only *P. anxia* is recognized as a white grub pest. In Massachusetts, five species of white grubs are identified that can cause damage. Therefore, any area suspected of having larval damage should be sampled to determine the species involved. Figures 364 and 365 show five grub species and their rastral patterns. In New Jersey (United States), *P. anxia* has not been identified as a problem species on cranberry.

Management of *P. anxia* infestations is difficult. The summer flood (mid-May through mid-July) may kill a large number of larvae but the current year's crop will be lost. The summer flood, however, is far less costly than complete bog renovation, a process in which the top 30 cm is scraped off (along with the larvae and plants) and hauled away and the bog replanted. This amounts to a crop loss of 3–5 years. The use of sex pheromones for use in mass trapping or disruption shows promise, although long-term research will be required to demonstrate their usefulness. Since the sex pheromone of *P. anxia* was identified, it has been demonstrated that *P. anxia* has three pheromone races, necessitating that infested sites must be characterized as to which of the races are present before pheromone management strategies are begun.

Fig. 364. Third-instar larvae of white grubs affecting cranberry or blueberry: cranberry white grub (*Phyllophaga anxia*), cranberry root grub (*Lichnanthe vulpina*), oriental beetle (*Anomala orientalis*), Japanese beetle (*Popillia japonica*), and hoplia (*Hoplia equina*). (Courtesy J. Ogrodnick; © APS)

Selected References

Dittl, T. G., and Kummer, L. D. 1997. Major cranberry insect pests of Wisconsin, 8-97. Wood County University Extension, Wisconsin Rapids, WI.

Evans, A. V., and Smith, A. B. T. 2009. An electronic checklist of the new world chafers (Coleoptera: Scarabaeidae: Melolonthinae) Version 3. Electronically published, Ottawa, Canada.

Franklin, H. J. 1950. Cranberry insects of Massachusetts. Bulletin 445. Parts II-VII. Massachusetts Agricultural Experiment Station, Amherst, MA.

Lim, K. P., Stewart, R. K., and Yule, W. N. 1980. A historical review of the bionomics and control of *Phyllophaga anxia* (LeConte) (Coleoptera: Scarabaeidae) with special reference to Quebec. Ann. Soc. Entomol. Quebec 25:163-178.

Luginbill, P., and Painter, H. R. 1953. May Beetles of the United States and Canada. U. S. Dep. Agric. Tech. Bull. No. 1040.

Robbins, P. S., Alm, S. R., Armstrong, C. D., Averill, A. L., Baker, T. C., Bauernfiend, R. J., Baxendale, F. P., Braman, S. K., Brandenburg, R. L., Cash, D. B., Couch, G. J., Cowles, R. S., Crocker, R. L., Delamar, Z. D., Dittl, T. G., Fitzpatrick, S. M., Flanders, K. L., Forgatsch, T., Gibb, T. J., Gill, B. D., Gilrein, D. O., Gorsuch, C. S., Hammond, A. M., Hastings, P. D., Held, D. W., Heller, P. R., Hiskes, R. T., Holliman, J. L., Hudson, W. G., Klein, M. G., Krischik, V. L., Lee, D. J., Linn, C. E., Jr., Luce, N. J., MacKenzie, K. E., Mannion, C. M., Polavarapu, S., Potter, D. A., Roelofs, W. L., Royals, B. M., Salsbury, G. A., Schiff, N. M., Sheltar, D. J., Skinner, M., Sparks, B. L., Sutschek, J. A., Sutschek, T. P., Swier, S. R., Sylvia, M. M., Vickers, N. J., Vittum, P. J., Weidman, R. B., Weber, D. C., Williamson, R. C., and Villani, M. G. 2006. Trapping *Phyllophaga* spp. (Coleoptera: Scarabaeidae: Melolonthinae) with sex attractants in the United States and Canada. J. Insect Sci. 6:1-124.

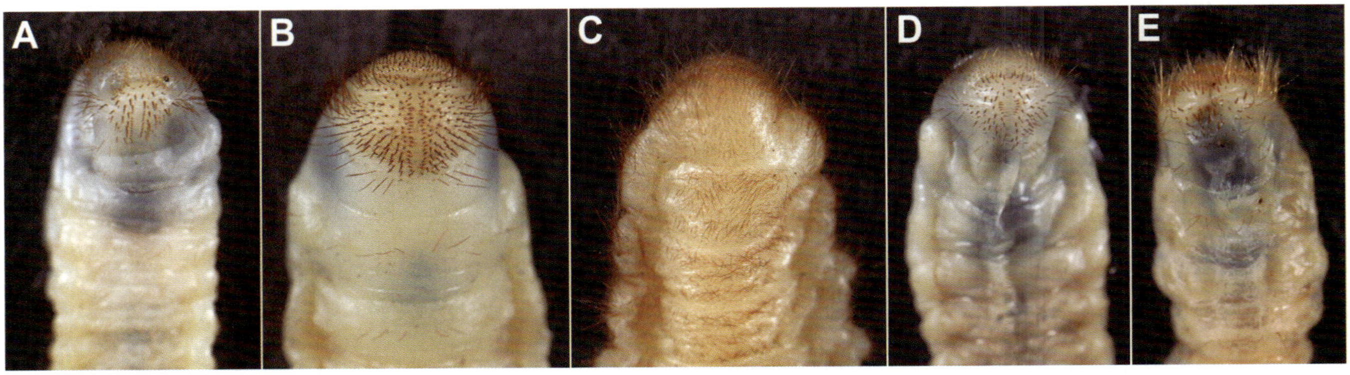

Fig. 365. Rastral patterns of the five species of white grubs affecting cranberry or blueberry. **A,** Oriental beetle (*Anomala orientalis*); **B,** cranberry white grub (*Phyllophaga anxia*); **C,** cranberry root grub (*Lichnanthe vulpina*); **D,** Japanese beetle (*Popillia japonica*); and **E,** hoplia (*Hoplia equina*). (Courtesy J. Ogrodnick; © APS)

Woodruff, R. E., and Beck, B. E. 1989. The scarab beetles of Florida (Coleoptera: Scarabaeidae). Part II. The May or June beetles (genus *Phyllophaga*). Arthropods Fla. Neighboring Land Areas 13:226.
Zhang, A., Robbins, P. S., Leal, W. S., Linn, C. E., Jr., Villani, M. G., and Roelofs, W. L. 1997. Essential amino acid methyl esters: Major sex pheromone components of the cranberry white grub, *Phyllophaga anxia* (Coleoptera: Scarabaeidae). J. Chem. Ecol. 23:231-245.

(Prepared by P. S. Robbins)

Red-Headed Flea Beetle

The red-headed flea beetle, *Systena frontalis* (Coleoptera: Chrysomelidae), is commonly found in low numbers in central and eastern cranberry regions in the latter part of the growing season. It also feeds on highbush blueberry and has been reported on lowbush blueberry. It is a univoltine, polyphagous species, utilizing a range of weed and crop species; it prefers some weeds over cranberry. Larvae reside in the soil and are slender and white, approximately 8 mm long when mature. In the absence of the roots of preferred weed hosts, they utilize cranberry. Adults emerge in July and feed on foliage through early September. The beetle is 3.5–5 mm long and is shiny black with reddish portions of the head (Fig. 366). The femur of the hind leg is enlarged for jumping. Damage occurs as a result of both larval and adult feeding. In Wisconsin (United States), girdled roots and vine death have been recorded. Evidence of larval feeding is similar to that of the cranberry girdler, but in the case of the red-headed flea beetle, a closer examination of the vines shows that the root is calloused and that weak upright growth may have taken place; these symptoms are diagnostic for the red-headed flea beetle but not the cranberry girdler. Adult populations are patchy and they prefer areas of lush cranberry growth. In areas of heavy feeding by adults, yield for the following year can be impacted. Sweep net sampling carried out postbloom picks up the adults and, if counts are high, foliar insecticide sprays have been used successfully to manage populations.

Fig. 366. Adult red-headed flea beetle (*Systena frontalis*) and injury as a result of its feeding on the tissue of the underside of a cranberry leaf. (Courtesy A. L. Averill; © APS)

Selected References

Jacques, R. L., Jr., and Peters, D. C. 1971. Biology of *Systena frontalis*, with special reference to corn. J. Econ. Entomol. 64:135-138.
Mahr, D. L. 2005. Redheaded flea beetle. Wisconsin Fruit, Cranberries, Insect Management, Insect Profiles. https://fruit.wisc.edu/wp-content/uploads/sites/36/2011/05/Redheaded-Flea-Beetle.pdf
Maltais, P. M., and Ouellette, M. C. 2000. A note on *Systena frontalis* (Coleoptera: Chrysomelidae) adults on lowbush blueberry, *Vaccinium angustifolium*. Phytoprotection 81:129-131.

(Prepared by A. L. Averill)

Root-Feeding Chrysomelids

Two species of beetles (Coleoptera: Chrysomelidae), cranberry rootworm, *Rhabdopterus picipes*, and striped colaspis, *Colaspis costipennis*, are pest problems in New Jersey and Massachusetts (United States), respectively. Larval feeding on the fine roots of cranberry plants causes dieback, and large areas can be impacted. Cranberry rootworm is widely distributed in eastern North America and has also been reported as an apple (New York, United States) and ornamental (Louisiana, United States) pest. Striped colaspis ranges from New England to Mississippi (United States) and has been reported on blueberry, grape, and pecan.

Life History and Description

Adult cranberry rootworm beetles are 5–6 mm long, dark brown, bronzed, and shiny (Fig. 367), while striped colaspis are 4–5 mm long with both the head and prothorax a metallic green and the elytra patterned with a yellow-black stripe (Fig. 368). Both species are univoltine. Adults emerge from the soil in June and can be found on the vines until late July. Larvae feed on roots until October. Full-grown cranberry rootworm larvae are about 8 mm long and striped colaspis are 1–2 mm smaller; both tend to be yellowish white with a light brown head. The legs are slender and small. Late-instar larvae overwinter under the soil in earthen cells. In the spring, larvae of both species feed for a short time; cranberry rootworm pupates in May and striped colaspis pupates in mid-June. Pupae are waxy white, are about 5 mm long, and tend to waggle the abdomen vigorously when disturbed.

Damage

Larval feeding on fine roots and the bark of larger roots leads to vine discoloration in the early stages and subsequent

Fig. 367. Cranberry rootworm adult (*Rhabdopterus picipes*) on cranberry. (Courtesy A. L. Averill; © APS)

vine death. Vines injured by larvae that are not yet dead tend to show dormant coloration earlier in the autumn than do surrounding healthy vines and to retain dormant coloration later in the spring. Injured vines may be easily pulled up like a carpet, although this is also found with most other root-feeding insects. Adult feeding of cranberry rootworm adults is rarely noticeable, but they may injure flower buds and flowers, as well as young foliage. Striped colaspis adult feeding is very characteristic in that the beetles chew the outer margins of leaves at the tip of the upright, resulting in a ragged notching of the leaves (Fig. 369).

Management

Cranberry rootworm infestations tend to take place on areas of the beds that are dry and near ditch edges. Striped colaspis infestations typically take place on upland beds in Massachusetts. Striped colaspis beetles are day active and are readily picked up in sweep netting. In contrast, cranberry rootworm adults are nocturnal, so scouting is best accomplished at night. For both species, the neonicotinoid imidacloprid has proven very effective when applied as a soil drench against young larvae in late July. Prior to registration of this compound, populations were suppressed by targeting adults, particularly with carbaryl or diazinon.

Fig. 368. Striped colaspis adult (*Colaspis costipennis*) on cranberry. (Courtesy A. L. Averill; © APS)

Fig. 369. Characteristic feeding injury to the cranberry tip by striped colaspis adult (*Colaspis costipennis*). (Courtesy A. L. Averill; © APS)

Selected References

Harmon, S. W. 1931. The cranberry rootworm as an apple pest. J. Econ. Entomol. 24:180-182.

Oliver, A. D., and Chapin, J. B. 1980. The cranberry rootworm: Adult seasonal history, and factors affecting its status as a pest of woody ornamentals in Louisiana. J. Econ. Entomol. 73:96-100.

Scammell, H. B. 1915. The cranberry rootworm. U. S. Dep. Agric. Tech. Bull. 263:1-8.

(Prepared by D. E. Schiffhauer and A. L. Averill)

Scale Insects

For more than a century, a number of scale insect species have been occasional pests on cranberry. This includes several species of armored scale (Hemiptera: Diaspididae), including oystershell scale (*Lepidosaphes ulmi*), greedy scale (*Hemiberlesia rapax*), Latania scale (*Hemiberlesia lataniae*), cranberry scale (*Hemiberlesia oxycoccus*), Putnam scale (*Diaspidiotus ancylus*), and Dearness scale (*Rhizaspidiotus dearnessi*). Two species of soft scale (Hemiptera: Coccidae), mainly reported in the Pacific Northwest, are the European fruit scale or Lecanium scale (*Parthenolecanium corni*) and brown soft scale (*Coccus hesperidum*). A final species, the azalea bark scale (*Acanthococcus azaleae*), is in the felt scale family (Hemiptera: Eriococcidae). Species and common names utilized here are in accordance with the online database, ScaleNet, and for accuracy, this must be checked for ongoing revisions. Reliable identification may require the assistance of an expert.

As of this writing, only two species, both of which are native to North America, are prevalent. Dearness scale is reported in all growing regions and Putnam scale has appeared at outbreak levels in Massachusetts (United States), beginning in 2010. Interestingly, Franklin observed in 1952 that the only scale insect of importance in Massachusetts cranberry was the cranberry scale, *Hemiberlesia oxycoccus*, but in the intervening years, this species has been completely absent. On the other hand, at the time of Franklin's survey, Putnam scale had been reported in Washington, New Jersey, and Long Island (United States) beds, but was absent from Massachusetts. Based on work by scale systematists and the injury patterns that were observed in current outbreaks, it is believed that Putnam scale has replaced cranberry scale as the primary scale pest in Massachusetts. Unfortunately, the Putnam scale is extremely variable and is probably a complex of several species.

Life History and Description

There is one generation of Dearness scale per year in all growing areas. Bright yellow-orange eggs remain under the cover of the female and tiny first instars (crawlers) begin to exit in early June in Massachusetts, Wisconsin (United States), and the Pacific Northwest. The crawlers are bright yellow, similar to the eggs, are quite active, and move over the vines, often settling at the base of a leaf stem or branch, typically on new growth. Crawlers settle within 1–4 days, lose their legs at the next molt, and reside under a bright white cover (Fig. 370). They continue to molt in place, building up cast skins plus wax to create a dingy white clamshell-shaped covering (Fig. 371). Upon completion of development, male covers are considerably smaller than the female covers, and they are flat and oblong with parallel sides. The 1-mm-long adult male is free living and winged. After mating, the adult female overwinters. Putnam scale undergoes two generations in Massachusetts, with crawlers appearing in mid-June and again in late July and into August. The second instar overwinters. Putnam scales are most dense on the old wood of the cranberry vine, with smaller numbers on new growth and rare appearances on leaves. For both species, females produce a few eggs per day over a span of several weeks and crawler emergence lasts through bloom. For several of the armored scale species, the 2-mm adult females are bright yellow, immobile soft-bodied sacks, beneath their cover (Fig. 372). The crawler stage is responsible for colonizing new plants and is possibly carried by the wind. See Blueberry Scales under the Blueberries (Highbush and Rabbiteye) section for additional general information on scale life history.

Fig. 370. Yellow crawlers and bright white covers of newly settled second instars of Dearness scale (*Rhizaspidiotus dearnessi*) on cranberry. (Courtesy A. L. Averill; © APS)

Fig. 371. Adult female Dearness scale (*Rhizaspidiotus dearnessi*) on cranberry. (Courtesy A. L. Averill; © APS)

Fig. 372. Yellow saclike females (with crawlers and eggs) of Putnam scale (*Diaspidiotus ancylus*) under lifted covers on cranberry. White scar areas remain on the vine when the scale is removed. (Courtesy A. L. Averill; © APS)

Damage

Scale feeding weakens the vines and, under high enough numbers, eventually causes reddening of the foliage and vine death. Some species form a crust over the vines (Fig. 373). For Putnam scale, this is most pronounced on the old wood. In contrast, Dearness scale is dotted over the vines. Apparently, it is often the case that scales are most abundant on the upright, but as infestations increase, they may be found on leaves and fruit as well.

The soft and felt scale species produce honeydew, fluid that is excreted by the scale as it ingests and processes large quantities of sap from the host plant. Several species of fungi (commonly called sooty mold) may grow on the honeydew, and although these fungi do not infect the plants, their layers may impact sunlight penetration to the leaf.

When Dearness scale populations are high enough, patches of vine injury become continuously weak and dead vine areas (Fig. 374). Upon inspection of the dead areas, the old white covers of last year's females are readily apparent on the brittle vines (Fig. 375). As the Putnam scale female matures, the bark grows over her cover, such that it may now only appear as a swelling along the vine. As a result, dead, red, and weakened vines, which become readily apparent when the winter water is removed, may be the first indication of a problem. In the spring, Putnam scale injury often appears as approximately 0.5-m circles, dotted here and there over the bog (Fig. 376).

Management

With the exception of cranberry scale, most of these scale species identified in cranberry are widespread and polyphagous. Thus, woodlands, tree fruit, or urban plantings may

Fig. 373. Crust of Putnam scales (*Diaspidiotus ancylus*) on an cranberry upright's stem. (Courtesy A. L. Averill; © APS)

Fig. 374. Dearness scale (*Rhizaspidiotus dearnessi*) injury in a Massachusetts cranberry bed. (Courtesy A. L. Averill; © APS)

be a source of new infection. It is important to note that extreme vigilance is required to ensure that new plantings are not started with infested cuttings.

Preliminary observations suggest that late water floods may impact Putnam scale, although this is not supported by old reports in the literature. Apparently, the soft and felt scales are rare on winter-flooded bogs; both *Parthenolecanium corni* and *Coccus hesperidum* are noted as occasional pests in the dry bogs of the Pacific Northwest and are absent in the Northeast and Wisconsin.

As a management program is assembled, encouragement of natural enemies is critical, since scales are host to numerous parasitic wasps as well as predators. It is widely held that scale outbreaks happen following foliar applications of broad-spectrum insecticides that eliminate key natural enemies. Indeed in Massachusetts, dichlorodiphenyltrichloroethane (DDT) was introduced in 1946, and six scale species were reported in 1952, three of which had rarely, if ever, been seen previously.

Fig. 375. Cranberry vines killed by Dearness scale (*Rhizaspidiotus dearnessi*) infestation. The vines are brittle and are dotted with last year's white adult scales. (Courtesy A. L. Averill; © APS)

Fig. 376. Putnam scale (*Diaspidiotus ancylus*) injury first appears as dead circular areas with reddish halos of cranberry vine. (Courtesy A. L. Averill; © APS)

However, across all the appearances and disappearances of high scale numbers in cranberry, insecticide use alone does not explain population dynamics. No doubt, populations are regulated by natural enemies (including fungal disease that may be impacted by fungicides), growing conditions, and plant resistance, along with commercial pesticide applications. Some experts have related Dearness and cranberry scale outbreaks to warm winters.

In both armored and soft scales, the cover protects the body from insecticides. However, it has been commonly observed that broad-spectrum insecticides, sprayed when the crawlers are active, can nearly eliminate infestations. At this point, no insecticides are labeled for scale management in cranberry. Dormant oils and insect growth regulators used in many other crops are targets for screening, and registrations of these, plus expansions to include scale where cranberry is already on the label, will increase options over time.

Selected References

Fitzpatrick, S. M., Mauza, B., and Robertson, M. 2004. Dearness scale found on cranberry in British Columbia. Cranberries Mag. 68(4):10-11.

Franklin, H. J. 1952. Cranberry insects in Massachusetts. Parts II-VII. Mass. Agric. Exp. Stn. Bull. No. 445.

Jackson, J. O., and Koval, C. F. 1977. Dearness scale, a vine pest of cranberry: Bionomics and spatial distribution. Cranberries Mag. 43(4):16-17 and 43(5):14-15.

Lacroix, D. S. 1926. Miscellaneous observations on a cranberry scale *Targionia dearnessi* (Ckll.) (Homp: Coccidae). Entomol. News 37:249-252.

Polavarapu, S., Davidson, J. A., and Miller, D. R. 2000. Life history of the Putnam scale, *Diaspidiotus ancylus* (Putnam) (Hemiptera: Coccoidea: Diaspididae) on blueberries (*Vaccinium corymbosum*, Ericaceae) in New Jersey, with a world list of scale insects in blueberries. Proc. Entomol. Soc. Wash. 102(3):549-560.

(Prepared by A. L. Averill and M. M. Sylvia)

Sparganothis Fruitworm

Sparganothis sulfureana (Lepidoptera: Tortricidae) is often very problematic in Massachusetts, Wisconsin, and New Jersey (United States) cranberry-growing regions and has appeared as a pest in eastern Canada and British Columbia (Canada). It is a native species and known as a general feeder; however, at least in Massachusetts cranberry habitats, it appears to favor cranberry and yellow loosestrife (*Lysimachia terrestris*), a widespread weed on bogs. Pheromone trapping in Massachusetts and observations in New Jersey suggest negligible moth activity outside of the cranberry bed. A sex pheromone blend was reported using extracts derived from female glands; in the course of these studies, a comparison of male responses (New Jersey versus Wisconsin) suggested that a pheromone polymorphism may exist.

Life History and Description

Sparganothis fruitworm is bivoltine and typically overwinters as a first instar. A partial third generation has been reported in New Jersey. Spring-generation larvae become active when new growth appears, webbing leaves and then uprights together as a feeding shelter. The first instar is a black-headed larva resembling a black-headed fireworm. Later instars have a yellow head and prothoracic shield with a whitish yellow, yellow, or drab greenish body with rows of white dots (Fig. 377). It is difficult to distinguish them from yellow-headed fireworm (*Acleris minuta*), although one could guess based on bog history. Yellow-headed fireworm typically appears on wild, abandoned, or neglected portions (often edges) of a commercial bog that were not flooded in the winter, while Sparganothis fruitworm is found on managed bogs. Sparganothis fruitworm larvae wriggle fiercely when disturbed, similar to their fireworm relatives. Mature larvae are approximately 10–15 mm long and pupation takes place within the silken retreat. First-generation moths fly in mid-June through early August, and in Massachusetts, peak flight is often in early July. The wings are vivid sulfur-yellow and 7–8 mm long with a brown "X" across the wings when they are held at rest (Fig. 378). Females lay eggs in masses of 10–30 eggs, often on the upper surface of the cranberry leaf, and just prior to hatch, the black head capsule of the neonate is readily apparent. Egg hatch occurs after 9–12 days and larvae complete development in 30–40 days. Second-generation larvae may pupate inside the fruit. Moths emerge again in August and flight may continue into September. The late-summer eggs hatch, and the resulting first instars construct a silken retreat attached to dead leaves and other debris and overwinter on the bog floor.

Fig. 377. Sparganothis fruitworm larva (*Sparganothis sulfureana*) on cranberry. (Courtesy E. de Lange; © APS)

Fig. 378. Sparganothis fruitworm moth (*Sparganothis sulfureana*) on cranberry. (Courtesy E. de Lange; © APS)

Damage

First-generation larvae feed on the new vine growth. The second-generation larvae feed on both leaves and berries, but prefer berries (Fig. 379). They do the most damage by partially feeding on many berries (Fig. 380), often creating a tent of uprights. Generally, it is possible to distinguish between the two fruitworms (cranberry fruitworm [*Acrobasis vaccinii*] versus Sparganothis fruitworm) by the type of damage. A cranberry fruitworm larva creates a neat hole as it enters the fruit and it remains there; once inside, it covers the entry with silk, and as it consumes the flesh, it fills the berry with frass. The green berry soon reddens. Sparganothis fruitworm larvae tend to move feeding among berries and eject frass from the berry's interior; injured berries typically remain green for some days, making an infestation very difficult to detect. On the other hand, if developing fruit are available on large-fruiting cultivars, such as Ben Lear, early Sparganothis fruitworm instars often enter the fruit, remain there, and fill it with frass in a manner similar to *Acrobasis vaccinii*. These populations are likely less vulnerable to both insecticide applications and natural enemies.

Management

Sparganothis fruitworm management is often difficult. The spring generation is monitored during sweep netting, but owing to their webbing behavior, only small numbers of larvae are

Fig. 379. Sparganothis fruitworm (*Sparganothis sulfureana*) feeding injury: larva in a cranberry. (Courtesy T. Dittl; © APS)

Fig. 380. Sparganothis fruitworm (*Sparganothis sulfureana*) larval feeding appears as gouging and ragged shot-hole injury in cranberries. (Courtesy A. L. Averill; © APS)

picked up. Visual inspections aimed at detecting webbed leaves and uprights are also an important scouting technique. In comparison to cranberry, larval retreats are easily detected in the webbed upper leaves of yellow loosestrife, a common weed. Because the larvae are so difficult to detect, the action threshold is set at very low counts. Sex pheromone traps are used to monitor first-generation flight and to time insecticide applications targeting second-generation larvae. For conventional sprays, an insecticide is recommended 2 weeks after peak flight; if using growth regulators, applications are based on the determination of biofix. For both generations, the most effective treatments target early-instar larvae. Larger larvae are very difficult to manage postbloom, but parasitoids are extremely common, and if conserved, can regulate populations. In areas where broad-spectrum conventional sprays are used, Sparganothis fruitworm outbreaks are more likely when these nontarget natural enemies are eliminated.

Resistance to organophosphate insecticides is common in Massachusetts populations, and alternatives such as spinosyn-based compounds and insect growth regulators must be used. Late water has not been shown to be effective, and although work suggests that a later 3- to 6-day flood aimed at larger larvae may be effective, an impact on the actively growing vines would likely be deleterious to yield. Similarly, an autumn harvest flood aimed at overwintering first instars must be held several days to realize even moderate levels of mortality, and this could also impact vine vigor required for growth and fruit set in the spring. Apparently, first instars inside hydrophobic silken retreats are kept dry and are afforded gas exchange in a manner similar to a plastron.

Selected References

Averill, A. L., and Sylvia, M. M. 1998. Cranberry insects of the Northeast. University of Massachusetts Extension, Amherst, MA.

Cockfield, S. D., Butkewich, S. L., Samoil, K. S., and Mahr, D. L. 1994. Forecasting flight activity of *Sparganothis sulfureana* (Lepidoptera: Tortricidae) in cranberries. J. Econ. Entomol. 87:193-196.

Polavarapu, S., Lonergan, G., Peng, H., and Neilson, K. 2001. Potential for mating disruption of *Sparganothis sulfureana* (Lepidoptera: Tortricidae) in cranberries. J. Econ. Entomol. 94:658-665.

Teixeira, L. A. F., and Averill, A. L. 2006. Evaluation of flooding for cultural control of *Sparganothis sulfureana* (Lepidoptera: Tortricidae) in cranberry bogs. Environ. Entomol. 35:670-675.

Vanden Heuvel, J. E., and Goffinet, M. C. 2008. The effects of flood initiation timing and water temperature during flooding on nonstructural carbohydrate concentration and anatomy of cranberry. HortScience 43:338-345.

Zhu, J., Polavarapu, S., Park, K.-C., Garvey, C., Mahr, D., Nojima, S., Roelofs, W. L., and Baker, T. 2009. Reidentification of pheromone composition of *Sparganothis sulfureana* (Clemens) and evidence of geographic variation in males' responses from two US states. J. Asia-Pacific Entomol. 12:247-252.

(Prepared by A. L. Averill)

Spotted Fireworm

Spotted fireworm, *Choristoneura parallela* (Lepidoptera: Tortricidae), is an important pest of cranberry in New Jersey (United States). Larvae cause damage both by foliar and fruit feeding. Spotted fireworm is highly omnivorous and has been found on a wide range of plants in the eastern United States and as far west as Minnesota (United States). Spotted fireworm was reported in Wisconsin (United States) cranberry in 1908 but was not considered to be a major pest of cranberry in any growing region until the 1980s. The advent of sprinkler irrigation systems for frost control allowed cranberry growers to remove the winter flood earlier in the season. This extended season enabled spotted fireworm to complete two generations per year and rise to pest status in New Jersey.

Life History and Description

There are two generations per year. Spotted fireworm overwinters as a second instar in the trash layer on the floor of cranberry beds. Overwintering larvae emerge in late April to early May and web cranberry uprights together. Larvae are yellowish white with brown heads when small; large larvae are olive green with conspicuous white spots along the whole length and a light reddish brown head (Fig. 381). Full-grown larvae are 20 mm long. The pupa is dark brown, approximately 13 mm long, and found among the webbed uprights. Adult moths are 19 mm wide and brown with two chocolate-colored stripes crossing each forewing diagonally (Fig. 382). First-generation adults emerge in early June; second-generation adults emerge in mid-August. Eggs are circular and flat and partly overlap each other. Eggs are laid in flat, shiny masses containing up to 350 eggs and are typically found on the upper surface of weed leaves. Eggs are lemon yellow at first, becoming orange, and as they near hatch, the brown heads of the larvae show through the shell.

Damage

First-generation (overwintering) larvae move to the tips of cranberry uprights and web them together for feeding. As lar-

Fig. 381. Spotted fireworm larva (*Choristoneura parallela*) on cranberry. (Courtesy D. Polk; © APS)

Fig. 382. Spotted fireworm adult (*Choristoneura parallela*) on cranberry. (Courtesy E. de Lange; © APS)

vae develop, they pull increasing numbers of uprights together with webbing; full-grown larvae may tie as many as 20 uprights together. First-generation larvae can cause significant damage to a cranberry bed if numbers are high, but this rarely happens. Second-generation larvae hatch during late bloom and also web uprights together. The second-generation larvae feed on foliage as well as fruit and can cause serious crop loss.

Management

Spotted fireworm has a large array of parasitoids and significant control by natural enemies is common. The development of newer, target-specific insecticides that have minimal impact on natural enemies has greatly aided the management of this pest. Two of the most common parasitoids of spotted fireworm are a braconid wasp (*Meteorus trachynotus*) and an egg parasitoid (*Trichogramma minutum*). The sex pheromone has been identified and pheromone trapping of adult male moths is possible. Since spotted fireworm females lay their egg masses almost exclusively on weeds, the eradication of weeds from cranberry beds is a cultural control tactic that can have a major impact.

Selected References

Polavarapu, S., and Lonergan, G. 1998. Sex pheromone of the *Choristoneura parallela* (Lepidoptera: Tortricidae): Components and development of a pheromone lure for population modeling. Environ. Entomol. 27(5):1242-1249.

Stuart, R. J., and Polavarapu, S. 1998. Oviposition preferences of the polyphagous moth *Choristoneura parallela* (Lepidoptera: Tortricidae): Effects of plant species, leaf size, and experimental design. Environ. Entomol. 27(1):102-109.

Stuart, R. J., and Polavarapu, S. 2000. Egg-mass variability and differential parasitism of *Choristoneura parallela* (Lepidoptera: Tortricidae) by endemic *Trichogramma minutum* (Hymenoptera: Trichogrammatidae). Ann. Entomol. Soc. Am. 93(5):1076-1084.

(Prepared by D. E. Schiffhauer)

Various Miscellaneous Cranberry Foliage Feeders

In the east, gypsy moth, *Lymantria dispar* (Lepidoptera: Erebidae), is an introduced species from Eurasia with an immense host range. Populations are highly variable among years. Eggs overwinter and normally hatch during the end of April to May. Early-instar larvae are black with long hairs and attack the terminal buds in the spring and, as new growth appears, feed on leaves and flower buds. Intermediate-instar larvae are black with orange dots along the back and the final instars have prominent blue and red tubercles in rows along each side of the back (Fig. 383). These later instars are heavy feeders and may eat all new growth, stripping the upright to the stem. If undetected, portions of the bed may be injured so severely that the vines yield poorly the following year. When mature, the larva is 2.5–5.0 cm long. The female moth is flightless and beds are originally infested by neonates ballooning in on silk threads or later instars crawling onto bogs. Egg masses (75–1,000 eggs) are able to overwinter on flooded bogs. Larvae are readily picked up in sweep net sampling but infestations are very patchy on a bed, often appearing on bog edges near forested uplands, so scouting must be thorough.

Cutworms (Lepidoptera: Noctuidae) are large-bodied larvae that can reach the same size as gypsy moth larvae and do comparable damage. Cranberry blossomworm *Epiglaea apiata* is covered in a separate section in this book. False armyworm, *Xylena nupera*, is of concern in the east and central North American growing regions. The adult (American swordgrass moth) emerges in the autumn and then overwinters. In the spring, although small larvae feed up on the vines, as they mature, feeding shifts to night. Larvae mature in late June, reaching up to 5 cm in length, and can range in color from grass green (Fig. 384) to dark brown to black. Gypsy moth and cutworm larvae, when small to medium-sized, are readily sampled with a sweep net and easily managed with either conventional or reduced-risk sprays. If late water floods are held, infestations are eliminated, so this is an excellent management tool for this group.

An array of spanworm (Lepidoptera: Geometridae) species is found in cranberry in all growing regions, but which ones are a problem varies by geographical area and by year. For all species discussed here, there is a single generation per year. The moths are small to medium in size (called geometers), with slender bodies and broad wings, and are often day active and can be seen flying over the vines. The larva is usually slender and twiglike (so as to be cryptic) and lacks pairs of prolegs toward the center of the body. Because they walk by extending the head end of the body as far forward as possible and then bringing the rear of the body up to front, they are called a looper or an inchworm. The larvae in this group do not silk up the vines, and they feed externally on the vines. In Wisconsin and Massachusetts (United States) and parts of eastern Canada, green spanworm (*Speranza sulphurea* [syn. *Itame sulphurea*]) is a reported problem. It overwinters as an egg and hatch begins in May and may be extended into June. The larva (Fig. 385) feeds on leaves and developing flower buds, reaching a mature length of approximately 2.5 cm. Moths (Fig. 386) emerge irregularly from late June through August. The wingspan of

Fig. 383. Gypsy moth larvae (*Lymantria dispar*) on cranberry. (Courtesy A. L. Averill; © APS)

Fig. 384. False armyworm larva (*Xylena nupera*) on cranberry. (Courtesy A. L. Averill; © APS)

Fig. 385. Green spanworm larva (*Speranza sulphurea*) on cranberry. (Courtesy A. L. Averill; © APS)

Fig. 386. Green spanworm moth (*Speranza sulphurea*) on cranberry. (Courtesy A. L. Averill; © APS)

the grayish yellowish male is 2.5 cm, and the yellow female is smaller, approximately 2.1 cm across. The wings of both have a few scattered brown markings. Rannoch looper, *Speranza brunneata* (syn. *Itame brunneata*), has been reported as a problem in Wisconsin and parts of eastern Canada and has a similar life history as the green spanworm. Larvae of another spanworm, spiny looper/halfwing geometer (*Phigalia titea*), appear in the early season. This species is better known as a forest pest where outbreaks have been recorded. Females are flightless and larvae disperse by ballooning on silk threads. The larva is dark gray with yellow or orange patches along the side of the body. In cranberry, damage can be severe in patchy areas as the larvae are aggregated and feed on the new growth, often clipping the new shoots so that they fall over. Another very early-season forest pest with flightless females that has been reported in both the east and west cranberry-growing regions is

Fig. 387. Brown spanworm larva (*Ematurga amitaria*) on cranberry. (Courtesy A. L. Averill; © APS)

the winter moth, *Operophtera brumata* (see Winter Moth under the Blueberries (Highbush and Rabbiteye) section above and Figs. 338 and 340). Finally, two spanworm species appear later in the growing season: brown spanworm (*Ematurga amitaria*) (Fig. 387) and big cranberry spanworm (*Eutrapela clemataria*). These species can be especially problematic to scout and manage since they are present during bloom. With the exception of the early instars of the winter moth, which invade the swelling bud of the upright, small to medium-sized spanworms are readily picked up in the sweep net during the day and can be managed with newer insecticides, such as spinosyn-based products and insect growth regulators.

Selected References

Averill, A. L., Sylvia, M. M., Kusek, C. C., and DeMoranville, C. J. 1987. Flooding in cranberry to minimize insecticide and fungicide inputs. Am. J. Altern. Agric. 12:50-54.

Butler, L. 1985. Biology of the half-wing geometer, *Phigalia titea* (Cramer) (Lepidoptera: Geometridae). J. Lepid. Soc. 39:177-186.

Ferguson, D. C. 2008. The Moths of North America: Geometroidea: Geometridae (part): Ennominae (part): Abraxini, Cassymini, Macariini. Fascicle 17.2. Wedge Entomological Research Foundation, Washington, DC.

Franklin, H. J. 1948. Cranberry insects in Massachusetts. Mass. Agric. Exp. Stn. Bull. 445, Part I.

Mahr, D. L. 2005. Brown cranberry spanworm. Wisconsin Fruit, Cranberries, Insect Management, Insect Profiles. https://fruit.wisc.edu/wp-content/uploads/sites/36/2011/05/Brown-Spanworm.pdf

Talerica, R. L. 1968. Life history of the looper *Phigalia titea* in Virginia. Ann. Entomol. Soc. Am. 61 557-561.

Wagner, D. L. 2005. Caterpillars of Eastern North America. Princeton University Press, Princeton, NJ.

(Prepared by A. L. Averill)

Lingonberries

Lingonberry Fruitworm

The lingonberry fruitworm, *Grapholita libertina* (Lepidoptera: Tortricidae), is a tortricid moth whose larvae infest the fruit of the wild lingonberry, *Vaccinium vitis-idaea* var. *minus*. It has been reported from eastern and western coastal provinces and states within North America and is potentially distributed throughout the lingonberry host range. As *G. libertina* has not been recorded in European lingonberry production areas, and there are few areas in North America in which lingonberries are commercially harvested, research on this species has been limited to the province of Newfoundland and Labrador in Canada. It is the primary insect pest species of wild lingonberries, feeding within corms of berries, reducing fruit volume and quality. Infestations were reported as high as 276 larvae per kilogram of unripe berries in a survey of sites in eastern Newfoundland.

Life History and Description

In Newfoundland, *G. libertina* is univoltine, with adults emerging from June to July and laying eggs directly on developing lingonberry fruit. Adult moths are nocturnal flyers, brown-gray, and 9–10 mm long (Fig. 388). Eggs are laid directly on the leaves; larvae hatch within a few days and bore into the berries. By late September, they exit the berries and enter a prepupal stage in a hibernaculum among ground litter. They overwinter as prepupae, with pupation taking place in the spring, and adults eclose by early summer.

Fig. 388. Lingonberry fruitworm moth (*Grapholita libertina*). (Courtesy N. K. Hillier; © APS)

Fig. 389. Lingonberry fruitworm larva (*Grapholita libertina*) on lingonberry. (Courtesy N. K. Hillier; © APS)

Final (fifth) instar larvae are approximately 10 mm long with a pale brown thoracic shield (Fig. 389). The remainder of the body is pale white to pink; however this coloration appears dependent on the maturity (and coloration) of the fruit cortex upon which they are feeding.

Damage

Damage by *G. libertina* is twofold, resulting in a direct reduction in viable fruit production and, more importantly, a significant decrease in the marketability of fruit because of the presence of larvae and frass within. Infested fruit may be identified by inspection to reveal bore holes or "dimpling" on the fruit surface. Damage is also evident by splitting open fruit, wherein accumulated frass is evident since the larvae do not eject frass from the fruit.

The primary host for *G. libertina* is lingonberry, although larvae have also been found infesting *V. angustifolium* and *V. macrocarpon*. Infestation of these secondary species is generally considered rare and only occurs where these plant species overlap with *V. vitis-idaea*. However, each of these *Vaccinium* spp. may function as an alternate host for *G. libertina*, making lingonberry fruitworm a possible threat to commercial development of several *Vaccinium* spp.

Management

In Newfoundland and Labrador, the lingonberry is primarily harvested from wild barrens and heaths. Thus, any means of managing *G. libertina* must be compatible with production in a heterogeneous plant environment. The use of pesticides would be prohibitive, since it would impact large areas of nontarget plant and animal species. As well, harvest and sales of lingonberry products relies, in part, on their marketability as wild, organic products. Therefore, management tools investigated to date have focused on noninsecticidal pest management alternatives.

Historically, control of lingonberry fruitworm has consisted of first delaying wild fruit harvest until most larvae have exited the fruit and, subsequently, placing berries into a barrel of water wherein larvae are skimmed from the surface after they exit the fruit. Neither method ensures a 100% pest-free product; therefore, in recent years, advances have been made in monitoring and controlling *G. libertina* using synthetic sex attractants and pheromone traps. Using such a system, it is possible to forecast lingonberry fruitworm population and damage levels for different regions.

A final prospect for *G. libertina* management is through the augmentation and preservation of a hymenopteran parasitoid, *Phanerotoma* sp., which parasitizes the larval stage. *Phanerotoma* parasitoids also present a potential method of augmentative biological control and could be useful to maintain reduced pest populations without the need for chemical applications.

Selected References

Dixon, P. L., and Hillier, N. K. 2003. Insect pests of wild cranberry, *Vaccinium macrocarpon*, in Newfoundland and Labrador. Phytoprotection 88:139-145.

Hillier, N. K. 2001. Quantitative chemical ecology of the lingonberry fruitworm, *Grapholita libertina*. Ph.D. thesis. Memorial University of Newfoundland, St. John's, Newfoundland, Canada.

Hillier, N. K., Dixon, P. L., Seabrook, W. D., and Larson, D. J. 2002. Field testing of synthetic attractants for male *Grapholita libertina* (Lepidoptera: Tortricidae). Can. Entomol. 134:657-665.

Hillier, N. K., Dixon, P. L., and Larson, D. J. 2003. Efficacy of five pheromone trap designs in trapping male *Grapholita libertina* (Lepidoptera: Tortricidae). Can. Entomol. 135:139-141.

Hillier, N. K., Dixon, P. L., and Larson, D. J. 2004. Trap captures of male *Grapholita libertina* (Lepidoptera: Tortricidae) moths: Relationship to larval numbers and damage in wild lingonberry. Environ. Entomol. 33:405-417.

Morris, R. F., Penney, B. G., Greenslade, G., Hendrickson, P. A., and McRae, K. B. 1988. Notes on the occurrence, distribution, population levels and control of *Grapholita libertina* Heinr. (Lepidoptera: Tortricidae), a pest of lingonberries in Nfld. Can. Entomol. 120:867-872.

(Prepared by N. K. Hillier and P. L. Dixon)

Appendix

Hosts

Vaccinium amoenum Aiton (syn. *V. virgatum*)—rabbiteye blueberry
Vaccinium angustifolium Aiton—lowbush blueberry
Vaccinium ashei J. M. Reade (syns. *V. amoenum*; *V. virgatum*)—rabbiteye blueberry
Vaccinium corymbosum L.—highbush blueberry
Vaccinium macrocarpon Aiton (syn. *Oxycoccus macrocarpus* (Aiton) Pers.)—American cranberry, large cranberry
Vaccinium myrtilloides Michx.—sour top blueberry
Vaccinium oxycoccos L.—small cranberry
Vaccinium virgatum Aiton (syn. *V. ashei*)—rabbiteye blueberry
Vaccinium vitis-idaea L.—lingonberry
Vaccinium vitis-idaea L. var. *minus* Lodd. et al.—wild lingonberry

Other Vaccinium spp. noted in the compendium
Vaccinium altomontanum Ashe (syn. *V. pallidum*)—hillside blueberry
Vaccinium arboreum Marshall—sparkleberry
Vaccinium australe Small (syn. *V. formosum* Andrews)—southern blueberry
Vaccinium caespitosum Michx.—dwarf bilberry
Vaccinium constablaei A. Gray (syn. *V. corymbosum*)—Constable's blueberry
Vaccinium darrowii Camp—Darrow's evergreen blueberry
Vaccinium deliciosum Piper—Cascade bilberry
Vaccinium elliottii Chapm.—Elliott's blueberry
Vaccinium fuscatum Aiton—black highbush blueberry
Vaccinium membranaceum Douglas ex Torr.—mountain bilberry
Vaccinium myrtillus L.—whortleberry, bilberry, European blueberry
Vaccinium oldhamii Miq.—natsuhaze
Vaccinium ovalifolium Sm.—Alaska blueberry
Vaccinium ovatum Pursh—evergreen huckleberry
Vaccinium pallidum Aiton (syn. *V. altomontanum*)—hillside blueberry
Vaccinium parvifolium Sm.—red huckleberry
Vaccinium scoparium Leiberg ex Coville—grouseberry
Vaccinium simulatum Small—upland highbush blueberry
Vaccinium stamineum L.—deerberry
Vaccinium uliginosum L.—bog blueberry

Diseases of Blueberry

The diseases/disorders and their causal organisms/agents are known for all blueberries unless indicated as occurring primarily on highbush/rabbiteye or lowbush blueberries.

Algal Disease

Algal stem blotch
Cephaleuros virescens Kunze ex Fries

Bacterial Diseases

Bacterial canker
Pseudomonas syringae van Hall
Bacterial leaf scorch
Xylella fastidiosa Wells et al.
Xylella fastidiosa Wells et al. subsp. *multiplex* Schaad et al.
Bacterial leaf spot
Burkholderia andropogonis (Smith) Gillis et al. (syn. *Pseudomonas andropogonis* (Smith) Stapp)
Crown gall
Agrobacterium rubi (Hildebrand) Starr & Weiss (syn. *Rhizobium rubi* (Hildebrand) Young et al.)
Agrobacterium tumefaciens (Smith & Townshend) Conn (syn. *Rhizobium radiobacter* (Smith & Townsend) Young et al.)
Stunt
'*Candidatus* Phytoplasma asteris' Lee et al., subgroup 16SrI-E

Fungal Diseases

Alternaria fruit rot/Alternaria leaf spot and fruit rot
Alternaria spp.
Alternaria tenuissima (Nees & T. Nees:Fr.) Wiltshire
Anthracnose/Anthracnose fruit rot
Colletotrichum fioriniae (J. A. P. Marcelino & S. Gouli) R. G. Shivas & Y. P. Tan
(syn. *Glomerella fioriniae* (J. A. P. Marcelino & S. Gouli) R. G. Shivas & Y. P. Tan)
(syn. *Colletotrichum acutatum* J. H. Simmonds)
(syn. *Glomerella acutata* Guerber & J. C. Correll)

Colletotrichum gloeosporioides (Penz.) Penz. & Sacc.
(syn. *Glomerella cingulata* (Stoneman) Spauld. & H. Schrenk)

Armillaria root rot
Armillaria mellea (Vahl:Fr.) P. Kumm.
Armillaria gallica Marxm. & Romagn.
Armillaria solidipes Peck
(syn. *Armillaria ostoyae* (Romagn.) Herink)

Botryosphaeria stem blight
Botryosphaeria dothidea (Moug.:Fr.) Ces & De Not.
(syn. *Fusicoccum aesculi* Corda)
Lasiodiplodia theobromae (Pat.) Griffon & Maubl.
Neofusicoccum ribis (Slippers, Crous & M. J. Wingf.) Crous, Slippers & A. J. L. Phillips

Botryosphaeria stem canker
Botryosphaeria corticis (Demaree & M. S. Wilcox) Arx & E. Müll.
(syn. *Physalospora corticis* Demaree & M. S. Wilcox)

Botrytis blight and fruit rot/Botrytis blossom blight
Botrytis cinerea Pers.:Fr.
(syn. *Botryotinia fuckeliana* (de Bary) Whetzel)

Brown leaf spot
Septoria sp.

Cylindrocladium rot
Calonectria ilicicola Boedijn & Reitsma
(syns. *Calonectria crotalariae* (Loos) D. K. Bell & Sobers; *Cylindrocladium parasiticum* Crous, M. J. Wingf. & Alfenas)
Cylindrocladium colhounii Peerally

Double spot
Dothichiza caroliniana Demaree & M. S. Wilcox

Exobasidium fruit and leaf spot (lowbush blueberry)
Exobasidium sp.

Exobasidium leaf and fruit spot (highbush and rabbiteye blueberries)
Exobasidium maculosum M. T. Brewer

Fusicoccum canker/Godronia canker
Godronia cassandrae Peck
(syn. *Fusicoccum putrefaciens* Shear)

Gibbera twig blight
Gibbera vacciniicola (Dearn. & House) M. E. Barr
(syn. *Dothidella vacciniicola* Dearn. & House)

Gloeocercospora leaf spot
Gloeocercospora inconspicua Demaree & M. S. Wilcox ex Deighton

Gloeosporium leaf spot and stem canker/Gloeosporium stem and leaf spot
Gloeosporium minus Shear

Leaf rust
In eastern North America:
Thekopsora minima P. Syd. & Syd.
(syn. *Pucciniastrum minimum* (Schwein.) Arthur)
In western North America:
Naohidemyces vaccinii (Jørst) S. Sato, K. Katsuya, & Y. Hirats. ex Vanderweyen & Fraiture
(syn. *Pucciniastrum vaccinii* (G. Winter) Jørst.)
(syn. *Pucciniastrum myrtilli* Arthur)

Mummy berry
Monilinia vaccinii-corymbosi (J. M. Reade) Honey
(syn. *Monilia vaccinii-corymbosi* J. M. Reade)

Penicillium fruit rot
Penicillium spp.

Pestalotia fruit rot
Pestalotia spp.
Pestalotia vaccinii (Shear) Guba

Phomopsis canker/Phomopsis twig blight and fruit rot
Phomopsis vaccinii Shear
(syn. *Diaporthe vaccinii* Shear)

Phyllosticta fruit rot/Phyllosticta leaf spot and fruit rot
Phyllosticta vaccinii Earle
(syn. *Botryosphaeria vaccinii* (Shear) M. E. Barr)

Phytophthora root rot (*see also* Oomycete Disease)
Phytophthora cinnamomi Rands

Powdery mildew
Erysiphe vaccinii Schwein.
(syns. *Microsphaera alni* (Wallr.:Fr.) G. Winter var. *vaccinii* (Schwein.) E. S. Salmon; *Microsphaera penicillata* (Wallr.:Fr.) Lév. f. *vaccinii* Jacz.; *Microsphaera penicillata* (Wallr.:Fr.) Lév. var. *vaccinii* (Schwein.) W. B. Cooke; *Microsphaera vaccinii* (Schwein.) Cooke & Peck)

Red leaf
Exobasidium sp.
Exobasidium vaccinii (Fuckel) Woronin

Septoria leaf spot (highbush, southern highbush, and rabbiteye blueberries)
Septoria albopunctata Cooke

Septoria leaf spot and stem canker (lowbush blueberry)
Septoria sp.

Silver leaf
Chondrostereum purpureum (Pers.: Fr.) Pouzar

Sooty blotch
Discostroma spp.
Hormonema spp.
Microcyclospora spp.
Pseudocercospora spp.
Rhizosphaera spp.

Stem blotch
Cercospora sp.

Trichoderma fruit rot
Trichoderma spp.

Valdensinia leaf spot
Valdensia heterodoxa Peyronel
(syn. *Valdensinia heterodoxa* Peyronel)

Witches'-broom
Pucciniastrum goeppertianum (J. G. Kühn) Kleb.

Nematodes (Parasitic)

Awl nematode
Dolichodorus heterocephalus Cobb

Cyst nematode
Meloidodera floridensis Chitwood, Hannon, & Esser

Dagger nematodes
Xiphinema americanum Cobb
Xiphinema chambersi Thorne
Xiphinema rivesi Dalmasso

Lance nematode
Hoplolaimus galeatus (Cobb) Thorne
(syn. *Hoplolaimus tylenchiformis* Andrassy)

Pin nematode
Paratylenchus projectus Jenkins

Ring nematodes
Bakernema inaequale (Taylor) Mehta & Raski
(syn. *Criconema inaequale* Taylor)
Crossonema menzeli (Stefanski) Schuurmans Stekhoven & Tenunissen
(syn. *Criconema menzeli* (Stefanski) Taylor)
Mesocriconema ornatum Raski
Mesocriconema xenoplax (Raski) Loof & De Grisse
(syns. *Criconemoides xenoplax* Raski; *Criconemella xenoplax* (Raski) Luc & Raski; *Macroposthonia xenoplax* (Raski) De Grisse & Loof)

Root-knot nematode
Meloidogyne carolinensis Eisenback

Root lesion nematodes
Pratylenchus crenatus Loof
(syn. *Pratylenchus pratensis* (de Man) Filipjev)
Pratylenchus penetrans (Cobb) Filipjev & Schuurmans Stekhoven
Sheath nematodes
Hemicycliophora ritteri Brizuela
Hemicycliophora uniformis Thorne
Hemicycliophora vaccinium Reed & Jenkins
(syn. *Hemicycliophora gracilis* Thorne)
Hemicycliophora vidua Raski
(syn. *Hemicycliophora similis* Thorne)
Hemicycliophora zuckermani Brzeski
Spiral nematodes
Helicotylenchus dihystera (Cobb) Sher
(syn. *Helicotylenchus nannus* Steiner)
Helicotylenchus erythrinae (Zimmermann) Golden
Helicotylenchus pseudorobustus Steiner
Rotylenchus goodeyi Loof & Oostenbrink
(syn. *Helicotylenchus goodeyi* (Loof & Oostenbrink) Perry)
Rotylenchus robustus (de Man) Filipjev
(syn. *Rotylenchus uniformis* (Thorne) Loof & Oostenbrink)
Stem nematode
Ditylenchus sp.
Stubby root nematodes
Paratrichodorus minor (Colbran) Siddiqi
(syn. *Trichodorus christei* Allen)
Paratrichodorus renifer Siddiqi
Trichodorus aequalis Allen
Stunt nematodes
Merlinius joctus (Thorne) Sher
(syn. *Tetylenchus joctus* Thorne
Tylenchorhynchus claytoni Steiner
Tylenchorhynchus dubius (Butschli) Filipjev
Tylenchorhynchus maximus Allen

Oomycete Disease

Phytophthora root rot
Phytophthora cinnamomi Rands

Parasitic Plant

Dodder
Cuscuta compacta Juss. ex Choisy

Pests

Aphids
Ericaphis fimbriata (Richards)
(syn. *Fimbriaphis fimbriata* Richards)
Illinoia azaleae (Mason)
Illinoia pepperi (MacGillivray)
Myzus persicae (Sulzer)
Blueberry blossom weevil
Anthonomus musculus Say
Blueberry bud mite
Acalitus vaccinii (Keifer)
Blueberry gall midge
Dasineura oxycoccana (Johnson)
Blueberry maggot fly
Rhagoletis mendax Curran
Blueberry sawfly
Neopareophora litura (Klug)
Pristiphora sp.
Pristiphora idiota Norton
Blueberry spanworm
Speranza argillacearia (Packard)
(syns. *Itame argillacearia* Packard; *Macaria argillacearia* (Packard))
Blueberry thrips
Catinathrips kainos O'Neill
Catinathrips similis Nakahara
Catinathrips vaccinophilus (Hood)
Frankliniella caudiseta Sakimura & O'Neill
Frankliniella vaccinii Morgan
Blueberry tip borer
Hendecaneura shawiana Kearfott
Bruce spanworm
Operophtera bruceata Hulst
Cherry fruitworm
Grapholita packardi Zeller
Cranberry fruitworm
Acrobasis vaccinii Riley
Flower thrips
Catinathrips kainos O'Neill
Frankliniella bispinosa (Morgan)
Frankliniella occidentalis (Pergande)
Frankliniella tritici (Fitch)
Frankliniella vaccinii Morgan
Japanese beetle
Popillia japonica Newman
Leafrollers
Argyrotaenia mariana (Fernald)—gray-banded leafroller
Argyrotaenia velutinana Walker—red-banded leafroller
Choristoneura rosaceana (Harris)—oblique-banded leafroller
Pandemis limitata Robinson—three-lined leafroller
Oriental beetle
Anomala orientalis (Waterhouse)
Plum curculio
Conotrachelus nenuphar (Herbst)
Red-striped fireworm
Aroga trialbamaculella Chambers
Scales
Acanthococcus azaleae Comstock—azalea bark scale
Acutaspis perseae (Comstock)—red bay scale
Ceroplastes japonicus Green—Japanese wax scale
Diaspidiotus ancylus (Putnam)—Putman scale
Eriococcus quercus (Comstock)—oak eriococcus
Lepidosaphes ulmi (Linnaeus)—oystershell scale
Mesolecanium nigrofasciatum (Pergande)—terrapin scale
Parthenolecanium corni (Bouché)—lecanium scale
Pinnaspis strachani (Cooley)—lesser snow scale
Pseudoparlatoria parlatorioides (Comstock)—parlatoria-like scale
Pulvinaria innumerabilis (Rathvon)—cottony maple scale
Sharp-nosed leafhopper
Scaphytopius magdalensis (Provancher)
Strawberry rootworm
Paria fragariae Wilcox
Winter moth
Operophtera brumata (Linnaeus)

Physiological Disorders

Environmental disorders
Cold injury
Salt injury
Water stress

Nutritional disorders
- Boron deficiency
- Boron toxicity
- Calcium deficiency
- Copper deficiency
- Copper toxicity
- Iron deficiency
- Manganese deficiency
- Manganese toxicity
- Magnesium deficiency
- Nitrogen deficiency
- Nitrogen toxicity
- Phosphorous deficiency
- Phosphorous toxicity
- Potassium deficiency
- Sulfur deficiency
- Zinc deficiency

Pesticide toxicity injury
- 2,4-D
- Diuron
- Glyphosate
- Hexazinone
- Norflurazon
- Paraquat
- Simazine
- Terbacil

Viral Diseases

Blueberry latent spherical virus (BLSV)—genus *Nepovirus*; family *Secoviridae*
Blueberry latent virus (BBLV)—genus *Amalgavirus*; family *Amalgaviridae*
Blueberry virus A (BVA)—genus *Closterovirus*; family *Closteroviridae*

Fruit drop
- Virus—possibly a new genus; family *Caulimoviridae*

Leaf mottle
- *Blueberry leaf mottle virus* (BLMV)—genus *Nepovirus*; family *Secoviridae*

Mosaic
- Blueberry mosaic associated virus (BlMaV)—genus *Ophiovirus*; family *Ophioviridae*

Necrotic ring blotch
- Associated with *Blueberry necrotic ring blotch virus* (BNRBV)—genus *Cilevirus*; family unassigned

Necrotic ringspot
- *Tobacco ringspot virus* (TRSV)—genus *Nepovirus*; family *Secoviridae*

Peach rosette mosaic
- *Peach rosette mosaic virus* (PRMV)—genus *Nepovirus*; family *Secoviridae*

Red ringspot
- *Blueberry red ringspot virus* (BRRSV)—genus *Soymovirus*; family *Caulimoviridae*

Scorch
- *Blueberry scorch virus* (BlScV)—genus *Carlavirus*; family *Betaflexiviridae*

Shock
- *Blueberry shock virus* (BlShV)—genus *Ilarvirus*; family *Bromoviridae*

Shoestring
- *Blueberry shoestring virus* (BSSV)—genus *Sobemovirus*; family unassigned

Tomato ringspot
- *Tomato ringspot virus* (ToRSV)—genus *Nepovirus*; family *Secoviridae*

Diseases of Cranberry

Bacterial Diseases

False blossom
- Phytoplasma

Stem gall
- Indole-3-acetic acid (IAA)-producing bacteria

Fungal Diseases

Berry speckle
- *Phyllosticta elongata* Weid.
  (syns. *Botryosphaeria vaccinii* (Shear) M. E. Barr; *Guignardia vaccinii* Shear)

Bitter rot
- *Colletotrichum fioriniae* (J. A. P. Marcelino & S. Gouli) R. G. Shivas & Y. P. Tan
  (syn. *Colletotrichum acutatum* J. H. Simmonds)
  (syn. *Glomerella acutata* Guerber & J. C. Correll)
- *Glomerella cingulata* (Stoneman) Spauld. & H. Schrenk
  (syns. *Colletotrichum gloeosporioides* (Penz.) Penz. & Sacc.; *Glomerella cingulata* (Stoneman) Spauld. & H. Schrenk var. *vaccinii* Shear)

Black rot
- *Allantophomopsis cytisporea* (Fr.:Fr.) Petr.
  (syns. *Apostrasseria lunata* (Shear) Nag Raj; *Ceuthospora lunata* Shear; *Phacidium lunatum* DiCosmo, Nag Raj & W. B. Kendr.)
- *Allantophomopsis lycopodina* (Höhn.) Carris
- *Strasseria geniculata* (Berk. & Broome) Höhn.
  (syn. *Strasseria oxycocci* (Shear) Shear)

Black spot
- *Mycosphaerella nigromaculans* Shear
  (syn. *Ramularia nigromaculans* Shear)

Blotch rot
- *Physalospora vaccinii* (Shear) Arx & E. Müll.
  (syn. *Acanthorhynchus vaccinii* Shear)

Cladosporium leaf spot
- *Cladosporium oxycocci* Shear

Cottonball
- *Monilinia oxycocci* (Woronin) Honey
  (syn. *Sclerotinia oxycocci* Woronin)

Early rot
- *Phyllosticta vaccinii* Earle

End rot
- *Godronia cassandrae* Peck
  (syn. *Godronia cassandrae* Peck f. *vaccinii* J. W. Groves)
  (syn. *Fusicoccum putrefaciens* Shear)

Fairy ring
- *Helicobasidium* sp.
  (syns. *Rhizoctonia* sp.; *Thanatophytum* sp.; *Tuberculina* sp.)

Phytophthora root and runner rot (*see also* Oomycete Disease)
Phytophthora spp.
Phytophthora cinnamomi Rands
Phytophthora megasperma Drechsler
Phytophthora taxon cranberry
Powdery mildew
Microsphaera vaccinii (Schwein.) Cooke & Peck
(syns. *Erysiphe vaccinii* Schwein.; *Microsphaera alni* (Wallr.:Fr.) G. Winter var. *vaccinii* (Schwein.) E. S. Salmon; *Microsphaera penicillata* (Wallr.:Fr.) Lév. var. *vaccinii* (Schwein.) W. B. Cooke
Podosphaera sp.
Protoventuria early leaf spot
Protoventuria barriae Carris & A. P. Poole
Protoventuria leaf spot and berry speckle
Protoventuria myrtilli (Cooke) M. E. Barr
(syn. *Gibbera myrtilli* (Cooke) Petr.)
Pyrenobotrys leaf spot
Pyrenobotrys compacta (Peck) B. Erikss.
(syn. *Gibbera compacta* (Peck) Shear)
Red gall
Synchytrium vaccinii F. Thomas
Red leaf spot
Exobasidium rostrupii Nannf.
(syn. *Exobasidium vaccinii* (Fuckel) Woronin)
Red shoot
Exobasidium perenne N. L. Nick.
Ripe rot
Coleophoma empetri (Rostr.) Petr.
(syn. *Sporonema oxycocci* Shear)
Rose bloom
Exobasidium oxycocci Rostr. ex Shear
Rust
Pucciniastrum vaccinii (G. Winter) Jørst.
(syn. *Pucciniastrum myrtilli* Arthur)
Twig blight
Lophodermium hypophyllum (Dearn. & House) Shear
(syn. *Lophodermium oxycocci* (Fr.) P. Karst. var. *hypophyllum* Dearn. & House)
Lophodermium oxycocci (Fr.) P. Karst.
Upright dieback
Colletotrichum fioriniae (J. A. P. Marcelino & S. Gouli) R. G. Shivas & Y. P. Tan
(syn. *Colletotrichum acutatum* J. H. Simmonds)
Colletotrichum gloeosporioides (Penz.) Penz. & Sacc.
Fusicoccum putrefaciens Shear
Phomopsis spp.
Phomopsis vaccinii Shear
(syn. *Diaporthe vaccinii* Shear)
Synchronoblastia crypta Uecker & F. L. Caruso
Viscid rot
Phomopsis vaccinii Shear
(syn. *Diaporthe vaccinii* Shear)
Yellow rot
Botrytis sp.
Botrytis cinerea Pers.:Fr.

Nematodes (Parasitic)

Ring nematodes
Mesocriconema spp.
Mesocriconema curvatum (Raski) Loof & De Grisse
(syns. *Criconemoides curvatus* Raski; *Criconemella curvatus* (Raski) Luc & Raski)
Mesocriconema xenoplax (Raski) Loof & De Grisse
(syns. *Criconemoides xenoplax* Raski; *Criconemella xenoplax* (Raski) Luc & Raski; and *Macroposthonia xenoplax* (Raski) De Grisse & Loof)
Root lesion nematode
Pratylenchus penetrans (Cobb) Filipjev & Schuurmans Stekhoven
Sheath nematodes
Hemicycliophora spp.
Hemicycliophora ritteri Brizuela
Hemicycliophora vidua Raski
Spiral nematodes
Helicotylenchus spp.
Stubby root nematodes
Paratrichodorus spp.
Paratrichodorus minor (Colbran) Siddiqi
Stunt nematodes
Merlinius spp.
Tylenchorhynchus spp.

Oomycete Disease

Phytophthora root and runner rot
Phytophthora spp.
Phytophthora cinnamomi Rands
Phytophthora megasperma Drechsler
Phytophthora taxon cranberry

Pests

Black-headed fireworm
Rhopobota naevana Hübner
Black vine weevil
Otiorhynchus sulcatus Fabricius
Blunt-nosed leafhopper
Limotettix vaccinii (Van Duzee)
(syn. *Scleroracus vaccinii* (Van Duzee))
Clay-colored weevil
Otiorhynchus singularis (Linnaeus)
Coptodisca leafminer
Coptodisca negligens Braun
Cranberry blossomworm
Epiglaea apiata (Grote)
Cranberry fruitworm
Acrobasis vaccinii Riley
Cranberry girdler
Chrysoteuchia topiaria (Zeller)
Cranberry root grub
Lichnanthe vulpina (Hentz)
Cranberry tipworm
Dasineura oxycoccana (Johnson)
Cranberry weevil
Anthonomus musculus Say
Hoplia
Hoplia equina LeConte
Phyllophaga
Phyllophaga anxia (LeConte)
Red-headed flea beetle
Systena frontalis (Fabricius)
Root-feeding chrysomelids
Colaspis costipennis Crotch—striped colaspis
Rhabdopterus picipes (Olivier)—cranberry rootworm
Scales
Acanthococcus azaleae Comstock—azalea bark scale
Coccus hesperidum Linnaeus—brown soft scale
Diaspidiotus ancylus (Putnam)—Putnam scale
Hemiberlesia lataniae (Signoret)—Latania scale
Hemiberlesia oxycoccus (Woglum)—cranberry scale
Hemiberlesia rapax (Comstock)—greedy scale
Lepidosaphes ulmi (Linnaeus)—oystershell scale

Parthenolecanium corni (Bouché)—European fruit scale/ lecanium scale
Rhizaspidiotus dearnessi (Cockerell)—Dearness scale
Sparganothis fruitworm
Sparganothis sulfureana (Clemens)
Spotted fireworm
Choristoneura parallela (Robinson)
Strawberry root weevil
Otiorhynchus ovatus Linnaeus
Various miscellaneous cranberry foliage feeders
Ematurga amitaria (Guenée)—brown spanworm
Eutrapela clemataria (J. E. Smith)—big cranberry spanworm
Lymantria dispar Linnaeus—gypsy moth
Operophtera brumata (Linnaeus)—winter moth
Phigalia titea (Cramer)—spiny looper/halfwing geometer
Speranza brunneata (Thunberg)—Rannoch looper
(syns. *Itame brunneata* (Thunberg); *Macaria brunneata* (Thunberg))
Speranza sulphurea (Packard)—green spanworm
(syns. *Itame sulphurea* (Packard); *Macaria sulphurea* (Packard))
Xylena nupera (Lintner)—false armyworm
Yellow-headed fireworm
Acleris minuta (Robinson)

Parasitic Plants

Dodder
Cuscuta cephalanthi Engelm.
Cuscuta compacta Juss. ex Choisy
Cuscuta gronovii Willd. ex Schult.

Physiological Disorders

Environmental disorders
Frost injury
Hail injury
Lightning injury
Oxygen deficiency injury
Salt injury
Sunscald and heat injury
Winter injury
Nutritional disorders
Boron deficiency
Calcium deficiency
Copper deficiency
Iron deficiency
Magnesium deficiency
Manganese deficiency
Nitrogen deficiency
Nitrogen toxicity
Phosphorus deficiency
Potassium deficiency
Sulfur deficiency
Yellow vine
Zinc deficiency
Pesticide toxicity injury
2,4-D
Chlorothalonil
Chlorpyrifos
Clopyralid
Copper sulfate
Dichlobenil
Glyphosate
Napropamide
Norflurazon
Simazine
Physiological (sterile) breakdown

Viral Diseases

Ringspot
Virus—genus *Soymovirus*; family *Caulimoviridae*
Tobacco streak
Tobacco streak virus (TSV)—genus *Ilarvirus*; family *Bromoviridae*

Diseases of Lingonberry

Fungal Diseases

Allantophomopsis leaf spot
Allantophomopsis cytisporea (Fr.:Fr.) Petr.
Allantophomopsis lycopodina (Höhn.) Carris
Coleophoma leaf spot
Coleophoma empetri (Rostr.) Petr.
Dwarfing redleaf
Exobasidium juelianum Nannf.
Foliar fungi (miscellaneous)
Godronia foliicola Schläpf.-Bernh.
Seimatosporium vaccinii (Fuckel) B. Erikss.
Gibbera canker
Gibbera vaccinii (Sowerby:Fr.) Fr.
(syn. *Brachysporium vaccinii* (Fr.) Sacc.)
Leaf rust
Pucciniastrum vaccinii (G. Winter) Jørst
(syn. *Naohidemyces vaccinii* (Jørst) S. Sato, K. Katsuya & Y. Hirats. ex Vanderweyen & Fraiture)
Little leaf
Phytoplasma
Mummy berry
Monilinia urnula (Weinm.) Whetzel
(syn. *Monilia urnula* (Weinm.) Grasso)
Mycosphaerella leaf spot
Mycosphaerella stemmatea (Fr.) Romell
Phacidium stem dieback
Phacidium vaccinii Fr.
(syn. *Ceuthospora anomala* Nag Raj)
Phomopsis leaf spot
Phomopsis columnaris D. F. Farr & Castl.
Phytophthora twig dieback and leaf necrosis
"*Phytophthora citricola*" complex
Phytophthora citricola Sawada
Phytophthora plurivora T. Jung & T. I. Burgess
Phytophthora ramorum Werres, De Cock & Man in 't Veld
Powdery mildew
Podosphaera myrtillina (C. Schub.:Fr.) Kunze
Red leaf
Exobasidium vaccinii (Fuckel) Woronin
Red shoot
Exobasidium splendidum Nannf.

Snow mold
Eupropolella vaccinii (Rehm) Höhn.
Witches'-broom rust
Pucciniastrum goeppertianum (J. G. Kühn) Kleb.
(syn. *Calyptospora columnaris* J. G. Kühn)

Pest

Lingonberry fruitworm
Grapholita libertina Heinrich

Glossary

a.i.—active ingredient
C—Celsius or centigrade
cm—centimeter (1 cm = 0.01 m = 0.3937 in.)
g—gram (1 g = 0.03527 oz)
gal—gallon (1 gal liquid (U.S.) = 3.785 L)
h—hour
ha—hectare (1 ha = 2.471 acres)
in.—inch (1 in. = 2.540 cm)
kg—kilogram (1 kg = 2.205 lb)
L—liter (1 L = 1.057 quarts liquid (U.S.))
lb—pound (1 lb = 453.59 g)
m—meter (1 m = 39.37 in.)
mg—milligram (1 mg = 0.001 g)
min—minute
ml—milliliter (1 ml = 0.001 L)
mm—millimeter (1 mm = 0.001 m = 0.03937 in.)
t—tonne; metric tonne
μg—microgram (1 μg = 10^{-6} g)
μm—micrometer (1 μm = 10^{-6} m)
nm—nanometer (1 nm = 10^{-9} m)
oz—ounce (1 oz = 28.35 g); fluid ounce (1 fl oz (U.S.) = 29.57 ml)
ppm—parts per million
sec—second

abaxial—directed away from the stem of a plant; pertaining to the lower surface of a leaf (*see* adaxial)
abdomen—in an insect, the third and most posterior region of the body, behind the thorax
abiotic—pertaining to the absence of life, as in disorders not caused by living organisms
abscise—to separate from a plant, as leaves, flowers, and fruits do when they fall
abscission—the shedding of leaves or other plant parts as the result of physical weakness in a specialized layer of cells (the abscission layer), which develops at the base of the structure
acervulus (pl. acervuli)—an erumpent, saucer-shaped, cushionlike fruiting body of a fungus bearing conidiophores, conidia, and sometimes setae
acid—(adj.) having a pH less than 7
acropetal—upward from the base to the apex of a shoot of a plant (*see* basipetal); in fungi, producing spores in succession in the direction of the apex, so that the apical spore is the youngest
acute—with reference to symptoms, developing suddenly (*see* chronic)
adaxial—directed toward the stem of a plant; pertaining to the upper surface of a leaf (*see* abaxial)
adventitious—arising from other than the usual place, as roots from a stem rather than branches of a root
aeciospore—a dikaryotic spore produced in the aecium of a rust fungus; in heteroecious rust fungi, the spore stage that infects the alternate host
aecium (pl. aecia; adj. aecial)—the fruiting body of a rust fungus in which the first dikaryotic spores (aeciospores) are produced
aerial—occurring in air
aerobic—living only in the presence of oxygen
aflatoxin—a chemical by-product of *Aspergillus flavus* and *A. parasiticus* that is harmful to humans and other animals
agar—a jellylike material derived from algae and used to solidify liquid culture media; a medium containing agar
alate—winged (*see* apterous)
alkaline—having basic (nonacidic) properties; having a pH greater than 7
alkaloid—any of various nitrogen-containing ring compounds produced by plants and having physiological effects on animals
allele—any of one or more alternative forms of a gene
alternate host—one of two kinds of plants on which a parasitic fungus (e.g., a rust fungus) must develop to complete its life cycle
alternative host—a plant other than the main host that can be colonized by a parasite but that is not required for completion of the developmental cycle of the parasite
anaerobic—living in the absence of oxygen
anamorph (adj. anamorphic)—the imperfect state or asexual form in the life cycle of certain fungi, producing asexual spores (e.g., conidia) or no spores (*see* holomorph; teleomorph)
annual—a plant that completes its life cycle and dies within 1 year (*see* biennial; perennial)
antagonist—an organism or substance that counteracts or limits the action of another organism or substance
anterior—situated toward the front or head (*see* posterior)
anther—the pollen-bearing portion of a flower
antheridium (pl. antheridia)—the male sexual organ (male gametangium) of certain fungi
anthesis—the period of the opening of a flower, during which pollination can occur
anthracnose—disease caused by acervuli-forming fungi (primarily in the order Melanconiales); characterized by sunken lesions and necrosis
antibiotic—any of various chemical compounds produced by microorganisms and killing or inhibiting the growth of other living organisms
antibody—a protein formed in the blood of warm-blooded animals in response to the presence of an antigen
antigen—any foreign chemical (normally, a protein) that induces antibody formation in warm-blooded animals
antiserum (pl. antisera)—blood serum containing an antibody
apex (pl. apices; adj. apical)—the tip of a root or shoot; contains the apical meristem
aphid—any of numerous species of small, sucking insects of the family Aphididae (order Hemiptera) that produce honeydew and injure plants when present in large populations
apothecium (pl. apothecia)—open, cuplike or saucerlike, ascus-bearing fruiting body (ascocarp) of ascomycetous fungi; often supported on a stalk
appressed—closely flattened down or pressed against a surface (syn. adpressed)
appressorium (pl. appressoria)—a swollen, flattened portion of a fungal filament that adheres to the surface of a higher plant, providing anchorage for invasion by the fungus
apterous—wingless (*see* alate)
arthropod—any member of the phylum Arthropoda, which consists of animals with articulated bodies and limbs, including insects, arachnids, and crustaceans
ascocarp—a sexual fruiting body (e.g., apothecium, ascostroma, cleistothecium, perithecium, pseudothecium) of ascomycetous fungi, producing asci and ascospores (syn. ascoma)
ascogenous—pertaining to ascus-producing hyphae
ascogonium (pl. ascogonia)—a specialized cell that gives rise to hyphae that produce asci
ascoma (pl. ascomata)—*see* ascocarp

ascomycetes (adj. ascomycetous)—a group of fungi in the phylum Ascomycota (kingdom Fungi) characterized by the production of sexual spores (ascospores) within an ascus

ascospore—a sexual spore borne in an ascus

ascostroma (pl. ascostromata)—a fruiting body of ascomycetous fungi containing bitunicate (double-walled) asci in locules (cavities); usually dark and containing multiple locules but sometimes containing only a single locule (*see* pseudothecium)

ascus (pl. asci)—a saclike structure containing ascospores (typically eight) and usually borne in a fruiting body

aseptate—lacking cross walls (septa); nonseptate; coenocytic

asexual—vegetative; lacking sex organs, gametes, or sexual spores (as in the imperfect or anamorphic stage of a fungus)

asexual reproduction—any type of reproduction not involving the union of gametes and meiosis

asexual stage—the imperfect (anamorphic) stage of certain fungi

autoecious—with reference to rust fungi, producing all spore forms on one species of host plant (*see* heteroecious)

autotroph—any of various organisms (e.g., plants capable of photosynthesis) that synthesize their nutritive substances from inorganic molecules

avirulent—nonpathogenic; unable to cause disease (*see* virulent)

axil—the angle formed by a leaf petiole and the stem

axillary—pertaining to or placed within an axil

bacilliform—shaped like a short rod with rounded ends

backcross—to cross (mate) an offspring with one of its parents

bactericide—a chemical or physical agent that kills bacteria

bacterium (pl. bacteria)—a prokaryotic, microscopic, single-celled organism having a cell wall and increasing by binary fission

basidiocarp—the sexual fruiting body of fungi in the group basidiomycetes (syn. basidioma)

basidiomycetes (adj. basidiomycetous)—a group of fungi in the phylum Basidiomycota (kingdom Fungi) characterized by the formation of external basidiospores on basidia

basidiospore—a haploid ($1n$) sexual spore produced on a basidium

basidium (pl. basidia; adj. basidial)—a structure on which basidiospores are produced externally

basipetal—situated down from the apex toward the base of a shoot of a plant; developing in the direction of the base, so that the apical part is oldest (*see* acropetal)

bicellular—two celled

biennial—a plant that produces seed and dies at the end of its second year of growth (*see* annual; perennial)

binucleate—having two nuclei per cell (*see* multinucleate; uninucleate)

bioassay—any test (assay) using a living organism

biocide—a compound toxic to all forms of life

biological control—exploitation of the natural competition, parasitism, or antagonism of organisms for the management of pests and pathogens (syn. biocontrol)

biotic—pertaining to living organisms; with reference to disease, caused by a living organism

biotroph—an organism that can live and multiply only on another living organism (syn. obligate parasite) (*see* necrotroph)

biotype—a subdivision of a species, subspecies, or race based on some identifiable physiological trait, such as a particular virulence pattern

bitunicate—having two walls

bivoltine—having two broods or generations per year (*see* multivoltine; univoltine)

blade—the flat portion of a grass leaf above the sheath

blight—the sudden, severe, and extensive spotting, discoloration, wilting, or destruction of leaves, flowers, stems, or entire plants

blotch—an irregularly shaped, usually superficial spot or blot

botryose—shaped like a bunch of grapes

bract—a reduced leaf associated with a flower or inflorescence; a modified leaf from the axil of which a flower arises

callus—specialized tissue that forms over a wound or cut in a plant (e.g., cork cambium may form, and the cells produced will gradually seal the wound)

canker—a plant disease characterized (in woody plants) by the death of cambium tissue and the malformation or loss of bark and (in nonwoody plants) by the formation of sharply delineated, dry, necrotic, localized lesions on the stem; a lesion caused by such a disease, particularly in woody plants

canopy—the expanded leafy top of a plant or plants

capsid—the protective layer of protein surrounding the nucleic acid core of a virus; the protein molecules that make up this layer (syn. coat protein)

carbohydrate—any of various chemical compounds composed of carbon, hydrogen, and oxygen, such as sugars, starches, and cellulose

carpel—the ovule-bearing structure of a flower in angiosperms

catenulate—with reference to the spores of certain fungi, formed in a chain

caudal—oriented toward, near, or at the posterior of the body

causal agent, causal organism—the agent or organism that produces a given disease

chelicera (pl. chelicerae)—in spiders, mites, and relatives, one of a pair of appendages in the anterior portion of the cephalothorax (where the head and thorax are fused together), often modified for food capture and defense

chlamydospore—a thick-walled or double-walled, asexual resting spore formed from hyphal cells (terminal or intercalary) or by transformation of conidial cells; functions as an overwintering stage

chlorophyll (adj. chlorophyllous)—any of a group of green pigments found in chloroplasts and critical to photosynthesis

chloroplast—a disklike structure containing chlorophyll in which photosynthesis occurs in green plants

chlorosis (adj. chlorotic)—the failure of chlorophyll development, caused by disease or a nutritional disturbance; the fading of green plant parts to light green, yellow, or white

chromosome—the structure that contains the genes of an organism (in eukaryotes, chromosomes are in the nucleus and can be visualized with an optical microscope as threads or rods during meiosis and mitosis; in bacteria, the chromosome is usually a single circle of DNA and cannot be visualized with an optical microscope)

chronic—with reference to symptoms, slow-developing, persistent, or recurring (*see* acute)

circulative—with reference to viruses, passing through the gut and circulating in the body of an insect vector before being transmitted to a host

circulative transmission—virus transmission in which the virus must accumulate within or pass through the lymphatic system of an insect vector before it can be transmitted to a plant

cirrhus (pl. cirrhi)—a curled, tendril-like mass of exuded spores held together by a slimy matrix

clavate—club shaped (syn. claviform)

cleistothecium (pl. cleistothecia)—a spherical ascocarp that is closed at maturity

coalesce—to grow together into one body, as enlarging lesions may join together to form a single spot

coenocytic—having multiple nuclei embedded in cytoplasm without cross walls; nonseptate

colonize—with reference to plant pathogens, to infect and ramify through plant tissue

colony—a microorganism growing in mass, especially as a pure culture

congener—an organism that is within the same genus as another

conidiogenous—producing and bearing conidia

conidiophore—a simple or branched hypha on which conidia are produced

conidium (pl. conidia)—an asexual, nonmotile fungal spore that develops externally or is liberated from the cell that formed it

cornicle—in aphids, a pair of hornlike tubes at the posterior end of the abdomen

cortex (adj. cortical)—the region of parenchyma tissue between the epidermis and phloem in stems and roots; the region beneath the rind of a sclerotium

cotyledon—the primary embryonic leaf within the seed, in which nutrients for the new plant are stored (monocots have one cotyledon; dicots have two)

crepuscular—active at twilight, dusk, and dawn

crown—the upper dome of a tree, bearing leaves, flowers, and fruits; the junction of the root and stem of a plant, usually at the soil line; in grafted woody plants, the rootstock portion of the plant near the soil surface

cultivar (abbr. cv.)—a cultivated variety of a plant species, resulting from deliberate genetic manipulation and having recognizable characteristics, such as color; shape of flowers, fruits, and seeds; height; and form (syn. variety)

cultural practices—the methods by which plants are grown (e.g., application of nutrients, irrigation, cultivation)

culture—the growth and propagation of microorganisms on nutrient media; the growth and propagation of living plants

cuticle (adj. cuticular)—the noncellular outer layer of an insect or nematode; the water-repellent, waxy layer of epidermal cells of plant parts, such as leaves, stems, and fruits

cyst—in fungi, a resting structure in a protective membrane or shell-like enclosure; in nematodes, the egg-laden carcass of a female nematode; in bacteria, a specialized cell enclosed in a thick wall; often dormant and resistant to environmental conditions

cytoplasm—the living protoplasm in a cell, not including the nucleus

damping-off—the death of a seedling before or shortly after emergence caused by decomposition of the root or lower stem (it is common to distinguish between preemergence and postemergence damping-off)

defoliation—the loss of leaves from a plant, whether normal or premature

dehiscent—opening by breaking into parts

desiccate—to dry out

diagnostic (n. diagnosis)—pertaining to a distinguishing characteristic important for the identification of a disease or disorder

diapause—a period of biological inactivity or dormancy

dichotomous—branching, often successively, into two more or less equal arms

dicot—a dicotyledon, or a plant having two cotyledons (*see* monocot)

dieback (v. die back)—the progressive death of shoots, leaves, or roots, beginning at the tips

differential host—a plant host that develops disease symptoms that distinguish between strains or races of a plant pathogen (syn. differential cultivar)

dikaryon (adj. dikaryotic)—an organism having two sexually compatible haploid nuclei per cell that divide simultaneously in the reproductive phase called dikaryophase

dimorphic—having two distinct shapes or forms

diploid—having two complete sets of chromosomes ($2n$ chromosomes) (*see* haploid; polyploid)

disease—the abnormal functioning of an organism

disease cycle—the succession of events and interactions among a host, a parasite, and the environment, from initial infection through pathogenesis to overseasoning, until another infection occurs

disinfest—to kill pathogens that have not yet initiated disease or to kill other contaminating microorganisms that occur in or on inanimate objects (e.g., soil or tools) or on the surface of plant parts (e.g., seed)

dispersal—the spread of infectious material (inoculum) from diseased to healthy plants (syn. dissemination)

dissemination—*see* dispersal

distal—situated away from the point of attachment or origin or away from the main body (*see* proximal)

DNA—deoxyribonucleic acid; the double-stranded, helical molecule that contains genetic code information; consists of repeating units, or nucleotides, each of which is composed of deoxyribose (a sugar), a phosphate group, and a purine (adenine or guanine) or a pyrimidine (thymine or cytosine) base

dominant—pertaining to a phenotypic trait that is expressed in hybrid progeny of diploid organisms even if contributed by only one of the parents (*see* recessive)

dormancy (adj. dormant)—a condition of suspended growth and reduced metabolism of an organism; generally induced by internal factors or environmental conditions as a mechanism of survival

dormant—resting; living in a state of reduced physiological activity

dorsal—situated toward the back or top of the body (ant. ventral)

dwarfing—underdevelopment of a plant or plant organs caused by disease, inadequate nutrition, or unfavorable environmental conditions

echinulate—having small spines projecting from the cell wall

eclosion—the emergence of an insect larva from the egg or of an adult from the pupal case

ecology—the study of the interactions between individual organisms, groups of organisms, and organisms and their environments

ectoparasite—a parasite that feeds from the exterior of its host (*see* endoparasite)

ectotrophic—pertaining to fungal development primarily over the root surface

edema—blisters produced on leaves or other plant parts under conditions of high moisture and restricted transpiration (syn. intumescence)

egg mass—a group of eggs held together by a gelatinous matrix

ELISA—enzyme-linked immunosorbent assay; a serological test in which the sensitivity of the reaction is increased by attaching an enzyme that produces a colored product in response to one of the reactants

ellipsoidal—elliptical in the plane section

elytron (pl. elytra)—one of the hardened forewings, characteristic of beetles, that typically serve to protect the underlying flight wings

embryo—an organism in the early stages of development, such as a young plant in the seed or a nematode before hatching from the egg

emergence—the growth of a seedling shoot through the surface of the soil

encapsidate—to enclose viral nucleic acid in a protein coat

encyst—to form a cyst or protective covering

endemic—native to a particular place; pertaining to a low and steady level of natural disease occurrence

endocarp—the inner layer of a fruit wall (*see* pericarp)

endoconidium (pl. endoconidia)—a conidium produced inside a hypha or conidiophore

endodermis—the layer of cells between the vascular tissue and the cortex of roots

endogenous—arising from inside (*see* exogenous)

endoparasite—a parasite that lives and feeds inside its host (*see* ectoparasite)

endophyte—a plant developing inside another organism; any of the various endoparasitic fungi associated with grass species and other plants

enzyme—a protein that catalyzes a specific biochemical reaction

enzyme-linked immunosorbent assay (abbr. ELISA)—a serological test in which the sensitivity of the reaction is increased by attaching an enzyme that produces a colored product in response to one of the reactants

epidemic—an increase of disease in a population; a general and serious outbreak of disease (*see* epiphytotic)

epidemiology (adj. epidemiological)—the study of factors influencing the initiation, development, and spread of infectious disease; the study of disease in populations

epidermis (adj. epidermal)—the surface layer of cells of leaves and other plant parts

epinasty—the abnormal downward curling of a leaf, leaf part, or stem

epiphyllous—located on the upper surface of a leaf

epiphyte—an organism growing on a plant surface but not as a parasite

epiphytotic—pertaining to the widespread and destructive outbreak of a plant disease

eradicant—a chemical used to eliminate a pathogen from a host or environment

eradication—the control of plant disease by eliminating the pathogen after it has become established or by eliminating the plants that carry the pathogen

erumpent—bursting or erupting through the surface of a substrate

ethylene—a plant hormone influencing vegetative growth, fruit ripening, abscission of plant parts, and senescence of flowers

etiolation—the elongation of stems caused by low levels of light intensity

etiology—the study of the causes of diseases

eukaryote (adj. eukaryotic)—any of various organisms whose cells contain a membrane-bound nucleus and other organelles, including all higher plants, animals, fungi, and protists (*see* prokaryote)

exogenous—originating from the outside (*see* endogenous)

extracellular—not occurring or located within a cell

exudate—a liquid excreted by or discharged from diseased tissues, roots, leaves, and fungi

f. sp. (pl. ff. spp.)—*forma specialis* (pl. *formae speciales*); a taxonomic group within a pathogenic species, defined by its host range (members of different *formae speciales* infect different groups of plants)

facultative—capable of changing lifestyle (e.g., from saprophytic to parasitic or the reverse)

facultative parasite—an organism that is normally saprophytic but capable of living as a parasite

facultative saprophyte—an organism that is normally parasitic but capable of living as a saprophyte

fallow—pertaining to cultivated land kept free from a crop or weeds during the normal growing season

fasciation—the malformation of shoots or floral organs manifested as enlargement and flattening, as if several parts were fused

fascicle (adj. fasciculate)—a small group, bundle, or cluster

fastidious—with reference to prokaryotic organisms, having special requirements for growth and nutrition

femur—a leg segment that is before the tibia and beyond the trochanter

filament (adj. filamentous)—a thin, flexible, threadlike structure

filamentous—threadlike (syn. filiform)

filiform—long, needlelike (syn. filamentous)

flagellum (pl. flagella)—a hairlike, whiplike, or tinsel-like appendage of a motile cell, bacterium, or zoospore that provides locomotion

fleck—a minute, discolored spot in green tissue

flexuous—having turns or windings; capable of bending

focus (pl. foci)—a small area of diseased plants within a larger plot or field

foliar—pertaining to leaves

forma specialis (pl. *formae speciales*)—a taxonomic group within a pathogenic species, defined by its host range (members of different *formae speciales* infect different groups of plants); abbr. f. sp. (pl. ff. spp.)

frass—insect excrement

fructification—any of the various spore-bearing organs formed by macro- and microfungi

fruiting body—any of the various complex, spore-bearing structures of fungi

fumigant (v. fumigate)—a gas or volatile substance used to kill or inhibit the growth of microorganisms or other pests

fungicide (adj. fungicidal)—a chemical or physical agent that kills or inhibits the growth of fungi

fungistat (adj. fungistatic)—a compound that inhibits fungal growth or sporulation but does not kill fungi

fusiform—spindle shaped; tapering at each end

gall—an abnormal swelling or localized outgrowth, often roughly spherical, produced by a plant as a result of attack by a fungus, bacterium, nematode, insect, or other organism (syn. tumor)

gametangium (pl. gametangia)—a cell containing gametes or nuclei that act as gametes

gamete—a sex cell

gelatinous—resembling gelatin or jelly

gene—a unit located on a chromosome and controlling the transmission of a heritable characteristic

genetic—pertaining to heredity or heritable characteristics

geniculate—bent, like a knee

genome—the complete genetic information of an organism or virus

genotype—the genetic constitution of an individual or group; a class or group of individuals sharing a specific genetic makeup (*see* phenotype)

genus (pl. genera)—a taxonomic category that consists of species that are closely related structurally or phylogenetically (the genus or generic name is the first name in a Latin binomial)

germ tube—a hypha resulting from an outgrowth of the spore wall and cytoplasm after germination

germinate (n. germination)—to begin growth, as of a seed, spore, sclerotium, or other reproductive body

germplasm—the bearer of heredity material (often loosely applied to cultivars and breeding lines)

girdle—to circle and cut through a stem or the bark and outer few rings of wood, disrupting the phloem and xylem

globose—nearly spherical

graft transmission—the transmission of a pathogen from one host plant to another through fusion of living tissue from the diseased host with living tissue of a healthy host

gram-negative bacteria—bacteria that are stained red or pink after treatment with Gram's stain

gram-positive bacteria—bacteria that are stained violet or purple after treatment with Gram's stain

guttulate—containing drops or droplike masses

guttule—drop, vesicle, or droplike mass

hair root—fine roots of some plants, typical of those in the family Ericaceae

haploid—having a single complete set of chromosomes (*see* diploid; polyploid)

hardiness—the ability to withstand stress

haustorium (pl. haustoria)—a specialized branch of a parasite formed inside a host cell to absorb nutrients

herbaceous—pertaining to primary, soft, nonwoody tissue of a plant or plant part; having the characteristics of an herb

herbicide—any of various chemicals that kill plants or inhibit their growth (e.g., a weed or grass killer)

heteroecious—with reference to rust fungi, passing different stages of the life cycle on alternate hosts (*see* autoecious)

heterokaryon (adj. heterokaryotic)—a cell with genetically different nuclei

heterothallism (adj. heterothallic)—a fungal life cycle in which sexual reproduction can occur only in the presence of genetically different mycelia (*see* homothallism)

holomorph—a fungus having sexual and asexual forms in its life cycle, considered in all its forms (*see* anamorph; teleomorph)

homothallism (adj. homothallic)—a fungal life cycle in which sexual reproduction occurs in a single, self-fertile thallus (*see* heterothallism)

honeydew—a sugary ooze or exudate often secreted by aphids; a characteristic symptom of ergot

host plant—a living plant attacked by or harboring a parasite or pathogen, from which the invader obtains part or all of its nourishment

host range—the range of plants on which an organism (particularly a parasite) feeds

hyaline—transparent or nearly so; translucent; colorless

hybrid (v. hybridize)—the offspring of two individuals of different genotypes

hydathode—an epidermal leaf structure specialized for the secretion or exudation of water; the opening at the terminus of a leaf vein

hymenium—the continuous, spore-bearing layer of a fungal fruiting body

hyperplasia (adj. hyperplastic)—an abnormal increase in the number of cells, often resulting in the formation of galls or tumors

hypersensitive—extremely or excessively sensitive, often with reference to an extreme reaction to a pathogen

hypertrophy (adj. hypertrophic)—an abnormal increase in the size of cells in a tissue or organ, often resulting in the formation of galls or tumors

hypha (pl. hyphae; adj. hyphal)—a tubular filament of a fungal thallus or mycelium; the basic structural unit of a fungus

immune—not capable of being infected by a given pathogen

immunogenic—able to induce the production of antibodies

imperfect state—the anamorph, or asexual form, in the life cycle of certain fungi, in which asexual spores (e.g., conidia) or no spores are produced (*see* perfect state)

in vitro—in glass, on an artificial medium, or in an artificial environment; not within a living host

in vivo—within a living organism

inclusion body—a structure developed within a plant cell as a result of infection by a virus; often useful in identifying the virus

indicator plant—a plant in which specific or distinctive symptoms develop in reaction to a pathogen or certain environmental conditions; used to detect or identify the pathogen or determine the effects of the environmental conditions

infection—the process in which an organism enters, invades, or penetrates and establishes a parasitic relationship with a host plant

infection court—a site in or on a host plant where infection can occur

infection pad—an organized mass of hyphae formed on the surface of a plant and producing numerous infective hyphae

infection focus—the initial site of infection, generally with reference to a population of plants

infection peg—the specialized, narrow hyphal strand located on the underside of an appressorium and penetrating host cells (syn. penetration peg)

infectious—with reference to disease, capable of spreading from plant to plant

infective—able to attack a host and cause infection; with reference to vectors, carrying or containing a pathogen and able to transfer it to a host plant

infest (n. infestation)—to attack as a pest (used especially for insects and nematodes); to contaminate (as with microorganisms); to be present in large numbers

inflorescence—a flower or flower cluster

initial inoculum—inoculum (usually from an overwintering source) that initiates disease in the field, as opposed to inoculum that spreads disease during the growing season (syn. primary inoculum)

injury—damage caused by transitory interaction with an agent such as an insect, chemical, or unfavorable environmental condition

inoculate (n. inoculation)—to place inoculum in an infection court; to insert a pathogen into healthy tissue

inoculum (pl. inocula)—a pathogen or its parts capable of causing infection when transferred to a favorable location

inoculum density—a measure of the number of propagules of a pathogenic organism per unit of area or volume

intercellular—between or among cells

internode (adj. internodal)—the portion of a stem between two successive nodes

interveinal—between veins (of a leaf)

intracellular—through or within cells

intumescence—blisters produced on leaves or other plant parts under conditions of high moisture and restricted transpiration (syn. edema)

isolate—(n.) a culture or subpopulation of a microorganism separated from its parent population and maintained under controlled conditions; (v.) to remove from soil or host material and grow in pure culture

isometric—with reference to virus particles, having an icosahedral structure and thus appearing approximately round

juvenile—an immature form that appears similar to the adult stage but is usually smaller and is not sexually mature, as in nematodes or insect species that undergo gradual metamorphosis

knot—a gall or localized abnormal swelling

lamina—the expanded part of a leaf (*see* petiole)

larva (pl. larvae)—the immature wormlike or caterpillar stage in the life cycle of an insect that undergoes complete metamorphosis

latent—present but not manifested or visible, as a symptomless infection

latent infection—an infection that does not have visible symptoms

latent period—the time elapsed between infection and the appearance of symptoms or production of new inoculum (sometimes synonymous with incubation period); the time elapsed after the acquisition of a pathogen by a vector but before the pathogen can be transmitted

leaf spot—a lesion typically restricted in development after reaching a characteristic size

leafhopper—any of various species of mobile insects with sucking mouthparts in the order Hemiptera

leaflet—one of the separate blades or divisions of a compound leaf

legume—a simple, dry, dehiscent fruit developing from a simple pistil and splitting at maturity along two seams; any plant in the family Fabaceae (formerly Leguminosae)

lesion—a localized diseased area or wound

life cycle—the cyclical stages in the growth and development of an organism

local lesion—a small, restricted lesion; often the characteristic reaction of a differential cultivar to a specific pathogen, especially in response to mechanical inoculation with a virus

lumen (pl. lumina)—the central cavity of a cell or other structure

macroconidium (pl. macroconidia)—the larger of two kinds of conidia formed by certain fungi (*see* microconidium)

macrocyclic—pertaining to the life cycle of rust fungi that typically exhibit all five spore stages (pycniospores, aeciospores, urediniospores, teliospores, and basidiospores) in the course of their development (*see* microcyclic)

manual transmission—the spread or introduction of inoculum to an infection court by hand

mating types—compatible strains (usually designated + and – or A and B), both of which are necessary for sexual reproduction in heterothallic fungi

mechanical injury—injury of a plant part by abrasion, mutilation, or wounding

mechanical transmission—the spread or introduction of inoculum to an infection court by human manipulation, accompanied by physical disruption of host tissues (wounding)

medium (pl. media)—a mixture of organic or inorganic chemical compounds and water providing the nutrients needed for the growth of a microorganism in vitro; with reference to higher plants, a mixture of fertilizers and other components in which plants are grown

meiosis—nuclear division in which the number of chromosomes per nucleus is halved (i.e., the diploid state is converted to the haploid state) (*see* mitosis)

melanin (adj. melanoid)—a brown-black pigment

melanized—having or containing melanin

meristem (adj. meristematic)—plant tissue characterized by frequent cell division, producing cells that become differentiated into specialized tissues

meristem culture—an aseptic culture of a plant or plant part from a portion of the meristem

mesophyll—the central, internal, nonvascular tissue of leaves, consisting of the palisade and spongy mesophyll

metabolite—any chemical participating in metabolism; a nutrient

microclimate—weather conditions on a small scale, as at the surface of a plant or within a field

microconidium (pl. microconidia)—the smaller of two kinds of conidia formed by certain fungi (*see* macroconidium)

microcyclic—pertaining to the life cycle of rust fungi that produce only teliospores and basidiospores (*see* macrocyclic)

microflora—the combination of all of the microorganisms in a particular environment

microorganism—an organism small enough that it can be seen only with the aid of a microscope (syn. microbe)

microsclerotium (pl. microsclerotia)—a microscopic, dense aggregate of darkly pigmented, thick-walled hyphal cells

midrib—the central, thickened vein of a leaf

migratory—moving from place to place on a plant or from plant to plant when feeding (*see* sedentary)

mildew—a thin coating of mycelial growth and spores on the surfaces of infected plant parts

mitochondrion (pl. mitochondria)—a cellular organelle outside the nucleus; functions in respiration

mitosis—nuclear division in which the number of chromosomes per nucleus remains the same (*see* meiosis)

MLO—mycoplasmalike organism (*see* phytoplasma)

mold—any microfungus with conspicuous, profuse, or woolly superficial growth (mycelium or spore masses) on various substrates; commonly grow on damp or decaying matter and on the surfaces of plant tissues, especially with reference to economically important saprobes

mollicute—any of a group of prokaryotic organisms bounded by a flexuous membrane and lacking a cell wall (*see* phytoplasma; spiroplasma)

molt—to shed a cuticle or body encasement during a phase of growth

monocot—a monocotyledon, or a plant having only one cotyledon, such as species of grasses, including grain crops and corn (*see* dicot)

monoculture—the cultivation of plants of the same species in close proximity and with few or no other plant species present

monoecious—having male and female reproductive organs on a single individual

morphology (adj. morphological)—the study of the form of organisms; the form and structure of organisms

mosaic—a disease symptom characterized by nonuniform coloration, with intermingled normal, light green, and yellowish patches; usually caused by a virus (often used interchangeably with mottle)

motile—capable of self-propulsion by means of flagella, cilia, or amoeboid movement

mottle—a disease symptom characterized by light and dark areas in an irregular pattern; usually caused by a virus (often used interchangeably with mosaic)

mucilaginous—viscous; slimy
mulch—a layer of material, such as organic matter or plastic film, applied to the surface of the soil for purposes such as retention of water and inhibition of weeds
multinucleate—having more than one nucleus per cell (*see* binucleate; uninucleate)
multiseptate—having many septa (cross walls)
multivoltine—having multiple (two or more) broods or generations per year (*see* bivoltine; univoltine)
muriform—having transverse and longitudinal septa
mutation—an abrupt heritable or genetic change in a gene or individual as a result of an alteration in a gene or chromosome or an increase in chromosome number
mycelium (pl. mycelia; adj. mycelial)—the mass of hyphae constituting the body (thallus) of a fungus
mycoplasmalike organism (abbr. MLO)—*see* phytoplasma
mycorrhiza (pl. mycorrhizae; adj. mycorrhizal)—a symbiotic association between a nonpathogenic or weakly pathogenic fungus and the roots of a plant
mycotoxin—any poisonous compound produced by a fungus

necrosis (adj. necrotic)—the death of cells or tissue; usually accompanied by blackening or browning
necrotroph—a parasite that typically kills and obtains its energy from dead host cells (*see* biotroph)
nematicide—an agent that kills nematodes; usually a chemical
nematode—any of various unsegmented roundworms (animals) that are parasitic on plants or animals or free living in soil or water
node (adj. nodal)—the enlarged portion of a shoot at which leaves or buds arise
noninfectious disorder—a disorder caused by an abiotic agent and not capable of being transmitted from plant to plant
nonpersistent transmission—virus transmission in which a virus is acquired and transmitted by a vector after a short feeding time and retained by the vector for only a short time (syn. stylet-borne transmission)
nonseptate—lacking cross walls; coenocytic
nymph—the juvenile stage in the life cycle of an insect with incomplete metamorphosis; superficially resembles the adult

obclavate—shaped like an upside-down club
obligate parasite—an organism that can grow only as a parasite in association with its host plant and cannot be grown in artificial culture media (syn. biotroph)
obovoid—shaped like an upside-down egg
obtuse—rounded or blunt; pertaining to an angle greater than 90°
oogonium (pl. oogonia)—a female gametangium of fungi in the group oomycetes containing one or more gametes
oomycetes (adj. oomycetous)—a group of funguslike organisms in the phylum Oomycota (kingdom Chromista); typically has nonseptate mycelium, asexual sporangia and zoospores, and sexual oospores
oospore—the thick-walled, sexually derived resting spore of organisms in the group oomycetes
organelle—any of various membrane-bound structures contained within cells and having a specialized function (e.g., mitochondria and chloroplasts)
ostiole (adj. ostiolate)—a pore; the opening in the papilla or neck of a perithecium, pseudothecium, or pycnidium through which spores are released
ovary—the female reproductive structure of an organism; in plants, the enlarged basal portion of the pistil, containing the ovules and developing into the fruit
overseason—to survive or persist from one planting season to the next
oversummer—to survive or persist through the summer
overwinter—to survive or persist through the winter
oviposit—to deposit or lay eggs with an ovipositor
ovule—an enclosed structure that after fertilization becomes a seed; an egg contained within an ovary

palisade parenchyma—the tissue located beneath the upper epidermis of leaves; composed of elongate, tubular cells arranged upright in the manner of posts in a palisade fortification
papilla (pl. papillae; adj. papillate)—a small, blunt projection
paraphysis (pl. paraphyses)—a hairlike cell within a fungal fruiting structure
parasite (adj. parasitic)—an organism that lives in intimate association with another organism, on which it depends for its nutrition; not necessarily a pathogen
parenchyma (adj. parenchymatous)—the soft tissue of living plant cells with undifferentiated, thin, cellulose walls
pathogen (adj. pathogenic)—a disease-producing organism or agent
pathogenesis—the initiation and development of disease
pathogenicity—the ability to cause disease
pathology—the study of diseases
pathotype—a subdivision of a pathogenic species characterized by its pattern of virulence or avirulence to differential host varieties
pathovar (abbr. pv.)—a subdivision of a plant-pathogenic bacterial species, defined by its host range (a pathovar of a bacterial species is analogous to a *forma specialis* of a fungal species)
pedicel—a small, slender stalk; a stalk bearing an individual flower, inflorescence, or spore
peduncle—the stalk or main stem of an inflorescence; part of an inflorescence or fructification
penetration—the initial invasion of a host by a pathogen
penetration peg—the specialized, narrow hyphal strand located on the underside of an appressorium and penetrating host cells (syn. infection peg)
perennial—a plant that survives for several to many years (*see* annual; biennial)
perfect—sexual; capable of sexual reproduction
perfect state—the teleomorph, or sexual form, in the life cycle of certain fungi (*see* imperfect state)
pericarp—the outer layer of a seed or fruit (*see* endocarp)
perithecium (pl. perithecia)—the flask-shaped or subglobose, thin-walled fruiting body (ascocarp) of an ascomycetous fungus, containing asci and ascospores and having a pore (ostiole) at its apex, through which ascospores are expelled or released
peritrichous—having hairs or flagella distributed over the whole surface
persistent transmission—virus transmission in which a virus is acquired and transmitted by a vector after a relatively long feeding time and remains transmissible for a prolonged period while in association with the vector (syn. circulative transmission)
pest—any organism that damages plants or plant products
pesticide—any of various chemicals used to control pests
petiole—the stalk portion of a leaf (*see* lamina)
pH—the negative logarithm of the effective hydrogen ion concentration; a measure of acidity (pH 7 is neutral; values less than pH 7 are acidic, and values greater than pH 7 are alkaline)
phenotype—the composite of observable physical qualities of an organism produced by the interaction of its genotype with the environment
phialide—the end cell of a conidiophore with one or more open ends, through which a basipetal succession of conidia develops
phialospore—a conidium produced on a phialide
phloem—the food-conducting, food-storing tissue in the vascular system of roots, stems, and leaves
photosynthate—a chemical product of photosynthesis
photosynthesis—the manufacture of carbohydrates from carbon dioxide and water in the presence of chlorophylls; uses light energy and releases oxygen
phyllody—a disorder in which floral organs are transformed into leaflike structures
phytoalexin—any of various substances produced by higher plants in response to chemical, physical, or biological stimuli and inhibiting the growth of certain microorganisms
phytopathology—the study of plant diseases (syn. plant pathology)
phytoplasma—any of various plant-parasitic pleomorphic mollicutes (prokaryotes lacking cell walls) found in phloem tissue and not capable of growth on artificial nutrient media (previously called mycoplasmalike organisms [MLOs])
phytotoxic—harmful to plants; usually with reference to chemicals
pigment—a colored compound, such as chlorophyll, in the cells of plants or fungi
pistil—the ovule-bearing organ of a plant, consisting of the ovary and its appendages (e.g., the style and stigma)
pith—parenchymatous tissue occupying the center of a stem
plant pathology—the study of plant diseases (syn. phytopathology)
pleomorphic—able to assume various shapes and perhaps sizes; having a life cycle characterized by a succession of distinctly different forms

polar—located at one end (pole) of the cell

pollen—male sex cells produced by the anthers of flowering plants or the cones of seed plants

pollination—the transfer of pollen from anther to stigma or from a staminate cone to an ovulate cone

polyploid—having three or more complete sets of chromosomes (*see* diploid; haploid)

positive-sense RNA—RNA that can serve directly as messenger RNA

posterior—situated toward the back or tail (*see* anterior)

predispose (n. predisposition)—to make prone to infection and disease

primary inoculum—inoculum (usually from an overwintering source) that initiates disease in the field, as opposed to inoculum that spreads disease during the season (syn. initial inoculum)

prokaryote (adj. prokaryotic)—any of various organisms lacking internal membrane-bound organelles and a distinct nucleus, such as bacteria and mollicutes (*see* eukaryote)

propagule—any part of an organism capable of independent growth

protectant—an agent (usually a chemical) applied to a plant surface in advance of exposure to a pathogen for the purpose of preventing infection

protein—any of numerous nitrogen-containing organic compounds containing combinations of amino acids

protoplasm—the living contents of a cell

protoplast—the living contents of a cell, exclusive of the cell wall

proximal—situated toward or near the point of attachment or main body (*see* distal)

pseudothecium (pl. pseudothecia)—a perithecium-like fruiting body containing asci and ascospores dispersed rather than in an organized hymenium; an ascostroma with a single locule (cavity) containing bitunicate asci

pupa (pl. pupae; v. pupate)—the quiescent stage between the larval and adult stages of certain insects

pustule—a small, blisterlike elevation of the plant epidermis that forms as spores emerge

pv.—pathovar; a subdivision of a plant-pathogenic bacterial species defined by its host range (a pathovar of a bacterial species is analogous to a *forma specialis* of a fungal species)

pycnidiospore—a spore (conidium) produced in a pycnidium

pycnidium (pl. pycnidia)—an asexual, globose or flask-shaped fruiting body of certain imperfect fungi, producing conidia

pycniospore—a haploid, sexually derived spore formed in the pycnium of a rust fungus (syn. spermatium)

pycnium (pl. pycnia)—the globose or flask-shaped haploid fruiting body of rust fungi, bearing receptive hyphae and pycniospores (syn. spermagonium)

pyriform—pear shaped

quarantine—an enforced isolation imposed to prevent the spread of disease; pertaining to plant disease, control of the transport of plants or plant parts to prevent the spread of pests or pathogens

quiescent—dormant; inactive

race—a subgroup or biotype within a species or variety that is distinguished from other races by virulence, symptom expression, or host range but not by morphology

radicle—the part of the plant embryo that develops into the primary root

recessive—pertaining to a phenotypic trait that is expressed in the progeny of diploid organisms only if contributed by both parents (*see* dominant)

resistant (n. resistance)—possessing properties that prevent or impede disease development (*see* susceptible)

respiration—a series of chemical reactions that make energy available through oxidation of carbohydrates and fat

resting spore—a spore (often thick walled) that can remain alive in a dormant state for some time, germinating later and capable of initiating infection

rhizome—a mostly horizontal, jointed, fleshy, often elongated, usually underground stem

rhizomorph—a macroscopic ropelike strand of compacted tissue formed by certain fungi

rhizosphere—the microenvironment in the soil immediately around a plant's roots

ribosome—a subcellular protoplasmic particle made up of one or more RNA molecules and several proteins; involved in protein synthesis

ring spot—a disease symptom characterized by yellowish or necrotic rings enclosing green tissue, as in some plant diseases caused by viruses

RNA—ribonucleic acid; any of several nucleic acids composed of repeating units of ribose (a sugar), a phosphate group, and a purine (adenine or guanine) or a pyrimidine (uracil or cytosine) base, transcribed from DNA and involved in translation to proteins

rogue—to remove and destroy individual plants that are diseased, infested by insects, or otherwise undesirable

root cap—a group of cells that protects the growing tip of a root

root hair—a threadlike, single-celled outgrowth from a root epidermal cell; note that blueberries and cranberries lack root hairs (*see* hair root)

rosette—a disease symptom characterized by short, bunchy growth habit caused by a shortening of internodes with no comparable reduction in leaf size

rot—softening, discoloration, and often disintegration of plant tissue as a result of fungal or bacterial infection

rotation—the practice of growing different kinds of crops in succession in the same field

runner—a slender, horizontal stem that grows close to the soil surface (syn. stolon)

runner plants—new plants produced asexually on a runner (stolon)

russet—a brownish, roughened area on the skin of fruit, resulting from cork formation

rust—any of several diseases caused by specialized fungi in the group basidiomycetes, some of the spores of which are a rusty color

sanitation—the destruction or removal of infected or infested plants or plant parts; the decontamination of tools, equipment, containers, work space, hands, and so on

saprobe—*see* saprophyte

saprophyte (adj. saprophytic)—an organism that obtains nourishment from nonliving organic matter (syn. saprobe)

scab—a roughened, crustlike diseased area on the surface of a plant organ

scald—a necrotic condition in which tissue is usually bleached and has the appearance of having been exposed to high temperature

sclerotium (pl. sclerotia)—a vegetative, resting body of a fungus composed of a compact mass of hyphae with or without included host tissue; usually has a darkened rind

scorch—a symptom that suggests the action of flame or fire on the affected tissue, often at the margins of leaves

scutellum (adj. scutellar)—the cotyledon of a grass embryo

secondary infection—an infection resulting from the spread of infectious material from a primary or secondary infection without an intervening inactive period

secondary inoculum—inoculum produced by an infection established earlier in the same growing season

secondary organism—an organism that multiplies in already diseased tissue but is not the primary pathogen

sedentary—remaining in a fixed location (*see* migratory)

seed—a ripened ovule consisting of an embryo and stored food enclosed by a seed coat

seed treatment—the application of a biological agent, chemical substance, or physical treatment to seed to protect the seed or resulting plant from pathogens or to stimulate germination or plant growth

seedborne—carried on or in seed

selective medium—a culture medium containing substances that inhibit or prevent the growth of certain species of microorganisms

senesce (adj. senescent; n. senescence)—to decline, as with maturation, age, or disease stress

sepal—one of the modified leaves composing a calyx

septate—having septa (cross walls)

septum (pl. septa; adj. septate)—a dividing wall; in fungi, a cross wall

serology (adj. serological)—a method using the specificity of the antigen–antibody reaction for the detection and identification of antigenic substances and the organisms that carry them

sessile—with reference to leaves, leaflets, flowers, florets, fruits, ascocarps, basidiocarps, and so on, lacking a stalk, petiole, pedicel, stipe, or stem; with reference to nematodes, permanently attached and not capable of moving about

seta (pl. setae)—a bristle- or hairlike structure, usually deep yellow or brown and thick walled
sexual spore—a spore produced during the sexual cycle of a fungus
shot hole—a symptom in which small lesions on a leaf fall out, giving the leaf the appearance of being hit by buckshot
sieve element—a tube-shaped living cell in the phloem, functioning in the transport of dissolved organic substances in the plant (syn. sieve tube element)
sign—an indication of disease from direct observation of a pathogen or its parts on a host plant (*see* symptom)
soft rot—a mushy disintegration of fleshy fruit or vegetables as a result of bacterial or fungal infection
soil drench—the application of a solution or suspension of a chemical to the soil, especially a pesticide to control soilborne pathogens
soilborne—carried on or beneath the soil surface
solarization—a disease management practice in which soil is covered with polyethylene sheeting to increase heating by sunlight and thus control soilborne plant pathogens
sooty mold—black, nonparasitic, superficial fungal growth; often develops on honeydew produced by aphids and other phloem-feeding insects
sorus (pl. sori)—a compact fruiting structure, especially with reference to spore masses of rust and smut fungi
sp. (pl. spp.)—species (the singular and plural abbreviations are used only after genus names; the singular form is used to refer to an undetermined species, and the plural form is used to refer to two or more species without naming them individually)
spermagonium (pl. spermagonia)—a structure in which male reproductive cells are produced; in rust fungi, the pycnium, a globose or flask-shaped haploid fruiting body composed of receptive hyphae and spermatia (pycniospores)
spermatium (pl. spermatia)—a male sex cell; a nonmotile male gamete; a haploid male gamete; in rust fungi, a pycniospore
spiroplasma—any of various spiral-shaped, plant-pathogenic mollicutes (prokaryotes lacking cell walls)
sporangiophore—the sporangium-bearing body of a fungus
sporangiospore—a nonmotile, asexual spore borne in a sporangium
sporangium (pl. sporangia)—a saclike fungal structure, the entire contents of which are converted into an indefinite number of asexual spores
spore—the reproductive structure of fungi and some other organisms, containing one or more cells; a bacterial cell modified to survive an adverse environment
sporidium (pl. sporidia)—the basidiospore of rust fungi, smut fungi, and other members of the group basidiomycetes
sporocarp—a spore-bearing fruiting body
sporodochium (pl. sporodochia)—a superficial, cushion-shaped, asexual fruiting body consisting of a cluster of conidiophores
sporulate—to produce spores
sporulating—producing and (often) liberating spores
spot—a small necrotic area occurring as a disease symptom on a leaf, flower, or stem
stamen (adj. staminal)—the male structure of a flower, composed of a pollen-bearing anther and a filament, or stalk
stele—the central cylinder of vascular tissue, especially with reference to roots
stem pitting—a viral disease symptom characterized by depressions on the stem
sterigma (pl. sterigmata)—a small, usually pointed projection that supports a spore
sterile—unable to reproduce sexually; free of living microorganisms
sterilization (adj. sterilized)—the total destruction of living organisms by various means, including heat, chemicals, and irradiation
stigma—the portion of a flower that receives pollen and on which pollen germinates
stipe—a stalk
stolon—a slender, horizontal stem that grows close to the soil surface (syn. runner); in fungi, a hypha that grows horizontally along a surface
stoma, **stomate** (pl. stomata; adj. stomatal)—a structure composed of two guard cells and the opening between them in the epidermis of a leaf or stem; functions in gas exchange
strain—a distinct form of an organism within a species or virus, differing from other forms of the species biologically, physically, or chemically
striate (n. striations)—to mark with delicate lines, grooves, or ridges
stroma (pl. stromata)—a compact mass of mycelium (with or without host tissue) that supports fruiting bodies or in which fruiting bodies are embedded
stunting—a reduction in the height of plants resulting from a progressive reduction in the lengths of successive internodes or a decrease in their number
stylet—the stiff, slender, hollow feeding organ of plant-parasitic nematodes or sap-sucking insects, such as aphids and leafhoppers
stylet-borne transmission—virus transmission in which a virus is acquired and transmitted by a vector after a short feeding time and is retained by the vector for only a short time (syn. nonpersistent transmission)
subepidermal—located or occurring just below the epidermis
subgenomic RNA—a piece of viral RNA, shorter than the entire genome of the virus, found in cells infected by the virus and sometimes encapsidated
substrate—the substance on which an organism lives or from which it obtains nutrients; a chemical substance acted upon, often by an enzyme
sunscald—the injury of plant tissues burned or scorched by direct sunlight
susceptible (n. susceptibility)—prone to develop disease when infected by a pathogen (*see* resistant)
symptom—an indication of disease by a reaction of the host (e.g., canker, leaf spot, wilt) (*see* sign)
symptomatology—the study of disease symptoms
symptomless carrier—a plant that is infected with a pathogen (usually a virus) but has no obvious symptoms
synergism (adj. synergistic)—an interaction in which the total effect of the interacting factors is greater than the additive effects of the factors acting individually
systemic—pertaining to a disease in which the pathogen (or a single infection) spreads generally throughout a plant; pertaining to chemicals that spread internally through a plant

taproot—a primary root that grows vertically downward and from which smaller lateral roots branch
taxonomy (adj. taxonomic)—the science of naming and classifying organisms
teleomorph—the perfect state, or sexual form in the life cycle of certain fungi (*see* anamorph; holomorph)
teliospore—a thick-walled resting spore produced by some fungi (notably rust and smut fungi) from which a basidium is produced
telium (pl. telia)—the fruiting body (sorus) that produces teliospores of a rust fungus
thallus—the vegetative body of a fungus
tissue—a group of cells, usually of similar structure, that perform the same or related functions
tissue culture—an in vitro method of propagating healthy cells from plant tissues
titer—a measure of the concentration of a substance in a solution
tolerance (adj. tolerant)—the ability of a plant to endure an infectious or noninfectious disease, adverse conditions, or chemical injury without serious damage or yield loss; with reference to pesticides, the amount of chemical residue legally permitted on an agricultural product entering commercial channels, usually measured in parts per million (ppm)
toxicity—the capacity of a substance to interfere with the vital processes of an organism
toxin—any poisonous substance of biological origin
translocation—the movement of water, nutrients, chemicals, or food materials within a plant
translucent—clear enough to allow the transmission of light
transmit (n. transmission)—to spread or transfer, as in spreading an infectious pathogen from plant to plant or from one generation of plants to another
transpiration—the loss of water by evaporation from leaf surfaces and through stomata
truncate—to end abruptly, as though an end has been cut off
tuber—an underground stem adapted for storage; typically produced at the end of a stolon
tumor—an abnormal swelling or localized outgrowth, often roughly spherical, produced by a plant as a result of attack by a fungus, bacterium, nematode, insect, or other organism (syn. gall)

turgid—swollen or inflated; plump or swollen as a result of internal water pressure
turgidity—the state of being rigid or swollen as a result of internal water pressure
tylosis (pl. tyloses)—a balloonlike extrusion of parenchyma cells into the lumina of contiguous vessels, partially or completely blocking them

unicellular—one celled
uniflagellate—having one flagellum
uninucleate—having one nucleus per cell (*see* binucleate; multinucleate)
univoltine—having a single brood or generation per year (*see* bivoltine; multivoltine)
urediniospore—the asexual, dikaryotic, often rust-colored spore of a rust fungus, produced in a uredinium; the repeating stage of a heteroecious rust fungus, in which it is capable of infecting the host plant on which it was produced
uredinium (pl. uredinia), **uredium** (pl. uredia)—the fruiting body (sorus) in which urediniospores of a rust fungus are produced

vacuole—a generally spherical organelle within a plant cell, bound by a membrane and containing dissolved materials, such as metabolic precursors, storage materials, and waste products
variegation—a pattern of two or more colors in a plant part, as in a green and white leaf
variety (adj. varietal)—a plant type within a species, resulting from deliberate genetic manipulation and having recognizable characteristics, such as color; shapes of flowers, fruit, and seeds; height; and form (syn. cultivar)
vascular—pertaining to fluid-conducting tissues (xylem and phloem) in plants
vascular bundle—a strand of conductive tissue; usually composed of xylem and phloem (in leaves, small bundles are called veins)
vascular cylinder—the cylinder of vascular tissue in stems and roots (syn. stele)
vector—a living organism (e.g., insect, mite, bird, higher animal, nematode, parasitic plant, human) that is able to carry and transmit a pathogen; in genetic engineering, a cloning vehicle, or self-replicating DNA molecule, such as a plasmid or virus, used to introduce a fragment of foreign DNA into a host cell
vegetative—pertaining to the somatic or asexual parts of a plant, which are not involved in sexual reproduction
vegetative propagation—asexual reproduction; in plants, the use of cuttings, bulbs, tubers, or other vegetative parts to grow new plants
vein—a small vascular bundle in a leaf
vein banding—a symptom of virus disease in which the regions along veins are darker green than the tissue between veins
vein clearing—disappearance of green color in or around leaf veins
vesicle—the thin sac in which zoospores are differentiated and from which they are released; the bulbous head terminating the conidiophores of *Aspergillus* spp.; a structure formed by an endomycorrhizal fungus within living cells of the root
vessel—the water-conducting structure of xylem tissue, with pit openings in end walls
viable (n. viability)—alive; capable of germination (e.g., of seeds, fungus spores, sclerotia, etc.); capable of growth
virion—a complete virus particle
viroid—any of various small, unencapsidated (naked), circular, single-stranded RNA molecules capable of causing disease in plants
virulence—the degree or measure of pathogenicity; the relative capacity to cause disease
virulent—pathogenic; having the capacity to cause disease (*see* avirulent)
viruliferous—virus laden, usually with reference to insects or nematodes as vectors
virus—any of numerous submicroscopic, intracellular, obligate parasites consisting of a core of infectious nucleic acid (either RNA or DNA) usually surrounded by a protein coat

water soaking—a disease symptom in which plant tissues or lesions appear wet, dark, and usually sunken and translucent
whorl—a circular arrangement of like parts
wild type—the phenotype characteristic of most individuals of a species under natural conditions
wilt—the drooping of leaves and stems caused by a lack of water (inadequate water supply or excessive transpiration); a vascular disease in which normal water uptake is interrupted
witches'-broom—a disease symptom in which many weak shoots arise at or close to the same point, producing abnormal massed, brushlike growth

xylem—the water- and mineral-conducting, food-storing, supporting tissue of a plant

zonate—pertaining to the targetlike development of a tree canker, characterized by successive perennial rings of callus; pertaining to any symptom appearing in concentric rings
zoosporangium—a sporangium (spore case) bearing zoospores
zoospore—a fungal spore with flagella; capable of locomotion in water

Index